Introduction to Classical Electrodynamics 1

Dedicated to the memory of my parents
Dedicated to the memory of Daniel Arnaudon
Dedicated to Prof. Amitabha Chakrabarti

Series Editor
Frédérique de Fornel

Introduction to Classical Electrodynamics 1

Boucif Abdesselam

iSTE

WILEY

First published 2025 in Great Britain and the United States by ISTE Ltd and John Wiley & Sons, Inc.

ISTE Ltd
27-37 St George's Road
London SW19 4EU
UK

www.iste.co.uk

John Wiley & Sons, Inc.
111 River Street
Hoboken, NJ 07030
USA

www.wiley.com

Library of Congress Control Number: 2025933055

British Library Cataloguing-in-Publication Data
A CIP record for this book is available from the British Library
ISBN 978-1-83669-000-9

Table of Contents

Preface

There is no place in this new kind of physics both for the field and matter, for the field is the only reality.

Albert Einstein

An *electric charge* is an abstract concept, comparable to the concept of *mass*, introduced to explain certain behaviors in physics. However, unlike mass, an electric charge can take two experimental forms that are arbitrarily called *positive* and *negative*. Two charges of the same nature, both positive or both negative, for instance, repel each other, while two opposite charges attract each other. This phenomenon is called *electrostatic interaction*.

The history of electric charges goes back to the ancient Greeks who observed that rubbing certain objects against fur produced an imbalance in the total electric charge. These "electrified" objects would also attract other light objects, like hair. What is more, if the objects are rubbed against each other for long enough, it is even possible to create a spark.

Electric charges can be experimentally measured. Particles observed in nature have charges that are integer multiples of an *elementary charge*, which is a *fundamental physical constant*. An elementary charge is the electric charge on a proton or, equivalently, the opposite of the electric charge on an electron. It is expressed in *Coulombs* (C), or in $A \times s$ in the international system of units. The value of an electric charge, considered to be invisible, and its *discrete nature* were first discovered in 1909 by Robert Andrews Millikan.

The elementary charge, denoted by e, has an approximate value $e = 1.602176634 \times 10^{-19}$C. The exact value of the elementary charge is currently defined by: $e = \sqrt{\frac{2h\alpha}{\mu_0 c}}$, where h is *Planck's constant*, with $h = 6.626070040 \times 10^{-34}$Js; α is the *fine structure constant*, with $\alpha = 7.2973525664 \times 10^{-3}$; μ_0 is the *vacuum magnetic permeability*, with $\mu_0 = 4\pi \times 10^{-7}\text{Hm}^{-1}$; and c is the speed of light in a vacuum, with $c = 299,792,458$ m/s. The existence of quarks was postulated in the beginning of 1960 and they are believed to possess a fractional electric charge $\left(\pm\frac{1}{3}e\right)$ or $\left(\pm\frac{2}{3}e\right)$, but they are confined within hadrons, whose charge and may be zero, and may be equal to the elementary charge, or may be an integer multiple of the elementary charge. Quarks have not yet been separately detected, but they are assumed to have existed in a free state in the very first instants of the Universe (the quark epoch). Charge is an invariant in the theory of special relativity. A particle of charge q, whatever its velocity, will always retain the same charge q.

There are two branches of study:

– *Electrostatics* is the branch of physics that studies the phenomena created by static electric charges for the observer. In physics, an electrically charged particle creates a vector field that is called an *electric field*. In the presence of a charged particle, the local properties of a space are therefore modified. This is therefore an accurate translation of the concept of a field. If there is another charge in this same field, it will be subject to the action of the electric force (one of the four forces of nature) exerted at a distance by the particle: the electric field is in a way the mediator of this distant action. If there is a movement of electric charges, generally electrons, within a conductive material under the effect of a potential difference at the ends of the material, an electric current is produced. The concept of electric current was first introduced by Benjamin Franklin. He was the first to imagine electricity as a type of invisible fluid present in all matter. He postulated that friction of an insulating surface causes the fluid to move and that a flow of this fluid constitutes an electric current. A material is made up of a large number of electric charges, but these charges balance out among themselves, that is, there is the same number of electrons (negative) as protons (positive). At normal temperatures, the material is electrically neutral. When a static electricity effect is produced, this means that there has been a displacement of charges from material A to material B. This is the *electrification phenomenon*. Excess or missing charges, that is, *uncompensated charges*, are responsible for electrical effects on a body (the example of the rubbed stick). There are two kinds of materials:

a) A material is said to be a *perfect conductor* if, when electrified, the uncompensated charges move freely through the material.

b) If the uncompensated charges do not move freely and remain fixed in the place where they were deposited, we say that it is a *perfect insulator (or dielectric)*.

In reality, a real material is obviously lies between these two limit states:

– *Magnetism* represents all physical phenomena in which objects exert attractive or repulsive forces on other materials. The electric currents and magnetic moments of the fundamental elementary particles are at the origin of the magnetic field generated by these forces. The magnetic field is therefore a vector quantity (characterized by an intensity and a direction) defined for each point in space. It is determined by the position and orientation of magnets, electromagnets and the displacement of electric charges. The presence of this field is seen in the existence of a force acting upon the moving electric charges (called the *Lorentz force*), and various effects that affect certain materials (paramagnetism, diamagnetism, or ferromagnetism, depending on the case). Magnetic susceptibility is the quantity that determines how much a magnetic field interacts with a material.

Electricity and magnetism are intimately related. In 1820, Ørsted discovered the relationship between electricity and magnetism: *a wire carrying an electric current is capable of deflecting the magnetized needle of a compass*. Subsequently, Ampère, based on Ørsted's work, discovered and formulated a number of laws governing the relationship between magnetism and electrodynamics. In 1831, Faraday discovered that if an electric current produces a magnetic field, the reverse is also true. An electric current can be produced by setting a magnetic field in motion, according to *Lenz's law*.

This marked the birth of electromagnetism. It is therefore the field of physics that studies the interactions between electrically charged particles, whether at rest or in motion, using the concept of the electromagnetic field. Electromagnetism can be defined as the branch that studies the electromagnetic field and its interactions with charged particles. The electric field and the magnetic field are the two components of the electromagnetic field described by electromagnetism. The equations that describe the evolution of the electromagnetic field are called *Maxwell equations*. Electric and magnetic fields waves can propagate freely in space and in most materials. These are called *electromagnetic waves* and correspond to all manifestations of light across wavelengths (radio waves, microwaves, infrared, visible domain, ultraviolet rays, X-rays and gamma rays). Maxwell wrote: *The agreement of the results seems to show that light and magnetism are affections of the same substance, and that light is an electromagnetic disturbance propagated through the field according to electromagnetic laws*. From this point of view, optics can be seen as an application of electromagnetism. The electromagnetic interaction is also one of the four fundamental interactions. Together with quantum mechanics, it enables us to understand the existence, cohesion and stability of atoms and molecules.

From the point of view of fundamental physics, the theoretical development of classical electromagnetism led to the theory of special relativity at the beginning of the 20th century. The need to reconcile electromagnetic theory and quantum mechanics led to the construction of quantum electrodynamics, in which electromagnetic interaction is interpreted as an exchange of particles called *photons*. In particle physics, *electromagnetic interaction* and *weak interaction* are unified within the *electroweak theory*.

Note to readers

Electromagnetic field theory is often poorly or insufficiently taught in university physics programs. The heavy dependence on mathematics, especially vector calculus, integral calculus, complex functions and special functions, etc., add to the difficulty in teaching this subject and may obscure certain phenomena such that students get mired in mathematical difficulties and lose sight of practical applications and vice versa. This modest contribution aims to lay out the essential ideas of the theory in the most logical order. Special importance is given to the mathematical developments of this theory. The experimental side of the theory is only lightly approached. We will not insist on units of measurement. These three volumes will contain many examples (the most classic). I have tried to give as much detail as possible in the calculations. Across the three volumes, the discussion is essentially divided into three main topics: (1) charges and/or dipoles, as the source of the electrostatic field; (2) currents and/or magnetization as sources of the magnetic field; (3) electrodynamics, where the electric field and magnetic field are of equal importance. This book discusses many applications of electromagnetic theory. For instance, there are the solutions of Laplace and Poisson equations in various coordinate systems, dielectric medias, magnetized medias, waveguides, transmission lines, electromagnetic radiation, radiation by point-like charges, Kirchhoff diffraction, special relativity, etc. For the reader's convenience, we have listed all of the mathematical tools used in the book into six appendices, which will be available to download online: vector analysis, tensors, special/hypergeometric functions, etc. References are presented at the end of each volume. Formulas are given here in the international system of units. We have tried to adopt the notations most commonly used in the literature. Graphs and 3D representations have been produced using Maple.

This three-volume set is meant for physics students, engineering students, as well as mathematics students who are interested in problems in electromagnetic theory. It can also be used by anyone interested in physics in general, or who simply wants to learn more out of intellectual curiosity. There are many books on the same subject and therefore it seems reasonable to repeat some classical developments that we find excellent, rather than striving for originality at any cost.

Volume 1 will cover the following topics:

– electrostatic field in vacuum (Gauss' theorem, etc.);

– conductor at equilibrium;

– electrokinetics;

– magnetic field in vacuum (Laplace force, Biot–Savart law, the Ampère theorem, etc.);

– electric and magnetic moments;

– boundary problems for static potentials in vacuum.

Volume 2 will study the following points:

– static potential with a source – Poisson equation (image method, Green function, etc.);

– electromagnetic induction phenomena;

– electromagnetic waves in vacuum;

– electromagnetic energies;

– polarization, polarized media (dielectric);

– magnetization, magnetized media;

– electromagnetic fields in matter.

Volume 3 will analyze the following concepts:

– waveguides;

– transmission lines;

– electromagnetic radiation;

– radiation from point-like charges;

– Kirchhoff diffraction;

– theory of special relativity;

– covariant formulation;

The online appendices will be available at the following URL:
www.iste.co.uk/abdesselam/electrodynamics1.zip

The appendices will cover:

– some vector analysis results (Appendix 1);

– tensors (Appendix 2);

– the Dirac function (Appendix 3);

– orthogonal functions (Appendix 4);

– beta and gamma functions (Appendix 5);

– special functions (Appendix 6).

Acknowledgments

This work would never have seen the light of day without the encouragement of many people. I am particularly grateful to my students, colleagues and many others. I thank Fréderique de Fornel for her patience and advice. Finally, I dedicate this book to my wife, Nadia, and my children, Abir, Ahmed Nour, Mohammed Sami and Samah. I would like to express my gratitude to them.

Boucif Abdesselam
March 2025

Chapter 1

Electrostatic Field in Vacuum

1.1. Electric charges

We now know that matter is made up of negatively charged electrons and positively charged protons, that is, the Universe contains two types of matter that we can call *negative* and *positive*. Particles of the same species repel each other and those of different species attract each other by a force called *electrostatic* or *Coulomb force*, similar to the gravitational force, which varies inversely with the square of the distance. However, the electrostatic force is much more intense than the gravitational force. The existence of these two types of charges was revealed in Franklin's experiments. For example, when a glass rod is rubbed with a dry cloth, some electrons from the glass shift onto the cloth. The cloth is then negatively charged, as it has more electrons than protons. The glass rod becomes positively charged because it has lost electrons (excess protons). The glass rod is then said to undergo an *electrification effect*. If this operation is repeated with a wax rod, the same electrification phenomenon is observed, except that this time, the electrons shift from the cloth to the wax. The wax rod is thus negatively charged. In addition, if two sufficiently electrified glass rods are brought together, they repel each other, that is, there is *repulsion*. The same repulsion is observed with two wax rods. On the other hand, if a glass rod is brought near a wax rod, they attract each other. It can thus be concluded that charges with the same sign repel each other and those with opposite signs attract each other. If the positively charged glass rod is brought close to a metallic ball attached to a wire (free to move), the electrons in the ball are attracted to the area on its surface that is close to the glass rod, leaving the other side of the ball positively charged: this is the phenomenon of *electrostatic induction*. An attractive force then appears between the ball and the glass rod. When it comes into contact with the ball, the positive charges on the rod are neutralized by part of the negative charges on the ball. This allows the ball-rod assembly to conserve a net positive charge. The ball, which is now positively charged, is then repelled by another positively charged rod. This transfer of charges is called the *conduction effect*. The metallic ball can also be called a *conductor*, as the charges on the ball are easily induced and conducted.

An electron bears an electric charge $-e$, whereas a proton has an electric charge $+e$. For a given quantity of ordinary matter, we can proceed with an algebraic sum of the elementary electric charges within the material. We therefore conclude that every body contains an *integer* quantity of an electric charge, denoted by e. This elementary charge represents the smallest possible value of an electric charge. This is the *charge quantification principle*. This quantization of electric charges means that any charged body is presented as a discontinuous distribution of its constituent charges. On a large scale, this distribution of constituent charges can be considered to be continuous (the elementary charges are very small). For an insulated system (no material can cross the system's boundaries), the total electric charge, that is, the algebraic sum of positive and negative charges present in the body at any given moment, remains constant. Charges are neither created nor destroyed.

1.2. Electrostatic force

Between two electric point-like charges q and q', located, respectively, at points $M(x, y, z)$ and $M'(x', y', z')$, a force is exerted which is inversely proportional to the square of the distance separating them and is carried by the line joining the two points (central force):

$$\overrightarrow{\mathbf{F}}_q = -\overrightarrow{\mathbf{F}}_{q'} = \frac{1}{4\pi\varepsilon_0} \frac{qq'\left(\overrightarrow{\mathbf{r}} - \overrightarrow{\mathbf{r}}'\right)}{\left|\overrightarrow{\mathbf{r}} - \overrightarrow{\mathbf{r}}'\right|^3} \qquad [1.1]$$

where $\overrightarrow{\mathbf{r}}$ (respectively, $\overrightarrow{\mathbf{r}}'$) is the vector that joins the origin O and the point $M(x, y, z)$ (respectively, $M'(x', y', z')$), as depicted in Figure 1.1. The forces applied to q and q' are of the same magnitude, but opposite in direction. The quantity ε_0 is called the *electric permittivity of vacuum* or *vacuum permittivity* and is related to the *speed of light* in vacuum c and the *magnetic permeability of vacuum* (or *vacuum permeability*) μ_0 through the relation $\varepsilon_0 = 1/\mu_0 c^2$. $\overrightarrow{\mathbf{F}}_q$ (respectively, $\overrightarrow{\mathbf{F}}_{q'}$) is the force applied on the charge q (respectively, q'). In particular, if $qq' > 0$, that is, of the same sign (respectively, $qq' < 0$, i.e. of the opposite sign), the force is attractive (respectively, repulsive). This is *Coulomb's law*.

According to the *superposition principle*, if there are several point-like charges $q'_1, \ldots, q'_n$ located, respectively, at $\overrightarrow{\mathbf{r}_1}', \ldots, \overrightarrow{\mathbf{r}_n}'$, the total force applied on the charge q is the resultant of all individual forces:

$$\overrightarrow{\mathbf{F}}_q = \frac{q}{4\pi\varepsilon_0} \sum_{i=1}^{n} \frac{q'\left(\overrightarrow{\mathbf{r}} - \overrightarrow{\mathbf{r}_i}'\right)}{\left|\overrightarrow{\mathbf{r}} - \overrightarrow{\mathbf{r}_i}'\right|^3} \qquad [1.2]$$

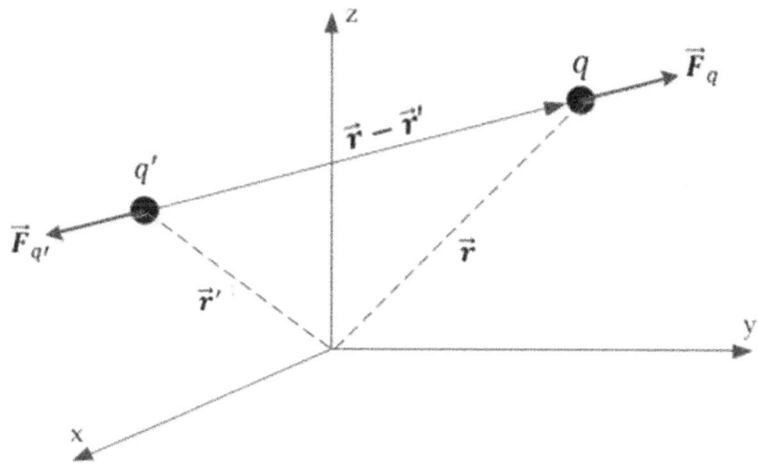

Figure 1.1. *Electrostatic force between two point-like charges q and q'*

EXAMPLE 1.1. (Multi-charge System).– Two identical electric charges, denoted here by q, located at A and B on the Oz axis, are separated by a distance $2d$. The origin O is chosen in the middle of the segment AB. Let us calculate the total force exerted on a test charge q' located at a point with coordinates $(x_0, 0, 0)$ on the Ox axis, as depicted in Figure 1.2. Using the superposition principle, we obtain:

$$\vec{\mathbf{F}} = \frac{q'q\left(x_0\,\vec{\mathbf{i}} - d\,\vec{\mathbf{k}}\right)}{4\pi\varepsilon_0\left(x_0^2 + d^2\right)^{3/2}} + \frac{q'q\left(x_0\,\vec{\mathbf{i}} + d\,\vec{\mathbf{k}}\right)}{4\pi\varepsilon_0\left(x_0^2 + d^2\right)^{3/2}} = \frac{2q'qx_0\,\vec{\mathbf{i}}}{4\pi\varepsilon_0\left(x_0^2 + d^2\right)^{3/2}} \qquad [1.3]$$

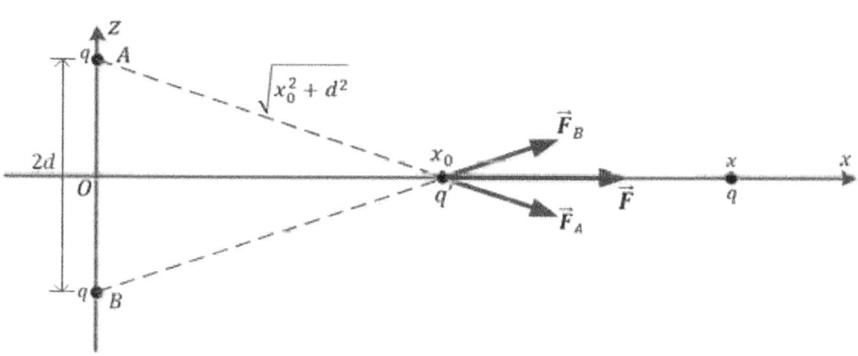

Figure 1.2. *Superposition principle for a system with multiple charges*

It is seen that the resultant force at $(x_0, 0, 0)$ only has a non-zero component along the Ox axis. This property can be explained by the fact that the resulting distribution is rotational symmetrical around Ox. In simpler terms, Ox is a symmetry axis. Note that if the charges at points A and B were of opposite polarities, then axis O would be an antisymmetry axis for the system and the total force exerted on charge q' would be normal to Ox. We will return to these properties in subsequent sections. At the origin, O ($x_0 = 0$), the force is zero (the forces applied by the charges are mutually annihilated); this is an unstable equilibrium position for the system: when moving away from $x_0 = 0$, q' tends to move further away. At a point that is very far from the origin, that is, when $x_0 \gg d$, the force $\vec{\mathbf{F}}$ takes the approximate value: $\vec{\mathbf{F}} \simeq \frac{q'(2q)\vec{\mathbf{i}}}{4\pi\varepsilon_0 x_0^2}$; this corresponds to a force applied by a charge $2q$ concentrated at O. The force [1.3] reaches its maximum at point $x_0 = \frac{d}{\sqrt{2}}$, that is, $\vec{\mathbf{F}}_{\max} = \frac{q'q\vec{\mathbf{i}}}{3\sqrt{3}\pi\varepsilon_0 d^2}$. Let us now add another charge q at the point $(x, 0, 0)$, such that $x > x_0$. The total force becomes:

$$\vec{\mathbf{F}} = \frac{q'q\vec{\mathbf{i}}}{4\pi\varepsilon_0}\left(\frac{2x_0}{\left(x_0^2 + d^2\right)^{3/2}} - \frac{1}{\left(x - x_0\right)^2}\right) \qquad [1.4]$$

This force cancels out for $x = x_0 + \frac{1}{\sqrt{2}}\frac{\left(x_0^2+d^2\right)^{3/4}}{\sqrt{x_0}}$. The position x_0 is here a stable equilibrium position for the system: when moving away from x_0, q' tends to return to x_0. ■

Consider a differential element of free charges dq' (located at $\vec{\mathbf{r}}'$) within a macroscopic distribution (of charges). The total force $\vec{\mathbf{F}}_q$ applied by the distribution on a test charge q (located at $\vec{\mathbf{r}}$) has the value:

$$\vec{\mathbf{F}}_q = \frac{q}{4\pi\varepsilon_0}\int_{\text{sys.}} \frac{\left(\vec{\mathbf{r}} - \vec{\mathbf{r}}'\right) dq'}{\left|\vec{\mathbf{r}} - \vec{\mathbf{r}}'\right|^3} \qquad [1.5]$$

The total charge in the distribution is obtained by summing over all differential elements, that is, by integrating over the whole system $Q' = \int_{\text{sys.}} dq'$. In general, there are three types of distribution, which are discussed below.

1.2.1. *Distribution along a line (linear) or a curve (curvilinear)*

When the electric charge is distributed along a line or a curve C', equation [1.5] is reduced to:

$$\vec{\mathbf{F}}_q = \frac{q}{4\pi\varepsilon_0}\int_{C'} \frac{\left(\vec{\mathbf{r}} - \vec{\mathbf{r}}'\right)\lambda\left(\vec{\mathbf{r}}'\right) dl'}{\left|\vec{\mathbf{r}} - \vec{\mathbf{r}}'\right|^3} \qquad [1.6]$$

where $\lambda\left(\overrightarrow{\mathbf{r}}'\right)$ is the *linear, line charge* or *curvilinear* (charge per unit length). Here, dl' represents an infinitesimal element of the line or curve. The total charge located along the line or curve is obtained from the integral $Q = \int_{C'} \lambda\left(\overrightarrow{\mathbf{r}}'\right) dl'$.

EXAMPLE 1.2. (Total Charge).– Let us determine the total charge in each of the following linear distributions: (a) a linear charge λ uniformly distributed along an infinite line; (b) a linear charge $\lambda(z) = \frac{qa}{\pi(a^2+z^2)}$ distributed along an infinite line; (c) a curvilinear charge (constant) λ distributed over a circle with radius R. For (a), we find $Q = \int_{-\infty}^{\infty} \lambda dz = \infty$; for (b), $Q = \int_{-\infty}^{\infty} \frac{qadz}{\pi(a^2+z^2)} = \frac{q}{\pi} \tan^{-1} \frac{z}{a}\Big|_{-\infty}^{\infty} = q$; for (c), $Q = \int_0^{2\pi} \lambda R d\varphi = 2\pi R\lambda$.

EXAMPLE 1.3. (Uniformly Charged Infinitely Long Wire).–

1) Consider an infinitely long wire that is uniformly charged (constant λ). For reasons of symmetry, we will calculate the force applied to a test charge q located in the Oxy plane, as shown in Figure 1.3. The charge element $\lambda dz'$ exerts the elementary force $d\overrightarrow{\mathbf{F}}_q = \frac{q}{4\pi\varepsilon_0} \frac{\left(x\overrightarrow{\mathbf{i}}+y\overrightarrow{\mathbf{j}}-z'\overrightarrow{\mathbf{k}}\right)\lambda dz'}{(x^2+y^2+z'^2)^{3/2}}$. After integration, this gives:

$$\overrightarrow{\mathbf{F}}_q = \frac{q}{4\pi\varepsilon_0} \int_{-\infty}^{\infty} \frac{\left(x\overrightarrow{\mathbf{i}} + y\overrightarrow{\mathbf{j}} - z'\overrightarrow{\mathbf{k}}\right)\lambda dz'}{\left(x^2+y^2+z'^2\right)^{3/2}} \qquad [1.7]$$

The third integral is zero, so $\int_{-\infty}^{\infty} \frac{\left(-z'\overrightarrow{\mathbf{k}}\right)\lambda dz'}{(x^2+y^2+z'^2)^{3/2}} = 0$, because the function under the sign of the integral is odd. Consequently, we obtain:

$$\overrightarrow{\mathbf{F}}_q = \frac{q}{4\pi\varepsilon_0} \int_{-\infty}^{\infty} \frac{\left(x\overrightarrow{\mathbf{i}} + y\overrightarrow{\mathbf{j}}\right)\lambda dz'}{\left(x^2+y^2+z'^2\right)^{3/2}} = \frac{q\lambda\left(x\overrightarrow{\mathbf{i}} + y\overrightarrow{\mathbf{j}}\right)}{2\pi\varepsilon_0\left(x^2+y^2\right)} \equiv \frac{q\lambda\overrightarrow{\mathbf{r}}}{2\pi\varepsilon_0 r^2} \qquad [1.8]$$

2) If, instead of λ, we take the linear charge (b) from Example 1.2, then:

$$\overrightarrow{\mathbf{F}}_q = \frac{q}{4\pi\varepsilon_0} \int_{-\infty}^{\infty} \frac{\left(\overrightarrow{\mathbf{r}} - z'\overrightarrow{\mathbf{k}}\right)}{\left(r^2+z'^2\right)^{3/2}} \frac{q'adz'}{\pi\left(a^2+z'^2\right)} = \frac{qq'a\overrightarrow{\mathbf{r}}}{4\pi^2\varepsilon_0} \int_{-\infty}^{\infty} \frac{dz'}{\left(r^2+z'^2\right)^{3/2}\left(a^2+z'^2\right)} \qquad [1.9]$$

To calculate this integral, let us first carry out the change variable $z' = r\tan\theta'$:

$$\overrightarrow{\mathbf{F}}_q = \frac{qq'\overrightarrow{\mathbf{r}}}{4\pi^2\varepsilon_0 ar^2} \int_{-\frac{\pi}{2}}^{\frac{\pi}{2}} \frac{\cos\theta' d\theta'}{1+\frac{r^2}{a^2}\tan^2\theta'} = \frac{qq'\overrightarrow{\mathbf{r}}}{4\pi^2\varepsilon_0 ar^2} \int_{-\frac{\pi}{2}}^{\frac{\pi}{2}} \frac{\cos^3\theta' d\theta'}{\cos^2\theta' + \frac{r^2}{a^2}\sin^2\theta'} \qquad [1.10]$$

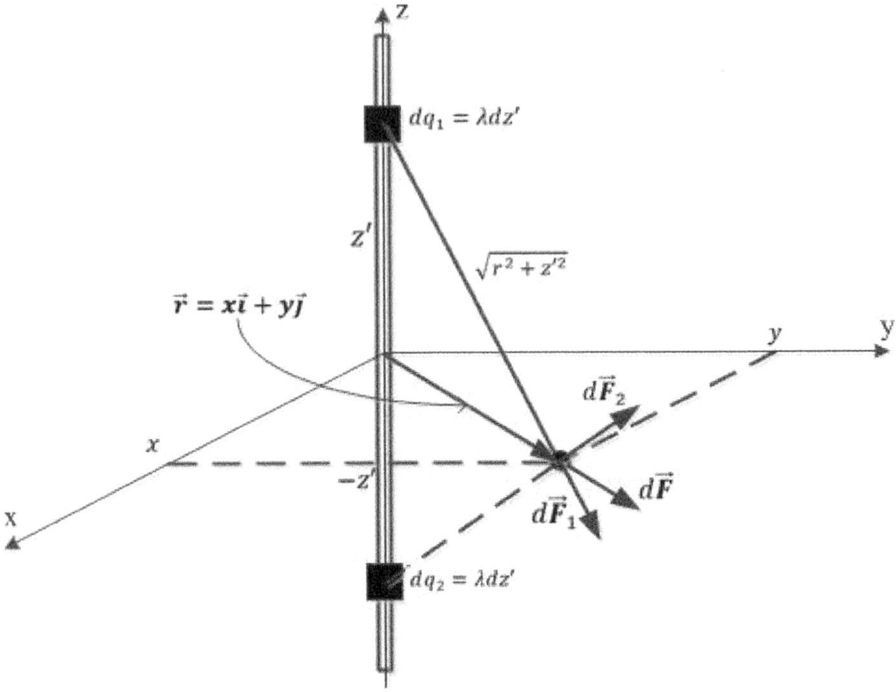

Figure 1.3. *Uniformly charged infinitely long wire dimensionless: by symmetry, the total force exerted on the test charge q is radial. Differential charges dq_1 and dq_2 located symmetrically at z' and $-z'$ produce forces whose projections add up in the Oxy plane and cancel each other out in the Oz direction*

Now taking $t' = \sin\theta'$, we obtain:

$$\vec{\mathbf{F}}_q = \frac{qq'\vec{\mathbf{r}}}{4\pi^2\varepsilon_0 ar^2}\int_{-1}^{1}\frac{\left(1-t'^2\right)dt'}{1+\left(\frac{r^2}{a^2}-1\right)t'^2}$$

$$\begin{cases} \dfrac{qq'\vec{\mathbf{r}}}{4\pi^2\varepsilon_0 ar^2}\displaystyle\int_{-1}^{1}\left(1-t'^2\right)dt' = \dfrac{qq'\vec{\mathbf{r}}}{3\pi^2\varepsilon_0 ar^2} & r=a \\ \dfrac{qq'a\,\vec{\mathbf{r}}}{2\pi^2\varepsilon_0 r^2\left(r^2-a^2\right)}\left[\dfrac{r^2}{a^2\sqrt{\frac{r^2}{a^2}-1}}\tan^{-1}\sqrt{\dfrac{r^2}{a^2}-1}-1\right] & r>a \\ \dfrac{qq'a\,\vec{\mathbf{r}}}{2\pi^2\varepsilon_0 r^2\left(a^2-r^2\right)}\left[1-\dfrac{r^2}{2a^2\sqrt{1-\frac{r^2}{a^2}}}\ln\left(\dfrac{1+\sqrt{1-\frac{r^2}{a^2}}}{1-\sqrt{1-\frac{r^2}{a^2}}}\right)\right] & r<a \end{cases} \quad [1.11]$$

EXAMPLE 1.4. (Uniformly Charged Wire of Finite Length L).– For a uniformly charged wire of finite length L, as in Figure 1.4, the total force experienced by a charge q located at the point $\left(\overrightarrow{\mathbf{r}}, z\right)$ has the value:

$$\begin{aligned}\overrightarrow{\mathbf{F}}_q(\overrightarrow{\mathbf{r}}, z) &= \frac{q}{4\pi\varepsilon_0}\int_{-\frac{L}{2}}^{\frac{L}{2}} \frac{\left[\overrightarrow{\mathbf{r}} + (z - z')\overrightarrow{\mathbf{k}}\right]\lambda dz'}{\left[r^2 + (z - z')^2\right]^{3/2}} \\ &= \frac{q\lambda}{4\pi\varepsilon_0}\left[\frac{\overrightarrow{\mathbf{r}}\left(\frac{L}{2} - z\right) + r^2\overrightarrow{\mathbf{k}}}{r^2\sqrt{r^2 + \left(\frac{L}{2} - z\right)^2}} + \frac{\overrightarrow{\mathbf{r}}\left(\frac{L}{2} + z\right) - r^2\overrightarrow{\mathbf{k}}}{r^2\sqrt{r^2 + \left(\frac{L}{2} + z\right)^2}}\right]\end{aligned} \qquad [1.12]$$

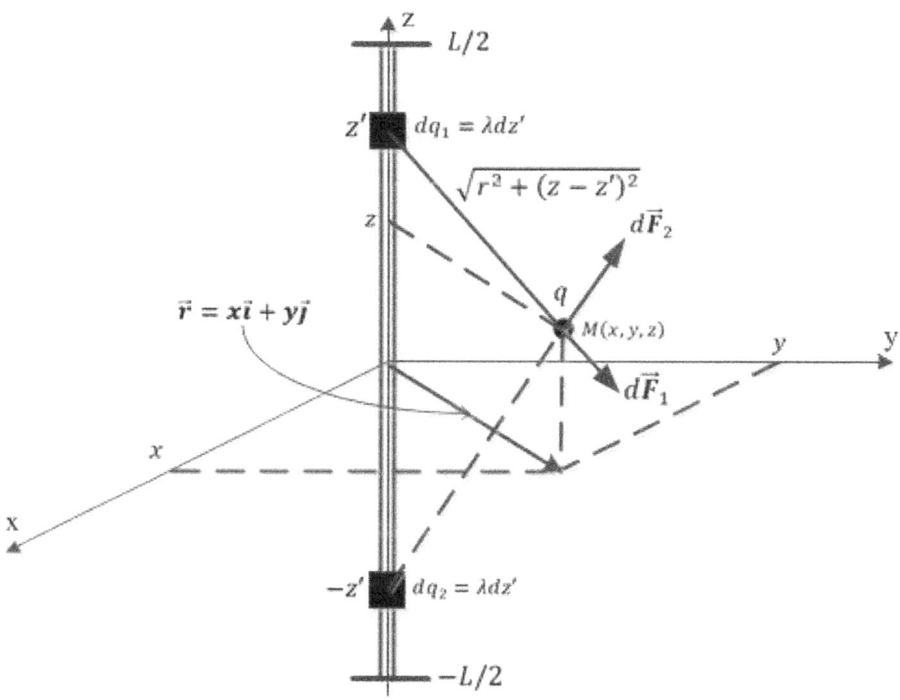

Figure 1.4. *For a uniformly loaded wire of finite length L, the force exerted by a test load q is not radial, except at the middle of the wire*

This is reduced in the middle of the wire, for $z = 0$, at $\overrightarrow{\mathbf{F}}_q\left(\overrightarrow{\mathbf{r}}, 0\right) = \frac{q\lambda}{4\pi\varepsilon_0}\frac{L\overrightarrow{\mathbf{r}}}{r^2\sqrt{r^2 + \frac{L^2}{4}}}$ When $L \longrightarrow \infty$, equation [1.8] is found.

1.2.2. *Distribution over a surface (surface charge density)*

When the charge is distributed over a surface S', the force experienced on the test charge q is written as:

$$\vec{\mathbf{F}}_q = \frac{q}{4\pi\varepsilon_0} \int_{S'} \frac{\left(\vec{\mathbf{r}} - \vec{\mathbf{r}}'\right) \sigma\left(\vec{\mathbf{r}}'\right) dS'}{\left|\vec{\mathbf{r}} - \vec{\mathbf{r}}'\right|^3} \qquad [1.13]$$

where $\sigma\left(\vec{\mathbf{r}}'\right)$ is the *surface charge density* (charge per unit area). dS' is an infinitesimal element of the surface. We have $dS' = dx'dy'$ in Cartesian coordinates, $dS' = r'dr'd\varphi'$ in polar coordinates, etc. The total charge on the surface S' is $Q = \int_{S'} \sigma\left(\vec{\mathbf{r}}'\right) dS'$.

EXAMPLE 1.5. (Infinite Plane Around a Disc).–

1) Consider an infinite plane of negligible thickness, which is charged as follows:

$$\sigma\left(\vec{\mathbf{r}}'\right) = \begin{cases} 0 & \text{for } x'^2 + y'^2 < R^2 \\ \sigma & \text{for } x'^2 + y'^2 \geq R^2 \end{cases} \qquad [1.14]$$

As shown in Figure 1.5, the force applied by $dq' = \sigma\left(r', \varphi'\right) dS' = \sigma r'dr'd\varphi'$ on a charge q located at $(0, 0, z)$ has the value $d\vec{\mathbf{F}}_q = \frac{q}{4\pi\varepsilon_0} \frac{\sigma\left(z\vec{\mathbf{k}} - r'\cos\varphi'\vec{\mathbf{i}} - r'\sin\varphi'\vec{\mathbf{j}}\right)}{\left(z^2+r'^2\right)^{3/2}}$ $r'dr'd\varphi'$. By summing, we obtain:

$$\vec{\mathbf{F}}_q = \frac{q}{4\pi\varepsilon_0} \int_R^\infty \int_0^{2\pi} \frac{\sigma\left(z\vec{\mathbf{k}} - r'\cos\varphi'\vec{\mathbf{i}} - r'\sin\varphi'\vec{\mathbf{j}}\right)}{\left(z^2 + r'^2\right)^{3/2}} r'dr'd\varphi' \qquad [1.15]$$

The integrals on $\sin\varphi$ and $\cos\varphi$ give 0. Therefore, it comes that:

$$\vec{\mathbf{F}}_q = \frac{q}{2\varepsilon_0} \int_R^\infty \frac{\sigma z\vec{\mathbf{k}}}{\left(z^2 + r'^2\right)^{3/2}} r'dr' = \frac{q\sigma z\vec{\mathbf{k}}}{2\varepsilon_0\sqrt{z^2 + R^2}} \qquad [1.16]$$

2) Instead of [1.14], we choose:

$$\sigma\left(\vec{\mathbf{r}}'\right) = \begin{cases} 0 & \text{for } x'^2 + y'^2 < R^2 \\ \frac{Rq'}{2\pi r'^3} & \text{for } x'^2 + y'^2 \geq R^2 \end{cases} \qquad [1.17]$$

We find:

$$\begin{aligned}\overrightarrow{\mathbf{F}}_q &= \frac{q}{4\pi\varepsilon_0}\int_R^{\infty}\int_0^{2\pi}\frac{\left(z\overrightarrow{\mathbf{k}} - r'\cos\varphi'\overrightarrow{\mathbf{i}} - r'\sin\varphi'\overrightarrow{\mathbf{j}}\right)}{\left(z^2+r'^2\right)^{3/2}}\frac{Rq'}{2\pi r'^3}r'dr'd\varphi' \\ &= \frac{qq'Rz\overrightarrow{\mathbf{k}}}{4\pi\varepsilon_0}\int_R^{\infty}\frac{dr'}{r'^2\left(z^2+r'^2\right)^{3/2}} = \frac{qq'R\overrightarrow{\mathbf{k}}}{4\pi\varepsilon_0 z^3}\left[\frac{z^2+2R^2}{R\sqrt{z^2+R^2}} - 2\right]\end{aligned} \quad [1.18]$$

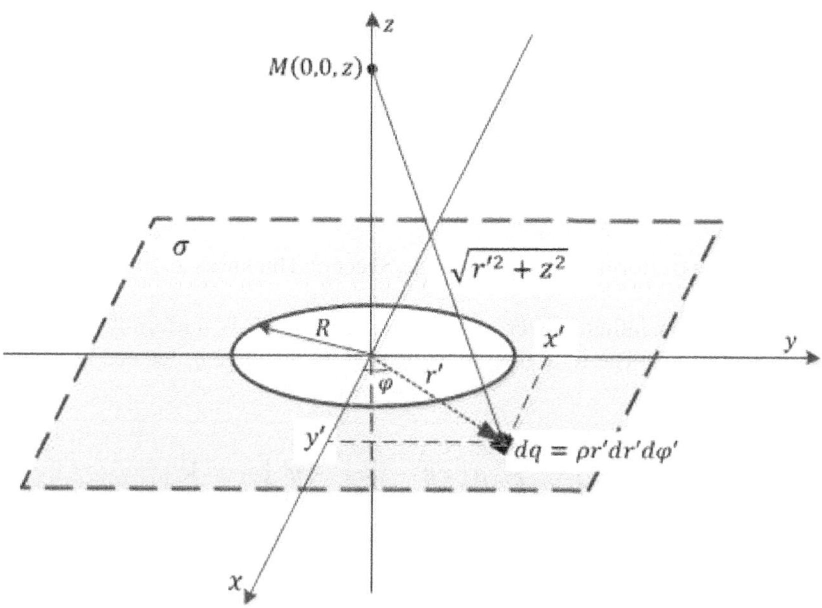

Figure 1.5. *Infinite plane distribution around a disc of radius R: By symmetry, the force experienced along the Oz axis is perpendicular to the plane*

1.2.3. *Distribution over a volume (volume charge density)*

When the charge is distributed within a volume τ', we get:

$$\overrightarrow{\mathbf{F}}_q = \frac{q}{4\pi\varepsilon_0}\int_{\tau'}\frac{\left(\overrightarrow{\mathbf{r}} - \overrightarrow{\mathbf{r}}'\right)\rho\left(\overrightarrow{\mathbf{r}}'\right)d\tau'}{\left|\overrightarrow{\mathbf{r}} - \overrightarrow{\mathbf{r}}'\right|^3} \quad [1.19]$$

where $\rho\left(\overrightarrow{\mathbf{r}}'\right)$ is the volume charge density (charge per unit volume) in τ'. $d\tau'$ is an infinitesimal element of the volume. In Cartesian coordinates, $d\tau'$ is equal to

$dx'dy'dz'$; in cylindrical coordinates, $d\tau'$ is $r'dr'd\varphi'dz'$; in spherical coordinates, it is $r'^2 \sin\theta' dr' d\theta' d\phi'$; etc. The total charge measured inside τ' is $Q = \int_{\tau'} \rho\left(\overrightarrow{\mathbf{r}}'\right) d\tau'$.

EXAMPLE 1.6. (Electron Cloud).– The electronic cloud of charge $+e$ around the hydrogen atom's nucleus is modeled by the spherically symmetric distribution:

$$\rho\left(\overrightarrow{\mathbf{r}}\right) = -\frac{e}{\pi r_0^3} e^{-\frac{2r}{r_0}} \qquad [1.20]$$

where r_0 is called the *Bohr radius*. The cloud's total charge is:

$$\begin{aligned} Q &= \int_\tau \rho d\tau = -\int_0^\infty \int_0^\pi \int_0^{2\pi} \frac{e}{\pi r_0^3} e^{-\frac{2r}{r_0}} r^2 \sin\theta dr d\theta d\varphi \\ &= -\int_0^\infty \frac{4e}{\pi r_0^3} e^{-\frac{2r}{r_0}} r^2 dr = \frac{2e}{\pi r_0^2} e^{-\frac{2r}{r_0}} \left[r^2 - \frac{r_0^2}{2}\left(-\frac{2r}{r_0} - 1\right)\right]\Bigg|_0^\infty = -e \end{aligned} \qquad [1.21]$$

EXAMPLE 1.7. (Uniformly Charged Infinite Sheet of Thickness $2a$).–

(1) Consider an infinite sheet of thickness $2a$, which is uniformly charged, as shown in Figure 1.6. The total force experienced by a charge q, located at the point $(0, 0, z)$, is written as:

$$\overrightarrow{\mathbf{F}}_q = \frac{q}{4\pi\varepsilon_0} \int_{-\infty}^{\infty} \int_{-\infty}^{\infty} \int_{-a}^{a} \frac{\rho\left(z\overrightarrow{\mathbf{k}} - x'\overrightarrow{\mathbf{i}} - y'\overrightarrow{\mathbf{j}} - z'\overrightarrow{\mathbf{k}}\right)}{\left[(z-z')^2 + x'^2 + y'^2\right]^{3/2}} dx'dy'dz' \qquad [1.22]$$

In cylindrical coordinates, it yields (we integrate with respect to r' and then with respect to z'):

$$\begin{aligned} \overrightarrow{\mathbf{F}}_q &= \frac{q\rho}{4\pi\varepsilon_0} \int_0^\infty \int_0^{2\pi} \int_{-a}^{a} \frac{(z-z')\overrightarrow{\mathbf{k}} - r'\cos\varphi\overrightarrow{\mathbf{i}} - r'\sin\varphi'\overrightarrow{\mathbf{j}}}{\left[(z-z')^2 + r'^2\right]^{3/2}} r'dr'd\varphi'dz' \\ &= \frac{q\rho}{2\varepsilon_0} \int_{-a}^{a} \frac{(z-z')\overrightarrow{\mathbf{k}}}{|z-z'|} dz' = \begin{cases} \dfrac{q\rho a}{\varepsilon_0}\overrightarrow{\mathbf{k}} & z \geq a \\ -\dfrac{q\rho a}{\varepsilon_0}\overrightarrow{\mathbf{k}} & z \leq -a \\ \dfrac{q\rho\overrightarrow{\mathbf{k}}}{2\varepsilon_0}\left[\displaystyle\int_{-a}^{z} dz' + \int_z^a (-dz')\right] = \dfrac{q\rho z}{\varepsilon_0}\overrightarrow{\mathbf{k}} & |z| < a \end{cases} \end{aligned} \qquad [1.23]$$

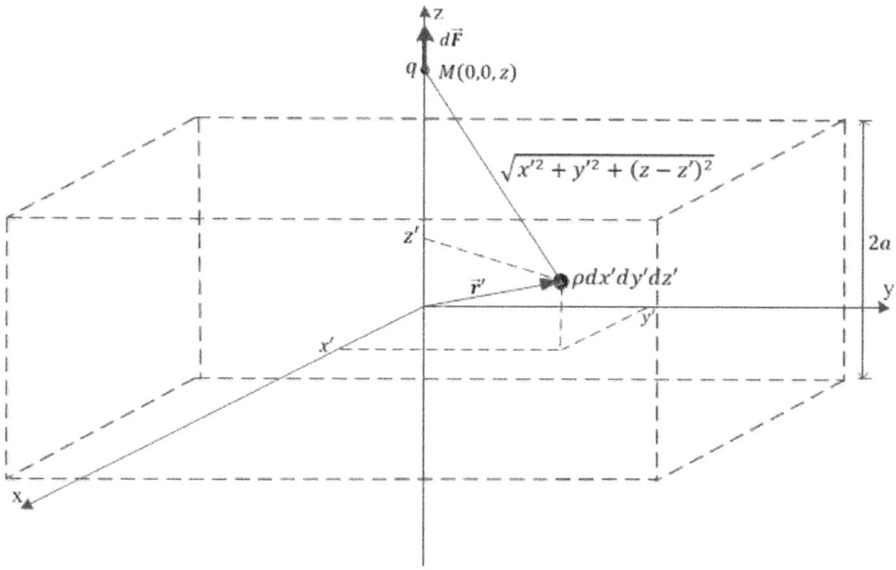

Figure 1.6. *Infinite sheet of thickness $2a$: the total force felt on the charge q is perpendicular to the sheet. Outside the sheet, the force does not depend on the distance between the charge and the sheet, while it depends on z inside*

We can see that $\vec{\mathbf{F}}_q$ is continuous on the boundaries $z = \pm a$.

(2) If the charge is now distributed over a maximum radius $r'_{\max}$ around Oz with the density $\rho(r', z') = \frac{\mu}{r'}$ (μ is a constant), the total force experienced by the charge q is:

$$\begin{aligned}
\vec{\mathbf{F}}_q &= \frac{q\mu\vec{\mathbf{k}}}{2\varepsilon_0}\int_{r'=0}^{r'_{\max}}\int_{z'=-a}^{a}\frac{(z-z')dr'dz'}{\left[(z-z')^2+r'^2\right]^{3/2}} \\
&= \frac{q\mu\vec{\mathbf{k}}}{2\varepsilon_0}\int_0^{r'_{\max}}\left[\frac{1}{\sqrt{(z-a)^2+r'^2}}-\frac{1}{\sqrt{(z+a)^2+r'^2}}\right]dr' \\
&= \frac{q\mu\vec{\mathbf{k}}}{2\varepsilon_0}\ln\left[\frac{r'+\sqrt{(z-a)^2+r'^2}}{r'+\sqrt{(z+a)^2+r'^2}}\right]\Bigg|_{r'=0}^{r'_{\max}} \\
&= \frac{q\mu\vec{\mathbf{k}}}{2\varepsilon_0}\ln\left[\frac{r'_{\max}+\sqrt{(z-a)^2+r'^2_{\max}}}{r'_{\max}+\sqrt{(z+a)^2+r'^2_{\max}}}\left|\frac{z+a}{z-a}\right|\right]
\end{aligned} \qquad [1.24]$$

For $r'_{\max} \to \infty$, equation [1.24] reduces $\overrightarrow{\mathbf{F}}_q = \frac{q\mu\overrightarrow{\mathbf{k}}}{2\varepsilon_0} \ln\left|\frac{z+a}{z-a}\right|$. The total charge held in the sheet is equal to $q' = \int_{-a}^{a}\int_0^{r'_{\max}} \frac{\mu}{r'} 2\pi r' dr' dz' = 4\pi\mu a r'_{\max}$; this gives $\mu = \frac{q'}{4\pi a r'_{\max}}$ and therefore $\overrightarrow{\mathbf{F}}_q = \frac{qq'\overrightarrow{\mathbf{k}}}{8\pi\varepsilon_0 a r'_{\max}} \ln\left[\frac{r'_{\max}+\sqrt{r'^2_{\max}+(z-a)^2}}{r'_{\max}+\sqrt{r'^2_{\max}+(z+a)^2}}\left|\frac{z+a}{z-a}\right|\right]$. ■

The Coulomb force [1.1] can be recovered from equation [1.19] using the Dirac distribution:

$$\begin{aligned}\rho\left(\overrightarrow{\mathbf{r}}'\right) &= q'\delta\left(\overrightarrow{\mathbf{r}}' - \overrightarrow{\mathbf{r}}'_{q'}\right) = q'\delta\left(x' - x'_{q'}\right)\delta\left(y' - y'_{q'}\right)\delta\left(z' - z'_{q'}\right) \\ &= q' r'^{-2} \sin^{-1}\theta' \delta\left(r' - r'_{q'}\right)\delta\left(\theta' - \theta'_{q'}\right)\delta\left(\varphi' - \varphi'_{q'}\right) \\ &= q' r'^{-2} \delta\left(r' - r'_{q'}\right)\delta\left(\cos\theta' - \cos\theta'_{q'}\right)\delta\left(\varphi' - \varphi'_{q'}\right)\end{aligned} \quad [1.25]$$

1.3. Electric field

The *electric field* $\overrightarrow{\mathbf{E}}\left(\overrightarrow{\mathbf{r}}\right)$ is defined as being the force exerted on a test charge equal to unity. Thus, the electric field created by a charge q' is equal to:

$$\overrightarrow{\mathbf{E}}\left(\overrightarrow{\mathbf{r}}\right) = \frac{\overrightarrow{\mathbf{F}}_q}{q} = \frac{q'}{4\pi\varepsilon_0} \frac{\overrightarrow{\mathbf{r}} - \overrightarrow{\mathbf{r}}'}{\left|\overrightarrow{\mathbf{r}} - \overrightarrow{\mathbf{r}}'\right|^3} \quad [1.26]$$

In the same way, the electric fields created by (a) several point-like charges $\{q'_i\}$, (b) a linear charge λ distributed over a curve C', (c) a surface charge density σ distributed over an area S' and (d) a volume charge density ρ distributed within a volume τ' are, respectively:

$$\overrightarrow{\mathbf{E}}\left(\overrightarrow{\mathbf{r}}\right) = \begin{cases} \frac{1}{4\pi\varepsilon_0}\sum_{i=1}^{n} \frac{q'_i\left(\overrightarrow{\mathbf{r}} - \overrightarrow{\mathbf{r}}'_i\right)}{\left|\overrightarrow{\mathbf{r}} - \overrightarrow{\mathbf{r}}'_i\right|^3} & \text{(dist. of charges)} \\ \frac{1}{4\pi\varepsilon_0}\int_{C'} \frac{\left(\overrightarrow{\mathbf{r}} - \overrightarrow{\mathbf{r}}'\right)\lambda\left(\overrightarrow{\mathbf{r}}'\right)dl'}{\left|\overrightarrow{\mathbf{r}} - \overrightarrow{\mathbf{r}}'\right|^3} & \text{(line dist.)} \\ \frac{1}{4\pi\varepsilon_0}\int_{S'} \frac{\left(\overrightarrow{\mathbf{r}} - \overrightarrow{\mathbf{r}}'\right)\sigma\left(\overrightarrow{\mathbf{r}}'\right)dS'}{\left|\overrightarrow{\mathbf{r}} - \overrightarrow{\mathbf{r}}'\right|^3} & \text{(area density dist.)} \\ \frac{1}{4\pi\varepsilon_0}\int_{\tau'} \frac{\left(\overrightarrow{\mathbf{r}} - \overrightarrow{\mathbf{r}}'\right)\rho\left(\overrightarrow{\mathbf{r}}'\right)d\tau'}{\left|\overrightarrow{\mathbf{r}} - \overrightarrow{\mathbf{r}}'\right|^3} & \text{(volume density dist.)} \end{cases} \quad [1.27]$$

If the charge distribution is known, the electric field is obtained by integrating one of the formulas [1.27]. A few simple rules must be observed:

– A distinction must be made between the coordinates of the field points and the source points (differential charges). The integration must always be carried out over the source coordinates.

– Equations [1.27] are vector equations: they generally have three components, that is, three integrals to be carried out. Certain symmetries can be used to show that certain field components are zero.

– The distance $\left|\vec{\mathbf{r}} - \vec{\mathbf{r}}'\right|$ is always positive. When square roots must be taken, it must be ensured that the positive root is taken.

– The solution to a particular problem can often be obtained by integrating the contributions of simpler differential systems (see Example 1.9).

EXAMPLE 1.8. (Uniformly Charged Infinitely Long Wire (see Figure 1.3)).– The electric field measured at the point (x, y, z) is obtained by dividing the force [1.8] by q:

$$\vec{\mathbf{E}} = \frac{\vec{\mathbf{F}}_q}{q} = \frac{\lambda \vec{\mathbf{r}}}{2\pi\varepsilon_0 r^2} \qquad [1.28]$$

The field $\vec{\mathbf{E}}$ has no component along the Oz axis, since the two charge elements dq'_1 and dq'_2, symmetrically located at a distance z on either side of the O point, produce fields of equal magnitude, oppositely directed in this direction and therefore cancel each other out, leaving only the radial components, which add to each other. If the wire has a length L (see Figure 1.4), formula [1.28] must be replaced, according to [1.12], by:

$$\vec{\mathbf{E}}\left(\vec{\mathbf{r}}, z\right) = \frac{\lambda}{4\pi\varepsilon_0}\left[\frac{\vec{\mathbf{r}}\left(\frac{L}{2} - z\right) + r^2\vec{\mathbf{k}}}{r^2\sqrt{r^2 + \left(\frac{L}{2} - z\right)^2}} + \frac{\vec{\mathbf{r}}\left(\frac{L}{2} + z\right) - r^2\vec{\mathbf{k}}}{r^2\sqrt{r^2 + \left(\frac{L}{2} + z\right)^2}}\right] \qquad [1.29]$$

EXAMPLE 1.9. (Uniformly Charged Infinite Plane).– Consider a uniformly charged infinite plane located at $z = 0$, as shown in Figure 1.7. The plane is divided into an infinite number of linear charges along the Oy axis of thickness dx with $d\lambda = \sigma dx$. It is also possible to divide the plane into linear charges along the Ox axis of thickness dy; this does not change the result. Each linear charge produces an infinitesimal electric field whose projection along the Oz axis has the value:

$$dE_z = \frac{\sigma dx}{2\pi\varepsilon_0\left(x^2 + z^2\right)^{1/2}}\cos\theta = \frac{\sigma z dx}{2\pi\varepsilon_0\left(x^2 + z^2\right)} \qquad [1.30]$$

We do not take into account the field component along the Ox axis, because the linear charge at $-z$ induces a field in this direction of the same amplitude but in the opposite direction, which thus cancels out the first field.

The total field is then obtained by integration over all linear charges:

$$\vec{\mathbf{E}} = \frac{\sigma z \vec{\mathbf{k}}}{2\pi\varepsilon_0} \int_{-\infty}^{\infty} \frac{dx}{(x^2 + z^2)} = \begin{cases} \dfrac{\sigma \vec{\mathbf{k}}}{2\varepsilon_0} & z \geq 0 \\ -\dfrac{\sigma \vec{\mathbf{k}}}{2\varepsilon_0} & z < 0 \end{cases} \quad [1.31]$$

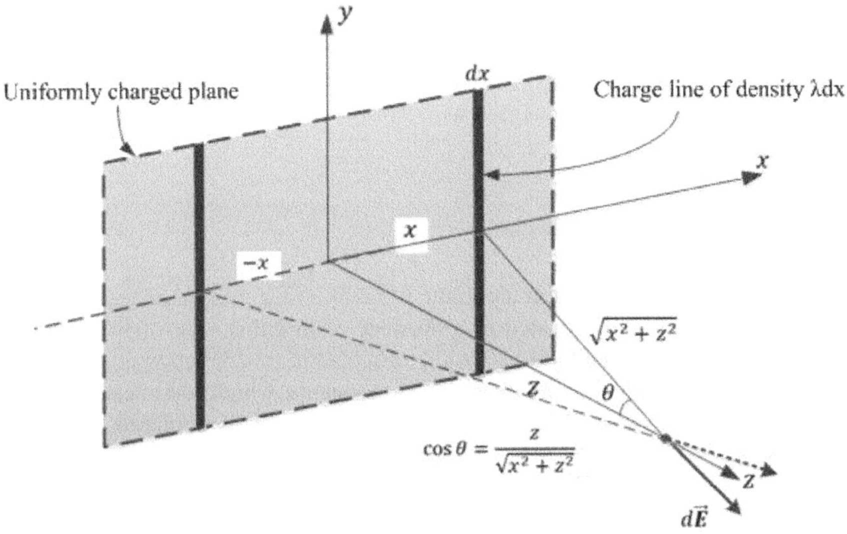

Figure 1.7. *Uniformly charged infinite plane composed of an infinite number of parallel charged lines (λdx) along Oy: the field created by lines symmetrically placed at x and $-x$ has components that cancel out in the Ox direction and components that add up in the Oz direction*

Note that the field does not decrease with distance from the plane.

EXAMPLE 1.10. (An Infinite Plane Around a Disc (see Figure 1.5)).– Along the Oz axis of the disc, a test charge q is subject to the force $\vec{\mathbf{F}}_q = \frac{q\sigma z \vec{\mathbf{k}}}{2\varepsilon_0\sqrt{z^2+R^2}}$. It therefore follows that:

$$\vec{\mathbf{E}} = \frac{\sigma z \vec{\mathbf{k}}}{2\varepsilon_0\sqrt{z^2 + R^2}} \quad [1.32]$$

When $R \longrightarrow 0$ (uniformly loaded infinite plane), we return to [1.31].

EXAMPLE 1.11. (Uniformly Charged Infinite Sheet of Thickness $2a$ (see Figure 1.6)).–

$$\vec{\mathbf{E}}(z) = \begin{cases} \frac{\rho a}{\varepsilon_0}\vec{\mathbf{k}} & z \geq a \\ \frac{\rho z}{\varepsilon_0}\vec{\mathbf{k}} & -a < z < a \\ -\frac{\rho a}{\varepsilon_0}\vec{\mathbf{k}} & z \leq -a \end{cases} \qquad [1.33]$$

EXAMPLE 1.12. (Uniformly Charged Disc with Negligible Thickness).– Let us calculate the electric field, created by a flat disk of negligible thickness, radius R and charge Q (uniformly distributed over the disk), at point M located along axis Oz of the disk at distance z from its center, as shown in Figure 1.8. The charge density is given by $\rho\left(\vec{\mathbf{r}}'\right) = \frac{Q}{\pi R^2}\delta\left(z'\right)$ for $x'^2 + y'^2 \leq R^2$ and 0 elsewhere. It therefore follows that:

$$\vec{\mathbf{E}}(0,0,z) = \frac{Q}{4\pi^2 R^2 \varepsilon_0}\int_{-R}^{R}\int_{-\sqrt{R^2-x'^2}}^{\sqrt{R^2-x'^2}} \frac{z\vec{\mathbf{k}} - x'\vec{\mathbf{i}} - y'\vec{\mathbf{j}}}{\left(x'^2+y'^2+z^2\right)^{3/2}}dx'dy' \qquad [1.34]$$

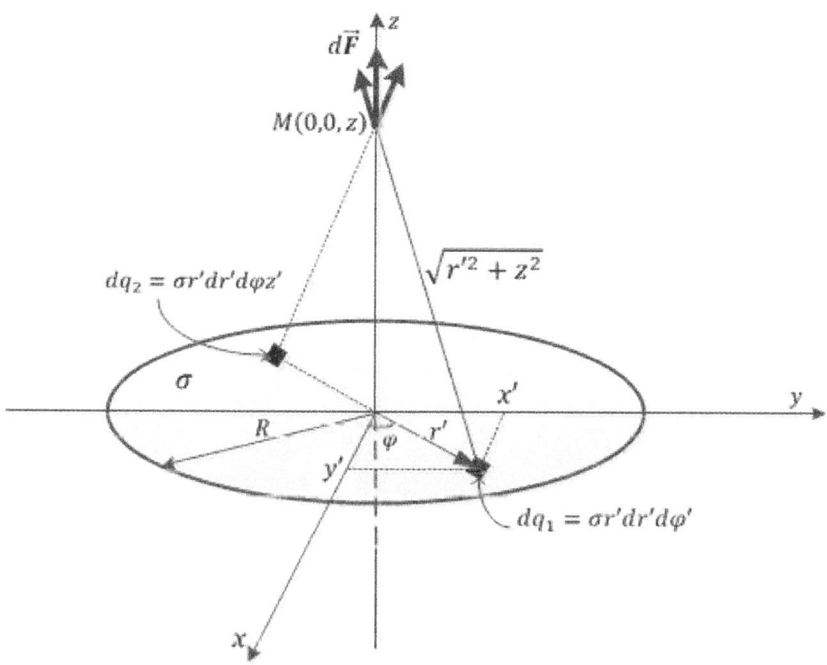

Figure 1.8. *By symmetry, the electric field created by a uniformly charged disk with radius R and no thickness is normal to the disk along points on the Oz axis*

By shifting to cylindrical coordinates, we obtain:

$$\vec{\mathbf{E}}\,(0,0,z) = \frac{Q}{4\pi^2 R^2 \varepsilon_0} \int_0^R \int_0^{2\pi} \frac{z\,\vec{\mathbf{k}} - r' cos\varphi'\,\vec{\mathbf{i}} - r' \sin\varphi'\,\vec{\mathbf{j}}}{\left(r'^2 + z^2\right)^{3/2}} r' dr' d\varphi' \qquad [1.35]$$

The integral on the variable φ' eliminates the terms in $\cos\varphi'$ and $\sin\varphi'$. Therefore, we have:

$$\vec{\mathbf{E}}\,(0,0,z) = \int_0^R \frac{Qz\,\vec{\mathbf{k}}\,r' dr'}{2\pi\varepsilon_0 R^2 \left(r'^2 + z^2\right)^{3/2}} = \frac{Q}{2\pi\varepsilon_0 R^2}\left[\frac{z\,\vec{\mathbf{k}}}{|z|} - \frac{z\,\vec{\mathbf{k}}}{\sqrt{R^2+z^2}}\right] \qquad [1.36]$$

When $z \ll R$, the electric field is reduced to $\frac{\sigma\vec{\mathbf{k}}}{2\varepsilon_0}\frac{z}{|z|}$ (i.e. to [1.31]).

1.3.1. *Distribution with spherical symmetry*

1.3.1.1. *Sphere charged in volume*

For simplicity's sake, let us start by considering a sphere of radius R uniformly charged in volume. Let us use a spherical coordinate system whose point of origin is the center of the sphere. We propose to calculate the electric field $\vec{\mathbf{E}}$ at a point M (inside or outside the sphere). By symmetry, it is clear that $\vec{\mathbf{E}}$ does not depend on the coordinates (θ, φ), as shown in Figure 1.9. We will start by writing the contribution to $\vec{\mathbf{E}}$ of the elementary charge $\rho d\tau'$ contained in a volume element $d\tau'$ ($\rho = \frac{Q}{\text{Volume of the sphere}} = \frac{3Q}{4\pi R^3}$, where Q is the total charge of the sphere) and then integrate over the sphere. The volume element is $r'^2 dr' \sin\theta' d\theta' d\varphi'$ (for simplicity, the point M is chosen along the Oz axis). The charge contained in this elementary volume creates the differential field $d\vec{\mathbf{E}} = \frac{1}{4\pi\varepsilon_0}\rho r'^2 dr' \sin\theta' d\theta' d\varphi' \frac{(\vec{\mathbf{r}} - \vec{\mathbf{r}}')}{\left|\vec{\mathbf{r}} - \vec{\mathbf{r}}'\right|^3}$ at point M. For reasons of symmetry, it is obvious that the electric field $\vec{\mathbf{E}}$ is radial. Consequently, we will consider only the radial component:

$$\begin{aligned}\vec{\mathbf{E}} &= \frac{1}{4\pi\varepsilon_0}\int_0^R \int_0^\pi \int_0^{2\pi} \frac{\rho r'^2 dr' \sin\theta' d\theta' d\varphi'}{\sqrt[3]{r^2 + r'^2 - 2rr'\cos\theta'}}\left(1 - \frac{r'}{r}\cos\theta'\right)\vec{\mathbf{r}} \\ &= \frac{1}{2\varepsilon_0}\int_0^R \int_0^\pi \frac{\rho r'^2 dr' \sin\theta' d\theta'}{\sqrt[3]{r^2 + r'^2 - 2rr'\cos\theta'}}\left(1 - \frac{r'}{r}\cos\theta'\right)\vec{\mathbf{r}}\end{aligned} \qquad [1.37]$$

In order to compute this integral, let us introduce the variable $s^2 = r^2 + r'^2 - 2rr'\cos\theta'$:

$$\vec{\mathbf{E}} = \frac{\rho}{4\varepsilon_0}\left[\int_{r'=0}^R \int_{s=|r-r'|}^{r+r'} r' dr' ds\left(1 + \frac{r^2 - r'^2}{s^2}\right)\right]\frac{\vec{\mathbf{r}}}{r^3} \qquad [1.38]$$

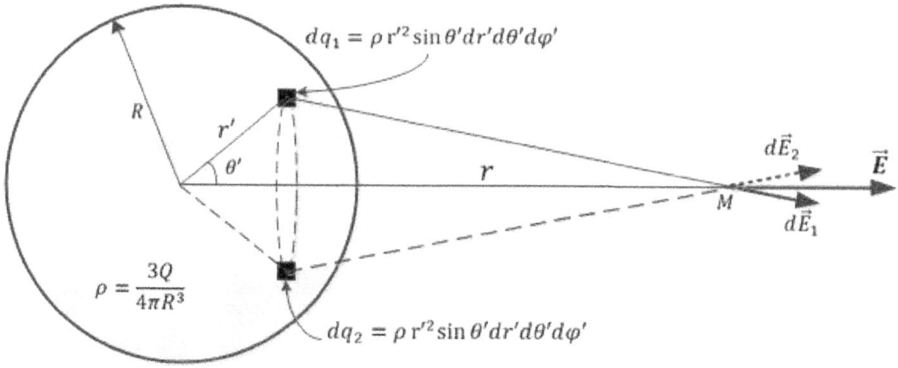

Figure 1.9. *Electric field created by a uniformly charged sphere in volume*

There are two possible cases:

– *Electric field outside the sphere*: when the point M lies outside the sphere, we always have $r > r'$:

$$\begin{aligned}\overrightarrow{\mathbf{E}}_{\text{ext}} &= \frac{\rho}{4\varepsilon_0}\left[\int_0^R \int_{r-r'}^{r+r'} r'dr'ds\left(1+\frac{r^2-r'^2}{s^2}\right)\right]\frac{\overrightarrow{\mathbf{r}}}{r^3} \\ &= \frac{\rho}{\varepsilon_0}\left[\int_0^R r'^2 dr'\right]\frac{\overrightarrow{\mathbf{r}}}{r^3} = \frac{\rho R^3}{3\varepsilon_0}\frac{\overrightarrow{\mathbf{r}}}{r^3} \equiv \frac{Q}{4\pi\varepsilon_0}\frac{\overrightarrow{\mathbf{r}}}{r^3}\end{aligned} \quad [1.39]$$

The electric field of a sphere of radius R and total charge Q coincides outside the sphere with the field of a point-like charge Q placed at the center of this sphere.

– *Electric field within the sphere*: when the point M lies within the sphere, the integration interval $\int_0^R dr'$ must be divided into two intervals, namely, $\int_0^r dr'$ and $\int_r^R dr'$:

$$\begin{aligned}\overrightarrow{\mathbf{E}}_{\text{int}} &= \frac{\rho}{4\varepsilon_0}\left[\int_0^r \int_{r-r'}^{r+r'}\left[1+\frac{r^2-r'^2}{s^2}\right] r'dr'ds \right. \\ &\left. + \int_r^R \int_{r'-r}^{r+r'}\left[1+\frac{r^2-r'^2}{s^2}\right] r'dr'ds\right]\frac{\overrightarrow{\mathbf{r}}}{r^3} \\ &= \frac{\rho}{\varepsilon_0}\left[\int_0^r r'^2 dr' + 0\right]\frac{\overrightarrow{\mathbf{r}}}{r^3} = \frac{\rho}{3\varepsilon_0}\overrightarrow{\mathbf{r}} + 0 \equiv \frac{Q}{4\pi\varepsilon_0 R^3}\overrightarrow{\mathbf{r}}\end{aligned} \quad [1.40]$$

EXAMPLE 1.13.– Let us now choose the spherically symmetric charge density described as follows: $\rho(r') = \rho_0\left(1-\frac{r'^2}{R^2}\right)$ if $r \leq R$ and 0 if $r > R$. Repeating the same calculation steps, we find the following:

– *Electric field outside the sphere*:

$$\vec{\mathbf{E}}_{\text{ext}} = \frac{\rho_0}{\varepsilon_0}\left[\int_0^R \left(1 - \frac{r'^2}{R^2}\right) r'^2 dr'\right] \frac{\vec{\mathbf{r}}}{r^3} = \frac{2\rho_0 R^3}{15\varepsilon_0} \frac{\vec{\mathbf{r}}}{r^3} \qquad [1.41]$$

– *Electric field inside the sphere*:

$$\vec{\mathbf{E}}_{\text{int}} = \frac{\rho_0}{\varepsilon_0}\left[\int_0^r \left(1 - \frac{r'^2}{R^2}\right) r'^2 dr'\right] \frac{\vec{\mathbf{r}}}{r^3} = \frac{\rho_0}{\varepsilon_0}\left(\frac{1}{3} - \frac{r^2}{5R^2}\right)\vec{\mathbf{r}} \qquad [1.42]$$

1.3.1.2. *Sphere charged on its surface*

Here, we will consider a charge Q distributed over the surface of the sphere, as shown in Figure 1.10. The radial component of the differential electric field due to the differential charge element $\sigma R^2 \sin\theta' d\theta' d\varphi'$ is $d\vec{\mathbf{E}} = \frac{\sigma R^2 (r - R\cos\theta') \sin\theta' d\theta' d\varphi'}{4\pi\varepsilon_0 [r^2 + R^2 - 2rR\cos\theta]^{3/2}} \frac{\vec{\mathbf{r}}}{r}$. Thus, the electric field created by the entire sphere is given as:

$$\begin{aligned}\vec{\mathbf{E}} &= \int_0^\pi \int_0^{2\pi} \frac{\sigma R^2 \sin\theta' \,(r - R\cos\theta')\, d\theta' d\varphi'}{4\pi\varepsilon_0 \left[r^2 + R^2 - 2rR\cos\theta\right]^{3/2}} \frac{\vec{\mathbf{r}}}{r} \\ &= \int_0^\pi \frac{\sigma R^2 \sin\theta' \,(r - R\cos\theta')\, d\theta'}{2\varepsilon_0 \left[r^2 + R^2 - 2rR\cos\theta\right]^{3/2}} \frac{\vec{\mathbf{r}}}{r}\end{aligned} \qquad [1.43]$$

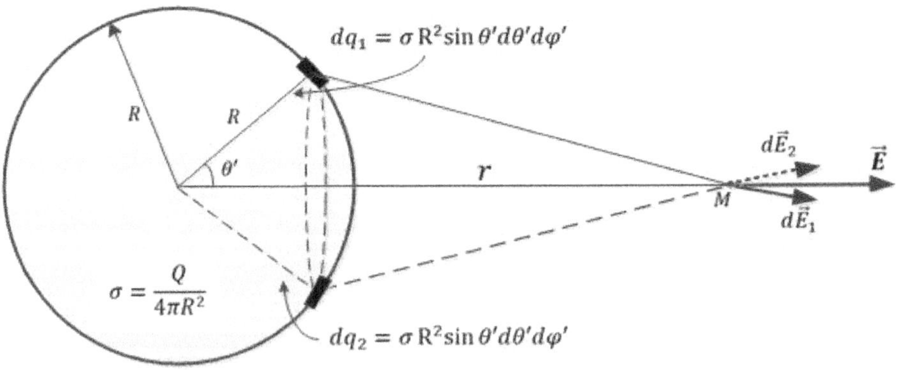

Figure 1.10. *A sphere of radius R with charge Q uniformly distributed on its surface. Symmetrically located differential charge elements indicate that the electric field is radial. The electric field inside the sphere is zero, while outside the sphere, the field is the same as if all the charge Q were concentrated in a point-like charge at the origin*

By introducing the variable change $s^2 = r^2 + R^2 - 2r\cos\theta$ in the same way, it is possible to rewrite the electric field integral as follows:

$$\begin{aligned}\overrightarrow{\mathbf{E}} &= \frac{\sigma R \overrightarrow{\mathbf{r}}}{4\varepsilon_0 r^3}\int_{|r-R|}^{r+R}\left[\frac{r^2-R^2}{s^2}+1\right]ds \\ &= \frac{\sigma R \overrightarrow{\mathbf{r}}}{4\varepsilon_0 r^3}\left[r+R-|r-R|-\left(r^2-R^2\right)\left(\frac{1}{r+R}-\frac{1}{|r-R|}\right)\right] \\ &= \begin{cases}\dfrac{\sigma R^2\overrightarrow{\mathbf{r}}}{\varepsilon_0 r^3} \equiv \dfrac{Q\overrightarrow{\mathbf{r}}}{4\pi\varepsilon_0 r^3} & \text{(outside the sphere } (r>R)) \\ 0 & \text{(inside the sphere } (r<R))\end{cases}\end{aligned} \quad [1.44]$$

1.3.2. *Thomson's model*

According to Thomson's model, a hydrogen atom is seen as a solid sphere with radius r_0 and a uniformly distributed positive charge inside. The force $\overrightarrow{\mathbf{F}}_e\left(\overrightarrow{\mathbf{r}}\right)$ acting on the electron placed at $\overrightarrow{\mathbf{r}}$ (inside the sphere) is:

$$\overrightarrow{\mathbf{F}}_e\left(\overrightarrow{\mathbf{r}}\right) = -e\overrightarrow{\mathbf{E}}\left(\overrightarrow{\mathbf{r}}\right) = -\frac{e^2}{4\pi\varepsilon_0 r_0^3}\overrightarrow{\mathbf{r}} = -k\overrightarrow{\mathbf{r}} \quad [1.45]$$

where $k = \frac{e^2}{4\pi\varepsilon_0 r_0^3}$. The electrical force acting on the electron is therefore analogous to the restoring force of a spring (of zero length in a vacuum): a central, attractive force proportional to the distance to the center (analogous to the elongation of the spring). The electron is thus elastically linked to the center of the atom. What is more, the moment of this force relative to O is zero. Using the angular momentum theorem, we can deduce that the electron's angular momentum $\overrightarrow{\mathbf{L}}$ with respect to O is constant. Since $\overrightarrow{\mathbf{L}} = m_e\overrightarrow{\mathbf{r}}\times\overrightarrow{\mathbf{v}}$, we can conclude that the vectors $\overrightarrow{\mathbf{r}}$ and $\overrightarrow{\mathbf{v}}$ are both perpendicular to a constant vector (i.e. they lie in the same plane): the electron's motion therefore takes place in a plane. Applying Newton's second law, that is, $m_e\frac{d^2\overrightarrow{\mathbf{r}}}{dt^2} = -k\overrightarrow{\mathbf{r}}$, we get:

$$\frac{d^2\overrightarrow{\mathbf{r}}}{dt^2} + \omega_0^2\overrightarrow{\mathbf{r}} = 0 \quad [1.46]$$

where $\omega_0^2 = \frac{k}{m_e}$. If ω_0 (or ν_0) corresponds to a frequency of one of the lines in the Lyman spectrum of the hydrogen atom λ_0, we write that $\omega_0 = 2\pi\nu_0 = \frac{2\pi c}{\lambda_0}$; which implies that $r_0 = \left(\frac{e^2\lambda_0^2}{16\pi^3\varepsilon_0 c^2}\right)^{1/3}$. Differential equation [1.46] has the solution:

$$\overrightarrow{\mathbf{r}}(t) = r_0 \cos\omega_0 t\, \overrightarrow{\mathbf{i}} + \frac{v_0}{\omega_0} \sin\omega_0 t\, \overrightarrow{\mathbf{j}} \qquad [1.47]$$

which corresponds to the ellipse $\frac{x^2}{r_0^2} + \frac{y^2}{\frac{v_0^2}{\omega_0^2}} = 1$ with center O and axis $\left(r_0, \frac{v_0}{\omega_0}\right)$.

In addition to the restoring force, if the electron is also subjected to a *frictional force* $\overrightarrow{\mathbf{f}} = -h\overrightarrow{\mathbf{v}}$, where $h > 0$ (an opposing force proportional to speed), the electron's equation of motion becomes:

$$m_e \frac{d^2\overrightarrow{\mathbf{r}}}{dt^2} = -k\overrightarrow{\mathbf{r}} - h\frac{d\overrightarrow{\mathbf{r}}}{dt} \qquad [1.48]$$

Let us put $\frac{h}{m_e} = 2\lambda$ and $\frac{k}{m_e} = \omega_0^2$. We obtain the *characteristic equation* $\alpha^2 + 2\lambda\alpha + \omega_0^2 = 0$ and its discriminant reduced to $\Delta = \lambda^2 - \omega_0^2$. There are three possible cases, depending on the sign of Δ:

a) $\Delta > 0$ *or* $\lambda > \omega_0$ *(high friction)*: the two solutions of the characteristic equation are real and negative: $\alpha_1 = -\lambda + \sqrt{\lambda^2 - \omega_0^2}$ and $\alpha_2 = -\lambda - \sqrt{\lambda^2 - \omega_0^2}$. The general solution of the differential equation is:

$$\overrightarrow{\mathbf{r}} = \frac{\overrightarrow{\mathbf{v}}_0 - \alpha_2 \overrightarrow{\mathbf{r}}_0}{\alpha_1 - \alpha_2} e^{\alpha_1 t} + \frac{\alpha_1 \overrightarrow{\mathbf{r}}_0 - \overrightarrow{\mathbf{v}}_0}{\alpha_1 - \alpha_2} e^{\alpha_2 t} \qquad [1.49]$$

b) $\Delta = 0$ *or* $\lambda = \omega_0$ *(critical regime)*: we have a double solution for the characteristic equation, $\alpha = -\lambda$. The general solution is:

$$\overrightarrow{\mathbf{r}} = \left[\overrightarrow{\mathbf{r}}_0 + \left(\overrightarrow{\mathbf{v}}_0 + \lambda\overrightarrow{\mathbf{r}}_0\right) t\right] e^{-\lambda t} \qquad [1.50]$$

c) $\Delta < 0$ *or* $\lambda < \omega_0$ *(low friction)*: we have two distinct complex solutions $\alpha_1 = -\lambda + \mathrm{i}\sqrt{\omega_0^2 - \lambda_0}$ and $\alpha_2 = -\lambda - \mathrm{i}\sqrt{\omega_0^2 - \lambda_0}$. The solution is therefore:

$$\overrightarrow{\mathbf{r}} = \left(\overrightarrow{\mathbf{r}}_0 \cos\sqrt{\omega_0^2 - \lambda_0} t + \frac{\left(\overrightarrow{\mathbf{v}}_0 + \lambda\overrightarrow{\mathbf{r}}_0\right)}{\sqrt{\omega_0^2 - \lambda_0}} \sin\sqrt{\omega_0^2 - \lambda_0} t\right) e^{-\lambda t} \qquad [1.51]$$

Even if the friction is small, the electron will inevitably whirl toward the center O along an elliptical trajectory with a smaller and smaller surface area. The atom described by Thomson's model is not stable.

1.3.3. *Symmetry in an electric field*

Since the field $\overrightarrow{\mathbf{E}}$ is created by a charge distribution τ', then if we know the symmetry properties of this distribution, we will know those of $\overrightarrow{\mathbf{E}}$. These properties simplify calculations considerably. If, in a homogeneous and isotropic medium, we apply a geometric transformation to a charge distribution, then its $\vec{\mathbf{E}}$ field will undergo the same transformation. If a charge distribution has a certain degree of symmetry, it is possible to deduce the field at one point from its value at another. The following can be observed:

– *Translation invariance*: a scalar function Φ or vector function $\vec{\mathbf{V}}$ is *translation invariant* along the Oz-axis, if, for any point, M and any translation parallel to Oz (regardless of displacement), we have $\Phi(M) = \Phi(M')$ and $\vec{\mathbf{V}}(M) = \vec{\mathbf{V}}(M')$, where M' is the point obtained from M through this translation. If the distribution of charges is translation invariant along the Oz-axis, the associated field $\vec{\mathbf{E}}$ does not depend on z. On the other hand, $\vec{\mathbf{E}}$ can have a component along Oz.

– *Axial symmetry*: a scalar function Φ or a vector function $\vec{\mathbf{V}}$ is said to be rotation invariant around an axis, denoted here by Oz; if for any point M and any rotation $\mathcal{R}_\varphi$ around Oz (regardless of the angle), we have $\Phi(M) = \Phi(M')$ and $\vec{\mathbf{V}}(M) = \vec{\mathbf{V}}(M')$, where $M' = \mathcal{R}_\varphi(M)$. If the distribution of charges is invariant through the rotation around the Oz axis, then the associated field, expressed here in cylindrical coordinates, does not depend on φ. We have:

$$\vec{\mathbf{E}}(M) = \int_{\tau(A)} \frac{\rho(A)\,\overrightarrow{\mathbf{AM}}d\tau(A)}{4\pi\varepsilon_0 AM^3} \qquad [1.52]$$

$$\vec{\mathbf{E}}(M') = \int_{\tau(A')} \frac{\rho(A')\,\overrightarrow{\mathbf{A'M'}}d\tau(A')}{4\pi\varepsilon_0 A'M'^3} \qquad [1.53]$$

For any point A', there exists a single point A such that $A' = \mathcal{R}_\varphi(A) \Longrightarrow \overrightarrow{\mathbf{A'M'}} = \mathcal{R}_\varphi\left(\overrightarrow{\mathbf{AM}}\right)$, $A'M' = AM$ and $d\tau(A) = d\tau(A')$. It then follows that:

$$\vec{\mathbf{E}}(M') = \int_{\tau(A)} \frac{\rho(\mathcal{R}_\varphi(A))\,\mathcal{R}_\varphi\left(\overrightarrow{\mathbf{AM}}\right)d\tau(A)}{4\pi\varepsilon_0 AM^3} \qquad [1.54]$$

Since the distribution ρ is rotation invariant (i.e. $\rho(\mathcal{R}_\varphi(A)) = \rho(A)$), it comes that:

$$\begin{aligned}\vec{\mathbf{E}}(M') &= \int_{\tau(A)} \frac{\rho(A)\,\mathcal{R}_\varphi\left(\overrightarrow{\mathbf{AM}}\right)\tau d(A)}{4\pi\varepsilon_0 AM^3} \\ &= \mathcal{R}_\varphi\left(\int_{\tau(A)} \frac{\rho(A)\,\overrightarrow{\mathbf{AM}}d\tau(A)}{4\pi\varepsilon_0 AM^3}\right) = \mathcal{R}_\varphi\left(\vec{\mathbf{E}}(M)\right)\end{aligned} \qquad [1.55]$$

– *Cylindrical symmetry*: if the charge distribution is *invariant by translation along Oz and by rotation about the same axis*, then the field depends only on the distance r from the axis Oz.

– *Spherical symmetry*: if the distribution of charges is *invariant through any rotation around a fixed point*, then $\vec{\mathbf{E}}$ only depends on the distance r measured between the center O and the field point.

– *Symmetry plane and antisymmetry plane*: we will look at symmetries and anti-symmetries with respect to a plane P. To do this, let us introduce a few useful definitions: (1) if M is any point in space, the point M' that is symmetric to M with respect to the plane P is denoted by $M' = \text{sym}_P M$ (P will appear here as the median plane of the segment MM'); (2) any vector $\vec{\mathbf{V}}$ can be written as $\vec{\mathbf{V}} = \vec{\mathbf{V}}_{\|} + \vec{\mathbf{V}}_{\perp}$, where $\vec{\mathbf{V}}_{\|}$ is the component parallel to P and $\vec{\mathbf{V}}_{\perp}$ is the normal component. The symmetric of $\vec{\mathbf{V}}$ with respect of the plane P is the vector:

$$\vec{\mathbf{V}}' = \text{sym}_P \vec{\mathbf{V}} = \vec{\mathbf{V}}_{\|} - \vec{\mathbf{V}}_{\perp} \qquad [1.56]$$

a) *Symmetry plane*: P is said to be a *symmetry plane* of the scalar function Φ if, for any point M, we have $\Phi(M') = \Phi(M)$, where $M' = \text{sym}_P M$. Similarly, P is a plane symmetry of the vector function $\vec{\mathbf{V}}$ if, for any M, we have: $\vec{\mathbf{V}}' = \vec{\mathbf{V}}(M') = \text{sym}_P \vec{\mathbf{V}}(M) = \vec{\mathbf{V}}_{\|} - \vec{\mathbf{V}}_{\perp}$, that is, $\vec{\mathbf{V}}'_{\|} = \vec{\mathbf{V}}_{\|}$ and $\vec{\mathbf{V}}'_{\perp} = -\vec{\mathbf{V}}_{\perp}$, as shown in Figure 1.11.

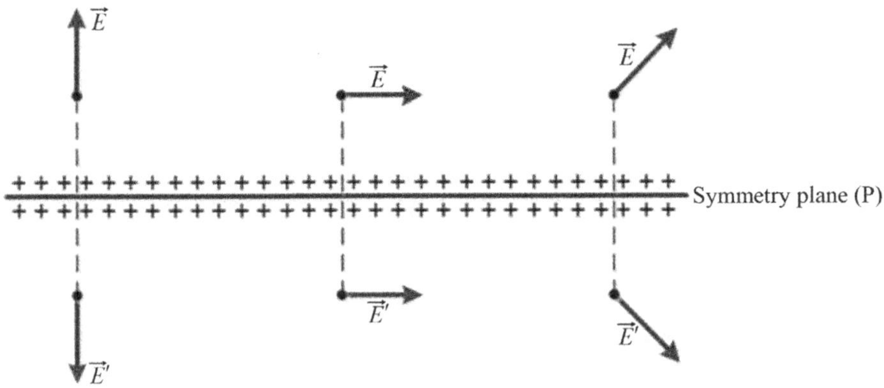

Figure 1.11. *Transformation of $\vec{\mathbf{E}}$ with respect to the symmetry plane (P): symmetry preserves the tangential component $\vec{\mathbf{E}}_{\|}$ and inverts the normal component $\vec{\mathbf{E}}_{\perp}$*

b) *Antisymmetry plane*: P is said to be a *antisymmetry plane* of Φ if, for any point M, we have $\Phi(M') = -\Phi(M)$. P is also a antisymmetry plane of $\vec{\mathbf{V}}$ if, for any

M, we have: $\vec{\mathbf{V}}' = \vec{\mathbf{V}}(M') = -\text{sym}_P \vec{\mathbf{V}}(M) = -\vec{\mathbf{V}}_{\parallel} + \vec{\mathbf{V}}_{\perp}$, that is, $\vec{\mathbf{V}}'_{\parallel} = -\vec{\mathbf{V}}_{\parallel}$ and $\vec{\mathbf{V}}'_{\perp} = \vec{\mathbf{V}}_{\perp}$, as given in Figure 1.12.

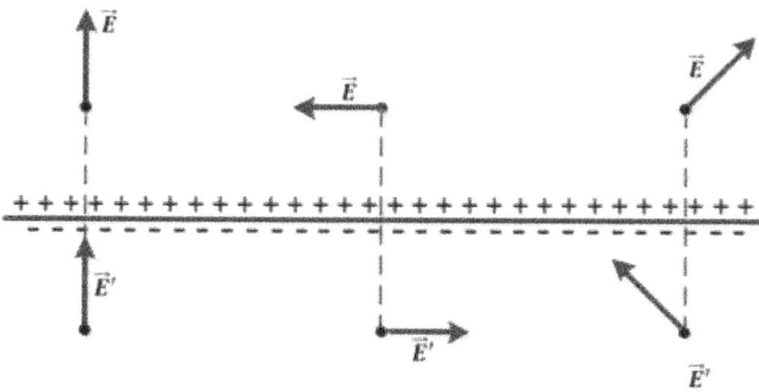

Figure 1.12. *Transformation of $\vec{\mathbf{E}}$ with respect to the antisymmetry plane (P): antisymmetry preserves the component normal $\vec{\mathbf{E}}_{\perp}$ and inverts the tangential component $\vec{\mathbf{E}}_{\parallel}$*

Consider two charges q and q' located at two points A and A' symmetrical with respect to a plane P, then calculate the fields created at two points M and M also symmetrical with respect to the same plane, as shown in Figure 1.13. The value of the electric field measured at point M is:

$$\vec{\mathbf{E}}(M) = \frac{1}{4\pi\varepsilon_0}\left(q\frac{\overrightarrow{\mathbf{AM}}}{AM^3} + q'\frac{\overrightarrow{\mathbf{A'M}}}{A'M^3}\right) = \vec{\mathbf{E}}_{\parallel}(M) + \vec{\mathbf{E}}_{\perp}(M) \qquad [1.57]$$

with:

$$\begin{aligned} \vec{\mathbf{E}}_{\parallel}(M) &= \frac{1}{4\pi\varepsilon_0}\left(q\frac{\overrightarrow{\mathbf{AM}_{\parallel}}}{AM^3} + q'\frac{\overrightarrow{\mathbf{A'M}_{\parallel}}}{A'M^3}\right) \\ \vec{\mathbf{E}}_{\perp}(M) &= \frac{1}{4\pi\varepsilon_0}\left(q\frac{\overrightarrow{\mathbf{AM}_{\perp}}}{AM^3} + q'\frac{\overrightarrow{\mathbf{A'M}_{\perp}}}{A'M^3}\right) \end{aligned} \qquad [1.58]$$

We also have

$$\vec{\mathbf{E}}(M') = \frac{1}{4\pi\varepsilon_0}\left(q\frac{\overrightarrow{\mathbf{AM'}}}{AM'^3} + q'\frac{\overrightarrow{\mathbf{A'M'}}}{A'M'^3}\right) = \vec{\mathbf{E}}_{\parallel}(M') + \vec{\mathbf{E}}_{\perp}(M') \qquad [1.59]$$

with:

$$\begin{aligned}\vec{\mathbf{E}}_{\parallel}(M') &= \frac{1}{4\pi\varepsilon_0}\left(q\frac{\overrightarrow{\mathbf{AM'}}_{\parallel}}{AM'^3} + q'\frac{\overrightarrow{\mathbf{A'M'}}_{\parallel}}{A'M'^3}\right)\\ \vec{\mathbf{E}}_{\perp}(M') &= \frac{1}{4\pi\varepsilon_0}\left(q\frac{\overrightarrow{\mathbf{AM'}}_{\perp}}{AM'^3} + q'\frac{\overrightarrow{\mathbf{A'M'}}_{\perp}}{A'M'^3}\right)\end{aligned} \qquad [1.60]$$

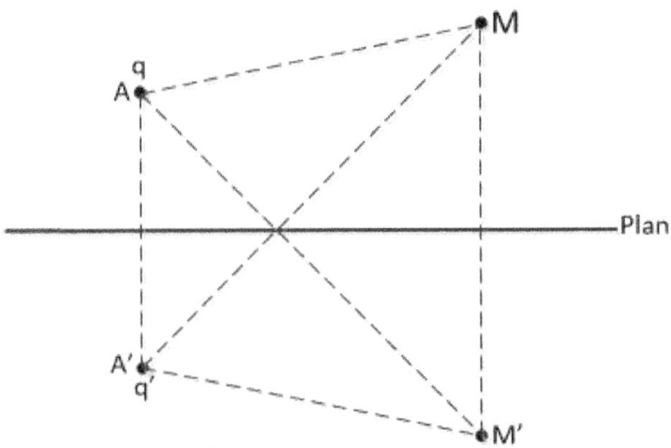

Figure 1.13. *Fields created by symmetric (respectively, antisymmetric) charges, $q' = q$ (respectively, $q' = -q$), at two symmetric points M and M' (with respect to a plane P). The symmetry (respectively, antisymmetry) of the charges implies the symmetry (respectively, antisymmetry) of the electric field with respect to the same plane. On the symmetry (respectively, antisymmetry) plane, the field $\vec{\mathbf{E}}$ is contained in this plane (respectively, is perpendicular to this plane)*

Further, since $AM' = A'M$, $\overrightarrow{\mathbf{AM'}}_{\parallel} = \overrightarrow{\mathbf{A'M}}_{\parallel}$, $\overrightarrow{\mathbf{AM'}}_{\perp} = -\overrightarrow{\mathbf{A'M}}_{\perp}$, $A'M' = AM$, $\overrightarrow{\mathbf{A'M'}}_{\parallel} = \overrightarrow{\mathbf{AM}}_{\parallel}$ and $\overrightarrow{\mathbf{A'M'}}_{\perp} = -\overrightarrow{\mathbf{AM}}_{\perp}$, it can be derived from this that:

$$\begin{aligned}\vec{\mathbf{E}}_{\parallel}(M') &= \frac{1}{4\pi\varepsilon_0}\left(q\frac{\overrightarrow{\mathbf{A'M}}_{\parallel}}{A'M^3} + q'\frac{\overrightarrow{\mathbf{AM}}_{\parallel}}{AM^3}\right)\\ \vec{\mathbf{E}}_{\perp}(M') &= \frac{1}{4\pi\varepsilon_0}\left(-q\frac{\overrightarrow{\mathbf{A'M}}_{\perp}}{A'M^3} - q'\frac{\overrightarrow{\mathbf{AM}}_{\perp}}{AM^3}\right)\end{aligned} \qquad [1.61]$$

It can therefore be concluded that:

1) If $q = q'$ (P is a symmetry plane), we therefore have $\vec{\mathbf{E}}_{\parallel}(M') = \vec{\mathbf{E}}_{\parallel}(M)$ and $\vec{\mathbf{E}}_{\perp}(M') = -\vec{\mathbf{E}}_{\perp}(M)$: the symmetry of the distribution of charges with respect to

P also implies that the field is symmetric with respect to the same plane. In particular, if M is in the plane, that is, $M' = M$, then we would have $\overrightarrow{\mathbf{E}}_{\perp}(M) = -\overrightarrow{\mathbf{E}}_{\perp}(M)$, that is, $\overrightarrow{\mathbf{E}}_{\perp}(M) = 0$ and $\overrightarrow{\mathbf{E}}(M) = \overrightarrow{\mathbf{E}}_{\parallel}(M)$. If the distribution of charges admits a plane of symmetry (P), then at all points in this plane, $\overrightarrow{\mathbf{E}}$ is contained in this field.

2) If $q' = -q$ (P is a antisymmetry plane), we thus have $\overrightarrow{\mathbf{E}}_{\parallel}(M') = -\overrightarrow{\mathbf{E}}_{\parallel}(M)$ and $\overrightarrow{\mathbf{E}}_{\perp}(M') = \overrightarrow{\mathbf{E}}_{\perp}(M)$: the antisymmetry of the distribution of charges implies the antisymmetry of the field. In particular, if $M \in P$, we will then have $\overrightarrow{\mathbf{E}}_{\parallel}(M) = -\overrightarrow{\mathbf{E}}_{\parallel}(M)$, that is, $\overrightarrow{\mathbf{E}}_{\parallel}(M) = 0$ and $\overrightarrow{\mathbf{E}}(M) = \overrightarrow{\mathbf{E}}_{\perp}(M)$. If the distribution of charges admits an antisymmetry plane (P), then at all points in this plane, $\overrightarrow{\mathbf{E}}$ is perpendicular to this plane.

These results can be easily generalized to more complex distributions.

1.3.4. *Field lines and tubes*

The concept of a *field line* or force line is very useful to form an idea of the general shape of the electric field or even a magnetic field in a region of space. It represents the orientation of the resultant electric field at a point in space. The field line of $\overrightarrow{\mathbf{E}}$ is a curve Γ defined in space so that at each of its points, the vector $\overrightarrow{\mathbf{E}}$ is tangent to this curve. It is conventionally oriented in the direction of $\overrightarrow{\mathbf{E}}$. Let $d\overrightarrow{\mathbf{l}}$ be an elementary displacement along a field line Γ. The fact that at every point on Γ, the field $\overrightarrow{\mathbf{E}}$ is parallel to $d\overrightarrow{\mathbf{l}}$ and translated by:

$$\overrightarrow{\mathbf{E}} \times d\overrightarrow{\mathbf{l}} = 0 \qquad [1.62]$$

This vector equation gives three coupled differential equations (these differential equations define the field line) that must be solved. For example, in Cartesian coordinates, we get:

$$\begin{array}{l} E_y(x,y,z)\,dz - E_z(x,y,z)\,dy = 0 \\ E_z(x,y,z)\,dx - E_x(x,y,z)\,dz = 0 \\ E_x(x,y,z)\,dy - E_y(x,y,z)\,dx = 0 \end{array} \qquad [1.63]$$

Solving these differential equations brings up three indeterminate constants. The values of these constants are fixed by choosing any point on the field line being considered. Another form of writing can also be adopted: if Γ is a curve that is parameterized by $\overrightarrow{\mathbf{r}}(t) = x(t)\,\overrightarrow{\mathbf{i}} + y(t)\,\overrightarrow{\mathbf{j}} + z(t)\,\overrightarrow{\mathbf{k}}$, with $t \in \mathbb{R}$, the vector that is tangent to Γ at the point $\overrightarrow{\mathbf{r}}(t)$ is the vector $\frac{d\overrightarrow{\mathbf{r}}(t)}{dt} = \frac{dx(t)}{dt}\overrightarrow{\mathbf{i}} + \frac{dy(t)}{dt}\overrightarrow{\mathbf{j}} + \frac{dz(t)}{dt}\overrightarrow{\mathbf{k}}$.

Thus, Γ is a field line for $\overrightarrow{\mathbf{E}} = E_x\left[x\left(t\right), y\left(t\right), z\left(t\right)\right]\overrightarrow{\mathbf{i}} + E_y\left[x\left(t\right), y\left(t\right), z\left(t\right)\right]\overrightarrow{\mathbf{j}} + E_z\left[x\left(t\right), y\left(t\right), z\left(t\right)\right]\overrightarrow{\mathbf{k}}$ if, for any t, we have:

$$\begin{aligned} \frac{dx\left(t\right)}{dt} &= E_x\left[x\left(t\right), y\left(t\right), z\left(t\right)\right] \\ \frac{dy\left(t\right)}{dt} &= E_y\left[x\left(t\right), y\left(t\right), z\left(t\right)\right] \\ \frac{dz\left(t\right)}{dt} &= E_z\left[x\left(t\right), y\left(t\right), z\left(t\right)\right] \end{aligned} \quad [1.64]$$

For any fixed point $\overrightarrow{\mathbf{r}}_0 = \overrightarrow{\mathbf{r}}\left(t_0\right)$, there is only one field line that passes through it.

EXAMPLE 1.14.– Let us determine the field lines of the vector field $\overrightarrow{\mathbf{V}} = -y\,\overrightarrow{\mathbf{i}} + x\,\overrightarrow{\mathbf{j}}$: $\overrightarrow{\mathbf{r}}\left(t\right) = x\left(t\right)\overrightarrow{\mathbf{i}} + y\left(t\right)\overrightarrow{\mathbf{j}} + z\left(t\right)\overrightarrow{\mathbf{k}}$ describes a field line if $\frac{d\overrightarrow{\mathbf{r}}(t)}{dt} = \frac{dx(t)}{dt}\overrightarrow{\mathbf{i}} + \frac{dy(t)}{dt}\overrightarrow{\mathbf{j}} + \frac{dz(t)}{dt}\overrightarrow{\mathbf{k}} = \overrightarrow{\mathbf{V}}\left[x\left(t\right), y\left(t\right), z\left(t\right)\right] = -y\left(t\right)\overrightarrow{\mathbf{i}} + x\left(t\right)\overrightarrow{\mathbf{j}}$, that is:

$$\begin{aligned} \frac{dx\left(t\right)}{dt} &= -y\left(t\right) \\ \frac{dy\left(t\right)}{dt} &= x\left(t\right) \\ \frac{dz\left(t\right)}{dt} &= 0 \end{aligned} \quad [1.65]$$

Thus, $\frac{dx(t)}{dt}x\left(t\right) + \frac{dy(t)}{dt}y\left(t\right) = \frac{1}{2}\frac{d}{dt}\left[x^2\left(t\right) + y^2\left(t\right)\right] = 0 \Longrightarrow x^2\left(t\right) + y^2\left(t\right) = C_1^2$ and $z\left(t\right) = C_2$, where C_1 and C_2 are constants. Γ describes a circle in the horizontal plane centered on the Oz axis.

EXAMPLE 1.15.– Let us find the field lines for the central field $f\left(r\right)\overrightarrow{\mathbf{e}}_r$: in spherical coordinates, a parameterized curve Γ is given by $r\left(t\right) \in \left]0, \infty\right[$, $\theta\left(t\right) \in \left[0, \pi\right]$ and $\varphi\left(t\right) \in \left[0, 2\pi\right[$. The points along the curve are marked by their position vectors $\overrightarrow{\mathbf{r}}\left(t\right) = r\left(t\right)\overrightarrow{\mathbf{e}}_r\left(t\right)$, where $\overrightarrow{\mathbf{e}}_r\left(t\right)$ also depends on t, since it changes direction with the point $\overrightarrow{\mathbf{r}}\left(t\right)$. The tangent vector to Γ at the point $\overrightarrow{\mathbf{r}}\left(t\right)$ is therefore $\frac{d\overrightarrow{\mathbf{r}}}{dt} = \frac{dr}{dt}\overrightarrow{\mathbf{e}}_r + r\frac{d\overrightarrow{\mathbf{e}}_r}{dt}$. Consequently, Γ is a field line if $\frac{dr}{dt}\overrightarrow{\mathbf{e}}_r + r\frac{d\overrightarrow{\mathbf{e}}_r}{dt} = f\left(r\right)\overrightarrow{\mathbf{e}}_r$ ($\frac{d\overrightarrow{\mathbf{e}}_r}{dt}$ is written as a function of $\overrightarrow{\mathbf{e}}_\theta\left(t\right)$ and $\overrightarrow{\mathbf{e}}_\varphi\left(t\right)$), that is:

$$\begin{aligned} \frac{dr\left(t\right)}{dt} &= f\left(r\right) \\ \frac{d\overrightarrow{\mathbf{e}}_r}{dt} &= 0 \end{aligned} \quad [1.66]$$

The second equation indicates that $\overrightarrow{\mathbf{e}}_r\,(t)$ is constant. The field lines are therefore radial lines centered on the source of the field. The first equation gives the distance $r\,(t)$ to the source:

$$\frac{dr\,(t)}{dt} = f\,(r) \Longrightarrow t = \int_{r_0}^{r} \frac{dr}{f\,(r)} \qquad [1.67]$$

where r_0 is the initial distance of the source point. In particular, for $f\,(r) = \frac{q}{4\pi\varepsilon_0 r^2}$, we will have $r^3 = r_0^3 + \frac{4\pi\varepsilon_0 t}{3q}$. For $q > 0$, the field lines are oriented toward the exterior: the electric field is repulsive. In contrast, if $q < 0$, the field lines converge toward the source: the electric field is attractive, as depicted in Figure 1.14. ■

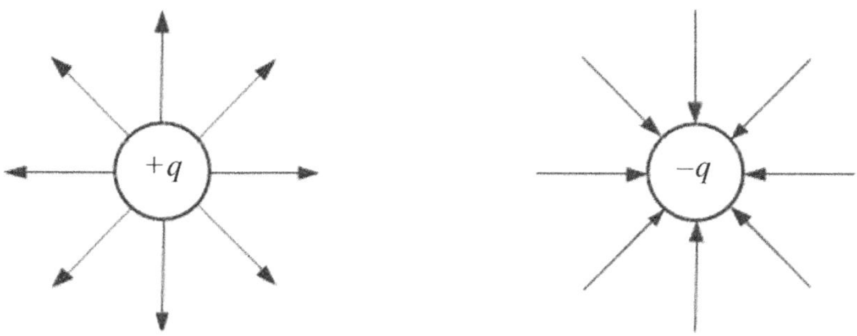

Field lines produced by positive charges **Field lines absorbed by negative charges**

Figure 1.14. *The field lines are oriented curves that translate the characteristics (direction, orientation and also intensity) of a vector field in a region in space: the electric field lines are produced by the positive charges (repulsion) and absorbed by the negative charges (attraction)*

A *field tube* is a bundle of field lines that go across a surface. Here are a few observations:

– The field lines are continuous between the positive and negative charges. The field lines are produced by the positive charges and absorbed by the negative charges.

– The number of field lines produced or absorbed by a charge is proportional to the value of the charge (the greater the charge, the more it produces or absorbs lines).

– The field lines must respect the symmetry of the distribution of charges.

– The field lines must not intersect.

Moving away from the charge distribution, the field lines appear to come from a point-like charge whose value is equal to the net charge of the distribution.

1.4. Electric potential

In the presence of two charges with opposite signs, work must be done against the force of attraction to separate them. This work can be recovered if the charges are allowed to approach each other. Similarly, if the charges are of the same sign, work must be done against the force of repulsion to bring them together. This work can be recovered if the charges are allowed to move away from each other. Therefore, a charge gains energy when it is moved in the opposite direction to the force that keeps it in a system (attraction) or pushes it away from a system (repulsion). In this case, we use the term *potential energy*, as the quantity of energy depends on the position of the charge in a force field.

Consider an electric field $\vec{\mathbf{E}}\left(\vec{\mathbf{r}}\right)$ created by a single point-like charge q' located at the coordinate origin ($\vec{\mathbf{E}}\left(\vec{\mathbf{r}}\right)$ and the field measured at position (x, y, z)). Let us take a test charge q small enough that its presence does not disturb the first charge q'. The work required to move this test charge at constant speed from M to another point N along a given trajectory will be:

$$W_{M \to N} = -\int_M^N q\vec{\mathbf{E}}\left(\vec{\mathbf{r}}\right) \cdot d\vec{\mathbf{r}} \qquad [1.68]$$

The negative sign indicates that the work carried out is against the field. The scalar product in [1.68] indicates that no work is required to displace the test charge perpendicular to the electric field (along spheres with a constant radius in our case). These surfaces are called *equipotential surfaces*. Non-zero work is required to move q to a different radius. Consequently, it can be seen that the work [1.68] depends only on the start and end positions (r_M and r_N) of the trajectory, and not on the form of the trajectory itself, as shown in Figure 1.15:

$$W_{M \to N} = -\frac{qq'}{4\pi\varepsilon_0}\int_{\vec{\mathbf{r}}_M}^{\vec{\mathbf{r}}_N} \frac{\vec{\mathbf{r}} \cdot d\vec{\mathbf{r}}}{r^3} = \frac{qq'}{4\pi\varepsilon_0}\left(\frac{1}{r_N} - \frac{1}{r_M}\right) \qquad [1.69]$$

It is easy to check the adequacy of the sign in [1.69] with the previous discussion: (1) if $qq' < 0$ (charge attraction) and $r_N > r_M$ (charge separation), then $W_{M \to N} > 0$ (positive work done on q to separate the two charges); (2) if $qq' > 0$ (chare repulsion) and $r_N < r_M$ (charges coming together), then $W_{M \to N} > 0$ (positive work done on q to bring the two charges together).

For a closed path C (i.e. starting and ending at the same point $r_N = r_M$), the work required to displace the charge is zero:

$$W|_C = -\oint_C q\vec{\mathbf{E}} \cdot d\vec{\mathbf{r}} = -\frac{qq'}{4\pi\varepsilon_0}\oint_C \frac{\vec{\mathbf{r}} \cdot d\vec{\mathbf{r}}}{r^3} = \frac{qq'}{4\pi\varepsilon_0}\oint_C d\left(\frac{1}{r}\right) = 0 \quad [1.70]$$

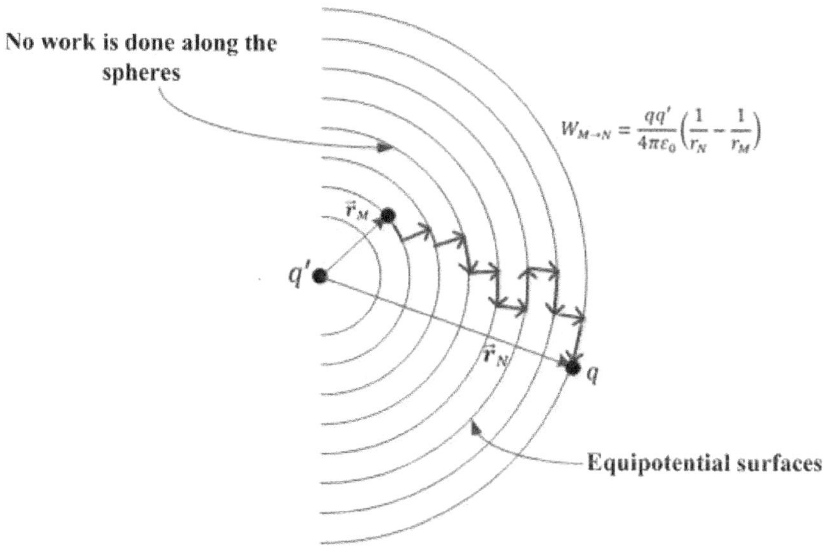

Figure 1.15. *The work required to displace a test charge q in the field of another point-like charge q' depends only on the starting and ending position, because it is zero on equipotential surfaces*

If the electric field is no longer created by a point-like charge q', but by a distribution of charges contained in τ', each curvilinear integral associated with a differential charge $\rho\left(\vec{\mathbf{r}}'\right) d\tau'$ is zero. We will therefore have for the entire distribution:

$$\oint \vec{\mathbf{E}}\left(\vec{\mathbf{r}}\right) \cdot d\vec{\mathbf{r}} = 0 \qquad [1.71]$$

The electric field is therefore *conservative*. This important property comes from the fact that the electrical force is a central force. Using Stokes's theorem, it is possible to convert the integral [1.71] into a surface integral of the curl of the electric field $\oint_C \vec{\mathbf{E}} \cdot d\vec{\mathbf{r}} = \int_S \left(\vec{\nabla} \times \vec{\mathbf{E}}\right) \cdot d\vec{\mathbf{S}}$. It follows that at any point in space, we will have:

$$\vec{\nabla} \times \vec{\mathbf{E}}\left(\vec{\mathbf{r}}\right) = 0 \qquad [1.72]$$

On the other hand, we also know that the curl of the gradient of a scalar function along a closed path is zero. This means that the electric field can be determined (or fully described) from the gradient of a scalar function of the position, called *electric potential*:

$$\vec{\mathbf{E}}\left(\vec{\mathbf{r}}\right)=-\vec{\nabla}V\left(\vec{\mathbf{r}}\right) \qquad [1.73]$$

It is easy to check that equations [1.72] and [1.73] are compatible: the field [1.73] satisfies equation [1.72]. The advantage of introducing potential is technical. If we can calculate the function V (a simple, double or triple integral to be realized), it is possible to directly obtain the three components of the electric field by applying the operator $-\vec{\nabla}$. This is often easier than calculating each component of the field separately (three integrals to be performed). It should be noted that:

1) The negative sign in equation [1.73] takes into account that the electric field is directed toward decreasing potentials.

2) According to the elementary properties of the gradient, the electric field is always perpendicular to surfaces with constant potential.

3) The potential V is determined to an additive constant. The potential V and $V+A$ (A is a constant here) give the same value for $\vec{\mathbf{E}}$. It is therefore necessary to specify a zero reference potential, which is often taken to infinity. In practice, zero potential is often attributed to the surface of the Earth (the "ground" reference point).

4) Since $\vec{\mathbf{E}}\cdot d\vec{\mathbf{r}}=-\vec{\nabla}V\cdot d\vec{\mathbf{r}}=-dV$, we deduce that electric potential is measured at a point $\vec{\mathbf{r}}$ by the integral $V\left(\vec{\mathbf{r}}\right)=-\int\vec{\mathbf{E}}\left(\vec{\mathbf{r}}\right)\cdot d\vec{\mathbf{r}}+A$ (A is an additive constant). Assuming that the potential is always zero at infinity, we finally obtain:

$$V\left(\vec{\mathbf{r}}\right)=\int_x^\infty\int_y^\infty\int_z^\infty\vec{\mathbf{E}}\left(\vec{\mathbf{r}}\right)\cdot d\vec{\mathbf{r}} \qquad [1.74]$$

The potential is a *continuous* function, except at points where the field tends to infinity. This follows directly from $dV=-\vec{\mathbf{E}}\cdot d\vec{\mathbf{r}}$: when $d\vec{\mathbf{r}}$ tends to zero, dV tends to zero if $\vec{\mathbf{E}}$ is finite in the neighborhood. We will see later that the electric field is discontinuous in the neighborhood of a charged surface.

5) The potential difference between two points $\vec{\mathbf{r}}_M$ and $\vec{\mathbf{r}}_N$ represents the work per unit charge required to move from $\vec{\mathbf{r}}_M$ to $\vec{\mathbf{r}}_N$ quasi-statically:

$$V\left(\vec{\mathbf{r}}_N\right)-V\left(\vec{\mathbf{r}}_M\right)=-\int_M^N q\vec{\mathbf{E}}\left(\vec{\mathbf{r}}\right)\cdot d\vec{\mathbf{r}}=\frac{W_{M\to N}}{q} \qquad [1.75]$$

It must be noted that equations [1.71], [1.72] and [1.75] are the *field version* of Kirchhoff's second law, which states that the algebraic sum of the potential differences around a closed loop in a circuit is zero. Using the usual convention that V is zero at infinity, it can finally be written that:

$$W_{\infty\to N}=qV\left(\vec{\mathbf{r}}_N\right)\equiv W_e\left(\vec{\mathbf{r}}_N\right) \qquad [1.76]$$

$W_e\left(\vec{\mathbf{r}}_N\right)$ is called the *electrostatic potential energy* of the charge q in the field $\vec{\mathbf{E}}\left(\vec{\mathbf{r}}\right)$ or, more accurately, it is the potential energy of the interaction between q and the sources of $\vec{\mathbf{E}}\left(\vec{\mathbf{r}}\right)$. We can adopt the following definition: *the electrostatic potential energy, W_e, of a charged particle placed in an electric field is equal to the work that must be carried out to bring this particle, quasi-statically, from infinity to its current position.*

6) The potential created at the point $\vec{\mathbf{r}}$ by a point-like charge q' placed at the origin is:

$$V\left(\vec{\mathbf{r}}\right) = \int_{\vec{\mathbf{r}}}^{\infty} \frac{q'dr}{4\pi\varepsilon_0 r^2} = \frac{q'}{4\pi\varepsilon_0 r} \qquad [1.77]$$

The potential $V\left(\vec{\mathbf{r}}\right)$ is then interpreted as the work per unit charge required to bring a charge from infinity to a certain distance r, from a point-like charge q'.

7) According to the superposition principle, the electric potential is additive. In fact, we have: $\vec{\mathbf{E}} = \vec{\mathbf{E}}_1 + \cdots + \vec{\mathbf{E}}_n = -\vec{\nabla}V_1 - \cdots - \vec{\nabla}V_n = -\vec{\nabla}\left(V_1 + \cdots + V_n\right)$, which gives $V = V_1 + \cdots + V_n$.

8) The net potential of multiple point-like charges is obtained by summing over the individual potentials of all of the charges:

$$V\left(\vec{\mathbf{r}}\right) = \sum_{i=1}^{n} \frac{q_i'}{4\pi\varepsilon_0 \left|\vec{\mathbf{r}} - \vec{\mathbf{r}}_i'\right|} \qquad [1.78]$$

If there is a continuous charge distribution, the summation becomes an integration over all of the differential charge elements dq':

$$V\left(\vec{\mathbf{r}}\right) = \int_{\text{Dist.}} \frac{dq'}{4\pi\varepsilon_0 \left|\vec{\mathbf{r}} - \vec{\mathbf{r}}'\right|} \qquad [1.79]$$

where the integral is a curvilinear integral for a linear charge, a surface integral for a surface charge density, and a volume integral for a volume charge density.

$$V\left(\vec{\mathbf{r}}\right) = \begin{cases} \displaystyle\int_{C'} \frac{\lambda\left(\vec{\mathbf{r}}'\right) dl'}{4\pi\varepsilon_0 \left|\vec{\mathbf{r}} - \vec{\mathbf{r}}'\right|} & \text{(linear charge)} \\ \displaystyle\int_{S'} \frac{\sigma\left(\vec{\mathbf{r}}'\right) dS'}{4\pi\varepsilon_0 \left|\vec{\mathbf{r}} - \vec{\mathbf{r}}'\right|} & \text{(surface charge density)} \\ \displaystyle\int_{\tau'} \frac{\rho\left(\vec{\mathbf{r}}'\right) d\tau'}{4\pi\varepsilon_0 \left|\vec{\mathbf{r}} - \vec{\mathbf{r}}'\right|} & \text{(volume charge density)} \end{cases} \qquad [1.80]$$

9) The electric field can be easily found again by applying the operator $-\overrightarrow{\nabla}$ over the potential V and by recalling that the derivatives must be taken with respect to the field positions (x, y, z), while the integration is carried out over the source positions (x', y', z'). The operator $\overrightarrow{\nabla}$ can therefore pass inside the integral and act on the quantity $\left|\overrightarrow{\mathbf{r}} - \overrightarrow{\mathbf{r}}'\right|$ only:

$$\overrightarrow{\mathbf{E}} = -\overrightarrow{\nabla} V = -\int_{\text{Dist.}} \frac{dq'}{4\pi\varepsilon_0} \overrightarrow{\nabla} \left(\frac{1}{\left|\overrightarrow{\mathbf{r}} - \overrightarrow{\mathbf{r}}'\right|} \right) = \int_{\text{Dist.}} \frac{dq' \left(\overrightarrow{\mathbf{r}} - \overrightarrow{\mathbf{r}}'\right)}{4\pi\varepsilon_0 \left|\overrightarrow{\mathbf{r}} - \overrightarrow{\mathbf{r}}'\right|^3} \qquad [1.81]$$

EXAMPLE 1.16. (Uniformly Charged Infinitely Long Wire).– Let us calculate the potential created by an infinite rectilinear wire carrying a uniform linear charge λ, as shown in Figure 1.3. The potential at point M is:

$$V = \int_{-\infty}^{\infty} \frac{\lambda dz'}{4\pi\varepsilon_0 \sqrt{r^2 + z'^2}} = \frac{\lambda}{4\pi\varepsilon_0} \ln\left(z' + \sqrt{z'^2 + r^2}\right)\Big|_{-\infty}^{\infty} \qquad [1.82]$$

This calculation is impossible because the potential cannot be zero at infinity in this case. This can be shown simply by calculating the potential difference between two points M and N from the field found in Example 1.8:

$$V(r_M) - V(r_N) = \int_{r_M}^{r_N} \overrightarrow{\mathbf{E}} \cdot d\overrightarrow{\mathbf{r}} = \int_{r_M}^{r_N} \frac{\lambda dr}{2\pi\varepsilon_0 r} = \frac{\lambda}{2\pi\varepsilon_0} \ln\left(\frac{r_N}{r_M}\right) \qquad [1.83]$$

We can clearly see that we cannot take $r_N = \infty$ and $V(r_N) = 0$. This is because the wire is infinitely long: the source is not localized and there are charges at infinity.

EXAMPLE 1.17. (Uniformly Charged Disc with Negligible Thickness).– Let us calculate the potential created by a flat disk, with no thickness, with radius R and charge Q (uniformly distributed over the disk), at a point M located on the axis of the disk denoted by Oz, a distance z from its center, as shown in Figure 1.8. The charge density is $\rho\left(\overrightarrow{\mathbf{r}}'\right) = \frac{Q}{\pi R^2}\delta(z')$ for $x'^2 + y'^2 \leq R^2$ and 0 elsewhere.

Consequently, the electric potential measured at the site M is:

$$\begin{aligned} V(0, 0, z) &= \frac{1}{4\pi\varepsilon_0} \int_{\text{Disc}} \frac{\rho\left(\overrightarrow{\mathbf{r}}'\right) d\tau'}{\left|\overrightarrow{\mathbf{r}} - \overrightarrow{\mathbf{r}}'\right|} = \frac{1}{4\pi\varepsilon_0} \int_0^R \int_0^{2\pi} \frac{Q}{\pi R^2} \frac{r' dr' d\varphi'}{\sqrt{r'^2 + z^2}} \\ &= \frac{Q}{2\pi\varepsilon_0 R^2} \left(\sqrt{R^2 + z^2} - |z|\right) \end{aligned} \qquad [1.84]$$

To once again arrive at the electric field $\vec{\mathbf{E}}\,(0,0,z)$, it is enough to write:

$$\vec{\mathbf{E}}\,(0,0,z) = -\frac{\partial V}{\partial z}\vec{\mathbf{k}} = \frac{Q}{2\pi\varepsilon_0 R^2}\frac{z}{|z|}\left(\frac{z}{|z|} - \frac{z}{\sqrt{R^2+z^2}}\right)\vec{\mathbf{k}} \qquad [1.85]$$

EXAMPLE 1.18. (Hollow Sphere with Uniform Surface Charge).– Consider a hollow sphere of radius R, with a charge Q that is uniformly distributed across its surface, as shown in Figure 1.10. Let us determine the potential created by the sphere at the point M. For reasons of symmetry, the field point M is chosen along the axis Oz. The differential charge $\sigma R^2 \sin\theta' d\theta' d\varphi'$ $\left(\sigma = \frac{Q}{4\pi R^2}\right)$ located on its surface produces the differential potential at the point M given by $dV = \frac{\sigma R^2 \sin\theta' d\theta' d\varphi'}{4\pi\varepsilon_0\sqrt{R^2+r'^2-2r'R\cos\theta'}}$. By integrating over the entire sphere, we obtain:

$$\begin{aligned} V(r) &= \int_{\theta'=0}^{\pi}\int_{\varphi'=0}^{2\pi} \frac{\sigma R^2 \sin\theta' d\theta' d\varphi'}{4\pi\varepsilon_0\sqrt{R^2+r'^2-2r'R\cos\theta'}} \\ &= \int_0^{\pi} \frac{\sigma R^2 \sin\theta' d\theta'}{2\varepsilon_0\sqrt{R^2+r'^2-2r'R\cos\theta'}} \end{aligned} \qquad [1.86]$$

Let us perform the change of variable $s^2 = R^2 + r'^2 - 2r'R\cos\theta'$

$$V(r) = \int_{|R-r|}^{R+r} \frac{\sigma R ds}{2r\varepsilon_0} = \frac{\sigma R}{2r\varepsilon_0}\left(R + r - |R-r|\right) = \begin{cases} \dfrac{\sigma R}{\varepsilon_0} = \dfrac{Q}{4\pi\varepsilon_0 R} & r \le R \\ \dfrac{\sigma R^2}{\varepsilon_0 r} = \dfrac{Q}{4\pi\varepsilon_0 r} & r > R \end{cases} \qquad [1.87]$$

and consequently:

$$\vec{\mathbf{E}}\,(r) = -\frac{dV}{dr}\frac{\vec{\mathbf{r}}}{r} = \begin{cases} 0 & r \le R \\ \dfrac{Q\vec{\mathbf{r}}}{4\pi\varepsilon_0 r^3} & r > R \end{cases} \qquad [1.88]$$

It can be seen that the potential is continuous along the boundary $r = R$, while the field is discontinuous.

EXAMPLE 1.19.– (Solid Sphere Uniformly Charged). Let us return to the sphere with radius R and charge Q uniformly distributed throughout its volume, as shown in Figure 1.9. It is possible to arrive at the expressions for the electric field inside and outside of the sphere from the potential $V\left(\vec{\mathbf{r}}\right)$. The differential charge

$\rho r'^2 dr' \sin\theta' d\theta' d\varphi'$, located inside the sphere, induces the potential $dV = \frac{\rho r'^2 dr' \sin\theta' d\theta' d\varphi'}{4\pi\varepsilon_0\sqrt{r^2+r'^2-2rr'\cos\theta'}}$ at the point M. Consequently, the potential created by the whole sphere can be obtained from the following integral (for reasons of symmetry, the potential will only depend on r):

$$\begin{aligned} V(r) &= \int_0^R \int_0^\pi \int_0^{2\pi} \frac{\rho r'^2 dr' \sin\theta' d\theta' d\varphi'}{4\pi\varepsilon_0\sqrt{r^2+r'^2-2rr'\cos\theta'}} \\ &= \frac{\rho}{2\varepsilon_0 r}\int_0^R r'dr' \int_{s=|r-r'|}^{r+r'} ds = \frac{\rho}{2\varepsilon_0 r}\int_0^R r'dr' \left(r+r'-|r-r'|\right) \end{aligned} \quad [1.89]$$

– *Electric potential outside the sphere* ($r > R$):

$$V_{\text{ext}}(r) = \frac{\rho}{\varepsilon_0 r}\int_0^R r'^2 dr' = \frac{\rho R^3}{3\varepsilon_0 r} = \frac{Q}{4\pi\varepsilon_0 r} \quad [1.90]$$

As with $\vec{\mathbf{E}}_{\text{ext}}$, the potential V_{ext} is the same as if the charge Q were concentrated at the center of the sphere. We arrive again at the electric field $\vec{\mathbf{E}}_{\text{ext}}$ using formula [1.77], that is, $\vec{\mathbf{E}}_{\text{ext}} = -\vec{\nabla}V_{\text{ext}} = -\frac{dV_{\text{ext}}}{dr}\frac{\vec{\mathbf{r}}}{r}$; this gives:

$$\vec{\mathbf{E}}_{\text{ext}} = \frac{Q}{4\pi\varepsilon_0}\frac{\vec{\mathbf{r}}}{r^3} \quad [1.91]$$

– *Electric potential inside the sphere* ($r \leq R$):

$$\begin{aligned} V_{\text{int}}(r) &= \frac{\rho}{\varepsilon_0 r}\int_{r'=0}^{r} r'^2 dr' + \frac{\rho}{\varepsilon_0}\int_{r'=r}^{R} r'dr' = \frac{\rho R^2}{6\varepsilon_0}\left[3 - \left(\frac{r}{R}\right)^2\right] \\ &= \frac{Q}{8\pi\varepsilon_0 R}\left[3 - \left(\frac{r}{R}\right)^2\right] \end{aligned} \quad [1.92]$$

and consequently:

$$\vec{\mathbf{E}}_{\text{int}} = -\vec{\nabla}V_{\text{int}} = -\frac{dV_{\text{int}}}{dr}\frac{\vec{\mathbf{r}}}{r} = \frac{Q}{4\pi\varepsilon_0 R^3}\vec{\mathbf{r}} \quad [1.93]$$

EXAMPLE 1.20.– If the sphere's volume charge density is governed by $\rho(r') = \rho_0\left(1 - \frac{r'^2}{R^2}\right)$ for $r' \leq R$ and 0 for $r' > R$, the electric potential has the value:

$$V(r) = \frac{\rho_0}{2\varepsilon_0 r}\int_0^R \left(1 - \frac{r'^2}{R^2}\right) r'\left(r + r' - |r - r'|\right) dr' \quad [1.94]$$

– *Electric potential outside the sphere*:

$$V_{\text{ext}} = \frac{\rho_0}{\varepsilon_0 r} \int_{r'=0}^{R} \left(1 - \frac{r'^2}{R^2}\right) r'^2 dr' = \frac{2\rho_0 R^3}{15\varepsilon_0 r} \qquad [1.95]$$

$$\vec{\mathbf{E}}_{\text{ext}} = -\vec{\nabla} V_{\text{ext}} = -\frac{dV_{\text{ext}}}{dr}\frac{\vec{\mathbf{r}}}{r} = \frac{2\rho_0 R^3}{15\varepsilon_0}\frac{\vec{\mathbf{r}}}{r^3} \qquad [1.96]$$

– *Electric potential inside the sphere*:

$$\begin{aligned} V_{\text{int}} &= \frac{\rho_0}{\varepsilon_0 r}\left[\int_0^r \left(1 - \frac{r'^2}{R^2}\right) r'^2 dr' + r \int_r^R \left(1 - \frac{r'^2}{R^2}\right) r' dr'\right] \\ &= \frac{\rho_0 R^2}{4\varepsilon_0}\left[1 - \frac{2}{3}\left(\frac{r}{R}\right)^2 + \frac{1}{5}\left(\frac{r}{R}\right)^4\right] \end{aligned} \qquad [1.97]$$

$$\vec{\mathbf{E}}_{\text{int}} = -\vec{\nabla} V_{\text{int}} = -\frac{dV_{\text{int}}}{dr}\frac{\vec{\mathbf{r}}}{r} = \frac{\rho_0}{\varepsilon_0}\left[\frac{1}{3} - \frac{1}{5}\left(\frac{r}{R}\right)^2\right]\vec{\mathbf{r}} \qquad [1.98]$$

1.5. Equipotential surfaces

The set of points in space whose potential V is equal to a certain constant value V_0 constitutes an *equipotential surface* V_0.

Consider an equipotential surface passed through a point $M(x, y, z)$. If we make an infinitesimal displacement on this surface toward a neighboring point $N(x + dx, y + dy + dz)$, the vector $\overrightarrow{\mathbf{MN}}$ is then in the plane tangent to the surface at M.

Moreover, $dV = -\vec{\mathbf{E}} \cdot \overrightarrow{\mathbf{MN}}$ and, by definition, the variation in potential during this displacement is zero, since $V(M) = V(N)$; this implies that $\vec{\mathbf{E}} \perp \overrightarrow{\mathbf{MN}}$. We can therefore conclude that the field $\vec{\mathbf{E}}$ is perpendicular to the equipotential at any point. Consequently, the field lines are perpendicular to equipotential surfaces.

EXAMPLE 1.21. (Equipotentials and Field Lines for Two Linear Charges of Opposite Polarity).–

1) Let us return to the example of a linear distribution with charge density λ and length L (located on Oz between $-\frac{L}{2}$ and $\frac{L}{2}$), as seen in Figure 1.16(a). The potential measured at the point $M(x, y, z)$ takes the value:

$$
\begin{aligned}
V(r,z) &= \int_{-\frac{L}{2}}^{\frac{L}{2}} \frac{\lambda dz'}{4\pi\varepsilon_0\sqrt{r^2+(z-z')^2}} = -\frac{\lambda \ln\left(\frac{z-\frac{L}{2}+\sqrt{r^2+\left(z-\frac{L}{2}\right)^2}}{z+\frac{L}{2}+\sqrt{r^2+\left(z+\frac{L}{2}\right)^2}}\right)}{4\pi\varepsilon_0} \\
&= -\frac{\lambda}{4\pi\varepsilon_0}\left[\sinh^{-1}\left(\frac{z-\frac{L}{2}}{r}\right)-\sinh^{-1}\left(\frac{z+\frac{L}{2}}{r}\right)\right]
\end{aligned}
\qquad [1.99]
$$

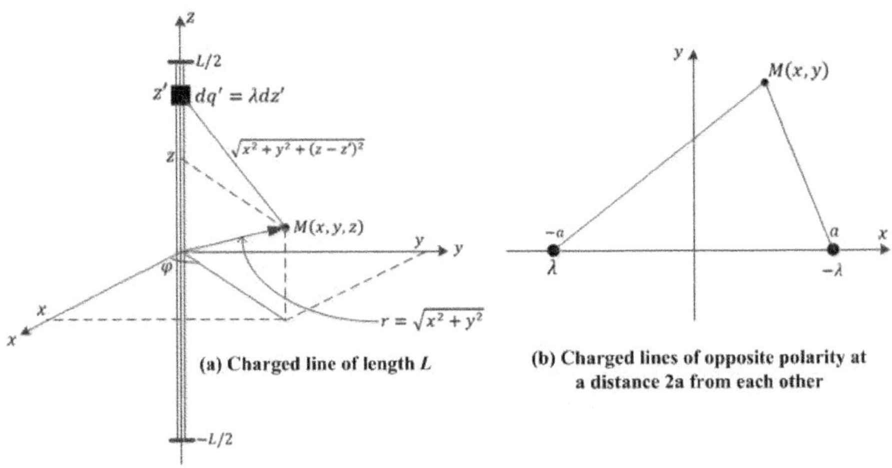

Figure 1.16. *Potential created by two linear charges with opposite polarities, which are parallel and separated by a distance of* $2a$

The components of $\overrightarrow{\mathbf{E}}$ can be obtained as follows:

$$
\begin{aligned}
E_z(r,z) &= -\frac{\partial V}{\partial z} = \frac{\lambda}{4\pi\varepsilon_0}\left[\frac{1}{\sqrt{r^2+\left(z-\frac{L}{2}\right)^2}}-\frac{1}{\sqrt{r^2+\left(z+\frac{L}{2}\right)^2}}\right] \\
E_r(r,z) &= -\frac{\partial V}{\partial r} = -\frac{\lambda}{4\pi\varepsilon_0 r}\left[\frac{z-\frac{L}{2}}{\sqrt{r^2+\left(z-\frac{L}{2}\right)^2}}-\frac{z+\frac{L}{2}}{\sqrt{r^2+\left(z+\frac{L}{2}\right)^2}}\right]
\end{aligned}
\qquad [1.100]
$$

When L becomes large, the expressions for the field and for the potential approach the expressions of an infinitely long line of charge:

$$\lim_{L\longrightarrow\infty}\begin{cases}E_z=0\\E_r=\dfrac{\lambda}{2\pi\varepsilon_0 r}\\V=-\dfrac{\lambda}{2\pi\varepsilon_0}\left[\ln r-\ln L\right]\end{cases}\qquad [1.101]$$

We observe that the potential has a constant term that becomes infinite as L tends toward infinity. The presence of this constant is explained by the fact that the potential [1.79] is not zero here for $r\longrightarrow\infty$, since the linear charge is infinitely long and, consequently, there is a non-zero charge at infinity. This additive and infinite constant is not important, as it does not contribute to the expression of the electric field and can therefore be ignored. Very far from the charge line, the potential [1.99] approaches the potential of a point charge λL:

$$\lim_{r^2+z^2\gg L^2} V=\frac{\lambda L}{4\pi\varepsilon_0\sqrt{r^2+z^2}}.\qquad [1.102]$$

2) If we are now in the presence of two infinite linear charges with opposite polarities, $\pm\lambda$, parallel and separated from each other by a distance $2a$ (the origin is chosen on the median), as shown in Figure 1.16(b), the total potential is just the superposition of these potentials [1.101]:

$$V(x,y)=-\frac{\lambda}{2\pi\varepsilon_0}\ln\sqrt{\frac{(x+a)^2+y^2}{(x-a)^2+y^2}}\qquad [1.103]$$

where the additive constants cancel each other out. Consequently, the equipotential surfaces (equipotential lines are observed in the Oxy plane) are described by the expressions:

$$\frac{(x+a)^2+y^2}{(x-a)^2+y^2}=e^{-4\pi\varepsilon_0 V/\lambda}\equiv K_1\qquad [1.104]$$

where K_1 is a constant (specific to each equipotential). This last equation can also be rewritten in the form:

$$\left(x-\frac{a(1+K_1)}{K_1-1}\right)^2+y^2-\frac{4K_1a^2}{(1-K_1)^2}\qquad [1.105]$$

This is the equation for a circle with radius $R=\frac{2\sqrt{K_1}a}{|1-K_1|}$ and center $\left(\frac{a(1+K_1)}{K_1-1},0\right)$, as seen in Figure 1.17.

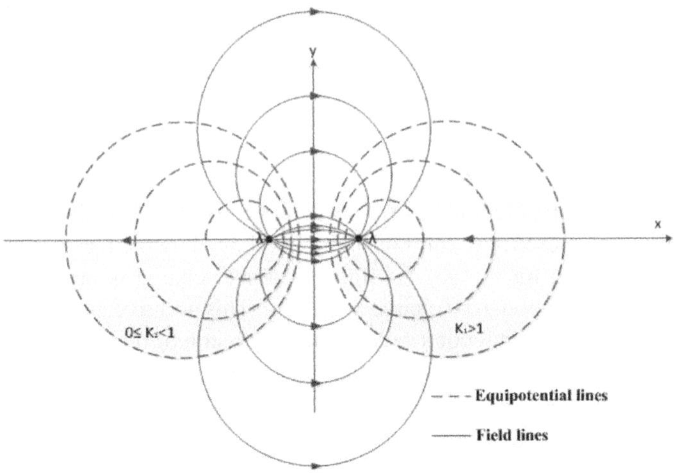

Figure 1.17. *Equipotential lines (dashed lines) and field lines (solid lines) forming a set of orthogonal circles*

For $K_1 = 1$, we obtain a circle of infinite radius whose center is at $(\pm\infty, 0)$: this therefore represents the plane $x = 0$. For $0 \leq K_1 < 1$, the equipotential circles are in the left half-plane, while for $1 < K_1 < +\infty$, the circles are in the right half-plane.

3) The electric field associated with the potential [1.103] is:

$$\vec{\mathbf{E}} = -\vec{\nabla} V = \frac{\lambda}{2\pi\varepsilon_0} \frac{2a\left(y^2 + a^2 - x^2\right)\vec{\mathbf{i}} - 4axy\,\vec{\mathbf{j}}}{\left[(x-a)^2 + y^2\right]\left[(x+a)^2 + y^2\right]} \qquad [1.106]$$

The following differential equation must be solved to obtain the field lines:

$$\frac{dy}{dx} = \frac{E_y}{E_x} = -\frac{2xy}{y^2 + a^2 - x^2} \qquad [1.107]$$

which can also be written as:

$$\frac{d\left(x^2 + y^2\right)}{a^2 - \left(x^2 + y^2\right)} + d\left(\ln y\right) = 0 \qquad [1.108]$$

After integrating the above equation, we get:

$$x^2 + \left(y - \frac{a}{\tan K_2}\right)^2 = \frac{a^2}{\sin^2 K_2} \qquad [1.109]$$

where K_2 is an integration constant that must be determined by specifying a point (x_0, y_0) along the chosen field line. The field lines are therefore also circles with radius $\frac{a}{\sin K_2}$ and centers at the points $\left(0, \frac{a}{\tan K_2}\right)$.

1.6. Gauss' law

Consider a closed surface S oriented toward the exterior and enclosing a volume τ. Let us take a differential surface element $d\vec{\mathbf{S}}$ (of the total surface) located around a point $M\ (x, y, z)$. The quantity:

$$d\Phi = \vec{\mathbf{E}} \cdot d\vec{\mathbf{S}} = \vec{\mathbf{E}} \cdot \vec{\mathbf{n}}\, dS = E \cos\theta dS = -\left[\frac{\partial V}{\partial x}\cos\alpha + \frac{\partial V}{\partial y}\cos\beta + \frac{\partial V}{\partial z}\cos\gamma\right] dS \quad [1.110]$$

where α, β and γ are the director angles of the outwardly directed normal $\vec{\mathbf{n}}$ to the surface dS, is called the *differential flux* of the field $\vec{\mathbf{E}}$ through the differential surface element $d\vec{\mathbf{S}}$ ($\vec{\mathbf{E}}$ is the field measured at the point $M\ (x, y, z)$). θ is the angle measured between $\vec{\mathbf{E}}$ and $\vec{\mathbf{n}}$, as shown in Figure 1.18. The expression $\frac{\partial V}{\partial x}\cos\alpha + \frac{\partial V}{\partial y}\cos\beta + \frac{\partial V}{\partial z}\cos\gamma$ is called the *directional derivative* V along $\vec{\mathbf{n}}$; it is denoted by $\frac{\partial V}{\partial n}$. Consequently, the total flux of $\vec{\mathbf{E}}$ across the entire surface S is equal to the sum of the elementary fluxes across the various elements, dS, of this surface, that is, the integral:

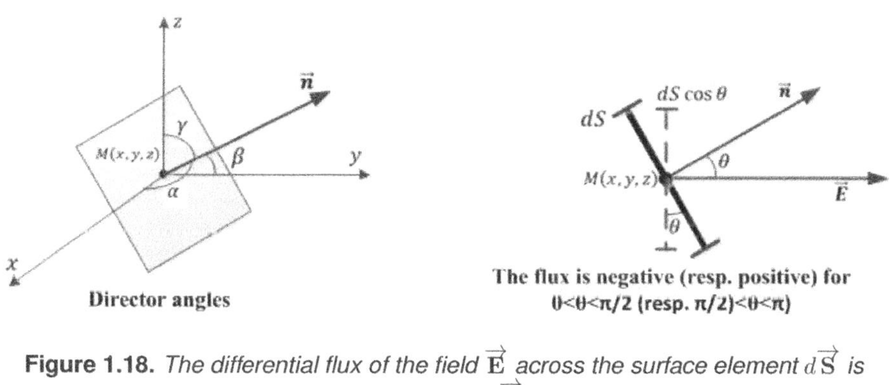

Figure 1.18. *The differential flux of the field $\vec{\mathbf{E}}$ across the surface element $d\vec{\mathbf{S}}$ is equal to the product of the magnitude of $\vec{\mathbf{E}}$ and $dS\cos\theta$ (the projection of dS onto the perpendicular plane to the field direction)*

$$\Phi = \oint_S \vec{\mathbf{E}} \cdot \vec{\mathbf{n}} dS = \oint_S E \cos\theta dS = -\oint_S \left[\frac{\partial V}{\partial x} \cos\alpha + \frac{\partial V}{\partial y} \cos\beta + \frac{\partial V}{\partial z} \cos\gamma \right] dS$$

[1.111]

1.6.1. *Integral form*

Let us now consider the two cases depicted in Figure 1.19, where the volume of an arbitrary shape τ (the surface surrounding τ is denoted S) surrounds a point-like charge q (i.e. q is located inside S) or is in the proximity of a point-like charge q (q is located outside the surface S). The point-like charge q is at position $M'(x', y', z')$. In both cases, the electric field emanates radially from the point-like charge. We wish to calculate the flux of the electric field across the surface S surrounding the volume:

$$\Phi = \oint_S \vec{\mathbf{E}} \cdot d\vec{\mathbf{S}} = \oint_S \frac{q\left(\vec{\mathbf{r}} - \vec{\mathbf{r}}'\right)}{4\pi\varepsilon_0 \left|\vec{\mathbf{r}} - \vec{\mathbf{r}}'\right|^3} \cdot d\vec{\mathbf{S}} = -\frac{q}{4\pi\varepsilon_0} \oint_S \vec{\nabla} \left(\frac{1}{\left|\vec{\mathbf{r}} - \vec{\mathbf{r}}'\right|} \right) \cdot d\vec{\mathbf{S}}$$

[1.112]

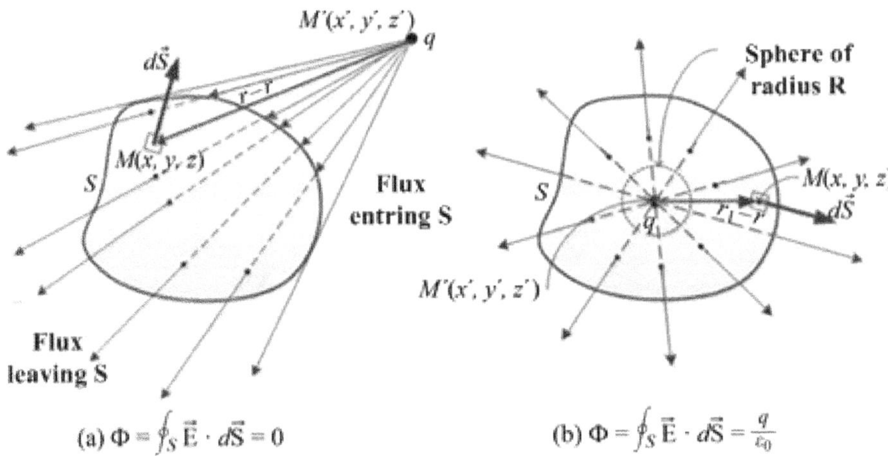

Figure 1.19. *(a) The net flux of the electric field across a closed surface due to a point-like charge outside the surface is zero. (b) The flux of the electric field from a point-like charge enclosed within a surface travels entirely across this surface*

Let us now use the divergence theorem to convert the surface integral [1.112] into a volume integral:

$$\begin{aligned}\Phi &= \oint_S \vec{\mathbf{E}} \cdot d\vec{\mathbf{S}} = -\frac{q}{4\pi\varepsilon_0} \int_\tau \vec{\nabla} \cdot \left[\vec{\nabla}\left(\frac{1}{\left|\vec{\mathbf{r}} - \vec{\mathbf{r}}'\right|}\right)\right] d\tau \\ &= -\frac{q}{4\pi\varepsilon_0} \int_\tau \Delta\left(\frac{1}{\left|\vec{\mathbf{r}} - \vec{\mathbf{r}}'\right|}\right) d\tau\end{aligned} \qquad [1.113]$$

1) If the point-like charge q is located outside the surface (i.e. outside the volume), the vectors $\vec{\mathbf{r}}$ and $\vec{\mathbf{r}}'$ never coincide (i.e. we always have $\vec{\mathbf{r}} - \vec{\mathbf{r}}' \neq 0$); this implies that $\Delta\left(\frac{1}{\left|\vec{\mathbf{r}} - \vec{\mathbf{r}}'\right|}\right) = 0$ (see Appendix A.1, which will be available online at: www.iste.co.uk/abdesselam/electrodynamics1.zip). It is thus seen that the net flux $\vec{\mathbf{E}}$ across the surface is equal to zero:

$$\Phi = \oint_S \vec{\mathbf{E}} \cdot d\vec{\mathbf{S}} = 0 \qquad [1.114]$$

This result can be explained qualitatively in Figure 1.19(a). The electric field coming from q on the part of surface S closest to q (i.e. the part of the surface facing q) has its normal component directed in the opposite direction to $d\vec{\mathbf{S}}$; this makes a negative contribution to the flux. On the other hand, on the other part of the surface, the electric field has its normal component in the same direction as $d\vec{\mathbf{S}}$; this is a positive contribution to the flux. These contributions are equal in magnitude but opposite in sign, so the total net flux is zero.

2) If the point-like charge q is now inside the surface, as shown in Figure 1.19(b), the electric field points outward when the charge q is positive, with a normal component in the direction of $d\vec{\mathbf{S}}$ at any point on S, so the flux is positive. On the contrary, if the charge q is negative, $\vec{\mathbf{E}}$ and $d\vec{\mathbf{S}}$ are oriented in opposite directions at any point of S, so that the flux is now negative. In both cases, the total net flux must therefore be non-zero. The function $\frac{1}{\left|\vec{\mathbf{r}} - \vec{\mathbf{r}}'\right|}$ becomes singular when $\vec{\mathbf{r}}$ and $\vec{\mathbf{r}}'$ coincide (because the charge is inside the surface). Consequently, the relation $\Delta\left(\frac{1}{\left|\vec{\mathbf{r}} - \vec{\mathbf{r}}'\right|}\right) = 0$ is no longer true when $\vec{\mathbf{r}} = \vec{\mathbf{r}}'$ and must be replaced by the Dirac distribution $-.4\pi\delta\left(\vec{\mathbf{r}} - \vec{\mathbf{r}}'\right)$ (see Appendix A.1, which will be available online at www.iste.co.uk/abdesselam/electrodynamics1.zip):

$$\Phi = \oint_S \vec{\mathbf{E}} \cdot d\vec{\mathbf{S}} = \frac{q}{4\pi\varepsilon_0} \int_\tau 4\pi\delta\left(\vec{\mathbf{r}} - \vec{\mathbf{r}}'\right) d\tau = \frac{q}{\varepsilon_0} \qquad [1.115]$$

It can clearly be seen that the sign of the flux depends on the polarity of the point-like charge. This result can also be obtained in the following way: since $\Delta\left(\frac{1}{\left|\vec{\mathbf{r}} - \vec{\mathbf{r}}'\right|}\right)$ is equal to zero everywhere, except when $\vec{\mathbf{r}} = \vec{\mathbf{r}}'$, it is possible to reduce the surface S to a small spherical surface S' with radius R surrounding the point-like charge (i.e. centered on the point $\overrightarrow{\mathbf{r}}'$). In the rest of the volume, we have $\overrightarrow{\mathbf{r}} \neq \overrightarrow{\mathbf{r}}'$ and, consequently, $\Delta\left(\frac{1}{\left|\overrightarrow{\mathbf{r}}-\overrightarrow{\mathbf{r}}'\right|}\right) = 0$:

$$\Phi = \oint_S \frac{q\left(\overrightarrow{\mathbf{r}} - \overrightarrow{\mathbf{r}}'\right)}{4\pi\varepsilon_0\left|\overrightarrow{\mathbf{r}} - \overrightarrow{\mathbf{r}}'\right|^3} \cdot d\overrightarrow{\mathbf{S}} = \oint_{\text{Spheres}S'} \frac{q\left(\overrightarrow{\mathbf{r}} - \overrightarrow{\mathbf{r}}'\right)}{4\pi\varepsilon_0\left|\overrightarrow{\mathbf{r}} - \overrightarrow{\mathbf{r}}'\right|^3} \cdot d\overrightarrow{\mathbf{S}}' \quad [1.116]$$

It is known that the electric field is purely radial over this sphere, that is, it is in the same direction as $d\overrightarrow{\mathbf{S}}'$. By performing the change of variable $\overrightarrow{\mathbf{R}} = \overrightarrow{\mathbf{r}} - \overrightarrow{\mathbf{r}}'$ and noting that $\overrightarrow{\mathbf{R}} \cdot d\overrightarrow{\mathbf{S}}' = R^3 d\Omega = R^3 \sin\theta d\theta d\varphi$, we obtain:

$$\oint_{S'} \frac{q\left(\overrightarrow{\mathbf{r}} - \overrightarrow{\mathbf{r}}'\right)}{4\pi\varepsilon_0\left|\overrightarrow{\mathbf{r}} - \overrightarrow{\mathbf{r}}'\right|^3} \cdot d\overrightarrow{\mathbf{S}}' = \frac{q}{4\pi\varepsilon_0}\int_0^{\pi}\int_0^{2\pi} \frac{R^3 \sin\theta d\theta d\varphi}{R^3} = \frac{q}{\varepsilon_0} \quad [1.117]$$

Thus, the flux across a closed surface S of the electric field produced by a point-like charge located within this surface is:

$$\Phi = \oint_S \overrightarrow{\mathbf{E}} \cdot d\overrightarrow{\mathbf{S}} = \frac{q}{\varepsilon_0} \quad [1.118]$$

If there are multiple point-like charges within the surface S, each charge q_i gives rise to a flux $\frac{q_i}{\varepsilon_0}$. It thus follows that the net flux of $\overrightarrow{\mathbf{E}}$ across a closed surface is equal to the product of $\frac{1}{\varepsilon_0}$ into the net charge enclosed by the surface:

$$\oint_S \overrightarrow{\mathbf{E}} \cdot d\overrightarrow{\mathbf{S}} = \frac{1}{\varepsilon_0} \sum_{q_i \text{ within } S} q_i \quad [1.119]$$

Charges located outside S do not contribute to the flux.

For a continuous distribution of charges, the flux, across any closed surface S, of the differential electric field $d\overrightarrow{\mathbf{E}}$ created by the differential element dq has the value:

$$d\Phi = \oint_S d\overrightarrow{\mathbf{E}} \cdot d\overrightarrow{\mathbf{S}} = \oint_S \frac{dq}{4\pi\varepsilon_0} \frac{\left(\overrightarrow{\mathbf{r}} - \overrightarrow{\mathbf{r}}'\right)}{\left|\overrightarrow{\mathbf{r}} - \overrightarrow{\mathbf{r}}'\right|^3} \cdot d\overrightarrow{\mathbf{S}} = \begin{cases} 0 & dq \text{ outside } S \\ \frac{dq}{\varepsilon_0} & dq \text{ inside } S \end{cases} \quad [1.120]$$

By integrating over the entire distribution, we obtain:

$$\Phi = \int_D \oint_S d\overrightarrow{\mathbf{E}} \cdot d\overrightarrow{\mathbf{S}} = \oint_S \left(\int_D d\overrightarrow{\mathbf{E}}\right) \cdot d\overrightarrow{\mathbf{S}} = \oint_S \overrightarrow{\mathbf{E}} \cdot d\overrightarrow{\mathbf{S}} = \frac{1}{\varepsilon_0}\int_{D\cap\tau} dq \quad [1.121]$$

where Dist. $\cap\ \tau$ indicates the part of the distribution located within the surface S. The quantity $\int_{\text{Dist.}\cap\tau} dq$ represents the total charge contained within S, as seen in Figure 1.20. We will use the notation:

$$Q_S = \int_{\text{Dist.}\cap\tau} dq \qquad [1.122]$$

Equation [1.127] therefore takes the form:

$$\oint_S \overrightarrow{\mathbf{E}} \cdot d\overrightarrow{\mathbf{S}} = \frac{Q_S}{\varepsilon_0} \qquad [1.123]$$

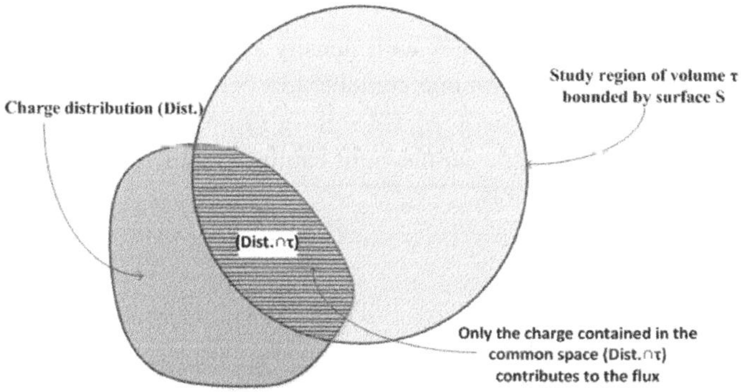

Figure 1.20. *Only the charge Q_S contained in the overlapping region between τ and Dist. contributes to the flux of the electric field across the surface S. The charge outside τ has no effect*

This is the *integral expression of Gauss's theorem* (or *Gauss's law*): *the electric flux across a closed surface, oriented from the interior to the exterior, is equal to the product of $\frac{1}{\varepsilon_0}$ multiplied by the total charge contained in the closed surface.* The surface S is sometimes called a *Gaussian surface*. Q_S includes all enclosed differential charge elements, so the total charge enclosed in this surface can be a line integral, a surface integral and/or a volume integral in addition to the sum of the point-like charges:

$$\begin{aligned}\oint_S \varepsilon_0 \overrightarrow{\mathbf{E}}\ d\overrightarrow{\mathbf{S}} &= \sum_{q_i \text{ within } S} q_i + \int_{q \text{ within } S} dq \\ &= \left(\sum q_i + \int_C \lambda dl + \int_S \sigma dS + \int_\tau \rho d\tau\right)_{\text{charge within } S}\end{aligned} \qquad [1.124]$$

Charges outside the volume do not contribute to the total flux across the surface S. Note that the quantities $\sum q_i$, $\int_C \lambda dl$ and $\int_S \sigma dS$ can also be expressed by triple integrals using Dirac distributions. Gauss' theorem [1.124] can be used to significantly simplify the calculation of the field for charges distributed in a spatially symmetric manner. The idea is to find a surface S with sections that are tangents to the electric field, so that the scalar product is zero, and sections perpendicular to the field and over which the field is constant, so that the scalar product and integration become simple multiplications. If such an appropriate surface is found, the surface integral becomes very easy to evaluate. Knowing the direction of the electric field often suggests the appropriate surface on which the integration can be performed [1.124].

1.6.2. *Differential form*

If a volume distribution of charges with density ρ and volume τ' is completely enclosed by a closed surface S (volume contained in S is greater than tau'), Gauss' theorem makes it possible to write $\oint_S \varepsilon_0 \overrightarrow{\mathbf{E}} \cdot d\overrightarrow{\mathbf{S}} = Q_S = \int_{\tau'} \rho d\tau'$. Since ρ is zero outside the volume τ', the integral on the right can be extended over the whole of τ without affecting the result. That is:

$$\oint_S \varepsilon_0 \overrightarrow{\mathbf{E}} \cdot d\overrightarrow{\mathbf{S}} = \int_\tau \rho d\tau \qquad [1.125]$$

The left-hand side of [1.125] can be transformed into a volume integral using the divergence theorem:

$$\int_\tau \overrightarrow{\nabla} \cdot \left(\varepsilon_0 \overrightarrow{\mathbf{E}}\right) d\tau = \int_\tau \rho d\tau \qquad [1.126]$$

Since [1.126] must be true for any volume τ, the quantities under the integral must therefore be equal at all points in space, that is, $\overrightarrow{\nabla} \cdot \left(\varepsilon_0 \overrightarrow{\mathbf{E}}\right) = \rho$, which gives:

$$\overrightarrow{\nabla} \cdot \overrightarrow{\mathbf{E}} = \frac{\rho}{\varepsilon_0} \qquad [1.127]$$

(As the permittivity of the vacuum, ε_0 is constant, it is possible to move it outside the divergence operator). This is the *differential expression for Gauss' theorem, the local form, or the Maxwell–Gauss equation*. This local form is not enough to determine $\overrightarrow{\mathbf{E}}$ from ρ, as we can add any vector function $\overrightarrow{\mathbf{A}}$ with zero divergence and obtain another solution for [1.127]. Considerations of symmetry can allow us to give more information on E, but in general, this local form of Gauss' theorem must be complemented by other constraints. The first constraint can be provided by equation [1.72], but this is not always sufficient to uniquely determine the electrostatic field because if $\vec{\mathbf{E}}$ satisfies equations [1.127] and [1.72], then $\vec{\mathbf{E}} + \vec{\mathbf{A}}$, where $\vec{\mathbf{A}}$ is constant, also satisfies these equations.

1.6.3. *The Helmholtz–Hodge theorem*

In order to uniquely determine the field $\vec{\mathbf{E}}$, we must use constraints related to the asymptotic behavior of the field. This can be done with the Helmholtz–Hodge theorem, which stipulates that any vector field can be separated into a longitudinal (*irrotational*, curl-free) component and a transverse (*solenoidal*, divergence-free) component, or the sum of the gradient of a scalar field and the curl of a vector field.

Suppose there exists a vector field $\vec{\mathbf{E}}$ whose curl $\vec{\nabla} \times \vec{\mathbf{E}}$ and divergence $\vec{\nabla} \cdot \vec{\mathbf{E}}$ we know. Using the Dirac distribution, $\vec{\mathbf{E}}$ can be expressed as follows:

$$\vec{\mathbf{E}}\left(\vec{\mathbf{r}}\right) = \int_{\tau'} \vec{\mathbf{E}}\left(\vec{\mathbf{r}}'\right) \delta\left(\vec{\mathbf{r}} - \vec{\mathbf{r}}'\right) d\tau' = -\frac{1}{4\pi} \int_{\tau'} \vec{\mathbf{E}}\left(\vec{\mathbf{r}}'\right) \Delta \left(\frac{1}{\left|\vec{\mathbf{r}} - \vec{\mathbf{r}}'\right|}\right) d\tau' \quad [1.128]$$

Further, since Δ acts only on $\vec{\mathbf{r}}$, we can derive the following (for simplicity's sake, we will write $R = \left|\vec{\mathbf{r}} - \vec{\mathbf{r}}'\right|$):

$$\vec{\mathbf{E}}\left(\vec{\mathbf{r}}\right) = -\frac{1}{4\pi}\Delta \int_{\tau'} \frac{\vec{\mathbf{E}}\left(\vec{\mathbf{r}}'\right)}{R} d\tau' = -\frac{1}{4\pi}\left[\vec{\nabla}\left(\vec{\nabla} \cdot \int_{\tau'} \frac{\vec{\mathbf{E}}\left(\vec{\mathbf{r}}'\right)}{R} d\tau'\right) - \vec{\nabla} \times \left(\vec{\nabla} \times \int_{\tau'} \frac{\vec{\mathbf{E}}\left(\vec{\mathbf{r}}'\right)}{R} d\tau'\right)\right] = -\frac{1}{4\pi}\left[\vec{\nabla}\left(\int_{\tau'} \vec{\mathbf{E}}\left(\vec{\mathbf{r}}'\right) \cdot \vec{\nabla}\left(\frac{1}{R}\right) d\tau'\right) + \vec{\nabla} \times \left(\int_{\tau'} \vec{\mathbf{E}}\left(\vec{\mathbf{r}}'\right) \times \vec{\nabla}\left(\frac{1}{R}\right) d\tau'\right)\right] = -\frac{1}{4\pi}\left[-\vec{\nabla}\left(\int_{\tau'} \vec{\mathbf{E}}\left(\vec{\mathbf{r}}'\right) \cdot \vec{\nabla}'\left(\frac{1}{R}\right) d\tau'\right) - \vec{\nabla} \times \left(\int_{\tau'} \vec{\mathbf{E}}\left(\vec{\mathbf{r}}'\right) \times \vec{\nabla}'\left(\frac{1}{R}\right) d\tau'\right)\right] \quad [1.129]$$

where the following relations have been used: $\Delta\vec{\mathbf{A}} = \vec{\nabla}\left(\vec{\nabla} \cdot \vec{\mathbf{A}}\right) - \vec{\nabla} \times \left(\vec{\nabla} \times \vec{\mathbf{A}}\right)$ and $\vec{\nabla}\left(\frac{1}{R}\right) = -\vec{\nabla}'\left(\frac{1}{R}\right)$. Now, by virtue of the identities $\vec{\mathbf{A}} \cdot \vec{\nabla}\psi = -\psi\left(\vec{\nabla} \cdot \vec{\mathbf{E}}\right) +$

$\vec{\nabla} \cdot \left(\psi \vec{\mathbf{A}}\right)$ and $\vec{\mathbf{A}} \times \vec{\nabla}\psi = \psi\left(\vec{\nabla} \times \vec{\mathbf{E}}\right) - \vec{\nabla} \times \left(\psi\vec{\mathbf{A}}\right)$, we find that:

$$\vec{\mathbf{E}}\left(\vec{\mathbf{r}}\right) = -\frac{1}{4\pi}\left[-\vec{\nabla}\left(-\int_{\tau'} \frac{\vec{\nabla}' \cdot \vec{\mathbf{E}}\left(\vec{\mathbf{r}}'\right)}{\left|\vec{\mathbf{r}} - \vec{\mathbf{r}}'\right|} d\tau' + \int_{\tau'} \vec{\nabla}' \cdot \left(\frac{\vec{\mathbf{E}}\left(\vec{\mathbf{r}}'\right)}{\left|\vec{\mathbf{r}} - \vec{\mathbf{r}}'\right|}\right) d\tau'\right) \right.$$
$$\left. -\vec{\nabla} \times \left(\int_{\tau'} \frac{\vec{\nabla}' \times \vec{\mathbf{E}}\left(\vec{\mathbf{r}}'\right)}{\left|\vec{\mathbf{r}} - \vec{\mathbf{r}}'\right|} d\tau' - \int_{\tau'} \vec{\nabla}' \times \left(\frac{\vec{\mathbf{E}}\left(\vec{\mathbf{r}}'\right)}{\left|\vec{\mathbf{r}} - \vec{\mathbf{r}}'\right|}\right) d\tau'\right)\right] \tag{1.130}$$

Using the divergence theorem. the preceding equation can also be rewritten in the form:

$$\vec{\mathbf{E}}\left(\vec{\mathbf{r}}\right) = -\frac{1}{4\pi}\left[-\vec{\nabla}\left(-\int_{\tau'} \frac{\vec{\nabla}' \cdot \vec{\mathbf{E}}\left(\vec{\mathbf{r}}'\right)}{\left|\vec{\mathbf{r}} - \vec{\mathbf{r}}'\right|} d\tau' + \oint_{S'} \vec{\mathbf{n}}' \cdot \frac{\vec{\mathbf{E}}\left(\vec{\mathbf{r}}'\right)}{\left|\vec{\mathbf{r}} - \vec{\mathbf{r}}'\right|} dS'\right) \right.$$
$$\left. -\vec{\nabla} \times \left(\int_{\tau'} \frac{\vec{\nabla}' \times \vec{\mathbf{E}}\left(\vec{\mathbf{r}}'\right)}{\left|\vec{\mathbf{r}} - \vec{\mathbf{r}}'\right|} d\tau' - \oint_{S'} \vec{\mathbf{n}}' \times \frac{\vec{\mathbf{E}}\left(\vec{\mathbf{r}}'\right)}{\left|\vec{\mathbf{r}} - \vec{\mathbf{r}}'\right|} dS'\right)\right]$$
$$= -\vec{\nabla}\left[\frac{1}{4\pi}\int_{\tau'} \frac{\vec{\nabla}' \cdot \vec{\mathbf{E}}\left(\vec{\mathbf{r}}'\right)}{\left|\vec{\mathbf{r}} - \vec{\mathbf{r}}'\right|} d\tau' - \frac{1}{4\pi}\oint_{S'} \vec{\mathbf{n}}' \cdot \frac{\vec{\mathbf{E}}\left(\vec{\mathbf{r}}'\right)}{\left|\vec{\mathbf{r}} - \vec{\mathbf{r}}'\right|} dS'\right]$$
$$+\vec{\nabla} \times \left[\frac{1}{4\pi}\int_{\tau'} \frac{\vec{\nabla}' \times \vec{\mathbf{E}}\left(\vec{\mathbf{r}}'\right)}{\left|\vec{\mathbf{r}} - \vec{\mathbf{r}}'\right|} d\tau' - \frac{1}{4\pi}\oint_{S'} \vec{\mathbf{n}}' \times \frac{\vec{\mathbf{E}}\left(\vec{\mathbf{r}}'\right)}{\left|\vec{\mathbf{r}} - \vec{\mathbf{r}}'\right|} dS'\right] \tag{1.131}$$

By defining:

$$V\left(\vec{\mathbf{r}}\right) = \frac{1}{4\pi}\int_{\tau'} \frac{\vec{\nabla}' \cdot \vec{\mathbf{E}}\left(\vec{\mathbf{r}}'\right)}{\left|\vec{\mathbf{r}} - \vec{\mathbf{r}}'\right|} d\tau' - \frac{1}{4\pi}\oint_{S'} \vec{\mathbf{n}}' \cdot \frac{\vec{\mathbf{E}}\left(\vec{\mathbf{r}}'\right)}{\left|\vec{\mathbf{r}} - \vec{\mathbf{r}}'\right|} dS',$$
$$\vec{\mathbf{A}}\left(\vec{\mathbf{r}}\right) = \frac{1}{4\pi}\int_{\tau'} \frac{\vec{\nabla}' \times \vec{\mathbf{E}}\left(\vec{\mathbf{r}}'\right)}{\left|\vec{\mathbf{r}} - \vec{\mathbf{r}}'\right|} d\tau' - \frac{1}{4\pi}\oint_{S'} \vec{\mathbf{n}}' \times \frac{\vec{\mathbf{E}}\left(\vec{\mathbf{r}}'\right)}{\left|\vec{\mathbf{r}} - \vec{\mathbf{r}}'\right|} dS' \tag{1.132}$$

it finally comes that:

$$\vec{\mathbf{E}} = -\vec{\nabla}V + \vec{\nabla} \times \vec{\mathbf{A}} \tag{1.133}$$

This is the so-called *Helmholtz–Hodge decomposition theorem.* In particular, if we assume that E satisfies the constraints [1.72] and [1.127], and we extend the previous calculations to the whole space (i.e. we tend S to infinity), the expressions [1.132] reduce to (as X is zero outside Y, the first term on the right remains the same):

$$V\left(\vec{\mathbf{r}}\right) = \frac{1}{4\pi\varepsilon_0}\int_{\tau'}\frac{\rho\left(\vec{\mathbf{r}}'\right)}{\left|\vec{\mathbf{r}} - \vec{\mathbf{r}}'\right|}d\tau' - \frac{1}{4\pi}\lim_{S'\longrightarrow\infty}\oint_{S'}\vec{\mathbf{n}}'\cdot\frac{\vec{\mathbf{E}}\left(\vec{\mathbf{r}}'\right)}{\left|\vec{\mathbf{r}} - \vec{\mathbf{r}}'\right|}dS'$$

$$\vec{\mathbf{A}}\left(\vec{\mathbf{r}}\right) = -\frac{1}{4\pi}\lim_{S'\longrightarrow\infty}\oint_{S'}\vec{\mathbf{n}}'\times\frac{\vec{\mathbf{E}}\left(\vec{\mathbf{r}}'\right)}{\left|\vec{\mathbf{r}} - \vec{\mathbf{r}}'\right|}dS' \quad [1.134]$$

It is clearly seen that if $\vec{\mathbf{E}}$ is zero at infinity, equations [1.72] and [1.127] admit a unique solution, given by:

$$\vec{\mathbf{E}} = -\vec{\nabla}V$$

$$V\left(\vec{\mathbf{r}}\right) = \frac{1}{4\pi\varepsilon_0}\int_{\tau'}\frac{\rho\left(\vec{\mathbf{r}}'\right)d\tau'}{\left|\vec{\mathbf{r}} - \vec{\mathbf{r}}\right|} \quad [1.135]$$

In fact, $\vec{\mathbf{E}}$ cannot be zero at infinity unless the charge Q is finite. The Helmholtz–Hodge theorem can also be stated as follows: if ρ tends to zero faster than $\frac{1}{r^2}$ (i.e. ρ is spatially bounded, $\int_{\text{space}}\rho d\tau \leq M$, where M is a finite constant), there exists a unique vector function $\vec{\mathbf{E}}$ of the position, which satisfies (1) $\vec{\nabla}\cdot\vec{\mathbf{E}} = \frac{\rho}{\varepsilon_0}$; (2) $\vec{\nabla}\times\vec{\mathbf{E}} = 0$ (this equation, valid only in electrostatics, is at the origin of the continuity equations for the electric field at the interface of an electrically charged sheet); and (3) $\lim_{r\longrightarrow\infty}\vec{\mathbf{E}} = 0$. This solution is given by [1.135].

By replacing $\vec{\mathbf{E}}$ with $-\vec{\nabla}V$ in [1.127], we obtain the *Poisson formula*:

$$\Delta V = -\frac{1}{\varepsilon_0}\rho \quad [1.136]$$

In a region where the charge density ρ is zero, the previous equation reduces to:

$$\Delta V = 0 \quad [1.137]$$

This is the *Laplace equation*. It appears in many areas of physics. These solutions obey the extreme value theorem: V has no extrema in any region where $\Delta V = 0$. A consequence of this theorem is the uniqueness theorem: the potential V is uniquely determined by the charge distribution in a volume τ and by the value of V or $\frac{\partial V}{\partial n} = \vec{\nabla}V\cdot\vec{\mathbf{n}}$ at any point on the surface surrounding this volume, where $\vec{\mathbf{n}}$ is the normal to the surface at the point under consideration. We will return to these results in detail.

1.6.4. *Applications*

The electric field is calculated as follows:

1) Using Gauss' theorem, when the charge distribution is symmetric so that the electric field is constant on certain hypothetical surfaces. This is generally the simplest method.

2) Find a solution to Poisson's equation or Laplace's equation, if the boundary conditions are known, to determine the integration constants.

EXAMPLE 1.22. (Solid Sphere Uniformly Charged in Volume).– Consider a sphere of radius R and center O, uniformly charged in volume (the total charge in the entire sphere $Q = \frac{4}{3}\pi\rho R^3$, where ρ is the charge density). Gauss' theorem is the simplest method here to calculate the electric field. The most appropriate Gaussian surface is a sphere centered on O and passing through the point of study M (this may be inside or outside the source).

– $\overrightarrow{\mathbf{E}}_{ext}$ *calculation*: the Gaussian surface in Figure 1.21(a), for $r > R$, contains the total charge Q. Moreover, by symmetry, we know that $\overrightarrow{\mathbf{E}}_{hboxext}$ must be radial and constant on this surface. Consequently, Gauss' theorem allows us to write $\oint_{S_{\text{ext}}} \overrightarrow{\mathbf{E}} \cdot d\overrightarrow{\mathbf{S}} = 4\pi r^2 \left|\overrightarrow{\mathbf{E}}_{\text{ext}}\right| = \frac{Q}{\varepsilon_0}$, which results in:

$$\overrightarrow{\mathbf{E}}_{\text{ext}} = \frac{Q}{4\pi\varepsilon_0} \frac{\overrightarrow{\mathbf{r}}}{r^3} \qquad [1.138]$$

– $\overrightarrow{\mathbf{E}}_{int}$ *calculation*: let us now calculate the electric field at a point M located inside the charged sphere. The Gaussian surface, passing through this point, has a radius that is smaller than R. It therefore contains a fraction of the total charge, namely, $\frac{4}{3}\pi\rho r^3 = \frac{Qr^3}{R^3}$, as shown in Figure 1.21(b). Similarly, the electric field observed on this surface is also radial and constant. In this case, Gauss' theorem gives us: $\oint_{S_{\text{int}}} \overrightarrow{\mathbf{E}} \cdot d\overrightarrow{\mathbf{S}} = 4\pi r^2 \left|\overrightarrow{\mathbf{E}}_{\text{int}}\right| = \frac{1}{\varepsilon_0}\rho\left(\frac{4}{3}\pi r^3\right)$. Therefore, it follows from this that:

$$\overrightarrow{\mathbf{E}}_{\text{int}} = \frac{\rho}{3\varepsilon_0}\overrightarrow{\mathbf{r}} = \frac{Q}{4\pi\varepsilon_0 R^3}\overrightarrow{\mathbf{r}} \qquad [1.139]$$

Note that if the charge distribution is not spherically symmetrical, the $\overrightarrow{\mathbf{E}}_{\text{ext}}$ and $\overrightarrow{\mathbf{E}}_{\text{int}}$ fields will also depend on the variables θ and φ, that is, they are no longer constant on the Gaussian surface. Consequently, Gauss' theorem will only give the mean value of the radial components of $\overrightarrow{\mathbf{E}}_{\text{ext}}$ and $\overrightarrow{\mathbf{E}}_{\text{int}}$ on this surface. Let us now try to obtain results [1.138] and [1.139] using Poisson's equation [1.136]. Given the spherical symmetry, we know that the potential can only depend on the radial coordinate r. Consequently, Poisson's equation is written as (in spherical coordinates):

$$\frac{1}{r^2}\frac{d}{dr}\left(r^2\frac{dV}{dr}\right)=\begin{cases}-\dfrac{\rho}{\varepsilon_0} & r\leq R\\ 0 & r>R\end{cases} \qquad [1.140]$$

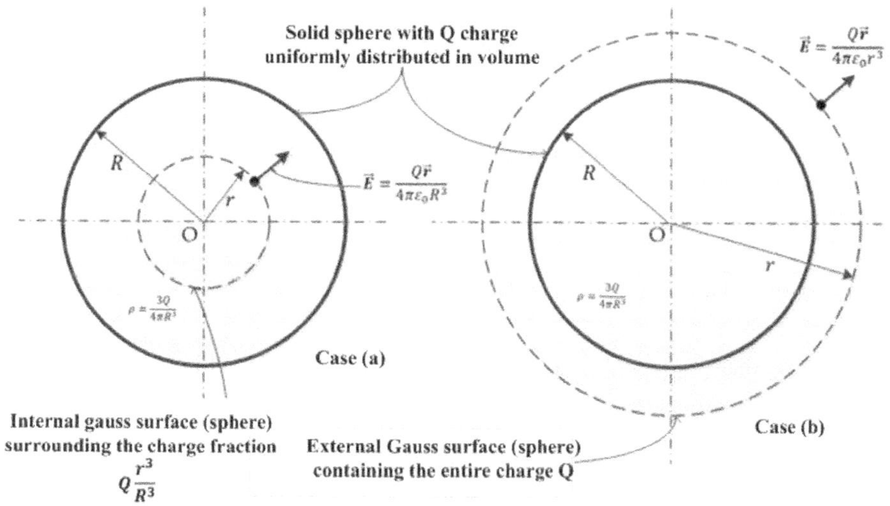

Figure 1.21. *The Gaussian surfaces for a sphere that is uniformly charged in its volume shows that the electric field outside it is equivalent to the field created by a point-like charge Q placed at the center O. Within the sphere, the electric field is linear, as the internal Gaussian surface contains only a fraction of the charge, proportional to r^3*

Let us begin by finding the form of the potential outside the sphere. By integrating twice with respect to r, it comes that:

$$V_{\text{ext}}(r)=\frac{A}{r}+B \qquad [1.141]$$

where A and B are integration constants. As the potential must be zero at infinity, we must choose $B=0$. The asymptotic behavior of the potential must be used to determine A. If we place ourselves very far from the charged sphere, we will see it essentially as a point-like charge. It is therefore necessary to have:

$$\lim_{r\longrightarrow\infty}V_{\text{ext}}(r)=\frac{Q}{4\pi\varepsilon_0 r}=\frac{\rho R^3}{3\varepsilon_0 r} \qquad [1.142]$$

By identification, it can be derived from this that $A = \frac{Q}{4\pi\varepsilon_0} = \frac{\rho R^3}{3\varepsilon_0}$. Within the sphere, the Poisson equation is written as:

$$\Delta V_{\text{int}} = \frac{1}{r^2}\frac{d}{dr}\left(r^2\frac{dV_{\text{int}}}{dr}\right) = -\frac{\rho}{\varepsilon_0} \qquad [1.143]$$

The first integration gives $r^2\frac{dV}{dr} = -\frac{\rho r^3}{3\varepsilon_0} + C$, or again, $\frac{dV}{dr} = -\frac{\rho r}{3\varepsilon_0} + \frac{C}{r^2}$ (C is an integration constant). By integrating again, we get:

$$V_{\text{int}} = -\frac{\rho r^2}{6\varepsilon_0} - \frac{C}{r} + D \qquad [1.144]$$

where D is a second integration constant. As the potential V_{int} must not diverge at $r = 0$, the constant C is necessarily zero. To determine D, let us use the continuity of the potential at $r = R$:

$$\lim_{r\longrightarrow R^-} V_{\text{int}}(r) = \lim_{r\longrightarrow R^+} V_{\text{ext}}(r) \qquad [1.145]$$

which shows that $-\frac{\rho R^2}{6\varepsilon_0} + D = \frac{\rho R^2}{3\varepsilon_0}$ and consequently, $D = \frac{\rho R^2}{2\varepsilon_0}$. Finally, the potential is written as:

$$V(r) = \begin{cases} \dfrac{\rho}{6\varepsilon_0}\left(3R^2 - r^2\right) = \dfrac{Q}{8\pi\varepsilon_0 R}\left[3 - \left(\dfrac{r}{R}\right)^2\right] & r \leq R \\ \dfrac{\rho R^3}{3\varepsilon_0 r} = \dfrac{Q}{4\pi\varepsilon_0 r} & r > R \end{cases} \qquad [1.146]$$

Now, to determine the $\overrightarrow{\mathbf{E}}$ field, we need to apply equation [1.73]. The same results are clearly obtained as in [1.39] and [1.40].

EXAMPLE 1.23.– If the sphere has a charge density with spherical symmetry described by $\rho(r') = \rho_0\left(1 - \frac{r'^2}{R^2}\right)$ for $r' \leq R$ and 0 for $r' > R$, the Gaussian surface for $r > R$ always encloses the total charge of the sphere $Q_{\text{tot}} = \int_0^R \rho(r')\,dr'$, while the Gaussian surface located at $r \leq R$ contains a part of the total charge, namely, $Q_r = \int_0^r \rho(r')\,dr'$. Since the field $\overrightarrow{\mathbf{E}}$ is radial and constant on these surfaces, Gauss' theorem is written as:

$$\begin{aligned} &\varepsilon_0 \oint_S \overrightarrow{\mathbf{E}} \cdot d\overrightarrow{\mathbf{S}} = \varepsilon_0 \left|\overrightarrow{\mathbf{E}}\right| 4\pi r^2 \\ &= \begin{cases} \displaystyle\int_0^r \rho_0\left(1 - \frac{r'^2}{R^2}\right) 4\pi r'^2 dr' = 4\pi\rho_0\left(\frac{r^3}{3} - \frac{r^5}{5R^2}\right) & r \leq R \\ \displaystyle\int_0^R \rho_0\left(1 - \frac{r'^2}{R^2}\right) 4\pi r'^2 dr' = \frac{8\pi\rho_0 R^3}{15} & r > R \end{cases} \end{aligned}$$

and consequently:

$$\vec{\mathbf{E}}(r) = \begin{cases} \dfrac{\rho_0}{\varepsilon_0}\left[\dfrac{1}{3} - \dfrac{1}{5}\left(\dfrac{r}{R}\right)^2\right]\vec{\mathbf{r}} & r \leq R \\ \dfrac{2\rho_0 R^3}{15\varepsilon_0}\dfrac{\vec{\mathbf{r}}}{r^3} & r > R \end{cases} \qquad [1.147]$$

In the same way, $\vec{\mathbf{E}}$ can be calculated from the expression for the potential, starting from the Poisson equation: $\frac{1}{r^2}\frac{d}{dr}\left(r^2\frac{dV}{dr}\right) = \begin{cases} \frac{\rho_0}{\varepsilon_0}\left(1 - \frac{r^2}{R^2}\right) & r \leq R \\ 0 & r > R \end{cases}$; after integration, this gives:

$$V(r) = \begin{cases} -\dfrac{\rho_0}{\varepsilon_0}\left(\dfrac{r^2}{6} - \dfrac{r^4}{20R^2}\right) - \dfrac{C}{r} + D & r \leq R \\ \dfrac{A}{r} + B & r > R \end{cases} \qquad [1.148]$$

Since the potential at infinity must be zero and does not diverge at the center of the sphere, the constants C and B must be chosen to be zero. To determine A, we must use the asymptotic behavior of the potential $V(r)$ at infinity. In fact, we must have $\lim_{r\longrightarrow\infty} V(r) = \frac{Q_{\text{tot}}}{4\pi\varepsilon_0 r} = \frac{2\rho R^3}{15\varepsilon_0 r}$ and, consequently, $A = \frac{2\rho R^3}{15\varepsilon_0}$. In addition, we must use the property of continuity of the potential on the surface of the sphere to calculate D: $\lim_{r\longrightarrow R^-} V_{\text{int}}(r) = \lim_{r\longrightarrow R^+} V_{\text{ext}}(r)$. It comes that: $-\frac{7\rho R^2}{60\varepsilon_0} + D = \frac{2\rho R^2}{15\varepsilon_0}$ and therefore, $D = \frac{\rho R^2}{4\varepsilon_0}$. Finally, the potential is written as:

$$V(r) = \begin{cases} \dfrac{\rho_0 R^2}{4\varepsilon_0}\left[1 - \dfrac{2}{3}\left(\dfrac{r}{R}\right)^2 + \dfrac{1}{5}\left(\dfrac{r}{R}\right)^4\right] & (r \leq R) \\ \dfrac{2\rho_0 R^3}{15\varepsilon_0 r} & (r > R) \end{cases} \qquad [1.149]$$

Using [1.73], we arrive again at expression [1.147].

EXAMPLE 1.24. (Hollow Sphere Charged on Surface).– Let us now assume that the total charge Q is uniformly distributed on the external surface of the sphere. Realizing from symmetry that the field $\vec{\mathbf{E}}$ is purely radial and depends only on r and not on θ and φ, let us start by drawing Gaussian spheres of radius r, as shown in Figure 1.22, inside the charged sphere ($r < R$) and outside the charged sphere ($r > R$). The Gaussian sphere inside the sphere encloses no charge, while the external sphere encloses the full charge $Q = \sigma 4\pi R^2$, that is, $\varepsilon_0 \oint_S \vec{\mathbf{E}} \cdot d\vec{\mathbf{S}} = \varepsilon_0 \left|\vec{\mathbf{E}}\right| 4\pi r^2 = \begin{cases} 0 & r < R \\ \sigma 4\pi R^2 & r > R \end{cases}$

and therefore, $\left|\vec{\mathbf{E}}\right| = \begin{cases} 0 & r < R \\ \frac{\sigma R^2}{\varepsilon_0 r^2} & r > R \end{cases}$. Thus, we find:

$$\vec{\mathbf{E}}(r) = \begin{cases} 0 & r < R \\ \dfrac{\sigma R^2 \vec{\mathbf{r}}}{\varepsilon_0 r^3} \equiv \dfrac{Q \vec{\mathbf{r}}}{4\pi\varepsilon_0 r^3} & r > R \end{cases} \qquad [1.150]$$

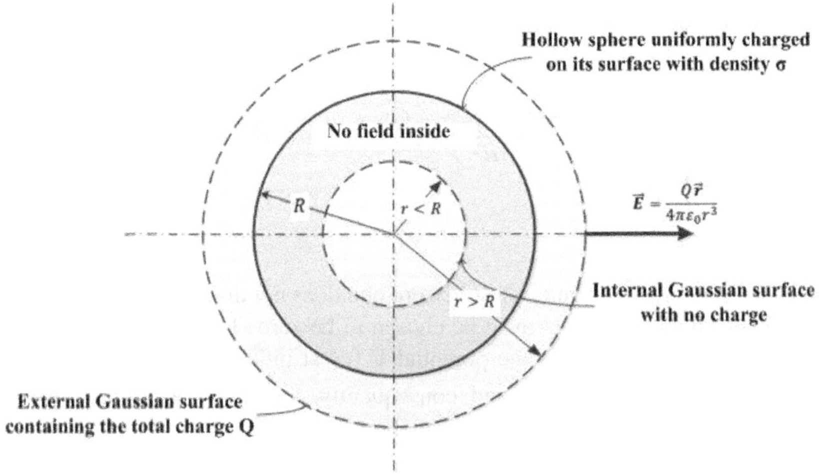

Figure 1.22. *Applying Gauss' theorem to concentric, spherical (Gaussian) surfaces inside ($r < R$) and outside ($r > R$) the charged sphere shows that the electric field inside is zero, while outside, it is as though the entire charge $Q = \sigma 4\pi R^2$ was concentrated into a point-like charge at the origin*

Note that the Poisson equation gives the solution $V(r) = A$ for $r < R$ and $\frac{B}{r}$ for $r > R$. The asymptotic behavior of the potential at infinity dictates that $B = \frac{Q}{4\pi\varepsilon_0} = \frac{\sigma R^2}{\varepsilon_0}$. The value of A is obtained using the continuity property of $V(r)$ on the boundary of the sphere, $\lim_{r \longrightarrow R^-} V(r) = \lim_{r \longrightarrow R^+} V(r)$, which shows that $A = \frac{B}{R} = \frac{Q}{4\pi\varepsilon_0 R} = \frac{\sigma R}{\varepsilon_0}$.

EXAMPLE 1.25. (Uniformly Charged Cylinder).–

1) An infinitely long cylinder with axis Oz and radius R carries the uniform charge density ρ within its volume, as seen in Figure 1.23. Let M be the point of coordinates (r, φ, z) (inside or outside the cylinder). For reasons of symmetry, the

electric field is purely radial and depends only on the distance r between M and the axis Oz. This can be demonstrated as follows: the plane (P_1) passing through M and containing the axis Oz is a plane of symmetry. The same applies to a plane(P_2) passing through M and perpendicular to Oz. The field must therefore be contained in both (P_1) and (P_2); therefore, it acts in the direction common to (P_1) and (P_2), that is, it is radial: $\overrightarrow{\mathbf{E}}(M) = E(r, \varphi, z)\frac{\overrightarrow{\mathbf{r}}}{r}$. Since the distribution of charges is translation invariant along Oz and rotation invariant around Oz, it can be concluded that the modulus of $\overrightarrow{\mathbf{E}}$ can only depend on the distance r, between M and the axis Oz, that is, $\overrightarrow{\mathbf{E}}(M) = E(r)\frac{\overrightarrow{\mathbf{r}}}{r}$. Let us apply Gauss' theorem to a cylindrical surface of height L, with axis Oz and radius r. The outgoing flux of $\overrightarrow{\mathbf{E}}$ across the bases of the Gaussian surface is zero, because $\overrightarrow{\mathbf{E}}$ is perpendicular to $d\overrightarrow{\mathbf{S}}_B$. At every point on the lateral section of the Gaussian surface, $\overrightarrow{\mathbf{E}}$ and $d\overrightarrow{\mathbf{S}}$ are collinear. Further, $E(r)$ is constant over this lateral part. The flux of $\overrightarrow{\mathbf{E}}$ across this lateral surface is therefore equal to $2\pi rLE(r)$. For a Gaussian surface located outside $(r > R)$, the charge enclosed within this surface is $Q = \pi R^2 L\rho$ (Figure 1.23(a)).

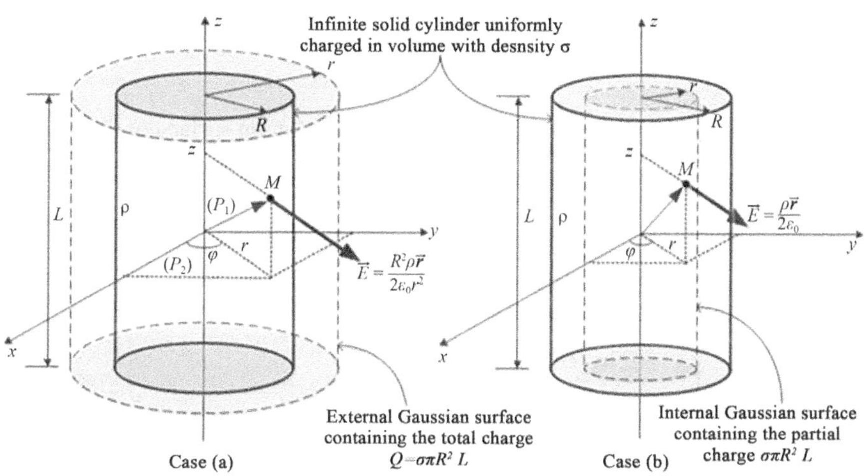

Figure 1.23. *Applying Gauss' theorem to concentric, cylindrical Gaussian surfaces shows that the external electric field behaves as though the charge on the cylinder was linearly distributed along the Oz axis, with the (linear) charge density $\lambda = \pi R^2\rho$. Inside the cylinder, the electric field is proportional to the distance, r, between M and the axis Oz*

Consequently, Gauss' theorem makes it possible to write that $2\pi rLE(r) = \frac{\pi R^2 L\rho}{\varepsilon_0}$; this gives $E(r) = \frac{R^2\rho}{2r\varepsilon_0}$ and consequently:

$$\overrightarrow{\mathbf{E}}(r) = \frac{R^2 \rho \overrightarrow{\mathbf{r}}}{2r^2 \varepsilon_0} \quad [1.151]$$

The electric field [1.151] is derived from the potential:

$$V(r) = -\frac{R^2 \rho_0 \ln r}{2\varepsilon_0} \quad [1.152]$$

For a Gaussian surface inside the cylinder, the enclosed charge is $Q_r = \pi r^2 L \rho$ (Figure 1.23(b)). It follows that $2\pi r L E(r) = \frac{\pi r^2 L \rho_0}{\varepsilon_0}$ and therefore:

$$\overrightarrow{\mathbf{E}}(r) = \frac{\rho_0 \overrightarrow{\mathbf{r}}}{2\varepsilon_0} \quad [1.153]$$

This field is derived from the potential:

$$V(r) = -\frac{\rho_0 r^2}{4\varepsilon_0} \quad [1.154]$$

2) If the charge is now uniformly distributed on the lateral surface of the cylinder, we obtain $\varepsilon_0 \oint_S \overrightarrow{\mathbf{E}} \cdot d\overrightarrow{\mathbf{S}} = \varepsilon_0 E_r 2\pi r L = \begin{cases} 0 & r < R \\ \sigma 2\pi R L & r > R \end{cases}$. This means that:

$$\overrightarrow{\mathbf{E}}(r) = \begin{cases} 0 & r < R \\ \dfrac{\sigma R \overrightarrow{\mathbf{r}}}{\varepsilon_0 r^2} & r > R \end{cases} \qquad V(r) = \begin{cases} -\dfrac{\sigma R \ln R}{\varepsilon_0} & r < R \\ -\dfrac{\sigma R \ln r}{\varepsilon_0} & r > R \end{cases} \quad [1.155]$$

Outside of the cylinder, the electric field resembles the electric field created by an infinitely long wire with charge density $\lambda = 2\pi R \sigma$.

EXAMPLE 1.26. (Uniformly Charged Plane).–

1) Let us calculate the field created at any point in space by a uniformly charged plane (charge density σ). The plane is assumed to be very large (infinitely large) to neglect edge effects. The axis Oz is oriented perpendicular to the plane as shown in Figure 1.24(a). For reasons of symmetry, the electric field $\overrightarrow{\mathbf{E}}$ can only be perpendicular to the plane, and its modulus can only depend on the distance between

the point M and the plane, that is, on z. In fact, $\overrightarrow{\mathbf{E}}$ is translation invariant along the axes Ox and Oy (i.e. it is independent of the variables x and y): $\overrightarrow{\mathbf{E}} = \overrightarrow{\mathbf{E}}(0,0,z)$. In addition, the Oxz and Oyz planes are symmetry planes. This implies that the field lies at their intersection, that is, along the Oz axis: $\overrightarrow{\mathbf{E}} = E(0,0,z)\overrightarrow{\mathbf{k}}$. y symmetry, $\overrightarrow{\mathbf{E}}$ is equal and opposite at distance $-z$. Given the symmetry of the problem, Gauss' theorem appears to be the best method for calculating $\overrightarrow{\mathbf{E}}$. An ideal Gaussian surface is a surface over which $\overrightarrow{\mathbf{E}}$ is constant and perpendicular or, on the contrary, $\overrightarrow{\mathbf{E}}$ is parallel. In this case, a cylinder whose lateral surface is perpendicular to the plane can be chosen as the Gaussian surface (i.e. parallel to $\overrightarrow{\mathbf{E}}$) and whose bases are parallel to the plane and equidistant from z on either side. On this basis, the field will be perpendicular to the surface of constant magnitude and outgoing. The total flux of $\overrightarrow{\mathbf{E}}$ across the cylinder is $\Phi = \Phi_1 + \Phi_{\text{lat.}} + \Phi_2$. On the lateral surface, we have $\overrightarrow{\mathbf{E}} \cdot d\overrightarrow{\mathbf{S}} = 0$, that is, $\Phi_{\text{lat.}} = 0$. On the surface S_1, we have $\Phi_1 = \int_{S_1} E(z)\,dS = E(z)\,S$. On the surface S_2, we have $\overrightarrow{\mathbf{E}} \cdot d\overrightarrow{\mathbf{S}} = -E(-z)\,dS$. Moreover, the charged plane is a symmetry plane, $E(-z) = -E(z)$, and therefore $\Phi_2 = \int_{S_2} E(z)\,dS = E(z)\,S$. We can therefore deduce that $\Phi = 2E(z)\,S$. The charge contained inside this Gaussian surface is σS. Applying Gauss' theorem, we write $\Phi = \frac{Q_{\text{int.}}}{\varepsilon_0}$ and, consequently, $2E(z)\,S = \frac{\sigma S}{\varepsilon_0}$. So, we obtain:

$$\overrightarrow{\mathbf{E}} = \begin{cases} \dfrac{\sigma}{2\varepsilon_0}\overrightarrow{\mathbf{k}} & \text{If } z > 0 \\ -\dfrac{\sigma}{2\varepsilon_0}\overrightarrow{\mathbf{k}} & \text{If } z < 0 \end{cases} \qquad [1.156]$$

2) If we now have two parallel planes A and B that are uniformly charged, with opposite polarities $-\sigma$ and $+\sigma$, and located, respectively, at $z_A = \frac{d}{2}$ and $z_B = -\frac{d}{2}$, we can draw Gaussian surfaces (i) passing through the two planes A and B, (ii) passing only through the A plane and (iii) passing only through the B plane, as shown in Figure 1.24(b). It can be seen that the plane A produces the field $-\frac{\sigma}{2\varepsilon_0}\overrightarrow{\mathbf{k}}$ in the subspace $z > \frac{d}{2}$ and $\frac{\sigma}{2\varepsilon_0}\overrightarrow{\mathbf{k}}$ in the subspace $z < \frac{d}{2}$. Similarly, B produces the field $\frac{\sigma}{2\varepsilon_0}\overrightarrow{\mathbf{k}}$ in the subspace $z > -\frac{d}{2}$ and $-\frac{\sigma}{2\varepsilon_0}\overrightarrow{\mathbf{k}}$ in the subspace $z < -\frac{d}{2}$. Using the principle of superposition, it can be concluded that the system of two planes produces the field $\frac{\sigma}{\varepsilon_0}\overrightarrow{\mathbf{k}}$ in the subspace $-\frac{d}{2} \leq z \leq \frac{d}{2}$ and 0 everywhere. It is verified that the electric field is indeed zero on the Gaussian surface (i). This electric field can be derived from the potential as follows:

$$V(z) = \begin{cases} -\dfrac{\sigma z}{\varepsilon_0} & \text{If } -\frac{d}{2} \leq z \leq \frac{d}{2} \\ -\dfrac{\sigma d\,|z|}{2\varepsilon_0 z} & \text{If } |z| > \frac{d}{2} \end{cases} \qquad [1.157]$$

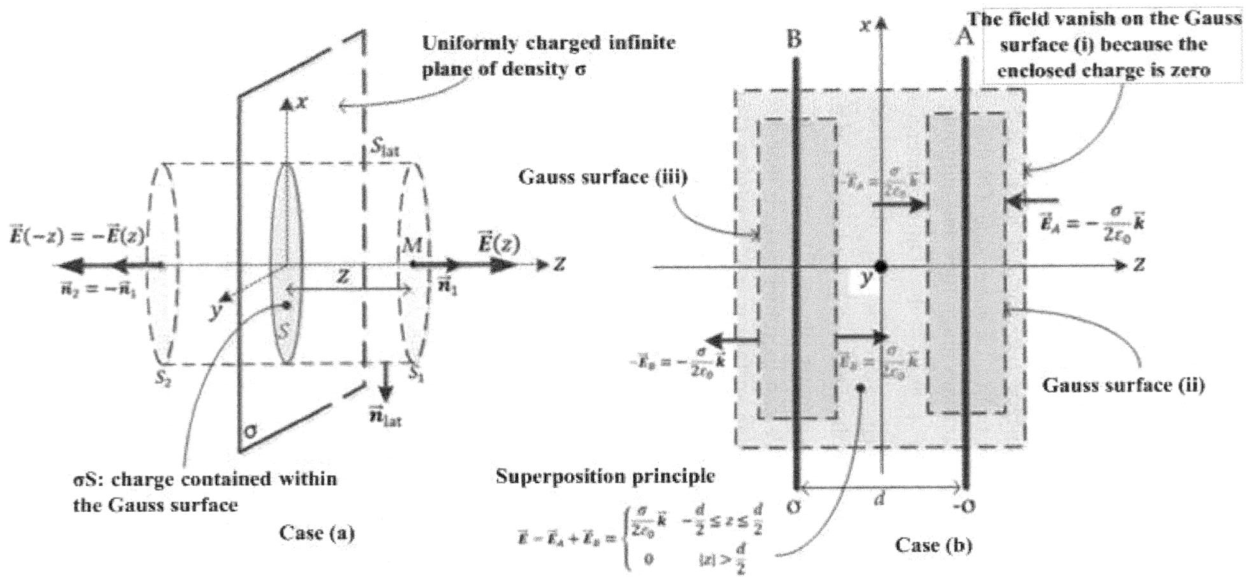

Figure 1.24. *Gauss theorem applied to a uniformly charged plane shows that the electric field is constant (on either side of the plane), perpendicular and outwardly directed (respectively, inwardly directed) if $\sigma > 0$ (respectively, $\sigma < 0$). For a double plane with opposite polarities, the field is constant between the planes (directed from the positive plate to the negative plate) and the external field is zero*

1.6.5. *Maximum and minimum potentials*

In a space with no electric charges, the potential does not admit any extreme values, that is, it does not go through a maximum or minimum. This is the *extreme value theorem*. In fact, if there is a maximum (respectively, a minimum) value for the potential, at a point M, the potential V would decrease (respectively, increase) in all directions around M. The electric field $\vec{\mathbf{E}} = -\vec{\nabla} V$ would therefore be directed from M outward all around M (respectively, from the outside toward M). Consequently, there would be a non-zero flux of $\vec{\mathbf{E}}$ across a closed surface surrounding M and Gauss' theorem would lead to the existence of a positive (respectively, negative) charge at this point, which contradicts the initial hypotheses. This result will be reviewed later. A direct consequence of what we have just seen is that a field line can only start from a place where charges exist, since the start of a field line corresponds to a maximum potential.

1.7. Electric field across a charged sheet

In the previous examples, we saw that (a) in the case of a volumetric charge distribution, the electrostatic field $\vec{\mathbf{E}}$ and the electrostatic potential V are continuous at all points, and (b) in the case of a surface distribution, there is a discontinuity in the field $\vec{\mathbf{E}}$ as it passes through the charged surface. This result is general. In fact, the tangential component of $\vec{\mathbf{E}}$ is continuous as it passes through the charged surface, whereas the normal component of $\vec{\mathbf{E}}$ is discontinuous; (c) the electrostatic potential V is always continuous.

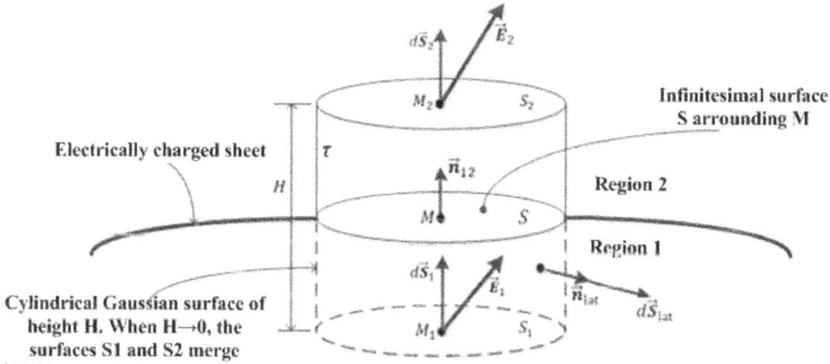

Figure 1.25. *Application of Gauss' theorem to an infinitely small Gaussian cylindrical surface containing a certain surface charge shows that the normal component $\varepsilon_0 \vec{\mathbf{E}}$ is discontinuous at the interface of a charged sheet*

To demonstrate result (b), consider an electrically charged sheet, that is, a surface distribution of charges σ, dividing the space into two regions 1 and 2. Let us draw a Gaussian surface, around a point M located on the sheet. This surface is infinitesimally small, cylindrical and closed, enclosing a volume τ and traversing the sheet, as shown in Figure 1.25. The surfaces S_1, S_2 and S are chosen parallel to the plane that is tangent to the sheet at the point M and are very small, so that $\vec{\mathbf{E}}$ and σ do not vary much over these planes. By applying Gauss' theorem:

$$\int_{S_2} \vec{\mathbf{E}}_2\Big|_{M_2} \cdot d\vec{\mathbf{S}}_2 + \int_{S_1} \vec{\mathbf{E}}_1\Big|_{M_1} \cdot d\vec{\mathbf{S}}_1 + \int_{S_{\text{lat}}} \vec{\mathbf{E}} \cdot d\vec{\mathbf{S}}_{\text{lat}} = \frac{1}{\varepsilon_0} \int_S \sigma(M)\, dS \tag{1.158}$$

where S_{lat} is the lateral surface of the cylinder. For a flattened cylinder, that is, when H tends toward zero, it is possible to eliminate the flow through the lateral surface from the previous expression, since S_{lat} becomes zero. Since, in addition, $d\vec{\mathbf{S}}_2 = \vec{\mathbf{n}}_{12} dS$ and $d\vec{\mathbf{S}}_2 = -\vec{\mathbf{n}}_{12} dS$, we obtain $\frac{1}{\varepsilon_0}\int_S \sigma(M)\, dS = \lim_{H\longrightarrow 0} \int_S \left[\vec{\mathbf{E}}(M_2) - \vec{\mathbf{E}}(M_1)\right] \cdot \vec{\mathbf{n}}_{12} dS$. This gives:

$$\frac{\sigma(M)\, S}{\varepsilon_0} = \left[\vec{\mathbf{E}}_2(M) - \vec{\mathbf{E}}_1(M)\right] \cdot \vec{\mathbf{n}}_{12} S \tag{1.159}$$

because the fields $\vec{\mathbf{E}}_1$ and $\vec{\mathbf{E}}_2$ are almost constants over the surfaces S_1 and S_2, respectively. $\vec{\mathbf{n}}_{12}$ here represents the normal to the sheet directed from medium 1 to medium 2. Upon simplifying by S, we finally find:

$$\frac{\sigma(M)}{\varepsilon_0} = \left[\vec{\mathbf{E}}_2(M) - \vec{\mathbf{E}}_1(M)\right] \cdot \vec{\mathbf{n}}_{12} = E_{n_{12}}(M)|_2 - E_{n_{12}}(M)|_1 \tag{1.160}$$

If we change the orientation, that is, direct it from medium 2 to medium 1, the above equation takes the form:

$$\frac{\sigma(M)}{\varepsilon_0} = \left[\vec{\mathbf{E}}_1(M) - \vec{\mathbf{E}}_2(M)\right] \cdot \vec{\mathbf{n}}_{21} = E_{n_{21}}(M)|_1 - E_{n_{21}}(M)|_2 \tag{1.161}$$

where $\vec{\mathbf{n}}_{21} = -\vec{\mathbf{n}}_{12}$ is the normal directed from medium 2 to medium 1. The normal component, E_n, of the electric field $\vec{\mathbf{E}}$ is therefore not continuous as it crosses a charged sheet.

Let us now draw an infinitesimal rectangular contour $ABCD$ around the point M on the charged sheet and passing across the sheet. The sides AB and CD are parallel to the plane tangent to the sheet at point M, while the sides BC and DA are perpendicular to this plane. The electric field $\vec{\mathbf{E}}$ is constant along the sides of the rectangle, as seen in Figure 1.26. Since $\vec{\nabla} \times \vec{\mathbf{E}} = 0$, the circulation of the field $\vec{\mathbf{E}}$ along the contour $ABCD$ is zero:

$$0 = \int_{AB} \vec{\mathbf{E}}\,(M_2)\cdot d\,\vec{\mathbf{l}} + \int_{BC} \vec{\mathbf{E}}\,(M_3)\cdot d\,\vec{\mathbf{l}} + \int_{CD} \vec{\mathbf{E}}\,(M_1)\cdot d\,\vec{\mathbf{l}} + \int_{DA} \vec{\mathbf{E}}\,(M_4)\cdot d\,\vec{\mathbf{l}} \quad [1.162]$$

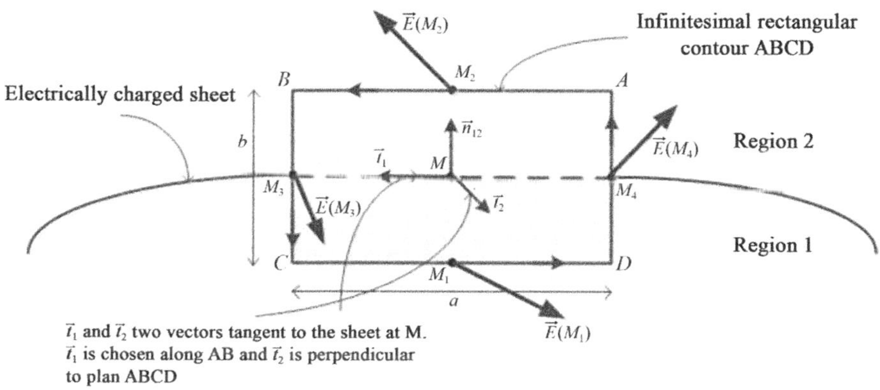

Figure 1.26. *Stokes' theorem, applied to an infinitesimal, rectangular contour and traversing a charge sheet separating two regions, 1 and 2, shows that the tangential component of $\vec{\mathbf{E}}$ is continuous along the interface*

By moving points B and C (respectively, A and D) to M_3 (respectively, M_4), that is, by applying the limit $b \longrightarrow 0$ (flattened contour) and noting that the circulation integrals $\lim_{b\longrightarrow 0} \int_{BC} bf\vec{E}\,(M_3)\cdot d\,\vec{\mathbf{l}}$ and $\lim_{b\longrightarrow 0} \int_{DA} \vec{\mathbf{E}}\,(M_4)\cdot d\,\vec{\mathbf{l}}$ become equal to zero, we obtain:

$$0 = \int_{AB} \left[\vec{\mathbf{E}}_2\,(M) - \vec{\mathbf{E}}_1\,(M)\right]\cdot \vec{\mathbf{t}}_1 dl \quad [1.163]$$

where $\vec{\mathbf{t}}_1$ is a unit vector oriented in the direction of AB. Since $\overrightarrow{\mathbf{AB}}$ is any vector, we deduce that:

$$\left[\vec{\mathbf{E}}_2\,(M) - \vec{\mathbf{E}}_1\,(M)\right]\cdot \vec{\mathbf{t}}_1 = 0 \quad [1.164]$$

By orienting the contour in the perpendicular direction, that is, toward $\overrightarrow{\mathbf{t}}_2$, we also find:

$$\left[\overrightarrow{\mathbf{E}}_2(M) - \overrightarrow{\mathbf{E}}_1(M)\right] \cdot \overrightarrow{\mathbf{t}}_2 = 0 \qquad [1.165]$$

We can therefore conclude that the projections of $\overrightarrow{\mathbf{E}}_2(M) - \overrightarrow{\mathbf{E}}_1(M)$ along $\overrightarrow{\mathbf{t}}_1$ and $\overrightarrow{\mathbf{t}}_2$ are zero. Consequently, the tangential part of $\overrightarrow{\mathbf{E}}_2(M) - \overrightarrow{\mathbf{E}}_1(M)$ is effectively zero. Equations [1.164] and [1.165] can be combined into a single equation:

$$\left[\overrightarrow{\mathbf{E}}_2(M) - \overrightarrow{\mathbf{E}}_1(M)\right] \times \overrightarrow{\mathbf{n}}_{12} = E_t(M)|_2 - E_t(M)|_1 = 0 \qquad [1.166]$$

The tangential component of the electric field is continuous across a charged sheet. Equations [1.160] and [1.166] are sometimes called *continuity equations of the electric field* at the interface of a charged sheet. They can be combined into a single equation of the form:

$$\overrightarrow{\mathbf{E}}_2(M) - \overrightarrow{\mathbf{E}}_1(M) = \frac{\sigma(M)}{\varepsilon_0}\overrightarrow{\mathbf{n}}_{12} \qquad [1.167]$$

where $\sigma(M)$ is the surface charge density at a point M on a charged sheet, $\overrightarrow{\mathbf{E}}_1(M)$ and $\overrightarrow{\mathbf{E}}_2(M)$ are fields in the neighborhood of M in the regions 1 and 2 separated by the sheet and $\overrightarrow{\mathbf{n}}_{12}$ is a unit vector normal to the sheet at M and oriented from 1 toward 2.

EXAMPLE 1.27. (Cylindrical and Spherical Sheets).–

(1) Let us write the continuity equations for the scalar potential V and the electric field $\overrightarrow{\mathbf{E}}$ through a sheet in the shape of an infinitely long cylinder, with axis Oz and radius R. The sheet carries a surface charge density $\sigma(\varphi, z)$, as seen in Figure 1.27(a). The scalar potentials (respectively, electric fields) outside and inside the cylinder are denoted by $V^{\text{ext}}(r, \varphi, z)$ (respectively, $\overrightarrow{\mathbf{E}}^{\text{ext}}(r, \varphi, z)$) and $V^{\text{int}}(r, \varphi, z)$ (respectively, $\overrightarrow{\mathbf{E}}^{\text{int}}(r, \varphi, z)$).

It has already been seen that the potential is continuous over the cylindrical interface when σ is finite, that is:

$$V^{\text{ext}}(R, \varphi, z) = V^{\text{int}}(R, \varphi, z) \qquad [1.168]$$

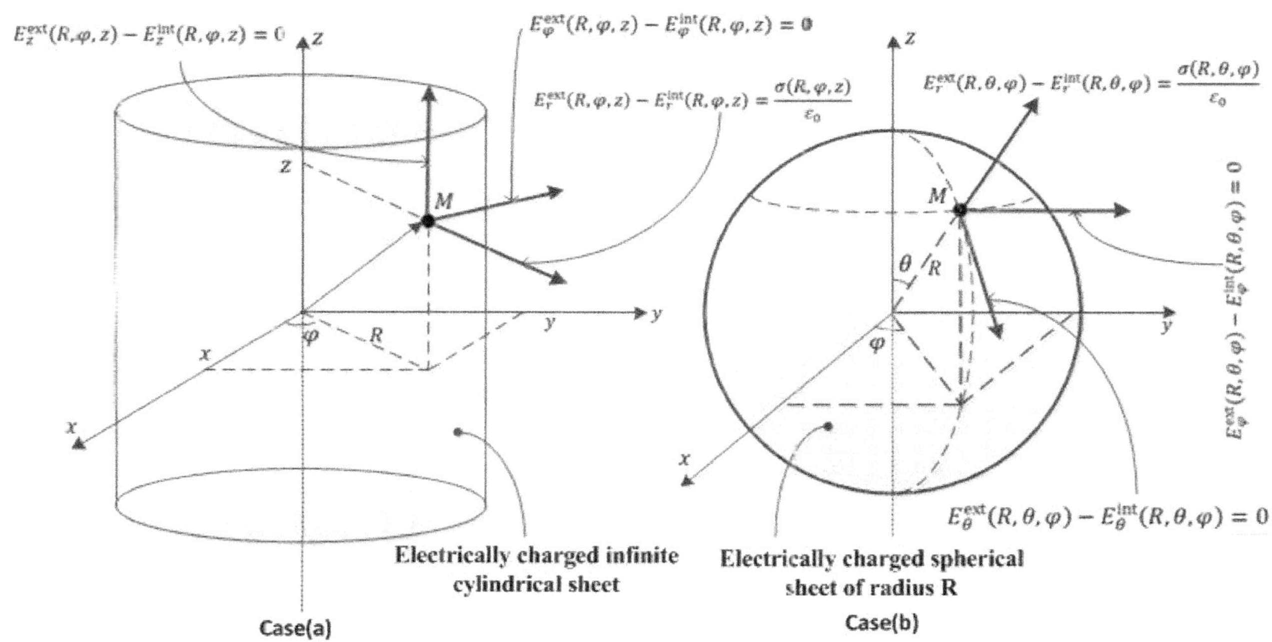

Figure 1.27. *Continuity equations for a cylindrical sheet (infinite, with axis Oz and radius R) and for a spherical sheet (radius R). Only the normal components are discontinuous, while the tangential components are continuous at the interfaces*

However, the situation is different for the electric field. According to the continuity equation [1.167], the normal (i.e. radial) component of the field across the interface obeys:

$$E_r^{\text{ext}}(R,\varphi,z) - E_r^{\text{int}}(R,\varphi,z) = \frac{\sigma(\varphi,z)}{\varepsilon_0} \tag{1.169}$$

while the two other tangential components verify the constraints:

$$E_\varphi^{\text{ext}}(R,\varphi,z) - E_\varphi^{\text{int}}(R,\varphi,z) = 0 \tag{1.170}$$

$$E_z^{\text{ext}}(R,\varphi,z) - E_z^{\text{int}}(R,\varphi,z) = 0 \tag{1.171}$$

In terms of the scalar potentials $V^{\text{ext}}(r,\varphi,z)$ and $V^{\text{int}}(r,\varphi,z)$, equations [1.169]–[1.171] are written as:

$$\begin{aligned}
\left.\frac{\partial V^{\text{int}}}{\partial r}\right|_{r=R} - \left.\frac{\partial V^{\text{ext}}}{\partial r}\right|_{r=R} &= \frac{\sigma(\varphi,z)}{\varepsilon_0} \\
\left.\frac{\partial V^{\text{int}}}{\partial \varphi}\right|_{r=R} - \left.\frac{\partial V^{\text{ext}}}{\partial \varphi}\right|_{r=R} &= 0 \\
\left.\frac{\partial V^{\text{int}}}{\partial z}\right|_{r=R} - \left.\frac{\partial V^{\text{ext}}}{\partial z}\right|_{r=R} &= 0
\end{aligned} \tag{1.172}$$

(2) For a spherical sheet with radius R, separating two regions 1 (internal) and 2 (external) and bearing a surface charge density $\sigma(\theta,\varphi)$, as in Figure 1.27(b), the continuity equations for the electric potential and the electric field are written as:

$$V^{\text{ext}}(R,\theta,\varphi) = V^{\text{int}}(R,\theta,\varphi) \tag{1.173}$$

and:

$$\begin{aligned}
E^{\text{ext}}(R,\theta,\varphi) - E^{\text{int}}(R,\theta,\varphi) &= \left.\frac{\partial V^{\text{int}}}{\partial r}\right|_{r=R} - \left.\frac{\partial V^{\text{ext}}}{\partial r}\right|_{r=R} = \frac{\sigma(\theta,\varphi)}{\varepsilon_0} \\
E^{\text{ext}}(R,\theta,\varphi) - E^{\text{int}}(R,\theta,\varphi) &= \frac{1}{R}\left(\left.\frac{\partial V^{\text{int}}}{\partial \theta}\right|_{r=R} - \left.\frac{\partial V^{\text{ext}}}{\partial \theta}\right|_{r=R}\right) = 0 \\
E^{\text{ext}}(R,\theta,\varphi) - E^{\text{int}}(R,\theta,\varphi) &= \frac{1}{R}\left(\left.\frac{\partial V^{\text{int}}}{\partial \varphi}\right|_{r=R} - \left.\frac{\partial V^{\text{ext}}}{\partial \varphi}\right|_{r=R}\right) = 0
\end{aligned} \tag{1.174}$$

1.8. Model of the hydrogen atom

Unlike Thomson's model, the atom can also be modeled by a fixed, point-like positive charge (the nucleus) and an electron cloud, that is, a zone surrounding the nucleus. This is the zone in which negatively charged particles are found: electrons, with charge $-e$. These electrons are distributed with a volume density ρ_e with spherical symmetry. To simplify, we shall restrict ourselves to the hydrogen atom. To calculate the electric field at a point M, we must take into account all charges found within the atom, that is, the point-like charge of the proton, $+e$, and the charge distributed in the electron cloud. For symmetry considerations, the field and electrostatic potential are actually functions of r, that is, $\vec{\mathbf{E}}(M) = E(r)\vec{\mathbf{e}}_r \equiv E(r)\frac{\vec{\mathbf{r}}}{r}$ and $V(M) = V(r)$. The field and potential resulting from the proton are given as:

$$\vec{\mathbf{E}}^+(M) = \frac{e}{4\pi\varepsilon_0 r^2}\vec{\mathbf{e}}_r \equiv \frac{e\vec{\mathbf{r}}}{4\pi\varepsilon_0 r^3} \qquad [1.175]$$

and:

$$V^+(M) = \frac{e}{4\pi\varepsilon_0 r} \qquad [1.176]$$

In addition, the electron density within the cloud can be chosen as $\rho_e(r) = \alpha e^{-\frac{r}{\beta}}$, where α is a negative constant with the dimension of a volume charge and β is a positive constant, so that the volume density converges to 0 when $r \to \infty$. β has the dimension of length. The elementary charge contained between two spheres with center O and having radii r and $r + dr$ will correspond to the product of $\rho_e(r)$ and the volume, $4\pi r^2 dr$, of the space between the two spheres. That is, $dq = 4\pi r^2 \alpha e^{-\frac{r}{\beta}} dr$; this corresponds to the radial density: $\lambda(r) = 4\pi r^2 \alpha e^{-\frac{r}{\beta}}$.

1) The radial density $\lambda(r)$ admits an extreme value at the point $r_0 = 2\beta$. In effect: $\lambda'(r) = 4\pi\alpha e^{-\frac{r}{\beta}}\left(2r - \frac{r^2}{\beta}\right)$; this shows that $\lambda'(r_0) = 0$.

2) The electron cloud contains a single electron: $\int_{r=0}^{\infty}\lambda(r)\,dr = \int_{r=0}^{\infty} e^{-\frac{r}{\beta}} \times 4\pi r^2 \alpha dr = 8\pi\alpha\beta^3 = -e$. Therefore, α and β can be expressed in terms of r_0 and $-e$: $\beta = \frac{r_0}{2}$ et $\alpha = -\frac{e}{\pi r_0^3}$. It becomes:

$$\lambda(r) = -\frac{e}{r_0}\left(\frac{2r}{r_0}\right)^2 e^{-\frac{2r}{r_0}} \qquad [1.177]$$

3) If $r \ll r_0$, we obtain $\lambda(r) \simeq -\frac{e}{r_0}\left(\frac{2r}{r_0}\right)^2$. In the neighborhood of $r = 0$, the curve is a parabola with a concave pointing downward. Similarly, we have

$\lim_{r\to\infty} \lambda(r) = 0$. Consequently, the function $\lambda(r)$ reaches its minimum at the point r_0, $\lambda(r_0) = -\frac{4e}{r_0}e^{-2}$.

The electric field associated with the electron cloud can be obtained using Gauss' theorem (the distribution has spherical symmetry), $\oint_S \vec{\mathbf{E}}^-(\vec{\mathbf{r}}) \cdot d\vec{\mathbf{S}} = \frac{Q(r)}{\varepsilon_0}$; this gives $4\pi r^2 E^-(r) = \frac{Q(r)}{\varepsilon_0}$, where $Q(r) = \int_{u=0}^{r} \lambda(u)\, du = -e{+}e\left[2\left(\frac{r}{r_0}\right)^2 + \frac{2r}{r_0} + 1\right]$ $\exp\left(-\frac{2r}{r_0}\right)$. We thus find:

$$\vec{\mathbf{E}}^-(\vec{\mathbf{r}}) = -\frac{e}{4\pi\varepsilon_0}\left[\frac{1}{r^2} - \left(\frac{2}{r_0^2} + \frac{2}{r_0 r} + \frac{1}{r^2}\right)\exp\left(-\frac{2r}{r_0}\right)\right]\vec{\mathbf{e}}_r \qquad [1.178]$$

If $r \ll r_0$, we have the approximation $\vec{\mathbf{E}}^-(\vec{\mathbf{r}}) \sim -\frac{er}{3\pi\varepsilon_0 r_0^3}\vec{\mathbf{e}}_r$. The electron charge appears as though it was uniformly distributed in an imaginary sphere with an effective radius $R = \left(\frac{3}{4}\right)^{1/3} r_0$. In the same way, if $r \gg r_0$, then $\vec{\mathbf{E}}^-(\vec{\mathbf{r}}) \sim -\frac{e}{4\pi\varepsilon_0 r^2}\vec{\mathbf{e}}_r$. At the point r_0, we find $\vec{\mathbf{E}}^-(r_0) = -\frac{e[1-5\exp(-2)]}{4\pi\varepsilon_0 r_0^2}\vec{\mathbf{e}}_r$.

To obtain the potential $V^-(\vec{\mathbf{r}})$ associated with $\vec{\mathbf{E}}^-(\vec{\mathbf{r}})$, we write:

$$V^-(\vec{\mathbf{r}}) = -\frac{e}{4\pi\varepsilon_0}\left[\frac{1}{r} + \left(f(r) - \frac{1}{r} - \frac{1}{r_0}\right)\exp\left(-\frac{2r}{r_0}\right)\right] \qquad [1.179]$$

where $f(r)$ is a function that must be determined. By calculating $\vec{\nabla}V^-(\vec{\mathbf{r}})$ and then comparing with the expression of $\vec{\mathbf{E}}^-(\vec{\mathbf{r}})$, we obtain the differential equation $f' - \frac{2}{r_0}f = 0$, which admits the solution $f = A\exp\left(\frac{2r}{r_0}\right)$, where A is a constant. We can derive from this that $V^-(\vec{\mathbf{r}}) = -\frac{e}{4\pi\varepsilon_0}\left[\frac{1}{r} + A - \left(\frac{1}{r} + \frac{1}{r_0}\right)\exp\left(-\frac{2r}{r_0}\right)\right]$. Since $V^-(\vec{\mathbf{r}})$ must be zero for $r \to \infty$, we will choose $A = 0$. The potential $V^-(\vec{\mathbf{r}})$ is therefore:

$$V^-(\vec{\mathbf{r}}) = -\frac{e}{4\pi\varepsilon_0}\left[\frac{1}{r} - \left(\frac{1}{r} + \frac{1}{r_0}\right)\exp\left(-\frac{2r}{r_0}\right)\right] \qquad [1.180]$$

For $r \ll r_0 : V^-(\vec{\mathbf{r}}) \sim \frac{e}{4\pi\varepsilon_0}\left(-\frac{1}{r_0} + \frac{2r^2}{3r_0^3}\right)$. For $r \gg r_0 : V^-(\vec{\mathbf{r}}) \sim -\frac{e}{4\pi\varepsilon_0 r}$.

By superposition, the field $\vec{\mathbf{E}}(\vec{\mathbf{r}})$ and the associated potential $V(\vec{\mathbf{r}})$, created by the hydrogen atom, are:

$$\vec{\mathbf{E}}(\vec{\mathbf{r}}) = \frac{e}{4\pi\varepsilon_0}\left(\frac{2}{r_0^2} + \frac{2}{r_0 r} + \frac{1}{r^2}\right)\exp\left(-\frac{2r}{r_0}\right)\vec{\mathbf{e}}_r \qquad [1.181]$$

and:

$$V\left(\overrightarrow{\mathbf{r}}\right) = \frac{e}{4\pi\varepsilon_0}\left(\frac{1}{r} + \frac{1}{r_0}\right)\exp\left(-\frac{2r}{r_0}\right) \qquad [1.182]$$

For $r \ll r_0$, we find (first-order expansion): $\overrightarrow{\mathbf{E}}\left(\overrightarrow{\mathbf{r}}\right) \sim \frac{e}{4\pi\varepsilon_0 r^2}\overrightarrow{\mathbf{e}}_r$ and $V\left(\overrightarrow{\mathbf{r}}\right) \sim \frac{e}{4\pi\varepsilon_0}\left(\frac{1}{r} - \frac{1}{r_0}\right)$. For $r \gg r_0$: $\overrightarrow{\mathbf{E}}\left(\overrightarrow{\mathbf{r}}\right) \sim 0$ and $V\left(\overrightarrow{\mathbf{r}}\right) \sim 0$. A discontinuity in the electrostatic field is therefore observed at $r = r_0$. The system behaves exactly as if the electron charge was uniformly distributed across the surface of a sphere of radius r_0, whose area is $4\pi r_0^2$. In effect, it has been verified that $E_r^{\text{ext.}} - E_r^{\text{int.}} = \frac{\sigma}{\varepsilon_0}$, which gives:

$$\sigma = -\frac{e}{4\pi r_0^2} \qquad [1.183]$$

Chapter 2

Conductors in Electrostatic Equilibrium

2.1. Conductors

A *conductor* is a medium in which charges are free to move in the presence of an external electric field $\overrightarrow{\mathbf{E}}_{\text{ext}}$. In the opposite case, that is, when $\overrightarrow{\mathbf{E}}_{\text{ext}} = 0$, they are immobile (in reality, the charges are mobile due to thermal excitation, but the mean vector velocities are zero (random displacements)). These charges are electrons or ions. Metals, electrolytes or plasmas (ionized gases) are conductive media. When the number of free charges tends to infinity or the friction forces (collision forces) inside the conductor are negligible (conductor with high conductivity), the conductor is said to be a *perfect* or *ideal*.

In an isolated atom, the electrons associated with the nucleus possess discrete negative energies. The electrons in the outermost shells are called *valence electrons*. When these electrons absorb sufficient energy, the atom becomes ionized and the electrons that are removed become "free electrons". The energy of these electrons can now take continuous values. In a solid, electrons do not have discrete energy, but the energy takes continuous values in a certain number of intervals because of interactions with neighboring atoms. This range of energies is known as the *energy band*. The highest band of electrons bound to the nucleus is called the *valence band*. Above this band is what is called the *conduction* band, whose electrons belong to all of the atoms. These electrons are therefore delocalized, that is, free to move throughout the solid. The valence and conduction bands are clearly separate in an insulating medium, as seen in Figure 2.1(a). In addition, the conduction band is almost empty. In a conductor, the two bands overlap (Figure 2.1(b)) and the electrons move from one band to the other: there are always *conduction electrons* in this medium. In a semiconductor medium, the two bands are disconnected (Figure 2.1(c)), but very close to each other. A small gain in energy is therefore enough for electrons to move from the valence band to the conduction band.

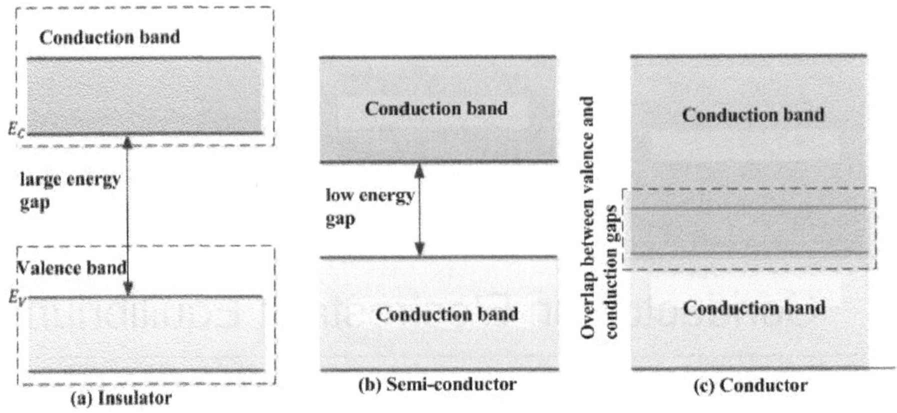

Figure 2.1. *Structure of bands in an insulator, a semi-conductor and a conductor. For a color version of this figure, see www.iste.co.uk/abdesselam/electrodynamics1.zip*

2.2. Electrostatic equilibrium

A conductor is said to be at *electrostatic equilibrium* if the free (conduction) charges are immobile, that is, no charge moves within the conductor. In metals, only electrons are mobile. The positive charge structure, called the *crystal lattice*, is almost fixed (except in the specific case of *holes* in semiconductors). In liquids and gases, ions are also displaced. The restoring forces on the surface of the conductor (electrostatic attraction of nuclei) are always considered sufficient to retain the electrons in a metal. If the external force (intense external electric field) is very strong, electrons may be discharged from the surface of the conductor: *electron emission* (*cathodic emission*) will occur. Electrons may also be discharged by applying high temperatures to the conductor. This is called *thermionic emission*.

Let there be a perfect neutral or charged conductor. The following are the essential properties of this conductor:

1) In equilibrium, the sum of the forces applied to any free charge must be zero. If we ignore the forces of gravity, friction (due to the crystalline structure) and in the absence of any other internal or external forces, this means that the electric force exerted on the free charge is zero. That is, $\vec{\mathbf{E}}_{\text{int}} = 0$ in the interior. If an external electric field $\vec{\mathbf{E}}$ is applied, the free charges move to the surface to create an induced electric field, which cancels out the external field exactly at any point on the conductor. This means that the total field $\vec{\mathbf{E}}_{\text{int}} = \vec{\mathbf{E}} + \vec{\mathbf{E}}_{\text{ind}}$ becomes zero inside the conductor, as shown in Figure 2.2. The field is therefore modified on the exterior.

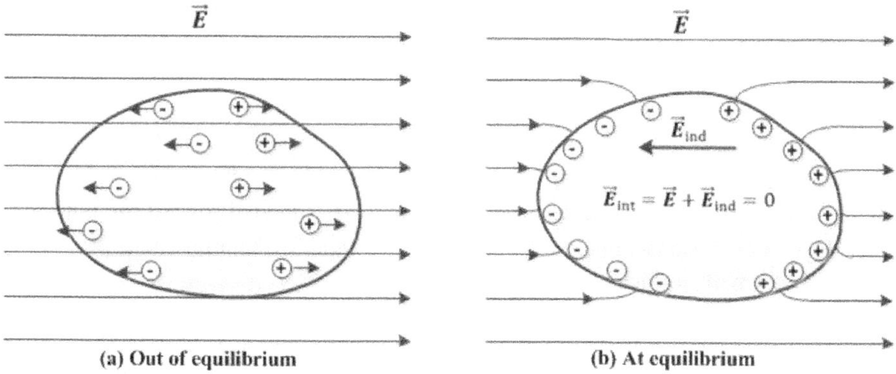

Figure 2.2. *The charges in a neutral conductor, placed in a uniform external field $\vec{\mathbf{E}}_{ext}$, move to create an induced field inside the conductor, which cancels out the external field*

2) Since $\vec{\mathbf{E}}$ is zero within the conductor (neutral or charged) and since $dV = \vec{\mathbf{E}} \cdot d\vec{\mathbf{r}}$ (V therefore appears as a continuous function at all points as long as $\vec{\mathbf{E}}$ is finite), it can be derived that V_{int} is uniform in any conductor, both inside and on its surface. Therefore, the conductor constitutes an equipotential surface. Consequently, the field $\vec{\mathbf{E}}$ must be normal to the surface (all points on the conductor's surface are at the same potential). The second consequence is that no field line can return to the conductor. That is, no field line can connect one point on a conductor to another point on the same conductor. In fact, the circulation of the field along this line dictates that $V(A) - V(B) = \int_A^B \vec{\mathbf{E}} \cdot d\vec{\mathbf{l}}$. Since A and B are two points on the surface of the same conductor, we must have $V(A) = V(B)$ and, consequently:

$$\int_A^B \vec{\mathbf{E}} \cdot d\vec{\mathbf{l}} = 0 \qquad [2.1]$$

which is impossible along a field line, because $\vec{\mathbf{E}}$ is parallel to $d\vec{\mathbf{l}}$, by definition.

3) Let us draw a closed Gaussian surface S, located within a conductor at equilibrium, and then apply Gauss's theorem to calculate the charge inside S:

$$Q_{int} = \varepsilon_0 \oint_S \vec{\mathbf{E}} \cdot d\vec{\mathbf{S}} = 0 \qquad [2.2]$$

since the field is zero within the conductor, and consequently in S. The sum of the charges inside this surface is therefore zero. In other words, excess charges can only be distributed on the surface of the conductor with a surface charge density designated by σ. The charge density ρ is zero inside a conductor.

4) *Coulomb's theorem*: we understand that the electric field is normal at any point M on the surface of the conductor. Additionally, from the continuity equation for the electric field at this point, it can be written that $\vec{\mathbf{E}}_{\text{ext}}(M) - \vec{\mathbf{E}}_{\text{int}}(M) = \frac{\sigma(M)}{\varepsilon_0}\vec{\mathbf{n}}$, where $\vec{\mathbf{n}}$ is the normal to the surface at the point M (directed toward the outside). Since $\vec{\mathbf{E}}_{\text{int}} = 0$, it can be concluded that the electrostatic field in immediate proximity to a conductor with surface charge density σ:

$$\vec{\mathbf{E}}_{\text{ext}}(M) = \frac{\sigma(M)}{\varepsilon_0}\vec{\mathbf{n}} \qquad [2.3]$$

This is *Coulomb's theorem.* A direct consequence of equation [2.3] is that the field lines arrive perpendicular to the surface of the conductor. They are outgoing (respectively, incoming) if $\sigma > 0$ (respectively, $\sigma < 0$). This equation also makes it possible to calculate σ from the field in the external neighborhood of the conductor's surface:

$$\sigma(M) = \varepsilon_0 \vec{\mathbf{E}}_{\text{ext}}(M) \cdot \vec{\mathbf{n}} \qquad [2.4]$$

5) *Hollow conductor*: consider a conductor in electrostatic equilibrium with an empty cavity (located entirely inside the conductor), as seen in Figure 2.3. Since it is at equilibrium, every point on the conductor is at the same electric potential. Any two points A and B on the surface of the cavity are therefore at the same potential. Now imagine there is an electric field $\vec{\mathbf{E}}$ inside the cavity, then calculate the potential difference $V_B - V_A = \int_B^A \vec{\mathbf{E}} \cdot d\,\vec{\mathbf{l}}$. Since $\vec{\mathbf{E}}$ is not zero, it is always possible to find a path between A and B for which the integral $\int_B^A \vec{\mathbf{E}} \cdot d\,\vec{\mathbf{l}}$ is not zero. However, $V_B - V_A = 0$, the circulation of $\vec{\mathbf{E}} \cdot d\,\vec{\mathbf{l}}$ is zero for any path between any two points on the conductor. As a result, the electric field is zero everywhere within the cavity. Upon applying the continuity equation at every point M on the surface of the cavity, it is seen that the surface charge density is zero throughout the cavity, since $\vec{\mathbf{E}}$ is zero on both sides. Therefore, there are no charges on the internal surface of the cavity. It can thus be concluded that if a hollow conductor is at an equilibrium, the cavity is automatically insulated from the outside world from an electrostatic point of view. Any hollow conductor maintained at constant potential is called an *perfect electrostatic screen* or *Faraday cage*.

6) *Tip point effect*: if a conductor (1) carrying a charge Q_1 is brought into contact with another conductor (2), which is neutral ($Q_2 = 0$) or charged ($Q_2 \neq 0$), we can observe a change in the charges of both conductors (1) and (2). In particular, if conductor (2) was initially neutral, it becomes charged. At the moment of contact,

electric charges instantaneously move from one to another, modifying the number of charges contained in each conductor, until an equilibrium is reached. Electrostatic equilibrium is attained when no more charges move from one toward the other (the system's internal electrostatic field is zero). To illustrate this, let us consider two conductive spheres (with radii R_1 and $R_2 > R_1$) at equilibrium, separated by a distance $D \gg R_1 + R_2$. To start with, they are insulated, with initial charges Q_1 and Q_2. They are then connected by a very long conducting wire. The total charge, $Q_1 + Q_2$, will be distributed between the two spheres, Q_1' for the first and Q_2' for the second. This creates a new equilibrium state of a single conductor with potential V_{int}^0, as shown in Figure 2.4. The potentials measured within these spheres are $V_{\text{int}}^{(1)} = \frac{Q_1'}{4\pi\varepsilon_0 R_1}$ for the first and $V_{\text{int}}^{(2)} = \frac{Q_2'}{4\pi\varepsilon_0 R_2}$ for the second. In equilibrium, we have $V_{\text{int}}^{(1)} = V_{\text{int}}^{(2)} = V_{\text{int}}^0$. It follows from this that:

$$V_{\text{int}}^0 = \frac{Q_1'}{4\pi\varepsilon_0 R_1} = \frac{Q_2'}{4\pi\varepsilon_0 R_2} \qquad [2.5]$$

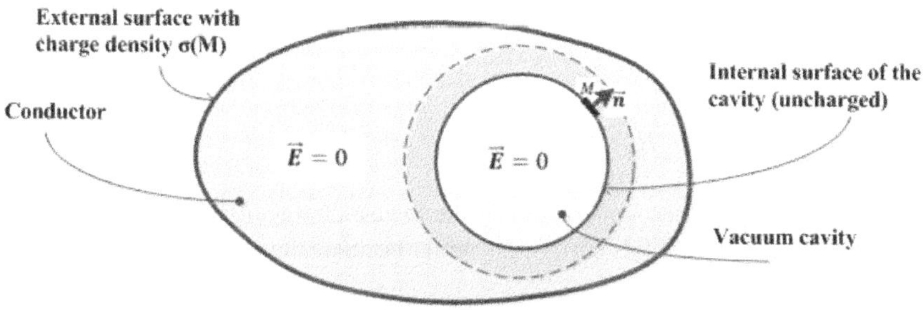

Figure 2.3. *Conductor containing a vacuum cavity. (1) $\vec{\mathbf{E}}$ is zero in the interior of the cavity. (2) There are no charges on the internal surface of the cavity. (3) The entire cavity is at the same potential as the conductor*

As the total charge in the system must be conserved, that is:

$$Q_1' + Q_2' = Q_1 + Q_2 \qquad [2.6]$$

the new charges on the two spheres can be calculated by simultaneously resolving equations [2.5] and [2.6]:

$$Q_1' = \frac{R_1 (Q_1 + Q_2)}{R_1 + R_2} \iff \sigma_1' = \frac{Q_1 + Q_2}{4\pi R_1 (R_1 + R_2)} \qquad [2.7]$$

$$Q_2' = \frac{R_2 (Q_1 + Q_2)}{R_1 + R_2} > Q_1' \iff \sigma_2' = \frac{Q_1 + Q_2}{4\pi R_2 (R_1 + R_2)} > \sigma_1' \qquad [2.8]$$

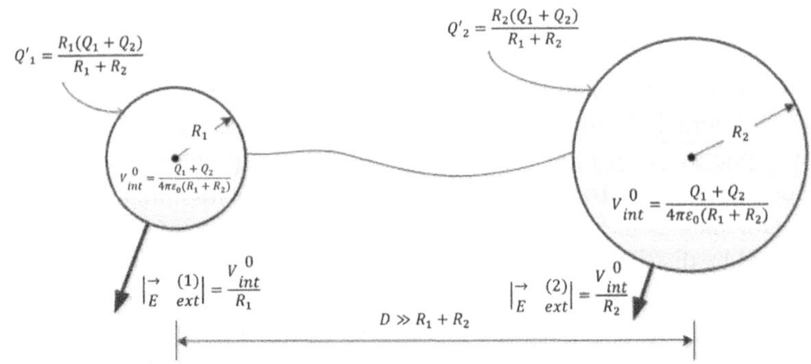

Figure 2.4. *Tip point effect observed in a system of two spheres connected by a metallic wire. The external electric field near the surface is greater on the smaller sphere field*

Consequently, the common equilibrium potential of the two spheres, V_{int}^0, becomes:

$$V_{\text{int}}^0 = \frac{Q_1 + Q_2}{4\pi\varepsilon_0 (R_1 + R_2)} \qquad [2.9]$$

Even though the smallest sphere carries the lowest charge, it is easy to see that the external electric field around its surface is greater because the charge density is higher. Indeed, we have:

$$\left|\vec{\mathbf{E}}_{\text{ext}}^{(1)}(r = R_1)\right| = \frac{Q'_1}{4\pi\varepsilon_0 R_1^2} = \frac{Q_1 + Q_2}{4\pi\varepsilon_0 R_1 (R_1 + R_2)} = \frac{V_{\text{int}}^0}{R_1} \qquad [2.10]$$

$$\left|\vec{\mathbf{E}}_{\text{ext}}^{(2)}(r = R_2)\right| = \frac{Q'_2}{4\pi\varepsilon_0 R_2^2} = \frac{Q_1 + Q_2}{4\pi\varepsilon_0 R_2 (R_1 + R_2)} = \frac{V_{\text{int}}^0}{R_2} < \left|\vec{\mathbf{E}}_{\text{ext}}^{(1)}(r = R_1)\right| \qquad [2.11]$$

It therefore follows that:

$$\frac{\left|\vec{\mathbf{E}}_{\text{ext}}^{(2)}(r = R_2)\right|}{\left|\vec{\mathbf{E}}_{\text{ext}}^{(1)}(r = R_1)\right|} = \frac{R_1}{R_2} < 1 \qquad [2.12]$$

The electric field $\vec{\mathbf{E}}_{\text{ext}}$ is therefore more intense at the level of the smaller sphere, that is, in proximity to the tip (corners and edges of the equipotential surfaces). In other words, for a greater curvature, the electric field $\overrightarrow{\mathbf{E}}_{ext}$ is more intense and greater in terms of surface charge density. This is called the *tip point effect*. In conclusion, it can be noted that at a large distance from a charged conductor, the field lines are radial; while in the vicinity of the conductor, they are perpendicular to the surface. They are more concentrated in proximity to a point where the conductor has a tip (i.e. high curvature) and more widely spaced in other regions. The electric field is very intense in the neighborhood of these points because of a strong accumulation of charges. If the field exceeds a certain threshold, the conductor releases charges into the external medium surrounding the tip, producing a luminous emission known as an *electric arc*.

7) *Earthing a conductor*: approximately, the earth can be considered a spherical conductor with a surface charge density of almost zero. The potential inside the Earth is, of course, zero. When connected to a conductor, the Earth can supply or absorb excess electrons without changing its own potential. We therefore conclude that the Earth imposes a zero potential on any conductor connected to Earth by a metal wire. If we place a voltage generator $V = V_1 - V_2$ between the ground and the conductor, where terminal (1) of the generator is connected to the conductor and terminal (2) is connected to the ground, the potential of the conductor becomes V and that of the ground remains zero.

8) *Capacitance of a conductor at equilibrium*: for an isolated conductor in electrostatic equilibrium, there is a relation between the conductor's potential and the charge distribution of its surface. The potential at any point $M(x, y, z)$ located within the conductor can be written as:

$$V_{\text{int}}(M) = \frac{1}{4\pi\varepsilon_0} \int_{S'} \frac{\sigma\left(\overrightarrow{\mathbf{r}}'\right) dS'}{\left|\overrightarrow{\mathbf{r}} - \overrightarrow{\mathbf{r}}'\right|} = \text{Const} \qquad [2.13]$$

where S', $\sigma\left(\overrightarrow{\mathbf{r}}'\right)$ and $\overrightarrow{\mathbf{r}}'$ are, respectively, the surface area of the conductor, the surface charge density and the position of the selected surface element dS'. The total charge on the conductor is the sum of elementary charges:

$$Q = \int_{S'} \sigma\left(\overrightarrow{\mathbf{r}}'\right) dS' \qquad [2.14]$$

By multiplying the density σ by a constant coefficient λ, we obtain a new total charge $Q' = \lambda Q$ and the conductor arrives at a new equilibrium state, defined by the potential $V'_{\text{int}} = \lambda V_{\text{int}}$. We therefore conclude that whatever the equilibrium state of a conductor, the ratio $\frac{Q}{V_{\text{int}}}$ remains constant. It depends only on the shape and size of the surface, S', that is, its geometrical characteristics: $\frac{Q}{V_{\text{int}}} = \frac{Q'}{V'_{\text{int}}} = \text{const}$. The proportionality coefficient is called *the conductor's capacitance*. We will use the following notation:

$$C = \frac{Q}{V_{\text{int}}} \qquad \Longleftrightarrow \qquad Q = CV_{\text{int}} \tag{2.15}$$

Capacitance is always a positive quantity and depends only on the geometric characteristics of the conductor.

EXAMPLE 2.1.(Capacitance of a Spherical Conductor).– Let us find the capacitance of a spherical conductor of radius R carrying a total charge Q on its surface. Inside the sphere, the potential is $V^0_{\text{int}} = \frac{Q}{4\pi\varepsilon_0 R}$ and, consequently:

$$C = \frac{Q}{V^0_{\text{int}}} = 4\pi\varepsilon_0 R \tag{2.16}$$

This is the capacitance of an isolated sphere (it depends only on the radius).

9) *Superposition of equilibrium states*: we have seen that an isolated conductor in electrostatic equilibrium is characterized by its total charge Q and its internal potential V_{int}, which are linked by the conductor's capacitance C. Inversely, if we know the surface charge distribution σ of a capacitance conductor C, we can completely determine its equilibrium state, since $Q = \int_{S'} \sigma dS'$ and $V_{\text{int}} = \frac{Q}{C}$. Let us now consider another equilibrium state for the same conductor defined by the surface charge density σ'. In this case, the conductor will carry a charge Q' and will have a potential V'_{int}. Due to the linear relationship between Q and V with σ, any linear combination of σ and σ' is again an equilibrium state:

$$\begin{aligned} \sigma'' &= a\sigma + b\sigma' \\ Q'' &= aQ + bQ' \\ V''_{\text{int}} &= aV_{\text{int}} + bV'_{\text{int}} \end{aligned} \tag{2.17}$$

with:

$$\frac{Q''}{V''_{\text{int}}} = \frac{aQ + bQ'}{aV_{\text{int}} + bV'_{\text{int}}} = C \tag{2.18}$$

This is *the principle of superposition of equilibrium states*: any superposition of equilibrium states (one or more conductors) is also an equilibrium state.

2.3. Electrostatic pressure

Consider a conductor in electrostatic equilibrium, bounded from the outside by a closed surface S, carrying a charge whose surface charge density (across S) is denoted σ. Let us choose an infinitesimal surface element dS around a point M on the charged surface S. The element dS with elementary charge $dq = \sigma dS$ is subject to an electrostatic force $d\vec{\mathbf{F}}$ by all charges other than those located in dS, as shown in Figure 2.5. To evaluate this force, we need to determine the electrostatic field applied to dq. In other words, the field created at the point M by the charges distributed outside dS. Let this field be denoted by $\vec{\mathbf{E}}_1(M)$. In addition, we know the field $\vec{\mathbf{E}}(M_1)$ near the surface, given by Coulomb's theorem:

$$\vec{\mathbf{E}}(M_1) = \frac{\sigma}{\varepsilon_0}\vec{\mathbf{n}} \qquad [2.19]$$

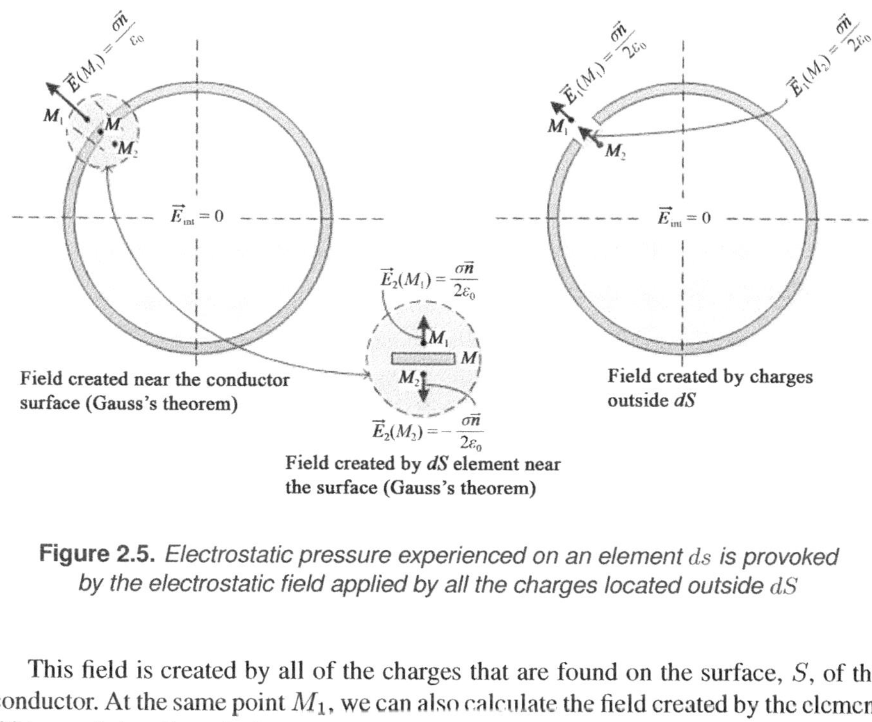

Figure 2.5. *Electrostatic pressure experienced on an element ds is provoked by the electrostatic field applied by all the charges located outside dS*

This field is created by all of the charges that are found on the surface, S, of the conductor. At the same point M_1, we can also calculate the field created by the element dS by applying Gauss's theorem:

$$\vec{\mathbf{E}}_2(M_1) = \frac{\sigma}{2\varepsilon_0}\vec{\mathbf{n}} \qquad [2.20]$$

By finally applying the principle of superposition, $\vec{\mathbf{E}}(M_1) = \vec{\mathbf{E}}_1(M_1) + \vec{\mathbf{E}}_2(M_1)$, we deduce that $\vec{\mathbf{E}}_1(M_1) = \frac{\sigma}{2\varepsilon_0}\vec{\mathbf{n}}$. This field $\vec{\mathbf{E}}_1(M_1)$ is practically the same as $\vec{\mathbf{E}}_1(M)$, since M and M_1 are very close, that is:

$$\overrightarrow{\mathbf{E}}_1(M) = \frac{\sigma(M)}{2\varepsilon_0}\overrightarrow{\mathbf{n}} \qquad [2.21]$$

Consequently, the charge dq contained in the element dS is subject to the force: $d\overrightarrow{\mathbf{F}}(M) = dq\overrightarrow{\mathbf{E}}_1(M) = \frac{\sigma^2}{2\varepsilon_0}\overrightarrow{\mathbf{n}}dS = \frac{\sigma^2}{2\varepsilon_0}d\overrightarrow{\mathbf{S}}$. It follows that *the electrostatic pressure* P (force per unit area) exerted on the element dS has the value:

$$P(M) = \left|\frac{d\overrightarrow{\mathbf{F}}}{dS}\right| = \frac{\sigma^2}{2\varepsilon_0} \qquad [2.22]$$

This quantity is always positive. In vector form, this is written as:

$$d\overrightarrow{\mathbf{F}}(M) = P(M)\,\overrightarrow{\mathbf{n}}dS = \frac{\sigma^2}{2\varepsilon_0}\overrightarrow{\mathbf{n}}dS \qquad [2.23]$$

Consequently, the force $d\overrightarrow{\mathbf{F}}$, at a point M on the surface, is directed outward, whatever the sign of the charge. In general, this force is too weak to remove charges from the conductor surface (except in the case of intense electrostatic fields). However, it can deform or displace the surface.

EXAMPLE 2.2. (Spherical Cap).– A hollow conductive sphere (radius R) is divided into two unequal parts along a horizontal plane perpendicular to the Oz axis. We obtain two spherical caps. The common base of these two caps is a circle of radius $R\sin\alpha$, as depicted in Figure 2.6. The sphere is taken to the potential V_0 and then isolated. The surface charge density that appears on the conductive sphere is:

$$\sigma = \frac{Q}{S} = \frac{4\pi\varepsilon_0 RV_0}{4\pi R^2} = \frac{\varepsilon_0 V_0}{R} \qquad [2.24]$$

Outside the sphere, $V = V_0\frac{R}{r}$ and $\overrightarrow{\mathbf{E}} = E_r\overrightarrow{\mathbf{e}}_r$ with $E_r = V_0\frac{R}{r^2}$. For $r = R$, we have $E_r(R+0) = \frac{\sigma}{\varepsilon_0}$ and, consequently, $\sigma = \frac{\varepsilon_0 V_0}{R}$. Let us calculate the force exerted by the lower (fixed) cap on the upper cap. As shown in Figure 2.6, each surface element dS of the upper cap located around the point M is subject to a force:

$$d\overrightarrow{\mathbf{F}}(M) = P\overrightarrow{\mathbf{n}}(M)\,dS = \frac{\varepsilon_0 V_0^2}{2R^2}\overrightarrow{\mathbf{n}}(M) \times R^2\sin\theta d\theta d\varphi \qquad [2.25]$$

where $\overrightarrow{\mathbf{n}}(M) = \sin\theta\left(\cos\varphi\,\overrightarrow{\mathbf{i}} + \sin\varphi\,\overrightarrow{\mathbf{j}}\right) + \cos\theta\,\overrightarrow{\mathbf{k}}$. After integrating over the entirety of the upper cap, the desired force is:

$$\begin{aligned}\overrightarrow{\mathbf{F}} &= \frac{\varepsilon_0 V_0^2}{2}\int_{\theta=0}^{\alpha}\sin\theta d\theta\int_{\varphi=0}^{2\pi}d\varphi\left[\sin\theta\left(\cos\varphi\,\overrightarrow{\mathbf{i}}+\sin\varphi\,\overrightarrow{\mathbf{j}}\right)+\cos\theta\,\overrightarrow{\mathbf{k}}\right] \\ &= \pi\varepsilon_0 V_0^2\int_0^{\alpha}\sin\theta\cos\theta\,\overrightarrow{\mathbf{k}} = \frac{1}{2}\pi\varepsilon_0 V_0^2\sin^2\alpha\,\overrightarrow{\mathbf{k}}\end{aligned} \quad [2.26]$$

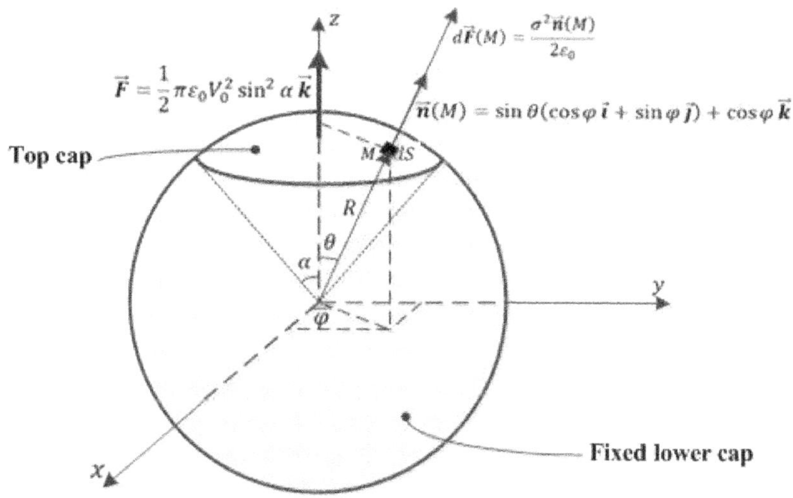

Figure 2.6. *Conductive sphere divided into two, unequal caps by a plane perpendicular to the Oz axis. For reasons of symmetry, the total force exerted on the upper cap by the rest of the sphere is directed upward along Oz. This is independent of the radius R*

The total force $\overrightarrow{\mathbf{F}}$ is directed upward, that is, in the positive direction of the Oz axis. Additionally, this force does not depend on the radius of the sphere. For $\alpha = \frac{\pi}{2}$ (the sphere is divided into two), we obtain:

$$\overrightarrow{\mathbf{F}} = \frac{1}{2}\pi\varepsilon_0 V_0^2\,\overrightarrow{\mathbf{k}} \quad [2.27]$$

2.4. Influence between conductors at electrostatic equilibrium

2.4.1. *Partial influence*

Consider a spherical conductor of radius R_1 denoted here by (1). The sphere (1) is taken to the positive potential V_0 and then isolated. At equilibrium, a uniformly distributed charge $Q_1 = 4\pi\varepsilon_0 R_1 V_0 > 0$ appears on the outer surface of the sphere. Inside, the field $\overrightarrow{\mathbf{E}}_{\text{int}}^{(1)}$ is zero, but outside of this, there is a field whose lines are perpendicular to the sphere. This external field is described by $V_{\text{ext}}^{(1)} = V_0 \frac{R_1}{r}$ and $\overrightarrow{\mathbf{E}}_{\text{ext}}^{(1)} = V_0 \frac{R_1}{r^2} \overrightarrow{\mathbf{e}}_r$. Consider another uncharged conductor, that is $Q_2 = 0$, designated here as (2), approaching (1) from the right. In conductor (2), fixed positive charges (ions) and free-moving negative charges (electrons) are uniformly distributed and compensate for each other at all points (electrically neutral conductor). We then observe a change in the field lines of conductor (1): the field lines running from the right-hand side of conductor (1) arrive on the left-hand side of conductor (2) (front zone) and this region becomes negatively charged through influence. The field lines from the other regions move away toward infinity or toward the ground, as shown in Figure 2.7. The positively charged conductor (1) will therefore influence conductor (2) by causing a displacement of charges on the latter: negative charges will migrate to the left side of conductor (2) and positive charges to the opposite side, creating an induced electric field. This movement will stop when the total field inside conductor (2) becomes zero, that is, when the induced field compensates for the external field:

$$\overrightarrow{\mathbf{E}}_{\text{int}}^{(2)} = \overrightarrow{\mathbf{E}}_{\text{ext}}^{(1)} + \overrightarrow{\mathbf{E}}_{\text{ind}}^{(2)} = 0 \qquad [2.28]$$

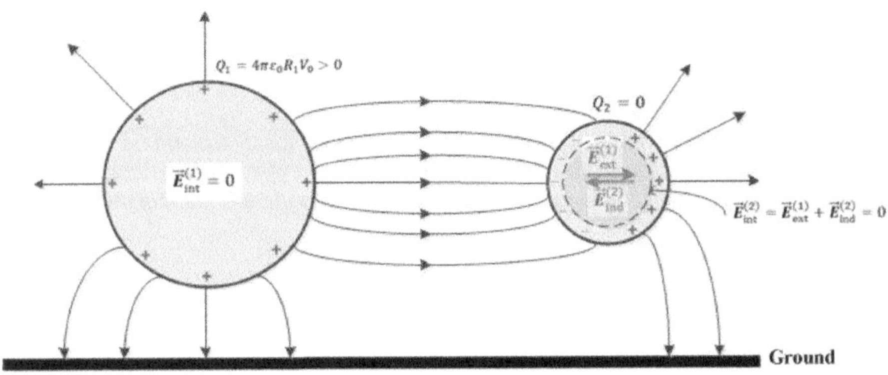

Figure 2.7. *Phenomenon of partial influence between conductors. The charged conductor (1) causes charges in (2) to be displaced so that an induced electric field is created to compensate for the external field experienced within the conductor: a new equilibrium state is thus established*

It is then said that conductors (1) and (2) are under *partial influence* and that (2) has been *electrified by influence* (electrification effect) (there was no creation of charge, just a displacement of the conductor's own charges).

What is the potential $V_{int}^{(2)}$ inside the conductor (2)? Let us recall that lines emerge from the conductor (1) toward (2) or toward infinity, or toward the ground. Since they are always directed toward the lower potentials, then $V_{\text{int}}^{(2)} < V_0$. In the same way, lines emerging from conductor (2) can only move toward infinity or the soil, which shows that $V_{\text{int}}^{(2)} > 0$. We therefore deduce that $0 < V_{\text{int}}^{(2)} < V_0$ while $Q^{(2)} = 0$. Without contact and through influence alone, the conductor has been brought a potential $V_{\text{int}}^{(2)} > 0$.

2.4.2. *Theorem of corresponding elements*

Let us sketch a small closed contour Γ_1 on the surface of the conductor (1). Let us assume that all field lines from Γ_1 (i.e. emerging from the interior of Γ_1) connect to the conductor (2) drawing the outline of a closed contour Γ_2. All of these field lines, together constitute what is called a *field tube* or *flux tube*. By construction, the flux in an electrostatic field across the lateral surface, S_{lat}, of the tube is zero, that is, $\int_{S_{\text{lat}}} \vec{\mathbf{E}} \cdot d\vec{\mathbf{S}} = 0$. Let us now consider the closed Gaussian surface $S = S_1 + S_{\text{lat}} + S_2$, where S_1 is a surface supported by Γ_1 and is immersed inside conductor (1), and S_2 is a similar surface for (2), as seen in Figure 2.8. The surfaces S_1 and S_2 are called *the corresponding elements*. Let us apply Gauss's theorem to the closed surface S:

$$\frac{q_1 + q_2}{\varepsilon_0} = \int_{S_1} \vec{\mathbf{E}}_{\text{int}}^{(1)} \cdot d\vec{\mathbf{S}} + \int_{S_{\text{lat}}} \vec{\mathbf{E}} \cdot d\vec{\mathbf{S}} + \int_{S_2} \vec{\mathbf{E}}_{\text{int}}^{(2)} \cdot d\vec{\mathbf{S}} = 0 \qquad [2.29]$$

where q_1 (respectively, q_2) is the charge enclosed in the loop Γ_1 (respectively, Γ_2) on the surface of the conductor (1) (respectively, conductor (2)). q_1 and q_2 are called *the charges on the corresponding elements*. It can be derived from this that:

$$q_2 = -q_1 \qquad [2.30]$$

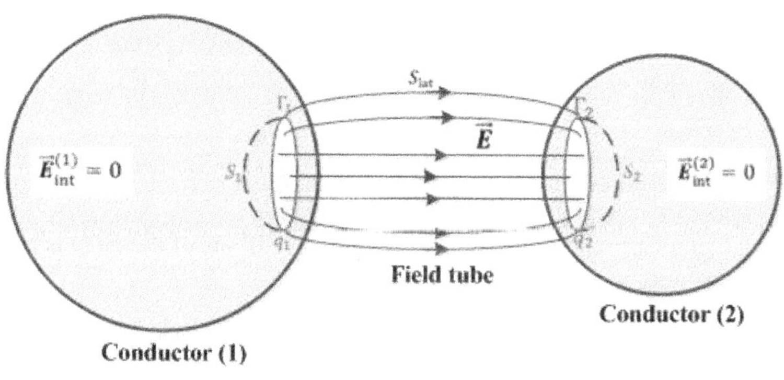

Figure 2.8. *Corresponding elements between two conductors (1) and (2) held within the contours Γ_1 and Γ_2. All field lines emerging from an element located on (1) arrive on an element on (2). Both of these elements have total charges with opposite signs*

We can therefore conclude that *the charges on two corresponding elements will be opposite in sign*. This is *the theorem of corresponding elements*.

2.4.3. *Earthing*

Let us return to the example in Figure 2.7. By connecting conductor (2) to the ground through a conductive wire, the positive charges located on the right side of the conductor will be neutralized by electrons from the earth. It can be said that the positive charges flowed into the earth. Consequently, no field line will be able to emerge from the conductor (2), either toward the ground (conductor (2) and the ground constitute a single conductor at the same potential) or toward another conductor, as shown in Figure 2.9.

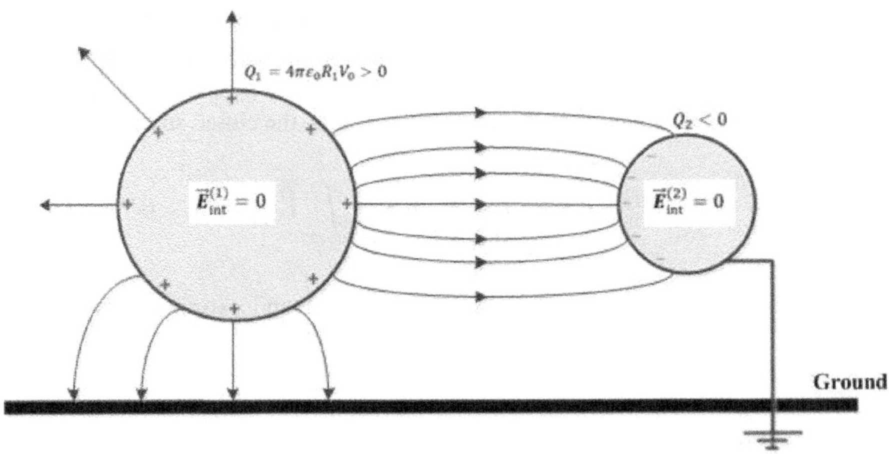

Figure 2.9. *If (2) is connected to the ground, the positive charges on (2) are canceled by the electrons from the ground. Therefore, no field line can emerge from (2)*

2.4.4. *Total influence*

Two conductors are said to be under total influence if one of them (1) is inside the other (2). Let us note some essential points:

(i) Now let us assume that conductor (2) is hollow. It is possible to create *conditions of total electrostatic influence* by placing the conductor (1) within (2). Since all field lines of (1) arrive on the inner surface of (2), the entire surface of (1) is then the corresponding element of the inner surface of (2), as shown in Figure 2.10(a). The charge $Q^{(2)}_{\text{int}} = -Q_1$ thus appears on the internal face of (2), regardless of the position of the conductor (1). This property is known as *Faraday's theorem*. By applying Gauss's theorem to calculate the field $\vec{\mathbf{E}}$ outside (2), it can be seen that only the charges on the external face of (2), $Q_{\text{ext}}^{(2)}$, come into play, as those on the internal surface, $Q_{\text{int}}^{(2)}$, are compensated by the charges on (1), that is, by Q_1. Therefore, the external electric field of (2) depends only on $Q_{\text{ext}}^{(2)}$. The same applies to the potential. If (2) is isolated during the experiment, its total charge cannot change. The outer surface of (2) will therefore see its surface charge increase by Q_1, in addition to any initial charge $Q_{\text{ini}}^{(2)}$. Consequently, the total charge on the external surface of (2) is simply:

$$Q_{\text{ext}}^{(2)} = Q_1 + Q_{\text{ini}}^{(2)} \quad [2.31]$$

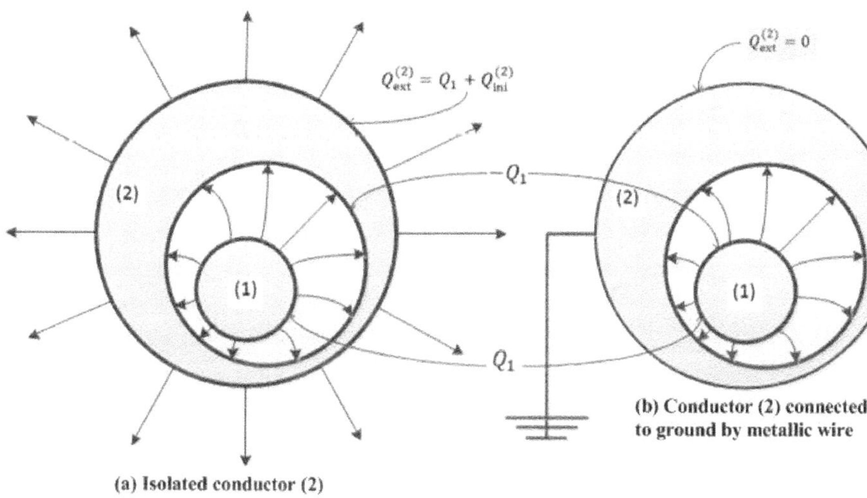

Figure 2.10. *Conductors (1) and (2) under total influence. All field lines emerging from (1) reach (2). The internal surface of (2) and the external surface of (1) have opposite charges. (a) On the external surface of (2), a charge Q_1 gets added to the initial charge on (2), $Q_{ext}^{(2)} = Q_{ini}^{(2)} + Q_1$. (b) By connecting the conductor (2) to the ground, the external charge flows into the ground, $Q_{ext}^{(2)} = 0$*

Note that the charge distribution on the inner surface of conductor (2) depends on the position of conductor (1) in the cavity. The charge distribution over the external surface does not depend on this.

(ii) When conductor (2) is connected to Earth, the total charge on the external surface of (2) flows to Earth by the connecting device and, as a result, $Q_{\text{ext}}^{(2)} = 0$. In this case, the electrostatic field measured outside (2) is zero, despite the presence of conductor (1) inside (2). Thus, the space outside (2) is protected against any electrostatic influence from within the cavity, as shown in Figure 2.10(b).

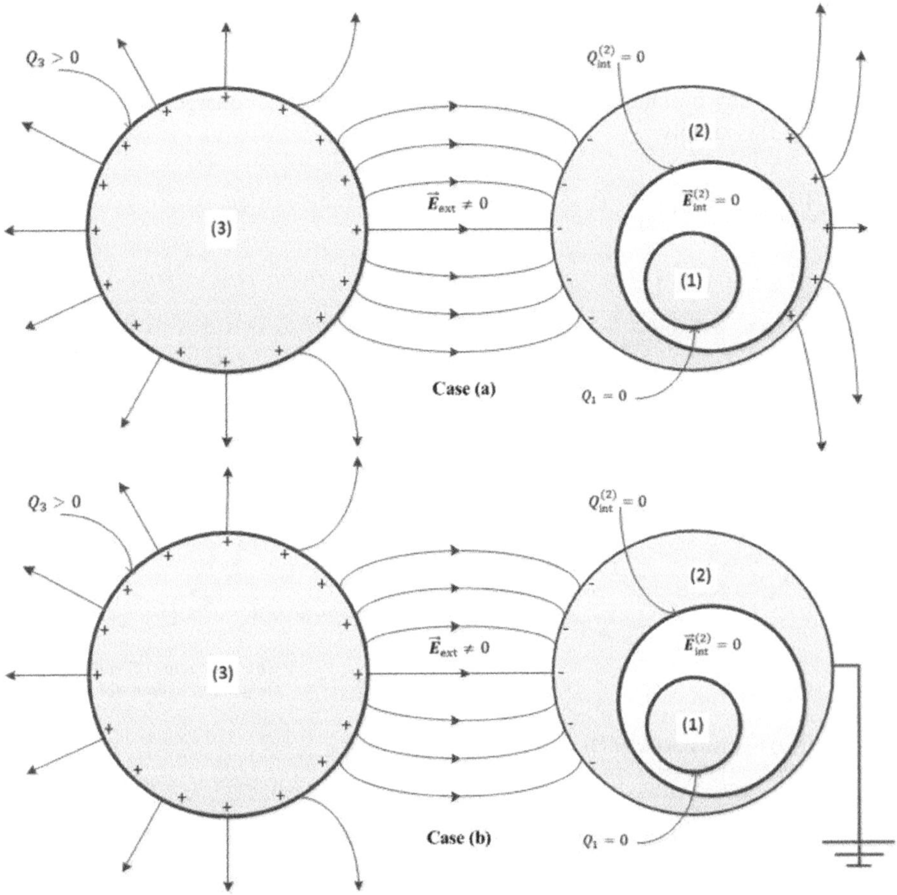

Figure 2.11. *Conductor (2) forms an electrostatic screen. Conductor (1), placed within (2), experiences no influence from (3)*

(iii) The reverse is also true, as shown in Figure 2.11(a). Let us assume that the hollow conductor (2) (containing a neutral conductor (1) inside) is close to a charged conductor (3). At equilibrium, the electrostatic field within the conductor (2) is zero, but outside it has no zero value, due to the presence of (3). In this case, too, the external field has no influence inside the cavity of (2). The state of conductor (1) does not change (its potential remains zero). We therefore conclude that conductor (2) forms an electrostatic screen. Note that this remains true even if the conductor (2) is connected to the ground, as shown in Figure 2.11(b). The inner and outer spaces of the conductor (2) are perfectly independent of each other.

2.4.5. *Electrostatic influence coefficients*

To fix ideas, let us consider two spherical conductors (1) and (2), carrying charges Q_1 and Q_2, respectively. The two spheres are separated by a distance $D \gg R_1 + R_2$, where R_1 and R_2 are, respectively, the radii of (1) and (2), as shown in Figure 2.12. According to the superposition principle, the potential observed at a point M_1 inside the conductor (1) is:

$$V_1(M_1) = \frac{1}{4\pi\varepsilon_0}\int_{M_1'\in S_1} \frac{\sigma_1(M_1')\,dS_1}{\left|\overrightarrow{\mathbf{M_1'M_1}}\right|} + \frac{1}{4\pi\varepsilon_0}\int_{M_2'\in S_2} \frac{\sigma_2(M_2')\,dS_2}{\left|\overrightarrow{\mathbf{M_2'M_1}}\right|} \qquad [2.32]$$

where the first term is due to the charges Q_1 and the second due to the charges on (2). When the distance, D, is much greater than the radii, the following approximation can be made for any point M_2' on the surface of (2): $\left|\overrightarrow{\mathbf{M_2'M_1}}\right| \simeq \left|\overrightarrow{\mathbf{O_2O_1}}\right| = D$ (O_1 and O_2 are the centers of the sphere). Consequently, we get:

$$V_{\text{int}}^{(1)} = \frac{Q_1}{4\pi\varepsilon_0 R_1} + \frac{Q_2}{4\pi\varepsilon_0 D} = \frac{Q_1}{C_1} + \frac{Q_2}{4\pi\varepsilon_0 D} \qquad [2.33]$$

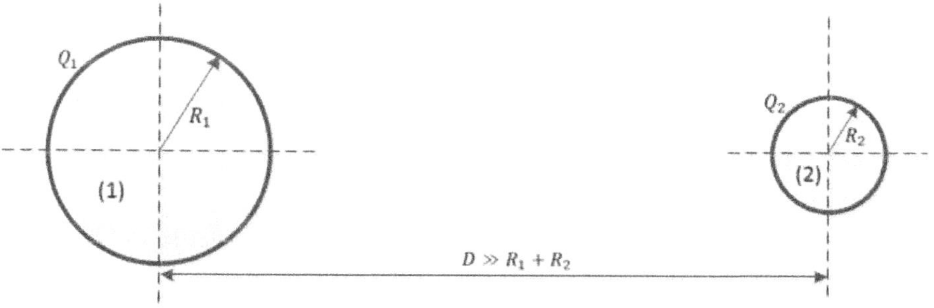

Figure 2.12. *Coefficients of electrostatic influence between two spherical conductors separated by a distance $D \gg R_1 + R_2$*

where C_1 is the capacitance of (1), that is, when it is isolated. By doing the same for conductor (2), we find:

$$V_{\text{int}}^{(2)} = \frac{Q_1}{4\pi\varepsilon_0 D} + \frac{Q_2}{4\pi\varepsilon_0 R_2} = \frac{Q_1}{4\pi\varepsilon_0 D} + \frac{Q_2}{C_2} \qquad [2.34]$$

where C_2 is the capacitance of (2). Equations [2.33] and [2.34] can also be written together in the matrix form:

$$\begin{pmatrix} V_{\text{int}}^{(1)} \\ V_{\text{int}}^{(2)} \end{pmatrix} = \begin{pmatrix} \frac{1}{C_1} & \frac{1}{4\pi\varepsilon_0 D} \\ \frac{1}{4\pi\varepsilon_0 D} & \frac{1}{C_2} \end{pmatrix} \begin{pmatrix} Q_1 \\ Q_2 \end{pmatrix} = \begin{pmatrix} p_{ij} \end{pmatrix} \begin{pmatrix} Q_1 \\ Q_2 \end{pmatrix} \qquad [2.35]$$

The potentials $V_{\text{int}}^{(1)}$ and $V_{\text{int}}^{(2)}$ measured inside and on the surface of the conductors (1) and (2) have a linear relation with their respective charges. The coefficients p_{ij} are called the *coefficients of potential* and depend only on the geometry of the system. By inverting the matrix (p_{ij}):

$$(p)_{ij}^{-1} = (C_{ij}) \qquad [2.36]$$

where:

$$\begin{aligned} C_{11} &= \frac{1}{C_2}\left[\frac{1}{\frac{1}{C_1 C_2} - \left(\frac{1}{4\pi\varepsilon_0 D}\right)^2}\right] \simeq C_1\left[1 + \frac{C_1 C_2}{(4\pi\varepsilon_0 D)^2}\right] \\ C_{22} &\simeq C_2\left[1 + \frac{C_1 C_2}{(4\pi\varepsilon_0 D)^2}\right] \\ C_{12} &= C_{21} = -\frac{1}{4\pi\varepsilon_0 D}\left[\frac{C_1 C_2}{1 - \frac{C_1 C_2}{(4\pi\varepsilon_0 D)^2}}\right] \simeq -\frac{C_1 C_2}{4\pi\varepsilon_0 D} \end{aligned} \qquad [2.37]$$

The coefficients C_{ii} are called the *capacitance coefficients* and C_{ij} $(i \neq j)$ are *coefficients of electrostatic induction*. It can clearly be seen that (in the presence of other conductors) the capacitance coefficients C_{ii} cannot be identified with the capacitances C_i of the isolated conductors and that the induction coefficients C_{ij} are indeed negative: *the capacitance coefficient of sphere (1) (respectively, (2)) increased in the presence of (2) (respectively, (1))*. In addition, the matrices (p_{ij}) and (C_{ij}) are symmetric.

Let us generalize this result. In a system of charges located on disconnected conductors, the potential at all points within a conductor, or on the surface of the conductor, has a linear relation with all the charges in question. Therefore, for a system of conductors $(1), \ldots, (n)$ which carry charges $Q_1, \ldots, Q_n$, the potential on conductor (i) can be written in terms of the charges Q_j located on the conductor (j) (including that on (i)):

$$V_i = \sum_j p_{ij} Q_j \qquad [2.38]$$

where p_{ij} are the *coefficients of potential*. They only depend on the geometry of the system (i.e. (a) the size and shape of the conductors, (b) the distance between them and (c) their arrangement in relation to each other). Although it is not always possible to calculate these coefficients, the system is entirely characterized by p_{ij}, which can be determined experimentally. These equations can also be inverted. That is:

$$Q_j = \sum_i \left(p^{-1}\right)_{ji} V_i \equiv \sum_i C_{ji} V_i \qquad [2.39]$$

where C_{ii} and C_{ij} $(i \neq j)$ are, respectively, *the capacitance coefficients* and *the induction coefficients* of the system. The coefficients C_{ij} $(\forall i, j)$ are also called *the influence coefficients*. Equation [2.39] can also be demonstrated in the following way: consider n conductors $(1), \ldots, (n)$ in electrostatic equilibrium, of charges $Q_1, \ldots, Q_n$ and potentials $V_1, \ldots, V_n$. They are in mutual influence with each other. If the conductor (1) is alone at the V_1 potential and the others are grounded (i.e. at a zero potential), the conductors $(1), \ldots, (n)$ will acquire, by influence, the loads $q_{11} = C_{11}V_1, q_{21} = C_{21}V_1, \ldots, q_{n1} = C_{n1}V_1$, respectively, as shown in Figure 2.13. Similarly, if the conductor (2) is now the only one with a potential of not zero, V_2 and all others are grounded, under influence, the drivers will receive charges $q_{12} = C_{12}V_2$, $q_{22} = C_{22}V_2, \ldots, q_{n2} = C_{n2}V_2$, and so on, respectively. Passing to the conductor (i) and assuming that he is taken at potential V_i, while the others are grounded, the conductors will then take, respectively, the charges $q_{1i} = C_{1i}V_i$, $q_{2i} = C_{2i}V_i, \ldots, q_{ni} = C_{ni}V_i$. We continue in this way until the end. After the superposition of all these equilibrium states, we find the most general state:

$$\begin{aligned} Q_1 &= C_{11}V_1 + C_{12}V_2 + \cdots + C_{1n}V_n \\ Q_2 &= C_{21}V_1 + C_{22}V_2 + \cdots + C_{2n}V_n \\ &\vdots \\ Q_n &= C_{n1}V_1 + C_{n2}V_2 + \cdots + C_{nn}V_n \end{aligned} \qquad [2.40]$$

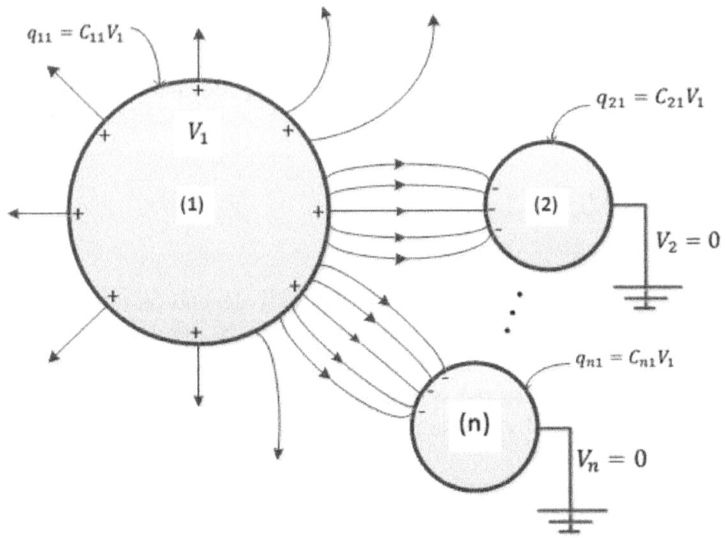

Figure 2.13. *Part of the field lines emerging from conductor (1) taken to the potential $V_1 > 0$ and, consequently, with the charge $q_{11} = C_{11}V_1 > 0$) reach the other conductors (the other lines go toward infinity) and, by influence, they cause the other conductors to acquire the negative charges $q_{21} = C_{21}V_1, \ldots, q_{n1} = C_{n1}V_1$. The coefficients C_{ii} are positive, while C_{ij} $(i \neq j)$ are all negative*

These equations show that there is a decomposition of the charge distribution for each conductor:

$$Q_i = \int_{S_i} \sigma_i dS_i = \sum_j \int_{S_i} \sigma_{ij} dS_i = \sum_j q_{ij} \equiv \sum_j C_{ij} V_j \qquad [2.41]$$

where σ_{ii} is the surface charge density appearing on the conductor (i), if all the others are earthed, and σ_{ij} $(i \neq j)$ is the surface charge density on the conductor (i), when all of the conductors (including conductor (i)) are earthed, except conductor (j).

The influence coefficients have the following properties:

a) The coefficients p_{ij} and C_{ij} are symmetric, that is, $p_{ij} = p_{ji}$ and $C_{ij} = C_{ji}$. To prove this, let us show that if (V_i, Q_i) and (V_i', Q_i') are two equilibrium states (of a system with n conductors at equilibrium), then:

$$\sum_{i=1}^{n} Q_i V_i' = \sum_{i=1}^{n} Q_i' V_i \qquad [2.42]$$

Indeed, for any point M_i on conductor (i), it is written that:

$$V_i = V(M_i) = \sum_{j=1}^{n} \int_{M_j \in S_j} \frac{\sigma_j(M_j)}{4\pi\varepsilon_0 \left|\overrightarrow{\mathbf{M_j M_i}}\right|} dS_j \qquad [2.43]$$

In addition, the total load on conductor (i) is $Q_i = \int_{M_i \in S_i} \sigma_i(M_i)\, dS_i$. It follows from this that:

$$\sum_{i=1}^{n} Q_i V_i' = \sum_{i=1}^{n} \sum_{j=1}^{n} \int_{M_i \in S_i} \int_{M_j \in S_j} \frac{\sigma_i(M_i)\, \sigma_j'(M_j)}{4\pi\varepsilon_0 \left|\overrightarrow{\mathbf{M_i M_j}}\right|} dS_i dS_j \qquad [2.44]$$

Since the expression is symmetric for σ_i and σ_j', it can be derived from this that $\sum_{i=1}^{n} Q_i V_i' = \sum_{i=1}^{n} Q_i' V_i$. Let us apply identity [2.42] to two different equilibrium states: one, where $V_i = V_0 \neq 0$ and $V_k = 0$ for all $k \neq i$; the other where $V_j' = V_0$ and $V_k' = 0$ for all $k \neq j$. Since $Q_k = C_{ki} V_0$, $Q_k' = C_{kj} V_0$, $\sum_{k=1}^{n} Q_k V_k' = C_{ji} V_0^2$ and $\sum_{k=1}^{n} Q_k' V_k = C_{ij} V_0^2$, according to [2.43], it is obtained to $C_{ji} V_0^2 = C_{ij} V_0^2$. This shows that $C_{ji} = C_{ij}$. Therefore, the coefficients C_{ij} are indeed symmetric. Additionally, the inverse of matrix (C_{ij}), that is, (p_{ij}), is also symmetric.

b) The coefficients C_{ii} are always positive. To demonstrate this, remember that two points on the same conductor cannot be connected by a field line. It can therefore be derived from this that the field lines emerging from conductor (i) at potential V_i: (1) either move toward the conductor with lower potential; (2) or move toward infinity, if $V_i > 0$; (3) emerge from a conductor at higher potential; (4) or from infinity, if $V_i < 0$. Let us keep all conductors at the potential 0, except the conductor (i) which is kept at the potential $V_i > 0$. We obtain $Q_i = C_{ii} V_i$. The field lines connecting (i) must necessarily start from the surface, therefore $\overrightarrow{\mathbf{E}}_{\text{ext}}$ is directed toward the exterior, in the external neighborhood of any point M_i on the surface of conductor (i). We therefore have $\sigma_i(M_i) > 0$ and after integration, $Q_i > 0$. The coefficient C_{ii} is therefore strictly positive, otherwise there is a contradiction.

c) The coefficients C_{ij} $(i \neq j)$ are all negative. Indeed, let us always examine the preceding configuration, that is, (b). For a conductor (j) with $j \neq i$, all field lines arriving on (j) come from (i) and no other field line leaves (j) toward any other conductor or to infinity. We therefore have $Q_j < 0$. Since $Q_j = C_{ji} V_i$ and $V_i > 0$, it can be derived from this that $C_{ij} < 0$.

d) $C_{ii} \geq -\sum_{j \neq i} C_{ji}$ Equality is only possible in the case of total influence. This inequality is a consequence of the theorem of corresponding elements. Keeping the conductor (i) at the potential V_0 and then earth all other conductors. Since all field lines emerging from conductor (i) do not necessarily arrive at the other conductors, except in the case of total influence, it can be written that $q_{ii} \geq \sum_{j \neq i} |q_{ji}| = -\sum_{j \neq i} C_{ji} V_0$. This results in the following:

$$C_{ii} \geq -\sum_{j \neq i} C_{ji} \qquad [2.45]$$

The influence coefficients given by [2.37] satisfy the assertions (a)–(d). It is of course verified that (1) $C_{11} > 0$ and $C_{22} > 0$, (2) $C_{12} = C_{21} < 0$ and (3) $C_{11} \geq C_{21} > 0$ and $C_{22} \geq C_{12} > 0$.

2.5. Capacitors

2.5.1. *Definition and capacitance of the capacitor*

Any system with two conductors under total electrostatic influence (i.e. all of the field lines from one of them all arrive to the other) is called *a capacitor* (under total influence). This property is realized if one of the conductors, referred to here by (2), completely surrounds the other, noted (1). Conductors (1) and (2) are *the armature* or *electrodes* of the capacitor. These two elements are separated by an insulating material or a vacuum. To simplify, suppose that there is only one vacuum:

– The electrode or armature (1) carries the surface charge Q_1.

– Under total influence, the internal surface of the armature (2) acquires the induced charge $Q_{\text{int}}^{(2)} = -Q_1$.

– The charge $Q_{\text{ext}}^{(2)}$ on the external surface of armature (2) will depend on its initial charge $Q_{\text{ini}}^{(2)}$ and on its electric state, as seen in Figure 2.14.

The potentials of (1) and (2) are V_1 and V_2, respectively. The charges Q_1 and Q_2 on the conductors (1) and (2) are expressed as linear combinations of V_1 and V_2 involving the influence coefficients:

$$\begin{aligned} Q_1 &= C_{11} V_1 + C_{12} V_2 \\ Q_2 &= C_{21} V_1 + C_{22} V_2 \end{aligned} \qquad [2.46]$$

First, assume that the conductors (1) and (2) are connected to the same source of potential V_0. In these conditions, it can be said that (1) and (2) form a single conductor, within which the electric field is zero. Therefore, Q_1 is necessarily zero, which leads to:

$$C_{11} = -C_{12} \qquad [2.47]$$

Suppose then that the armature (2) is connected to the Earth by a grounding device and that (1) carries the charge Q. The potential of (2) is therefore zero. In this case, both the potential and the field are null in the outer space of (2). This conductor has no charge on its outer surface (Gauss' theorem). However, due to the total influence, the internal surface of (2) must carry the charge $-Q$. This results in $Q_2 = -Q = C_{21}V_1 = -C_{11}V_1$, which gives:

$$C_{12} = -C_{11} = C_{21} \quad [2.48]$$

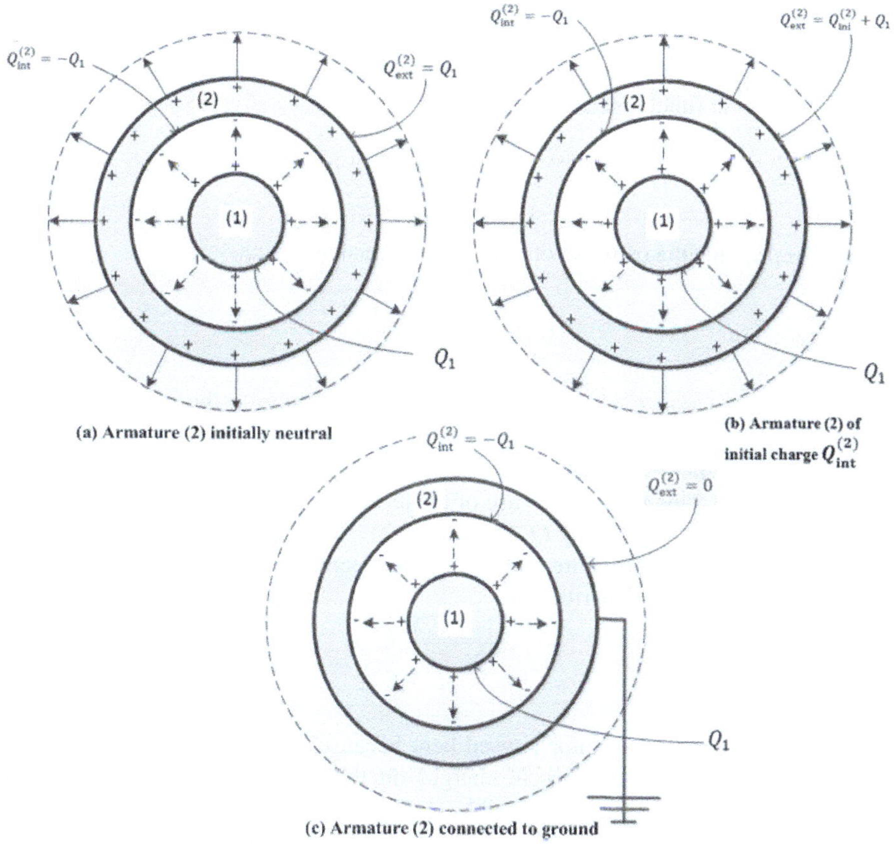

Figure 2.14. *Capacitor under total influence: (a) armature (1) carries the charge Q_1. (b) Due to total influence, the internal surface of armature (2) will acquire the induced charge $Q_{int}^{(2)} = -Q_1$. (c) The charge $Q_{ext}^{(2)}$ on the external surface of (2) is equal to the sum of Q_1 and the initial charge $Q_{ini}^{(2)}$. If armature (2) is earthed, the external charge $Q_{ext}^{(2)}$ flows into the ground*

Therefore, for any potentials V_1 and V_2, it can be written as:

$$\begin{aligned} Q_1 &= C_{11}\left(V_1 - V_2\right) \\ Q_2 &= \left(C_{22} - C_{11}\right)V_2 - C_{11}\left(V_1 - V_2\right) \end{aligned} \qquad [2.49]$$

Since (1) carries the charge Q_1, the internal surface of (2) must carry the charge $-Q_1$ by total influence. The charge carried by the outer surface of (2) is therefore:

$$Q_{\text{ext}}^{(2)} = Q_2 + Q_1 = \left(C_{22} - C_{11}\right)V_2 \qquad [2.50]$$

It is independent of the electrical state of the armature (1) and remains even if (2) is at zero potential (uncharged). The quantity:

$$C_2 = C_{22} - C_{11} \qquad [2.51]$$

therefore represents the capacity of (2) alone. The value:

$$C = C_{11} \qquad [2.52]$$

is called the *capacitance of the capacitor*. It depends only on the geometry of the system and the permittivity of the insulation between the armatures. In general, C_2 is much smaller than C and the external charge $Q_{\text{ext}}^{(2)}$ of (2) is therefore negligible compared to $|Q_1|$. This justifies the use of the term *charge capacitor* to charge Q_1 of the internal armature. By setting $Q = Q_1$ and $V = V_1 - V_2$ (difference between the potential of the internal armature and that of the external armature), the fundamental relation of the capacitors is written as:

$$Q = CV \qquad [2.53]$$

Note that the name capacitor is used here because the system has the ability to condense (or accumulate) electrical charges on the armatures. By constructing a high-capacity capacitor, C, it is possible to obtain high condensation of electrical charges even at low voltages V. To make a very high-capacity capacitor, imagine a full-influence capacitor with two conductors (1) and (2), separated by a thin insulating material (e.g. vacuum) of thickness d, as shown in Figure 2.15. The external conductor (2) carries the potential V_2, while the internal conductor (1) is at V_1. Since the field lines are perpendicular to the surfaces and the insulating layer is very thin, it can be inferred that the field lines between the two frames are all of length d. Thus, two infinitesimal element surfaces dS facing each other are corresponding elements and carry equal and opposite loads. Therefore, we deduce that the charge densities carried by the two surfaces are equal and opposite: σ and $-\sigma$. In addition, the electric field is $E = \frac{\sigma}{\varepsilon_0}$ on these field lines, since it is the field in the immediate vicinity of the surface of a conductor. By integrating along a line of length d, we obtain:

$$V = V_1 - V_2 = Ed = \frac{\sigma d}{\varepsilon_0} \quad [2.54]$$

The charge density is therefore uniform. If S is the area of the surfaces opposite each other on either side of the insulating material (the internal surface of the conductor (2) is equal to the surface of the conductor (1)), then the total charge carried by the capacitor can be calculated as follows:

$$Q = \sigma S = \frac{\varepsilon_0 VS}{d} \quad [2.55]$$

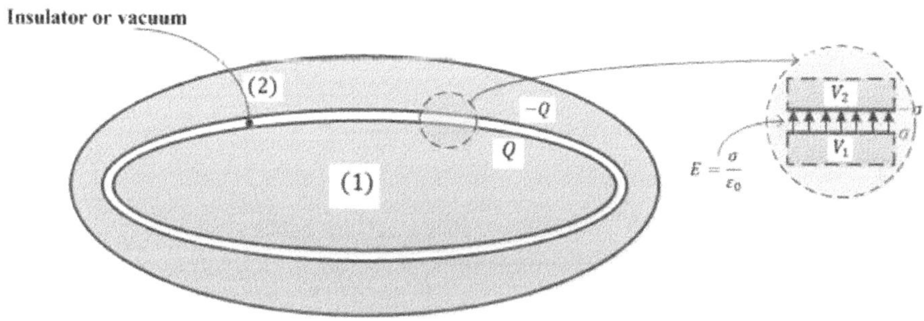

Figure 2.15. *Creating a high-capacity capacitor. When the contact area S is large and the distance between the two conductors is small, it produces enormous capacities*

Consequently, we have the following capacitance for the capacitor:

$$C = \frac{Q}{V} = \frac{\varepsilon_0 S}{d} \quad [2.56]$$

If the two surfaces facing each other are large and the distance between them is very small, the capacitances are much higher than those of a capacitor standing alone in space and, consequently, the charges on the armatures are high. The charges therefore tend to accumulate in regions that are very close to each other, in order to annihilate each other, if possible. A capacitor is represented by the symbol in Figure 2.16 for a plane capacitor. This symbol is used to recall the fact that the expression for capacitance of a capacitor is almost the same as that for a plane capacitor when the distance between the conductors is very small.

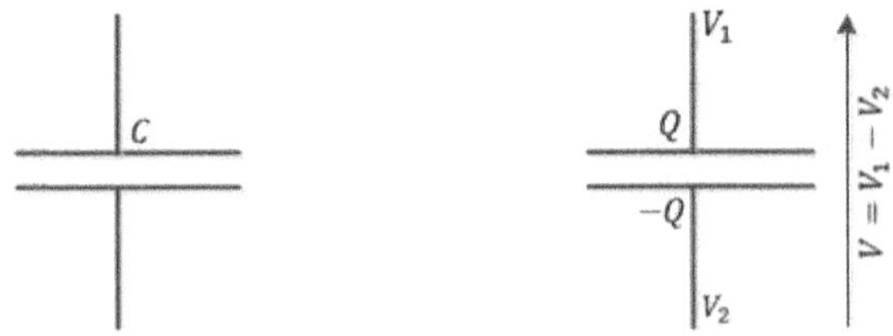

Figure 2.16. *Symbolic representation of a capacitor*

2.5.2. *Capacitor with closely spaced armatures*

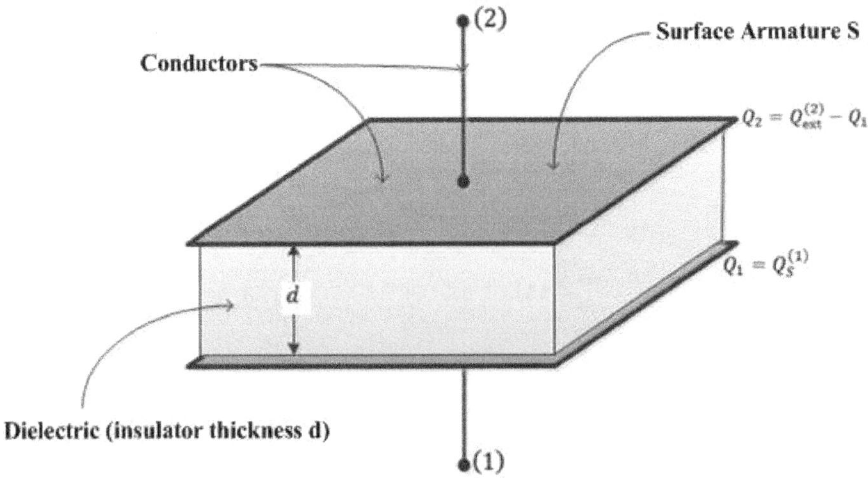

Figure 2.17. *Capacitor with closely spaced armatures. When the thickness d is very small with respect to the dimensions of the armatures, the influence may be considered to be quasi-total and, consequently, it is possible to apply the fundamental relation for capacitors $Q = CV$ with high precision. The thickness of the dielectric is chosen such that the charge transfer through the tunnel effect can be neglected*

Typical capacitors do not really have the structure described above. In general, they are made up of two conductive surfaces (armatures) separated by an insulating material (dielectric). Electric contact occurs in each of the armatures, as shown in Figure 2.17. However, total influence does generally occur; that is, charges on the facing armatures are opposite in nature and the relation $Q = CV$ can be applied with high precision. These capacitors are called *capacitors with closely spaced armatures* (almost total influence). In order to create almost total influence, it is sufficient to take armatures that are very wide with respect to the separation distance, so that edge effects can be neglected, that is, to not have any loss of field. In these conditions, we will also have $C_{11} = -C_{12} = -C_{21} \neq C_{22}$. In this type of capacitor, the charges Q_1 and Q_2 correspond to the charges that are found to be distributed over the entire surface of each conductor. However, if the distance separating the two armatures is very small compared to the size of the armatures, the electrostatic influence will condense the charges onto the surfaces facing each other, such that the following hypothesis or approximation can be made:

$$\left[\begin{array}{l} Q_1 = Q_{\text{ext}}^{(1)} + Q_S^{(1)} \simeq Q_S^{(1)} \\ Q_2 = Q_{\text{ext}}^{(2)} + Q_S^{(2)} \simeq Q_{\text{ext}}^{(2)} - Q_S^{(1)} = Q_{\text{ext}}^{(2)} - Q_1 \end{array}\right. \qquad [2.57]$$

which leads to expressions that are identical to the expression of a capacitor under total influence.

2.5.3. *Examples of capacitors*

In what follows, we will look at several examples of capacitance calculations. To obtain the capacitance C of a capacitor, we need to calculate the relationship between the capacitor's charge Q and its voltage $V_1 - V_2$, that is, $V_1 - V_2 = \int_{\Gamma_{12}} \overrightarrow{\mathbf{E}} \cdot d\overrightarrow{\mathbf{l}} = \frac{Q}{C}$. This gives:

$$C = \frac{Q}{V_1 - V_2} = \frac{Q}{\displaystyle\int_{\Gamma_{12}} \overrightarrow{\mathbf{E}} \cdot d\overrightarrow{\mathbf{l}}} \qquad [2.58]$$

where Γ_{12} is any path that connects armature (1) to armature (2). In other words, the circulation of the electrostatic field from armature (1) to armature (2) must be computed, as well as the charge on the capacitor Q. To determine Q, we need to take a Gaussian surface surrounding the initial armature (in this case (1)) and apply Gauss's theorem:

$$C = \frac{\displaystyle\oint_{S_1} \varepsilon_0 \overrightarrow{\mathbf{E}} \cdot d\overrightarrow{\mathbf{S}}}{\displaystyle\int_{\Gamma_{12}} \overrightarrow{\mathbf{E}} \cdot d\overrightarrow{\mathbf{l}}} = \frac{\displaystyle\oint_{S_2} \varepsilon_0 \overrightarrow{\mathbf{E}} \cdot d\overrightarrow{\mathbf{S}}}{\displaystyle\int_{\Gamma_{21}} \overrightarrow{\mathbf{E}} \cdot d\overrightarrow{\mathbf{l}}} \qquad [2.59]$$

where S_2 is a fictitious surface that completely surrounds the armature (2) and Γ_{21} is the opposite path Γ_{12} in direction.

EXAMPLE 2.3. (Spherical Capacitor).–

a) Consider the total influence capacitor shown in Figure 2.18. The conductor (1) inside is a sphere with radius R_1 and center O. The inner wall of (2) is a concentric sphere of radius R_2, while the outer wall can be of any form. The conductor (1) carries a charge Q that is uniformly distributed on its surface. The space between the two conductors is empty. Under total influence, a charge $-Q$ is induced on the internal wall of the conductor (2). Let us draw a fictional concentric Gaussian sphere Σ with radius r ($R_1 < r < R_2$). For reasons of symmetry, the field $\vec{\mathbf{E}}$ is radial and its modulus is the same everywhere over Σ. By applying Gauss's theorem, the electrostatic field has the value $\vec{\mathbf{E}}(M) = \frac{Q\vec{r}}{4\pi\varepsilon_0 r^3}$, where M is a point located between the armatures. This field will induce the following voltage between the armatures: $V_1 - V_2 = \int_{R_1}^{R_2} \vec{\mathbf{E}} \cdot d\vec{\mathbf{l}} = \frac{Q}{4\pi\varepsilon_0}\int_{R_1}^{R_2}\frac{dr}{r^2} = \frac{Q}{4\pi\varepsilon_0}\left(\frac{1}{R_1} - \frac{1}{R_2}\right)$ and will therefore provide the following capacitance:

$$C = \frac{Q}{V_1 - V_2} = \frac{4\pi\varepsilon_0 R_1 R_2}{R_2 - R_1} \tag{2.60}$$

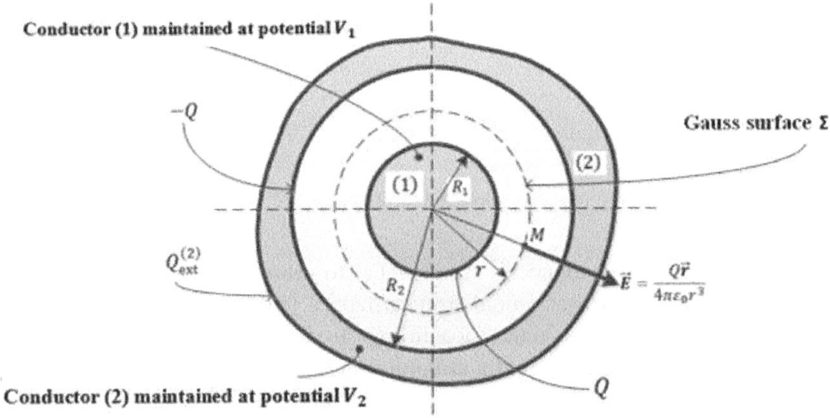

Figure 2.18. *Spherical capacitor*

In particular, if the distance $d = R_2 - R_1$ between the armatures is small compared to R_1 and R_2, we obtain:

$$C = \frac{4\pi\varepsilon_0 R_1 (R_1 + d)}{d} \simeq \frac{4\pi\varepsilon_0 R_1^2}{d} = \frac{\varepsilon_0 S}{d} \tag{2.61}$$

When $V_2 = 0$, there is a screen effect from the outside and we do not know what happens inside. The external wall of conductor (2) does not carry any charge, that is,

$Q^{(2)}_{\text{ext}} = 0$. Consequently, the total charge on the conductor (2) has the value $Q_2 = -Q = -CV_1$. If $V_2 \neq 0$, there exists an external charge $Q^{(2)}_{\text{ext}}$. To calculate the total charge on the armature (2), let us assume this on the exterior, (2) appears to be a sphere of radius R_3. In this case, the external charge will be $Q^{(2)}_{\text{ext}} = 4\pi\varepsilon_0 R_3 V_2 \simeq 4\pi\varepsilon_0 R_2 V_2$. Therefore, it follows that the total charge on the armature (2) is $Q_2 = Q^{(2)}_{\text{ext}} - Q \simeq 4\pi\varepsilon_0 R_3 V_2 - \frac{4\pi\varepsilon_0 R_1 R_2}{R_2 - R_1}(V_1 - V_2)$. For $d \ll R_1, R_2$, we find approximately $Q_2 \simeq 4\pi\varepsilon_0 R_3 V_2 - \frac{4\pi\varepsilon_0 R_1^2}{d}(V_1 - V_2)$. We can see that $Q^{(2)}_{\text{ext}}$ is small compared to the charge $|Q|$, that is, the entire charge on the armature (2) tends to be located on its internal surface.

b) It is possible to proceed differently to calculate C by directly evaluating the potentials outside and between the armatures. It is known that the function $V(r) = \frac{A}{r} + B$, where A and B are constants, satisfies the Laplace equation and its equipotential surfaces are spheres of center O. This function makes it easy to find potentials in the desired regions. Let us first take the region outside (2) (the external surface of (2) is assumed to be a sphere of radius R_3). As it must be zero at infinity, the potential in this region takes the form $V(r) = \frac{A}{r}$. By imposing the condition $V = V_2$ for $r = R_3$, it is found that:

$$V(r \geq R_3) = \frac{V_2 R_3}{r} \tag{2.62}$$

Let us now consider the region between the armatures. Looking for a solution of the form $V(M) = \frac{A}{r} + B$ and imposing the boundary conditions $V(R_1) = V_1$ and $V(R_2) = V_2$, we obtain $A = \frac{(V_1 - V_2) R_1 R_2}{R_2 - R_1}$ and $B = \frac{R_2 V_2 - R_1 V_1}{R_2 - R_1}$. It therefore follows that:

$$V(R_1 \leq r \leq R_2) = \frac{(V_1 - V_2) R_1 R_2}{R_2 - R_1}\frac{1}{r} + \frac{R_2 V_2 - R_1 V_1}{R_2 - R_1} \tag{2.63}$$

Equations [2.62] and [2.63] define the desired solution. The electrostatic field is therefore radial:

$$E_r(r) = \begin{cases} \dfrac{V_2 R_3}{r^2} & r \geq R_3 \\ \dfrac{(V_1 - V_2) R_1 R_2}{R_2 - R_1}\dfrac{1}{r^2} & R_1 \leq r \leq R_2 \end{cases} \tag{2.64}$$

Applying Gauss's theorem makes it possible to find the charges on the conductors. Consequently, the charge on the external surface of (2) is $Q^{(2)} = 4\pi r^2 \varepsilon_0 E_r(r \geq R_3) = 4\pi\varepsilon_0 V_2 R_3$, while that on (1) is $Q_1 = 4\pi r^2 \varepsilon_0 E_r(R_1 \leq r \leq R_2) = (V_1 - V_2)\, 4\pi\varepsilon_0 \frac{R_1 R_2}{R_2 - R_1}$. From this, it can be derived, first of all, that the capacitance of conductor

(2) alone is $C' = 4\pi\varepsilon_0 R_3$, and second that the capacitance of the capacitor itself is $C = 4\pi\varepsilon_0 \frac{R_1 R_2}{R_2 - R_1}$.

EXAMPLE 2.4. (Spherical Capacitor with Three Armatures).– A capacitor is made up of three concentric spherical armatures with center O. The central armature, (1), is a conductive sphere of radius R_1. The second armature (2), surrounding (1), is a spherical ring of inner radius $R_2 > R_1$ and outer radius R_3. The third armature, (3), surrounding (1) and (2) is a spherical ring with an inner radius $R_4 > R_3$ and an outer radius R_5, as shown in Figure 2.19. The potentials on (1) and (3) are, respectively, fixed at V_1 and V_3, while (2) is initially neutral, that is, it has a total charge of zero. This system has spherical symmetry around the center O and, consequently, it is possible to repeat the same reasoning used in the previous example. Let us begin with the area outside the conductors, defined by $r > R_5$. In this region, the expression for the potential is $V(r > R_5) = \frac{A}{r} + B$. Conditions $\lim_{r \longrightarrow \infty} V(r) = 0$ and $V(R_5) = V_3$ give, respectively, $B = 0$ and $A = V_3 R_6$. It therefore follows that:

$$V(r \geq R_5) = \frac{V_3 R_5}{r} \qquad [2.65]$$

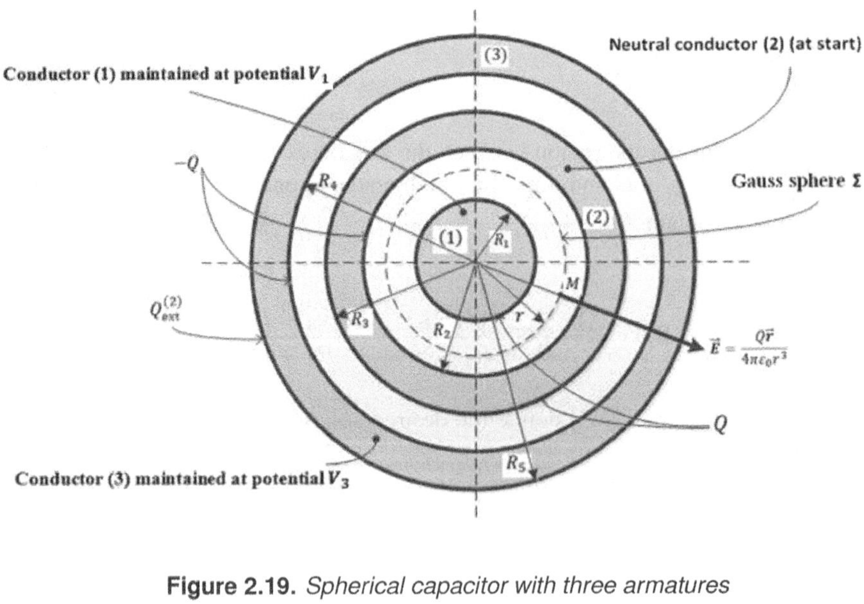

Figure 2.19. *Spherical capacitor with three armatures*

In the region defined by $R_3 \leq r \leq R_4$, the potential also takes the form $V(R_3 \leq r \leq R_4) = \frac{A'}{r} + B'$ with the constraint $V(R_4) = V_3$, which implies that $B' = V_3 - \frac{A'}{R_4}$. Consequently, we obtain:

$$V(R_3 \le r \le R_4) = A'\left(\frac{1}{r} - \frac{1}{R_4}\right) + V_3 \qquad [2.66]$$

because the potential is not specified a priori on (2). Over the sphere $r = R_3$, the previous expression gives:

$$V_2 = A'\left(\frac{1}{R_3} - \frac{1}{R_4}\right) + V_3 \qquad [2.67]$$

In the region defined by $R_1 \le r \le R_2$, the potential is written as $V(R_1 \le r \le R_2) = \frac{A''}{r} + B''$, with the condition $V(R_1) = V_1$, which gives $B'' = V_1 - \frac{A''}{R_1}$. Consequently, it can be derived that:

$$V(R_1 \le r \le R_2) = A''\left(\frac{1}{r} - \frac{1}{R_1}\right) + V_1 \qquad [2.68]$$

By writing $r = R_3$, we arrive at a second formula for V_2:

$$V_2 = A''\left(\frac{1}{R_2} - \frac{1}{R_1}\right) + V_1 \qquad [2.69]$$

In the various regions, the expression for the electrostatic field is:

$$E_r = \begin{cases} \dfrac{A''}{r^2} & R_1 \le r \le R_2 \\ \dfrac{A'}{r^2} & R_3 \le r \le R_4 \\ \dfrac{V_3 R_5}{r^2} & r \ge R_5 \end{cases} \qquad [2.70]$$

and therefore the flux in this field moving outwards across the closed, Gaussian surface with center O and radius r is equal to:

$$\Phi = \begin{cases} 4\pi A'' & R_1 \le r \le R_2 \\ 4\pi A' & R_3 \le r \le R_4 \\ 4\pi V_3 R_5 & r \ge R_5 \end{cases} \qquad [2.71]$$

Since (2) is neutral, the fluxes in the regions $R_1 \le r \le R_2$ and $R_3 \le r \le R_4$ must be equal, because the Gauss spheres contain the same charge Q (Q is carried by (1)), that is, $A' = A'' = \frac{Q}{4\pi\varepsilon_0}$. Comparing the two expressions [2.67] and [2.69], we obtain:

$$V_1 - V_3 = A'\left(\frac{1}{R_3} - \frac{1}{R_4} + \frac{1}{R_1} - \frac{1}{R_2}\right) \qquad [2.72]$$

Let us write:

$$C_1 = \frac{4\pi\varepsilon_0 R_1 R_2}{R_2 - R_1} \qquad C_2 = \frac{4\pi\varepsilon_0 R_3 R_4}{R_4 - R_3} \tag{2.73}$$

C_1 and C_2 represent, respectively, the capacitances of the capacitors formed by the armatures. Equation [2.72] thus leads to the following:

$$V_1 - V_3 = Q\left(\frac{1}{C_1} + \frac{1}{C_2}\right) \tag{2.74}$$

This system of conductors is equivalent to a single capacitor whose capacitance, C, is given by the relation $\frac{1}{C} = \frac{1}{C_1} + \frac{1}{C_2}$, or again:

$$C = \frac{C_1 C_2}{C_1 + C_2} \tag{2.75}$$

We will come back to these assemblies later. The potential of (2) is calculated from [2.69]:

$$V_2 = V_1 - \frac{Q}{C_1} = V_1 - (V_1 - V_3)\frac{C_2}{C_1 + C_2} \tag{2.76}$$

EXAMPLE 2.5. (Cylindrical Capacitor).– Consider the cylindrical capacitor shown in Figure 2.20. The internal armature and the internal wall of the external armature are coaxial cylinders. The external surface may take any shape. If the cylinders are infinitely long, the field $\overrightarrow{\mathbf{E}}$, at the point M, is oriented along the OM axis for reasons of symmetry. Take the part of the cylinder of length L, situated between two planes (P) and (P') perpendicular to the cylinder axis, carrying a charge Q. Let us apply Gauss' theorem to the surface formed by the fictional cylinder of revolution with radius $OM = r$, bounded by (P) and (P'): $\Phi = 2\pi r L E = \frac{Q}{\varepsilon_0}$. This gives $E = \frac{Q}{2\pi\varepsilon_0 L r}$. Since $E = -\frac{dV}{dr}$, it can be derived from this that $-dV = \frac{Q}{2\pi\varepsilon_0 L}\frac{dr}{r}$. By integrating between the limits R_1 and R_2, we obtain: $V_1 - V_2 = \frac{Q}{2\pi\varepsilon_0 L}\int_{R_1}^{R_2}\frac{dr}{r} = \frac{Q}{2\pi\varepsilon_0 L}\ln\frac{R_2}{R_1}$ and consequently:

$$C = \frac{2\pi\varepsilon_0 L}{\ln\frac{R_2}{R_1}} \tag{2.77}$$

The capacitance per unit length (along the axis of the cylinder) is therefore:

$$C_1 = \frac{2\pi\varepsilon_0}{\ln\frac{R_2}{R_1}} \tag{2.78}$$

In reality, cylinders 1 and 2 are not infinite, and perturbations in the field occur near their ends. These perturbations modify the expressions of E and C. However, if the length of the armatures is very large compared to $d = R_2 - R_1$, it is still possible to use [2.77] with a high degree of accuracy. If $d = R_2 - R_1 \ll R_1, R_2$, it is written that $\ln\frac{R_2}{R_1} = \ln\left(\frac{R_1+d}{R_1}\right) = \ln\left(1+\frac{d}{R_1}\right) \simeq \frac{d}{R_1}$ and, consequently, $C = \frac{2\pi\varepsilon_0 L R_1}{d} = \frac{\varepsilon_0 S}{d}$, where S represents the lateral surface of the cylinder. If the two armatures do not have the same axis, this problem becomes more difficult to solve. To solve this problem, we need to use the method *of images*, which is simpler to use in this situation. We will come back to this later in the text.

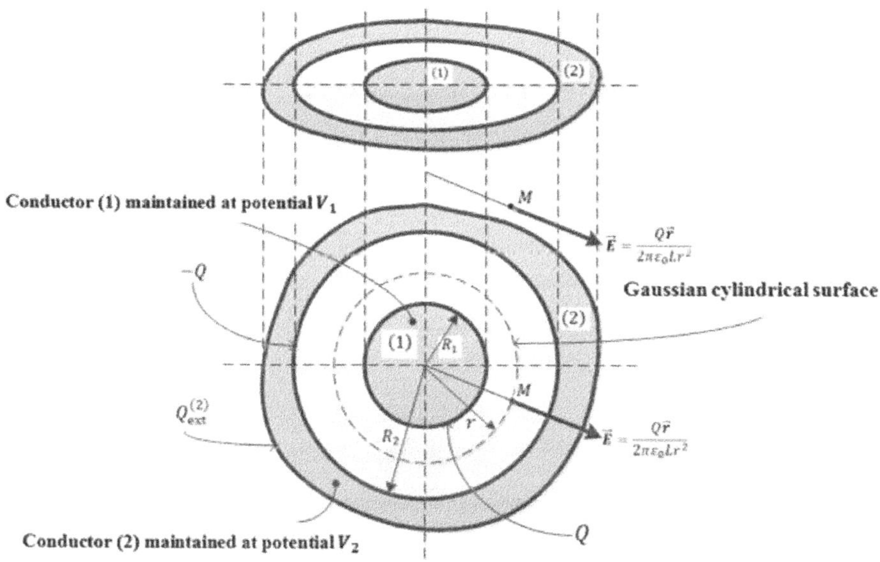

Figure 2.20. *Cylindrical capacitor*

EXAMPLE 2.6.(Plane Capacitor).– The idea behind the planar capacitor is to imagine two identical and parallel conducting planes (1) and (2) with a surface area S, separated by a very thin dielectric d. They therefore form a set of two conductors with almost total influence, as shown in Figure 2.21. It will be assumed that d is very small compared to the dimensions of the armatures (the armatures are therefore infinite to eliminate edge effects and ensure almost total influence). The Oz-axis is assumed to be perpendicular to the armatures. Armature (1) carries a surface charge density $+\sigma$, while armature (2) inductively acquires a surface charge density $-\sigma$ under influence effect. An electrostatic field is therefore created between these armatures: $\vec{\mathbf{E}} = \vec{\mathbf{E}}^{(1)} + \vec{\mathbf{E}}^{(2)} = \frac{\sigma}{2\varepsilon_0}\vec{\mathbf{k}} + \frac{-\sigma}{2\varepsilon_0}\left(-\vec{\mathbf{k}}\right) = \frac{\sigma}{\varepsilon_0}\vec{\mathbf{k}}$ (the field is zero outside the armatures). This internal electrostatic field will induce a potential difference between the two plates, (1) and (2), $V_1 - V_2 = \int_0^d \overrightarrow{\mathbf{E}} \cdot d\overrightarrow{\mathbf{l}} = \frac{\sigma}{\varepsilon_0} \int_0^d dz = \frac{\sigma d}{\varepsilon_0} = \frac{Qd}{\varepsilon_0 S}$. Finally, it gives:

$$C = \frac{\varepsilon_0 S}{d} \tag{2.79}$$

The capacitance per unit area is therefore: $C_1 = \frac{\varepsilon_0}{d}$.

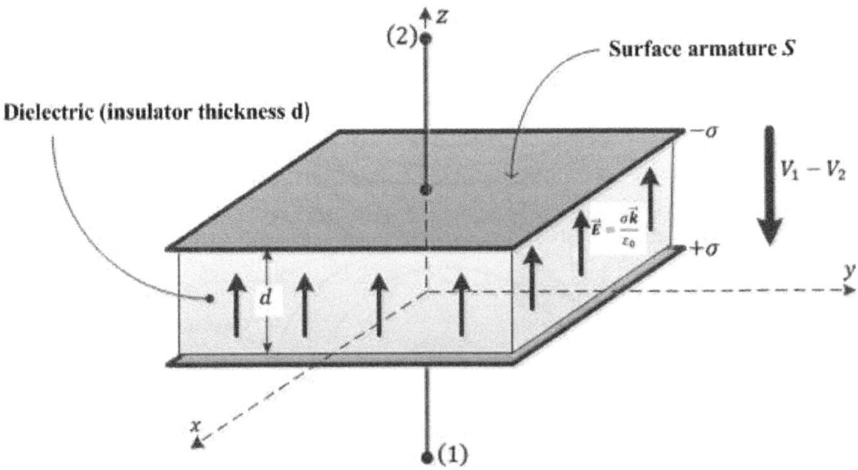

Figure 2.21. *Plane capacitor: when the plates are brought sufficiently close to one another, the charges tend to accumulate on the internal walls (which is conducive to total influence)*

2.5.4. *Association of capacitors*

There are two possibilities:

1) *Capacitors in series*: when two capacitors are connected in series before being charged, as in Figure 2.22(a), the central conductor (within the rectangle in dotted lines), made up of two armatures joined together, bears a total charge of zero. When a potential difference is applied between the external armatures, through influence, the charges $+Q$, $-Q$, $+Q'$ and $-Q'$ appear on the plates. Since the central conductor is insulated, its total charge remains zero. This implies that $Q' - Q = 0$ and, consequently, $Q' = Q$. Both capacitors therefore have the same charge. Applying the fundamental law of capacitors, we obtain $Q = C_1 (V_1 - V_2)$ for the first capacitor and $Q = C_2 (V_2 - V_3)$ for the second capacitor. From these equations, it can be

derived that: $(V_1 - V_2) + (V_2 - V_3) = (V_1 - V_3) = Q\left(\frac{1}{C_1} + \frac{1}{C_2}\right)$. Two capacitors connected in series behave like a single capacitor whose capacitance C is given by:

$$\frac{1}{C} = \frac{1}{C_1} + \frac{1}{C_2} \quad \Longleftrightarrow \quad C = \frac{C_1 C_2}{C_1 + C_2} \qquad [2.80]$$

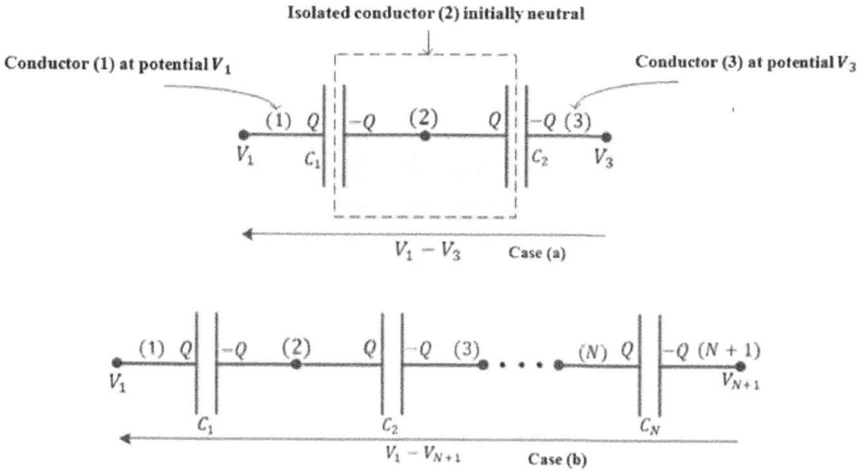

Figure 2.22. *Capacitors in series*

This result can easily be generalized to N capacitors connected in series, with capacitances $C_1, C_2, \ldots, C_N$ (Figure 2.22(b)): if the two ends of the chain are taken to the potentials V_1 and V_{N+1}, respectively, this will result in a charge Q appearing on the first capacitor. Assuming that all capacitors are initially neutral, the armatures of the following capacitors will develop the charge $\pm Q$ by influence, and so on down the chain. The voltage difference across the capacitor chain is written as: $V_1 - V_{N+1} = (V_1 - V_2) + \cdots + (V_N - V_{N+1}) = \frac{Q}{C_1} + \cdots + \frac{Q}{C_N} = \left(\sum_{i=1}^{N} \frac{1}{C_i}\right) Q$. The equivalent capacitance of the association is therefore calculated using the formula:

$$\frac{1}{C} = \sum_{i=1}^{N} \frac{1}{C_i} \qquad [2.81]$$

Note that the equivalent capacity is smaller than the smallest capacities in the series. In fact, we always have $\frac{C_1 C_2}{C_1 + C_2} < C_1$ and $\frac{C_1 C_2}{C_1 + C_2} < C_2$. This shows that $\frac{C_1 C_2}{C_1 + C_2} < \inf(C_1, C_2)$.

2) *Capacitors in parallel*: let us connect two capacitors in parallel, as shown in Figure 2.23(a). Here, we have only two conductors, (1) and (2), in total influence, boxed in dotted lines. This influence is exerted on several surfaces: (A_1, B_1) on the one hand and (A_2, B_2) on the other hand. If Q_1 is the charge carried by A_1, then B_1 must carry the charge $-Q_1$ and consequently, we obtain $Q_1 = C_1(V_1 - V_2)$ (for the first capacitor). Similarly, if A_2 carries the charge Q_2, then B_2 must carry the charge $-Q_2$, and consequently, $Q_2 = C_2(V_1 - V_2)$ (for the second capacitor). The total charge on (1) is $Q = Q_1 + Q_2$. We can therefore deduce that $Q = (C_1 + C_2)(V_1 - V_2)$. This equation shows that the combination behaves like a single capacitor whose capacitance, C, is given by:

$$C = C_1 + C_2 \qquad [2.82]$$

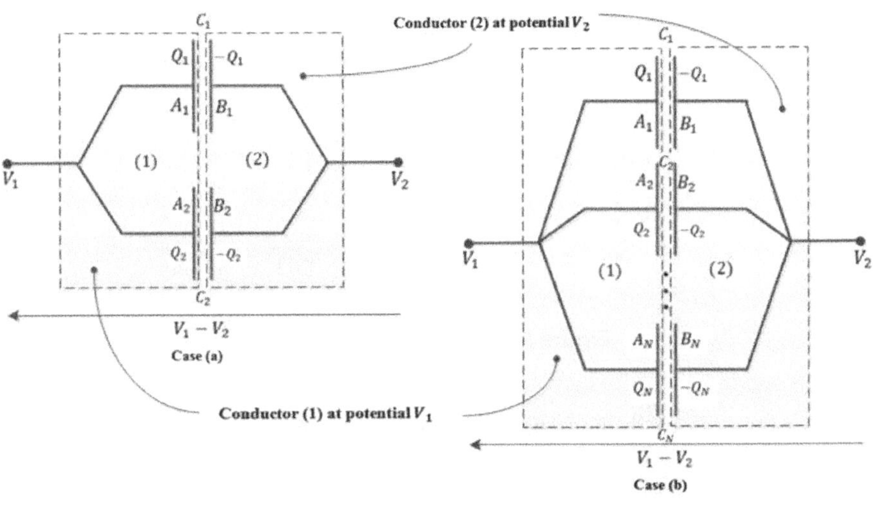

Figure 2.23. *Capacitors connected in parallel*

This formula can be easily generalized to N capacitors $C_1, C_2, \ldots, C_N$, connected in parallel as shown in Figure 2.23(b). In this case too, we have only two conductors (1) and (2) with several surfaces under total influence. The electric charge on each capacitor is given by $Q_i = C_i(V_1 - V_2)$. Therefore, it follows that the total electric charge is simply: $Q = \sum_{i=1}^{N} Q_i = \left(\sum_{i=1}^{N} C_i\right)(V_1 - V_2)$. This corresponds to a single equivalent capacitance:

$$C = \sum_{i=1}^{N} C_i \qquad [2.83]$$

We can conclude that if the capacitors are connected in parallel, the equivalent capacitance is equal to the sum of the individual capacitances.

EXAMPLE 2.7.– Consider capacitors (all with identical capacitance C) connected, as shown in Figure 2.24. Since the capacitors are in series along the branches, and furthermore, since the equivalent capacitances calculated on these branches are in parallel, we obtain:

$$\begin{aligned} C_{\text{tot.}} &= C + \left(\frac{1}{C}+\frac{1}{C}\right)^{-1} + \left(\frac{1}{C}+\frac{1}{C}+\frac{1}{C}\right)^{-1} + \cdots + \\ &\left(\frac{1}{C}+\cdots+\frac{1}{C}\right)^{-1} = C\sum_{n=1}^{N}\frac{1}{n} \end{aligned} \quad [2.84]$$

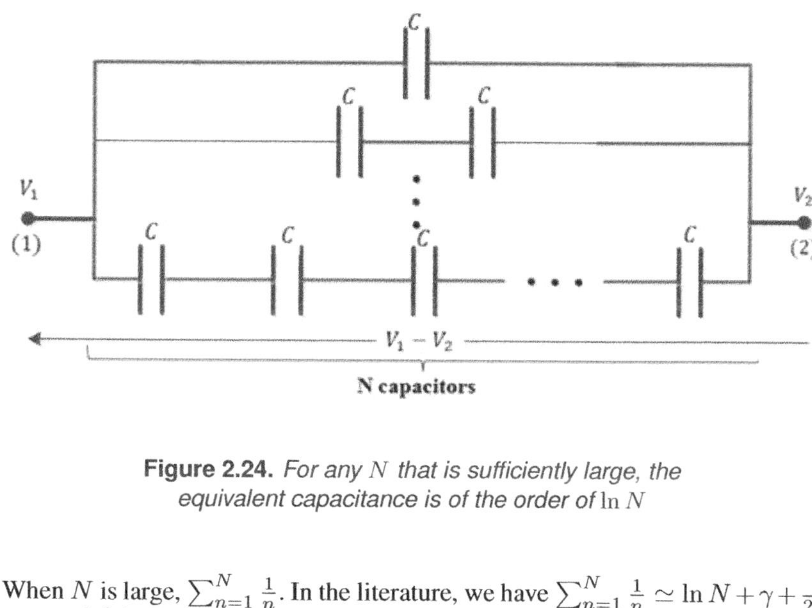

Figure 2.24. *For any N that is sufficiently large, the equivalent capacitance is of the order of $\ln N$*

When N is large, $\sum_{n=1}^{N}\frac{1}{n}$. In the literature, we have $\sum_{n=1}^{N}\frac{1}{n} \simeq \ln N + \gamma + \frac{1}{2N} - \frac{1}{12N^2} + O\left(\frac{1}{N^4}\right)$, where $\gamma = 0.577215$ is Euler's constant.

2.5.5. *Capacitors with dielectrics*

It has already been seen that the capacitances of capacitors depended on their geometric characteristics, as well as the permittivity of the vacuum. In some cases, the vacuum between the armatures can be replaced by another medium. However, this medium must be an insulator to prevent charges flowing between the two armatures (which would turn the capacitor into a single conductor) and to allow the armatures to influence each other. This type of medium is called a *dielectric*. For these capacitors, the capacitance is calculated from formulas similar to those already obtained, except that the vacuum permittivity ε_0 by the medium permittivity $\varepsilon = \varepsilon_r\varepsilon_0$ (ε_r is called *relative permittivity*).

It is therefore clear that the capacitance of a capacitor using a dielectric whose relative permittivity is ε_r will increase according to the relationship:

$$C = \varepsilon_r C_0 \quad [2.85]$$

where C_0 is the capacitance calculated with a vacuum between the armatures.

EXAMPLE 2.8. (Capacitor with Two Dielectrics).–

1) Two dielectrics with relative permittivity $\varepsilon_{r,1}$ and $\varepsilon_{r,2}$, respectively, are arranged side by side, as shown in Figure 2.25(a), within a plane capacitor with surface area S and armature spacing d. Each dielectric occupies half the surface area of the capacitor. Each part containing a single dielectric can be regarded as a capacitor. The first, with capacitance C_1 and relative permittivity, $\varepsilon_{r,1}$ covers the area $S_1 = S/2$, while the second with capacitance C_2 and relative permittivity $\varepsilon_{r,2}$ covers the second half of the surface $S_2 = S/2$. The armatures of each capacitor are at the same potential on either side of the dielectrics because they form the same conductor on both sides. Consequently, the two capacitors are maintained at the same potential difference. It is clearly seen that the charge of the capacitor is divided between the surfaces S_1 and S_2. Therefore, it is possible to write $V_1 - V_2 = \frac{Q_1}{C_1} = \frac{Q_2}{C_2}$. However, since $Q = \sigma\,(S_1 + S_2) = Q_1 + Q_2 = (C_1 + C_2)(V_1 - V_2)$, it can be derived that $C = C_1 + C_2$. These are therefore two capacitors in parallel. The equivalent capacity is given by:

$$C = C_1 + C_2 = \frac{\varepsilon_{r,1}\varepsilon_0\,(S/2)}{d} + \frac{\varepsilon_{r,2}\varepsilon_0\,(S/2)}{d} = \frac{\varepsilon_0 S}{2d}\left(\varepsilon_{r,1} + \varepsilon_{r,2}\right) \quad [2.86]$$

Everything happens as if we are dealing with a single dielectric of relative permittivity $\varepsilon_r = \frac{1}{2}\left(\varepsilon_{r,1} + \varepsilon_{r,2}\right)$.

2) The two dielectrics are now arranged one on top of the other and have the same thickness, $\frac{d}{2}$, as shown in Figure 2.25(b). To calculate the capacitance, we need to calculate the potential difference for each dielectric. To do this, simply note that the electric field inside each dielectric is given by $E_i = \frac{\sigma}{\varepsilon_r\varepsilon_0}$ for $i = 1, 2$. The potential difference for the entire capacitor is: $V_1 - V_2 = E_1\frac{d}{2} + E_2\frac{d}{2} = \frac{\sigma d}{2\varepsilon_0}\left(\frac{1}{\varepsilon_{r,1}} + \frac{1}{\varepsilon_{r,2}}\right) =$

$\frac{Qd}{2\varepsilon_0 S}\left(\frac{1}{\varepsilon_{r,1}}+\frac{1}{\varepsilon_{r,2}}\right)$, where Q is the charge on the entire capacitor. Consequently, we obtain:

$$C=\frac{2\varepsilon_0 S}{d}\left(\frac{\varepsilon_{r,1}\varepsilon_{r,2}}{\varepsilon_{r,1}+\varepsilon_{r,2}}\right) \quad [2.87]$$

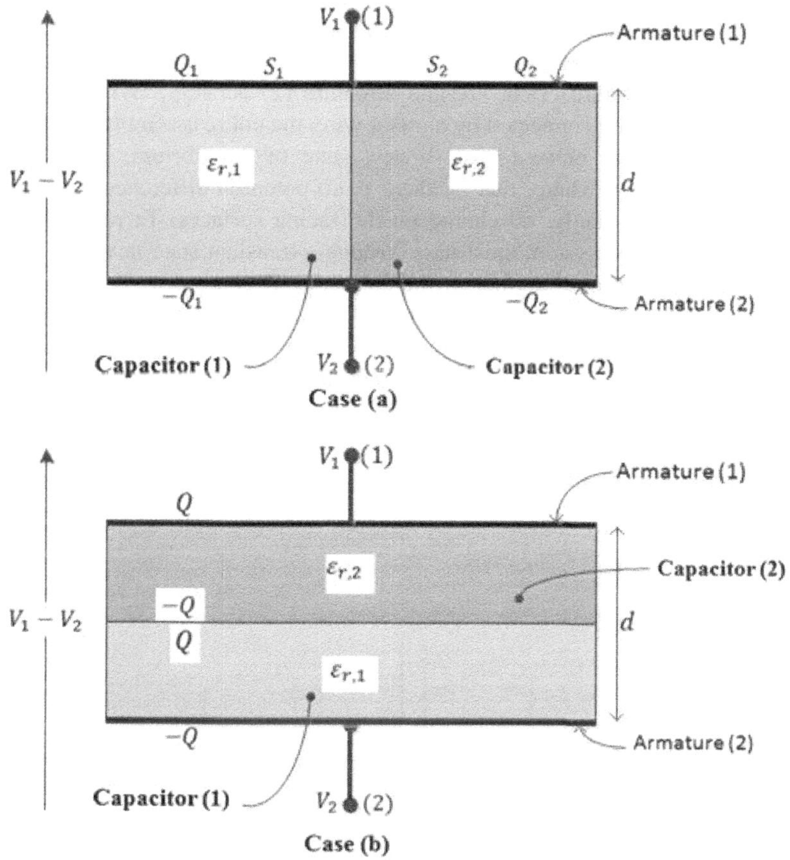

Figure 2.25. *a) For two dielectrics arranged side by side, the equivalent capacitance corresponds to a series association of the constituent capacitances C_1 and C_2. b) If they are disposed on atop the other, the equivalent capacitance corresponds to a connection in parallel of capacitances C_1 and C_2*

This equation can also be rewritten in the following form:

$$\frac{1}{C}=\frac{d}{2\varepsilon_0 S}\left(\frac{1}{\varepsilon_{r,1}}+\frac{1}{\varepsilon_{r,2}}\right)=\frac{1}{C_1}+\frac{1}{C_2} \quad [2.88]$$

This shows that in this configuration, the capacitor is equivalent to two capacitors, C_1 and C_2, connected in series.

2.6. Energy stored in a capacitor

Consider a capacitor charged to Q_0 and V_0 (with $Q_0 = CV_0$). One of the armatures, (2), is earthed. The other, (1), is taken to V_0 by an external source. We then cut off the connection between (1) and this source. Therefore, the capacitor remains charged (armature (1) at Q_0 and armature (2) at $-Q_0$). When these two armatures (1) and (2) are connected by a metal wire, the entire assembly will become a single conductor and move toward a new state of equilibrium. The potential becomes the same everywhere, that is, there is no potential difference between the armatures and, consequently, no charge on the facing surfaces. To reach this new state of equilibrium, the system must pass through a transient state in which charges flow from one armature to the other. If the conductor is charged to V, a quantity of charge $|dQ|$, which passes from armature (1) to armature (2), sees its potential change from V to 0. Therefore, the quantity of charges receives a quantity of energy $dW_e = V\,|dQ|$, as shown in Figure 2.26. During this time, the charge capacitor decreases by $|dQ|$ and, consequently, its potential decreases by $|dV| = \frac{|dQ|}{C}$. It then follows that $dW_e = V\,|dQ| = -CVdV$. Therefore, when V changes from V_0 to 0, the charges will receive a total quantity of energy:

$$W_e = -\int_{V_0}^{0} CVdV = \frac{1}{2}CV_0^2 \qquad [2.89]$$

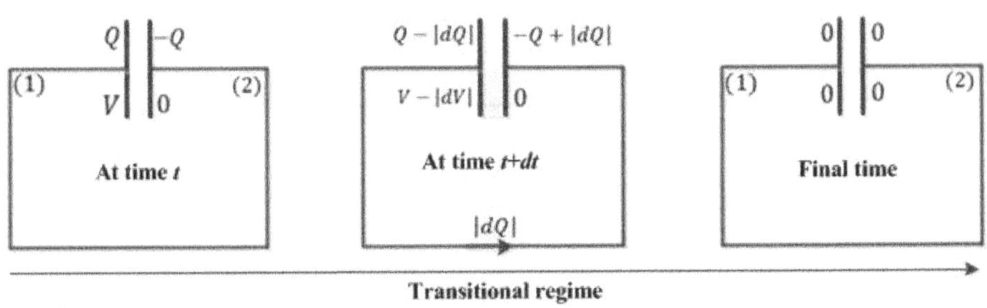

Figure 2.26. *During the transient phases, any elementary charge $|dQ|$ that leaves armature (1) to move toward armature (2) sees a change in potential from V to 0 and, consequently, it will receive a potential energy $dW_e = Vd\,|dQ|$*

Since the armatures of the capacitor are no longer connected to any source, it can be concluded that the energy received by the charges was supplied by the capacitor itself. A charged capacitor therefore has, in reserve, a quantity of energy equal to $\frac{1}{2}CV_0^2$. This is the energy stored within a capacitor. This also corresponds to the electrostatic potential energy of the interaction between the charges on the conductor (1) and those on the internal surface of the conductor (2). It is given by:

$$W_e = \frac{1}{2}Q_1 V_1 + \frac{1}{2}Q_2^{(2)} V_2 = \frac{1}{2}Q_1 (V_1 - V_2) = \frac{1}{2}CV_0^2 = \frac{Q^2}{2C} \qquad [2.90]$$

where $V_0 = V_1 - V_2$ and $Q = Q_1$. The potential energy [2.90] is additive. For example, if we have several capacitors, $C_1, \ldots, C_N$, connected in series, the total potential energy is the sum of individual potential energies associated with each capacitor. Since the capacitors carry the same charge:

$$W_{e,\mathrm{tot}} = \sum_{i=1}^{N} W_i = \sum_{i=1}^{N} \frac{Q^2}{2C_i} = \frac{1}{2}\left(\sum_{i=1}^{N} \frac{1}{C_i}\right) Q^2 = \frac{Q^2}{2C} \qquad [2.91]$$

where C is the equivalent capacitance. For the capacitors $C_1, \ldots, C_N$ connected in parallel, the total potential energy is given by:

$$W_{e,\mathrm{tot}} = \sum_{i=1}^{N} W_i = \sum_{i=1}^{N} \frac{1}{2}C_i V^2 = \frac{1}{2}\left(\sum_{i=1}^{N} C_i\right) V^2 = \frac{1}{2}CV^2 \qquad [2.92]$$

since the capacitors have the same potential difference V.

EXAMPLE 2.9.– Two square conductor plates with side a and negligible thickness form a plane-parallel capacitor. They are separated by a distance d. A third neutral conductive plate, identical to the first two, but $\frac{d}{3}$ thick, is inserted between the first two plates, equidistant from both, at a depth x, as shown in Figure 2.27. Armature charge densities located in the zones A and B are denoted by σ_A and σ_B, respectively. For zone A, it is possible to write $V_1 - V_2 = E_A d = \frac{\sigma_A d}{\varepsilon_0}$, while in zone B, we will have $V_1 - V_2 = (V_1 - V_3) + (V_3 - V_2) = E_B \frac{d}{3} + E_B \frac{d}{3} = \frac{\sigma_B d}{3\varepsilon_0} + \frac{\sigma_B d}{3\varepsilon_0}$. By comparing these two expressions, we find $\frac{\sigma_A d}{\varepsilon_0} = \frac{2\sigma_B d}{3\varepsilon_0}$. This shows that $\sigma_A = \frac{2}{3}\sigma_B$. All that now remains is to calculate the total charge on the external plates:

$$Q - \sigma_A a\,(a - x) + \sigma_B a x = \sigma_A a\left(a + \frac{x}{2}\right) \implies \begin{cases} \sigma_A = \dfrac{Q}{a\left(a + \frac{x}{2}\right)} \\ \sigma_B = \dfrac{3Q}{2a\left(a + \frac{x}{2}\right)} \end{cases} \qquad [2.93]$$

and, consequently, the electric field is:

$$E_A = \frac{Q}{\varepsilon_0 a\left(a + \frac{x}{2}\right)} \qquad E_B = \frac{3Q}{2\varepsilon_0 a\left(a + \frac{x}{2}\right)} \qquad [2.94]$$

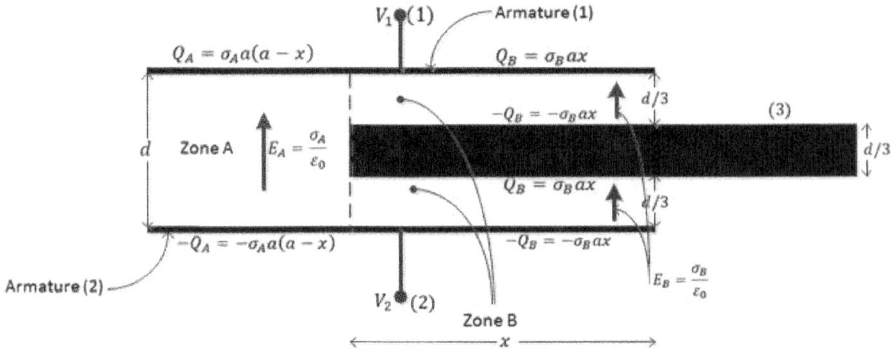

Figure 2.27. *By inserting a sheet of thickness $\frac{d}{3}$ inside a plane capacitor at equal distances from the two armatures, the whole system will behave like a association of two identical capacitors in series, each of capacitance $C_1 = \frac{3\varepsilon_0 xa}{d}$, in parallel with another capacitor of capacitance $C_2 = \frac{\varepsilon_0(a-x)a}{d}$. The equivalent capacitance is $C(x) = \left(\frac{1}{C_1} + \frac{1}{C_1}\right)^{-1} + C_2 = \frac{C_1}{2} + C_2 = \frac{\varepsilon_0 a}{d}\left(a + \frac{x}{2}\right)$*

The capacitance of the system can be calculated by calculating the ratio $C(x) = \frac{Q}{V_1 - V_2}$, that is:

$$C(x) = \frac{Q}{V_1 - V_2} = \frac{Q}{\frac{\sigma_A d}{\varepsilon_0}} = \frac{\varepsilon_0 a}{d}\left(a + \frac{x}{2}\right) \qquad [2.95]$$

It is clearly seen that $C_{\text{min}} = \frac{\varepsilon_0 a^2}{d}$ and $C_{\text{max}} = \frac{2\varepsilon_0 a^2}{2d} = \frac{3}{2}C_{\text{min}}$. The quantity of energy stored within the capacitor is given by:

$$W_e(x) = \frac{1}{2}C(x)(V_1 - V_2)^2 = \frac{\varepsilon_0 a}{2d}\left(a + \frac{x}{2}\right)(V_1 - V_2)^2 \qquad [2.96]$$

■

The potential energy stored within a plane capacitor is written $W_e = \frac{Q^2}{2C} = \frac{\sigma^2 Sd}{2\varepsilon_0}$, or again:

$$W_e = \left(\frac{\varepsilon_0}{2}E^2\right) Sd \qquad [2.97]$$

because $E = \frac{\sigma}{\varepsilon_0}$ (E is the field measured between the two armatures). Since Sd here represents the internal volume of the capacitor (i.e. the volume contained within the two armatures), the term $w_E = \frac{\varepsilon_0 E^2}{2}$ can be interpreted as being the *electrostatic energy density* within the capacitor. We can therefore conclude that the energy stored in the capacitor is contained in the electric field, that is, it is electrostatic energy. Therefore, to obtain W_e, it suffices to integrate the density $w_E = \frac{\varepsilon_0 E^2}{2}$ over the internal volume of the capacitor. We will return to these results later on. To verify what we have seen, let us return to Example 2.3 with the spherical capacitor. The electric field within the two spherical armatures is given by $E_r = \frac{(V_1-V_2)R_1R_2}{R_2-R_1}\frac{1}{r^2}$. This corresponds to the electrostatic energy density $w_E = \frac{\varepsilon_0 E^2}{2} = \frac{\varepsilon_0}{2}\frac{(V_1-V_2)^2 R_1^2 R_2^2}{(R_2-R_1)^2}\frac{1}{r^4}$. By integrating into the volume contained within the two spheres of radius R_1 and R_2, we obtain:

$$W_e = \int_{R_1}^{R_2}\left(\frac{\varepsilon_0 E^2}{2}\right)4\pi r^2 dr = 2\pi\varepsilon_0\frac{(V_1 - V_2)^2 R_1 R_2}{R_2 - R_1} = \frac{1}{2}C(V_1 - V_2)^2 \qquad [2.98]$$

where $C = 4\pi\varepsilon_0\frac{R_1R_2}{R_2-R_1}$ is the capacitance of the spherical capacitor.

2.7. Forces of attraction between the armatures of a capacitor

2.7.1. *Calculating force from pressure*

It has already been seen that any charged conductor experiences the following electrostatic pressure at any point M on its surface: $P(M) = \frac{\sigma^2}{2\varepsilon_0}$. Consequently, this point experiences an electrostatic force $d\vec{\mathbf{F}} = P(M)\,d\vec{\mathbf{S}} = \frac{\sigma^2}{2\varepsilon_0}d\vec{\mathbf{S}}$ directed toward the outside, regardless of the sign of the charge. Therefore, the total force acting on the conductor is:

$$\vec{\mathbf{F}} = \int P(M)\,d\vec{\mathbf{S}} = \int \frac{\sigma^2}{2\varepsilon_0}d\vec{\mathbf{S}} \qquad [2.99]$$

This force is exerted on the charges and tends to extract these charges from the conductor. However, because of the cohesion of matter, this force will be applied to the conductor itself and will tend to deform or displace it. Therefore, it is possible to observe the displacement of the conductor. This phenomenon is observed with capacitors where one of the armatures is mobile. The mobile armature will tend to move closer to the other, as the forces applied to the armatures are attractive (the charges on the armatures have opposite signs), as shown in Figure 2.28. Calculating the force acting on a conductor using formula [2.99] requires prior knowledge of the surface density of charges on the conductor's surface. This calculation is easier than in the case of uniform charge distribution, but is not possible in most cases. Note that there is an alternative method, based on equation [2.90], which allows these forces to be obtained by direct calculation, that is, based on energy considerations, as we will see later. However, before examining this method, it is interesting to note that we can give the force [2.99] a general expression that does not explicitly involve this density. To do this, let us consider a constant unit vector $\vec{\mathbf{u}}$ and then form the vector:

$$\vec{\mathbf{W}} = \left(\vec{\mathbf{u}} \cdot \vec{\mathbf{E}}\right) \vec{\mathbf{E}} - \frac{\left|\vec{\mathbf{E}}\right|^2}{2} \vec{\mathbf{u}} \qquad [2.100]$$

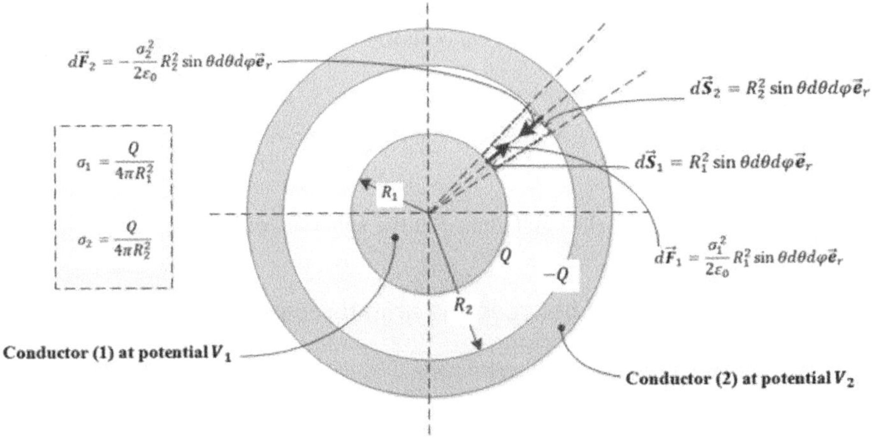

Figure 2.28. *The surface elements $d\vec{\mathbf{S}}_1$ and $d\vec{\mathbf{S}}_2$ on the spherical armatures (1) and (2), respectively, experience the attractive forces $d\vec{\mathbf{F}}_1 = P(M_1)\, d\vec{\mathbf{S}}_1$ and $d\vec{\mathbf{F}}_2 = P(M_2)\, d\vec{\mathbf{S}}_2$ which makes them move closer to one another, where $P(M_1) = \frac{\sigma_1^2}{2\varepsilon_0}$ and $P(M_2) = \frac{\sigma_2^2}{2\varepsilon_0}$ are, respectively, the electrostatic pressures experienced by armatures (1) and (2)*

To simplify, let us take $\vec{\mathbf{u}} = \vec{\mathbf{k}}$ and then calculate the divergence of $\vec{\mathbf{W}}$: $\vec{\nabla} \cdot \vec{\mathbf{W}} = \vec{\nabla} \cdot \left(E_z \vec{\mathbf{E}}\right) - \frac{\partial}{\partial z}\left(\frac{\left|\vec{\mathbf{E}}\right|^2}{2}\right) = E_z\left(\vec{\nabla} \cdot \vec{\mathbf{E}}\right) + \left(\vec{\mathbf{E}} \cdot \vec{\nabla}\right) E_z - E_x \frac{\partial E_x}{\partial z} - E_y \frac{\partial E_y}{\partial z} - E_z \frac{\partial E_z}{\partial z}$. Furthermore, since the curl of $\vec{\mathbf{E}}$ is zero, it can be derived that $\frac{\partial E_x}{\partial z} = \frac{\partial E_z}{\partial x}$ and $\frac{\partial E_y}{\partial z} = \frac{\partial E_z}{\partial y}$. Consequently, the preceding formula can be transformed as follows: $\vec{\nabla} \cdot \vec{\mathbf{W}} = E_z\left(\vec{\nabla} \cdot \vec{\mathbf{E}}\right) + \left(\vec{\mathbf{E}} \cdot \vec{\nabla}\right) E_z - E_x \frac{\partial E_z}{\partial x} - E_y \frac{\partial E_z}{\partial y} - E_z \frac{\partial E_z}{\partial z} = E_z\left(\vec{\nabla} \cdot \vec{\mathbf{E}}\right)$, because $E_x \frac{\partial E_z}{\partial x} + E_y \frac{\partial E_z}{\partial y} + E_z \frac{\partial E_z}{\partial z} = \left(\vec{\mathbf{E}} \cdot \vec{\nabla}\right) E_z$. Finally, it comes to:

$$\vec{\nabla} \cdot \vec{\mathbf{W}} = \left(\vec{\mathbf{E}} \cdot \vec{\mathbf{u}}\right)\left(\vec{\nabla} \cdot \vec{\mathbf{E}}\right) \qquad [2.101]$$

Since there are no charges in the space between the two conductors, that is, $\vec{\nabla} \cdot \vec{\mathbf{E}} = 0$, we have $\vec{\nabla} \cdot \vec{\mathbf{W}} = 0$. We can therefore conclude that outside the conductors, $\vec{\mathbf{W}}$ has a conservative flux. Let us take a closed Σ surface that completely surrounds the conductor (1), without surrounding the others. According to what we have discussed above, the outgoing flux from $\vec{\mathbf{W}}$ through Σ is equal to the outgoing flux through the surface S_1 of (1) (approached from the outside). Furthermore, at any point on the surface S_1, the electric field is given by $\frac{\sigma_1}{\varepsilon_0}\vec{\mathbf{n_1}}$, where σ_1 and $\vec{\mathbf{n}}_1$ are, respectively, the surface charge density and normal to S_1 at that point. It therefore follows that:

$$\vec{\mathbf{W}} = \frac{\sigma_1^2}{\varepsilon_0^2}\left[\left(\vec{\mathbf{u}} \cdot \vec{\mathbf{n}}_1\right)\vec{\mathbf{n}}_1 - \frac{1}{2}\vec{\mathbf{u}}\right] \Longrightarrow \vec{\mathbf{W}} \cdot \vec{\mathbf{n}}_1 = \frac{\sigma_1^2}{2\varepsilon_0^2}\left(\vec{\mathbf{u}} \cdot \vec{\mathbf{n}}_1\right) \quad [2.102]$$

The conservation of flux imposes:

$$\int_{S_1} \frac{\sigma_1^2}{2\varepsilon_0}\left(\vec{\mathbf{u}} \cdot \vec{\mathbf{n}}_1\right) dS_1 = \varepsilon_0 \int_{\Sigma}\left[\left(\vec{\mathbf{u}} \cdot \vec{\mathbf{E}}\right)\vec{\mathbf{E}} - \frac{\left|\vec{\mathbf{E}}\right|^2}{2}\vec{\mathbf{u}}\right] \cdot \vec{\mathbf{n}}_1 d\Sigma \quad [2.103]$$

Clearly, the first term is none other than the component along $\vec{\mathbf{u}}$ of the force [2.99]. Therefore, it is found that:

$$\vec{\mathbf{F}}^{(1)} = \varepsilon_0 \int_{\Sigma}\left[\left(\vec{\mathbf{n}} \cdot \vec{\mathbf{E}}\right)\vec{\mathbf{E}} - \frac{\left|\vec{\mathbf{E}}\right|^2}{2}\vec{\mathbf{n}}\right] d\Sigma \quad [2.104]$$

EXAMPLE 2.10. (Plane Capacitor).– Let us calculate the forces exerted on the armatures of a plane capacitor with capacitance $C = \frac{\varepsilon_0 S}{z_2 - z_1}$, where S is the surface area of the armatures located at z_1 and $z_2 > z_1$ (the Oz axis, perpendicular to the plates, points upwards), to which the potential difference V is applied. The following forces are applied to plates (1) and (2), respectively:

$$\begin{aligned} \vec{\mathbf{F}}^{(1)} &= \frac{\sigma^2 S}{2\varepsilon_0}\vec{\mathbf{k}} = \frac{Q^2}{2\varepsilon_0 S}\vec{\mathbf{k}} = \frac{\varepsilon_0 S V^2}{2\left(z_2 - z_1\right)^2}\vec{\mathbf{k}} \\ \vec{\mathbf{F}}^{(2)} &= -\frac{\sigma^2 S}{2\varepsilon_0}\vec{\mathbf{k}} = -\frac{Q^2}{2\varepsilon_0 S}\vec{\mathbf{k}} = -\frac{\varepsilon_0 S V^2}{2\left(z_2 - z_1\right)^2}\vec{\mathbf{k}} \end{aligned} \quad [2.105]$$

EXAMPLE 2.11. (Flat Capacitor with Moving Armature Attached to a Spring).– Consider a flat, initially uncharged capacitor with surface area S and distance d (the initial distance between the plates of the capacitor). The capacitor is attached to a spring with stiffness k through one of its armatures, designed here by (2). The external ends of the spring and the capacitor (armature (1)) are fixed. We wish to calculate the extension of the spring when the capacitor is taken to a potential difference V. In equilibrium, the sum of forces applied to the point of spring-armature contact is zero. At the point of contact, there is the spring return force $\mathsf{k}z\overrightarrow{\mathbf{k}}$ and the attractive force $-\frac{\varepsilon_0 SV^2}{2(d-z)^2}\overrightarrow{\mathbf{k}}$ experienced by the armature (2), where $(d-z)$ is the new distance between the plates, as shown in Figure 2.29. Consequently, it can be derived that $\frac{\varepsilon_0 SV^2}{2(d-z)^2} = \mathsf{k}z$. In particular, when $z \ll d$, the preceding expression is reduced to $\frac{\varepsilon_0 SV^2}{2d^2}\left(1 + 2\frac{z}{d}\right) = \mathsf{k}z$. This gives $z = \frac{\frac{\varepsilon_0 SV^2}{2d^2}}{\mathsf{k} - \frac{\varepsilon_0 SV^2}{d^3}}$.

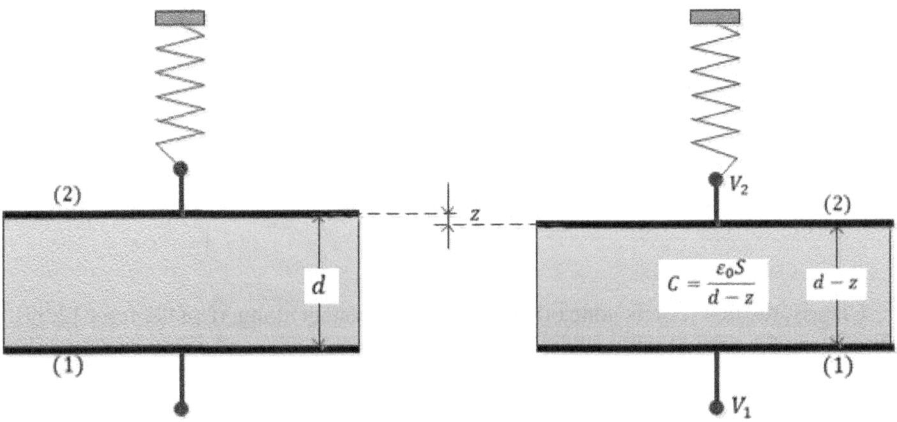

Figure 2.29. *When a potential difference is applied across the capacitor, armature (2) tends to move closer to armature (1), which changes the capacitor's capacitance and causes an extension of the spring*

2.7.2. *Force calculation from the energy expression*

It is also possible to calculate this attractive force acting between the armatures using the principle of virtual work, which can be seen as a special application of the law of conservation of energy to the case of statics. In this case, the kinetic energy is zero. As a result, the work carried out by external forces must be equal to the work carried out by internal forces. It should also be noted that *the virtual displacements* chosen must be compatible with each other, or take into account the connections between solids and the possibilities of motion. In addition, they must be chosen small enough so that the force $\overrightarrow{\mathbf{F}}$ can be considered constant. However, the energy balance in a virtual displacement of the armatures differs depending on whether the displacement is carried out with a constant load (electrically insulated armatures) or with a constant potential difference (armatures connected to a source capable of supplying loads while keeping V constant).

(a) In the case where the capacitor is insulated, the charge Q remains constant. The only energy received is the work of external forces (which must compensate for the work of internal forces). The virtual displacement $d\vec{\mathbf{l}}$ of the system leads a variation of the potential energy: $dW_e = \vec{\mathbf{F}}_{\text{ext}} \cdot d\vec{\mathbf{l}} = -\vec{\mathbf{F}}_{\text{int}} \cdot d\vec{\mathbf{l}}$. Moreover, according to the definition of the gradient, we have $dW_e = \vec{\nabla} W_e \cdot d\vec{\mathbf{l}}$. Consequently, the internal force is:

$$\vec{\mathbf{F}}_{\text{int}} = -\left.\vec{\nabla} W_e\right|_Q \qquad [2.106]$$

For example, in the case of a plane capacitor with capacitance $C = \frac{\varepsilon_0 S}{z_2 - z_1}$, where z_1 and $z_2 > z_1$ indicate the positions of armatures (1) and (2), we imagine that the coordinates z_1 and z_2 have changed by dz_1 and dz_2. Therefore, the work carried out by the external forces to displace the armatures is, on the one hand, equal to:

$$dW_e = F_{\text{ext}}^{(1)} dz_1 + F_{\text{ext}}^{(2)} dz_2 = -F_{\text{int}}^{(1)} dz_1 - F_{\text{int}}^{(2)} dz_2 \qquad [2.107]$$

where $F_{\text{int}}^{(1)}$ and $F_{\text{int}}^{(2)}$ are the forces of attraction experienced on the armatures. On the other hand, the variation in potential energy (if the charge Q is not allowed to vary) is:

$$dW_e = -\frac{Q^2}{2C^2} dC = -\frac{Q^2}{2C^2}\left(\frac{\partial C}{\partial z_1} dz_1 + \frac{\partial C}{\partial z_2} dz_2\right) \qquad [2.108]$$

By comparing the two equations [2.107] and [2.108], we deduce that:

$$\begin{cases} F_{\text{int}}^{(1)} = \dfrac{Q^2}{2C^2}\dfrac{\partial C}{\partial z_1} = \dfrac{V^2}{2}\dfrac{\varepsilon_0 S}{(z_2 - z_1)^2} \Longrightarrow \vec{\mathbf{F}}_{\text{int}}^{(1)} = \dfrac{V^2}{2}\dfrac{\varepsilon_0 S}{(z_2 - z_1)^2}\vec{\mathbf{k}} \\ F_{\text{int}}^{(2)} = \dfrac{Q^2}{2C^2}\dfrac{\partial C}{\partial z_2} = -\dfrac{V^2}{2}\dfrac{\varepsilon_0 S}{(z_2 - z_1)^2} \Longrightarrow \vec{\mathbf{F}}_{\text{int}}^{(2)} = -\dfrac{V^2}{2}\dfrac{\varepsilon_0 S}{(z_2 - z_1)^2}\vec{\mathbf{k}} \end{cases} \qquad [2.109]$$

We can clearly see that these forces are attractive (i.e. they tend to attract the armatures toward each other).

(b) We now consider a capacitor where the thickness of the dielectric between the two plates is variable and a constant potential difference, V, is maintained between the plates. The first armature (1) is fixed, while the second (2) is brought closer by a quantity $d\vec{\mathbf{l}}$. During this elementary displacement, the energy received by the capacitor is constituted by the work of external forces, as well as the electrical work $dW = PdQ$, supplied by the source. Consequently, it is possible to write:

$$dW_e = \vec{\mathbf{F}}_{\text{ext}} \cdot d\vec{\mathbf{l}} + VdQ = -\vec{\mathbf{F}}_{\text{int}} \cdot d\vec{\mathbf{l}} + PdQ \qquad [2.110]$$

In addition, the energy stored in the capacitor $W_e = \frac{1}{2}QV$ has changed by $dW_e = \frac{1}{2}VdQ$ and, consequently, $PdQ = 2dW_e$. By replacing VdQ by its expression in [2.124], we finally obtain $dW_e = -\vec{\mathbf{F}}_{\text{int}} \cdot d\vec{\mathbf{l}} + 2dW_e$ and, consequently:

$$dW_e = \vec{\mathbf{F}}_{\text{int}} \cdot d\vec{\mathbf{l}} \qquad [2.111]$$

The electric force acting on the capacitor's armature is therefore given by:

$$\vec{\mathbf{F}}_{\text{int}} = \vec{\nabla} W_e\Big|_V \qquad [2.112]$$

For the plane capacitor with capacitance $C = \frac{\varepsilon_0 S}{z_2 - z_1}$, the attractive forces experienced on the armatures, maintained at a constant potential difference, are:

$$\begin{cases} F_{\text{int}}^{(1)} = \dfrac{V^2}{2}\dfrac{\partial C}{\partial z_1} \\ F_{\text{int}}^{(2)} = \dfrac{V^2}{2}\dfrac{\partial C}{\partial z_2} \end{cases} \Longrightarrow \begin{cases} \vec{\mathbf{F}}_{\text{int}}^{(1)} = \dfrac{V^2}{2}\dfrac{\varepsilon_0 S}{(z_2 - z_1)^2}\vec{\mathbf{k}} \\ \vec{\mathbf{F}}_{\text{int}}^{(2)} = -\dfrac{V^2}{2}\dfrac{\varepsilon_0 S}{(z_2 - z_1)^2}\vec{\mathbf{k}} \end{cases} \qquad [2.113]$$

We obtain the same results as in the previous case. Since the forces depend only on the state of the capacitor, the results should be identical for both configurations.

EXAMPLE 2.12.– Let us return to Example 2.9. According to the preceding discussion, the force applied along x on the plate (of thickness $\frac{d}{3}$) introduced in the middle of the capacitor will have the following value (this force attracts the plate towards the interior of the capacitor):

$$F_x = +\frac{dW_e}{dx}\bigg|_V = \frac{\varepsilon_0 a V^2}{4d} \qquad [2.114]$$

EXAMPLE 2.13. (Thin Sheet between the Armatures of a Capacitor).– A plane capacitor is made up of two parallel armatures of width l, both of which are maintained at the same potential V, as shown in Figure 2.30. The distance between the armatures is here denoted by d. A wide, thin and grounded conductive sheet is partially inserted at a distance x from the armature (2) and at a depth s. Let us find the forces applied on the sheet within the capacitor. As the electric field is zero everywhere, except in the regions lying between the sheet and the two armatures, the electrostatic energy of the capacitor-sheet system is:

$$W_e = \frac{\varepsilon_0}{2}\left(\frac{V}{x}\right)^2 lsx + \frac{\varepsilon_0}{2}\left(\frac{V}{d-x}\right)^2 ls\,(d-x) = \frac{\varepsilon_0 V^2 lsd}{2x\,(d-x)} \qquad [2.115]$$

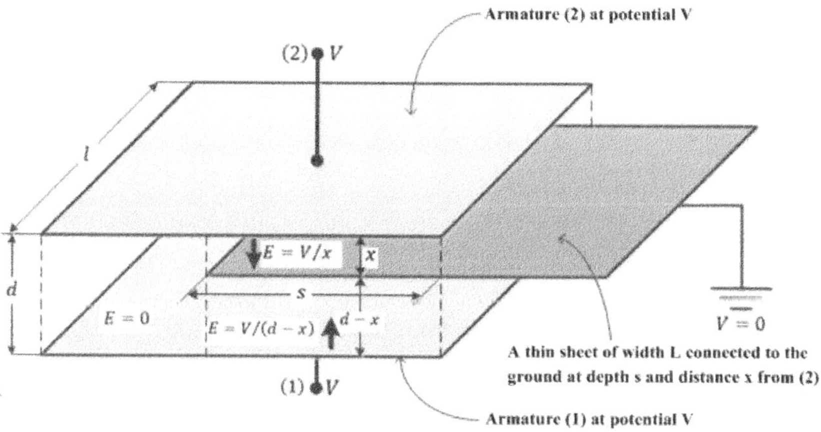

Figure 2.30. *Sheet inserted between the armatures of a capacitor, both maintained at the same potential V: the sheet is drawn toward the inside of the conductor by a force that reaches its maximum when it is inserted in the middle*

Consequently, the forces applied along the directions s and x, respectively, are given by:

$$\begin{aligned} F_s &= +\left.\frac{dW_e}{ds}\right|_V = \frac{\varepsilon_0 V^2 ld}{2x\,(d-x)} \\ F_x &= +\left.\frac{dW_e}{dx}\right|_V = -\frac{\varepsilon_0 V^2 lsd\,(d-2x)}{2x^2\,(d-x)^2} \end{aligned} \qquad [2.116]$$

The force F_s tends to draw the sheet toward the inside of the capacitor. It is clearly seen that the force F_x is zero at the half-way point, that is $x = \frac{d}{2}$, and F_s reaches its maximum:

$$F_s|_{\text{max.}} = \frac{2\varepsilon_0 V^2 l}{d} \qquad [2.117]$$

Chapter 3

Electrokinetics

3.1. Conduction in solids

Up to now, we have only considered stationary or static charges, which are sources of electrostatic fields. Stationary charges are studied in electrostatics. Electrokinetics, on the other hand, is the branch of electromagnetism devoted to the study of *electric currents* (i.e. moving charges) in a given medium, their evolution in time and their distribution in this medium. An *electric current* is therefore defined as a displacement of charge carriers. When electric charges are in motion, the expression for the electric field $\vec{\mathbf{E}}$ is modified. The electric currents produce not only an electric field but also a magnetic field that is manifested through a magnetic force exerted upon a charge in motion. An electric current can pass through a medium more or less easily, depending on the nature of this medium. This characteristic, called *resistance*, depends on the nature of the material and its temperature.

First, let us give a simple description of the conduction effect in solids:

– *Conductors*: conductive medium have electrons that are loosely bound to their atoms. Their *binding energy* (to atoms) is relatively low. They are easily released from this bond by an external disturbance (e.g. presence of an electric field). Consequently, these electrons are no longer attached to any atom and can freely move in the crystal lattice of the medium (ionic lattice), producing an electric current. These free electrons are called *conduction electrons*. Conductive media therefore allow an electric current to flow easily

– *Insulators*: in insulators, electrons are very strongly bound to the nucleus, unless the electric field is highly intense (this is known as *electrical breakdown*). It is therefore impossible to observe an electric current passing through these mediums. In reality, there is no perfect insulator (i.e. a material with infinite resistivity). In general, an insulator has high, but finite, resistivity.

– *Semiconductors*: semiconductors are mediums that behave like almost perfect insulators at very low temperatures, but have a resistivity that is intermediate between that of conductors and insulators at ordinary temperatures.

The conduction mechanism (in all three cases) can be explained coherently and simply using band theory (in quantum mechanics) already mentioned in the previous chapter.

3.2. Current density and intensity

We know that charged particles move under the effect of an electric field. Additionally, the motion of charged particles is accelerated in vacuum. In general, in a solid, charge carriers are electrons that are more or less strongly bonded to the atoms in the crystal lattice, depending on whether the solid is a conductor, an insulator or some other medium. As a result, in addition to the external electrical force, electrons are also subject to frictional forces of varying magnitude, depending on the medium under consideration. This displacement produces an *electric current*. The current intensity is always defined with respect to a surface. Consider an oriented surface S inside a solid. The *current intensity* or *electric intensity*, I_S, is the quantity of charges that crosses S per unit time, counted positively if it is in the same direction as the orientation of S and negatively if it is in the opposite direction. If dQ_S is the charge crossing S during dt, then:

$$I_S = \frac{dQ_S}{dt} \qquad [3.1]$$

In other words, the intensity of the electric current is equal to the flow of charges across the surface S. By convention, the direction of the current corresponds to the direction of movement of the positive charges.

Suppose that a given medium contains several types of charge carriers α and $d\vec{\mathbf{S}}$ a surface element oriented inside this medium. In general, each medium (conductor, ionized gas, insulator, semiconductor, etc.) is seen as a plasma with two types of carriers (i.e. $\alpha = \pm$) with the same charge but opposite signs ($\pm q$) with masses $m_\pm$ and densities $n_\pm$ (number of α carriers per unit volume), respectively. In principle, a plasma is an electrically neutral medium made up of positive ions (atoms or molecules that have lost one or more electrons) and negative electrons, which interact with each other. We will come back to this in more detail later, but for now, let us consider the most general case possible. Let n_α denote the number of α carriers per unit volume, q_α the individual charge and $\vec{\mathbf{v}}_\alpha$ the velocity (here $\vec{\mathbf{v}}_\alpha$ designates the mean value of the velocities of the α carriers) relative to a reference frame in which $d\vec{\mathbf{S}}$ is fixed.

The charges of type α passing through $d\vec{\mathbf{S}}$ (oriented along the unit vector $\vec{\mathbf{n}}$) between t and $t + dt$ are all contained inside the cylinder of volume:

$$d^2\tau_\alpha = \left| d\vec{\mathbf{S}} \cdot \vec{\mathbf{v}}_\alpha dt \right| = dS_\perp v_\alpha dt \qquad [3.2]$$

where $dS_\perp$ is the projection of $d\vec{\mathbf{S}}$ onto the plane perpendicular to $\vec{\mathbf{v}}_\alpha$ and $v_\alpha = \left|\vec{\mathbf{v}}_\alpha\right|$, as depicted in Figure 3.1. The volume contains the quantity of charges:

$$d^2Q_\alpha = n_\alpha q_\alpha d^2\tau_\alpha \qquad [3.3]$$

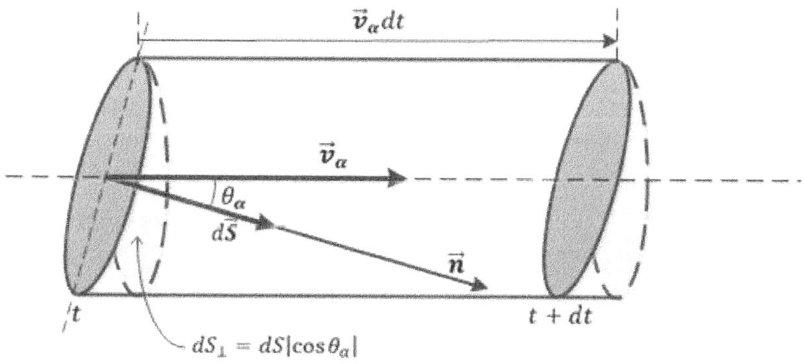

Figure 3.1. *Quantity of charges of type α, passing through an oriented surface dS during a time interval dt*

If θ_α represents the angle measured between $\vec{\mathbf{v}}_\alpha$ and $\vec{\mathbf{n}}$, then $dS_\perp = dS\left|\cos\theta_\alpha\right|$ and, consequently:

$$d^2Q_\alpha = n_\alpha q_\alpha v_\alpha dt dS \left|\cos\theta_\alpha\right| \qquad [3.4]$$

In particular, for $\cos\theta_\alpha > 0$, that is, $\theta_\alpha \in \left[0, \frac{\pi}{2}\right[$ ($\vec{\mathbf{v}}_\alpha$ and $\vec{\mathbf{n}}$ are in the same direction), we have: $dI_\alpha = \frac{d^2Q_\alpha}{dt} = n_\alpha q_\alpha v_\alpha dS\cos\theta_\alpha$. Inversely, for $\cos\theta_\alpha < 0$, that is, $\theta_\alpha \in \left]\frac{\pi}{2}, \pi\right[$ ($\vec{\mathbf{v}}_\alpha$ and $\vec{\mathbf{n}}$ are in opposite directions), we write $dI_\alpha = -\frac{d^2Q_\alpha}{dt} = n_\alpha q_\alpha v_\alpha dS\cos\theta_\alpha$. In both cases, we find:

$$dI_\alpha = n_\alpha q_\alpha v_\alpha dS\cos\theta_\alpha = n_\alpha q_\alpha \vec{\mathbf{v}}_\alpha \cdot d\vec{\mathbf{S}} \equiv \vec{\mathbf{J}}_\alpha \cdot d\vec{\mathbf{S}} \qquad [3.5]$$

where:

$$\vec{\mathbf{J}}_\alpha = n_\alpha q_\alpha \vec{\mathbf{v}}_\alpha \qquad [3.6]$$

is the *current vector of α-type carriers*. By summing over all types of carriers, we finally obtain:

$$dI = \sum_{\alpha} dI_{\alpha} = \vec{\mathbf{J}} \cdot d\vec{\mathbf{S}} = \left(\vec{\mathbf{J}} \cdot \vec{\mathbf{n}}\right) dS \qquad [3.7]$$

where:

$$\vec{\mathbf{J}} = \sum_{\alpha} \vec{\mathbf{J}}_{\alpha} = \sum_{\alpha} n_{\alpha} q_{\alpha} \vec{\mathbf{v}}_{\alpha} = \sum_{\alpha} \rho_{\alpha} \vec{\mathbf{v}}_{\alpha} \qquad [3.8]$$

$\vec{\mathbf{J}}$ is the *total current vector*. The quantity $\rho_{\alpha} = n_{\alpha} q_{\alpha}$ represents here *the volume density of α-type carriers*. A direct consequence of this equation is that it is very possible to have an electric current in a neutral body. In conclusion, it should be noted that not all α carriers have the same $\vec{\mathbf{v}}_{\alpha}$ velocity. In fact, equation [3.6] must be replaced by $\vec{\mathbf{J}}_{\alpha} = \sum_{i=1}^{n_{\alpha}} q_{\alpha} \vec{\mathbf{v}}_{\alpha,i} = q_{\alpha} n_{\alpha} \frac{\sum_{i=1}^{n_{\alpha}} \vec{\mathbf{v}}_{\alpha,i}}{\sum_{i=1}^{n_{\alpha}} 1} = q_{\alpha} n_{\alpha} \langle \vec{\mathbf{v}}_{\alpha} \rangle$, where $\vec{\mathbf{v}}_{\alpha,i}$ is the velocity of the ith α charge carrier. $\langle \vec{\mathbf{v}}_{\alpha} \rangle$ is the mean velocity of the α-type carriers. So, the *current intensity* through a finite oriented surface S is:

$$I = \int_S dI = \int_S \vec{\mathbf{J}} \cdot d\vec{\mathbf{S}} = \int_S \left(\vec{\mathbf{J}} \cdot \vec{\mathbf{n}}\right) dS \qquad [3.9]$$

In other terms, the current intensity is simply the flux of the current vector $\vec{\mathbf{J}}$ through the surface S.

EXAMPLE 3.1. (Change of Reference Frames).– Let us show that the current vector $\vec{\mathbf{J}}$ is independent of the reference frame, at any point M, for a neutral medium. Consider a Cartesian system $O'x'y'z'$ attached to the medium, which cannot only move rectilinearly relative to a immobile Galilean reference frame $Oxyz$ but can also rotate, as shown in Figure 3.2. $Oxyz$ is called the absolute reference frame, while $O'x'y'z'$ (attached to the medium) is called the *mobile or relative reference frame* (non-Galilean). The absolute velocity at any point M is equal to the sum of the relative and the training velocities, that is, $\vec{\mathbf{v}}_{\alpha} = \vec{\mathbf{v}}'_{\alpha} + \vec{\mathbf{v}}_{e}$, where $\vec{\mathbf{v}}_{e} = \frac{d\vec{\mathbf{R}}}{dt} + \vec{\omega} \times \overrightarrow{\mathbf{O'M}}$. $\vec{\omega} = \omega_{x'} \vec{\mathbf{i}}\,' + \omega_{y'} \vec{\mathbf{j}}\,' + \omega_{z'} \vec{\mathbf{k}}\,'$ (expressed in $O'x'y'z'$) is here the rotational velocity of $O'x'y'z'$ (i.e. of the medium) with respect to $Oxyz$. Consequently, we obtain:

$$\vec{\mathbf{J}} = \sum_{\alpha} n_{\alpha} q_{\alpha} \vec{\mathbf{v}}_{\alpha} = \sum_{\alpha} n_{\alpha} q_{\alpha} \vec{\mathbf{v}}'_{\alpha} + \left(\sum_{\alpha} n_{\alpha} q_{\alpha}\right) \vec{\mathbf{v}}_{e} = \sum_{\alpha} n_{\alpha} q_{\alpha} \vec{\mathbf{v}}'_{\alpha} = \vec{\mathbf{J}}\,' \qquad [3.10]$$

because the medium is neutral ($\sum_\alpha n_\alpha q_\alpha = 0$). We therefore conclude that the current vector is independent of the reference frame. ■

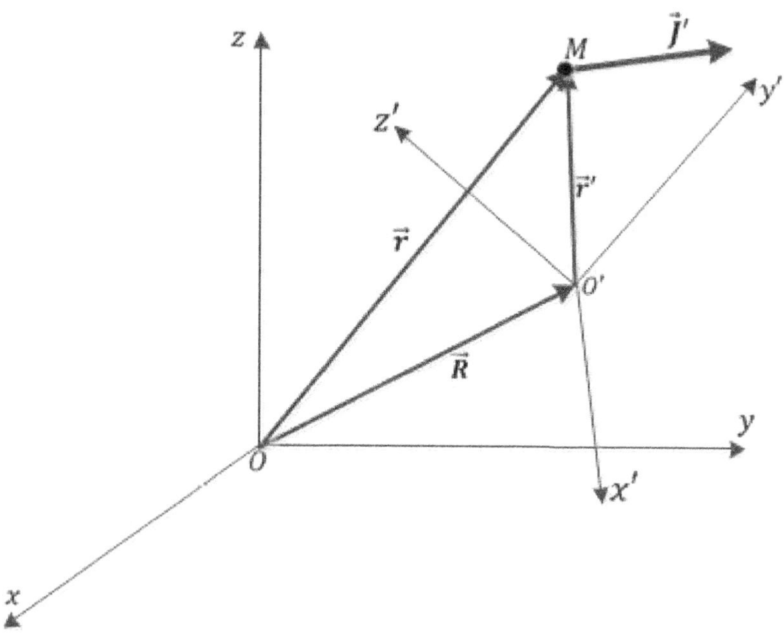

Figure 3.2. *The current vector is independent of the frame of reference*

In a neutral metal conductor, the current vector is written as:

$$\vec{J} = \rho_- \vec{v}_- + \rho_+ \vec{v}_+ \quad [3.11]$$

where $\vec{v}_-$ and $\vec{v}_+$ are, respectively, the velocity of conduction electrons and the velocity of the ions in the crystalline lattice. Since the ions are static with respect to each other inside the metal, the velocity $\vec{v}_+$ (of the ions with respect to the crystalline lattice of the metal) is zero. The current vector is therefore reduced to:

$$\vec{J} = \rho_- \vec{v}_- = -en_- \vec{v}_- \quad [3.12]$$

In a metal, electrons move in the opposite direction to the current vector.

By default, a current is a volume current. When an electric current flows essentially on the surface of an object, we speak about a *surface current density* (or *surface current*), which is also vectorial. If we consider a medium in which one of the three dimensions is very weak compared with the other two (e.g. a plane of thickness d), we can compare this medium to a sheet of current of negligible thickness (surface current) and, consequently, we can define surface current in this way:

$$\vec{\mathbf{K}} = \int_0^d \vec{\mathbf{J}} \, dz \qquad [3.13]$$

where z denotes, here, a coordinate measured perpendicular to the surface.

In general, if the current moves over a sheet $z = \Sigma(x, y)$, the volume current density can be written as: $\vec{\mathbf{J}}(x, y, z) = \vec{\mathbf{K}}(x, y)\,\delta(z - \Sigma(x, y))$. Let us imagine an infinitesimal rectangular surface element $ABCD$, centered at point $M(x, y)$ on sheet $z = \Sigma(x, y)$ and going perpendicularly through the sheet in a symmetrical manner (AB and CD are tangent to the sheet), as shown in Figure 3.3(a). The current intensity going across the differential surface element $ABCD$ is equal to $\Delta I = \int_{ABCD} \vec{\mathbf{J}} \cdot \vec{\mathbf{n}}\, dS \approx \left(\vec{\mathbf{K}} \cdot \vec{\mathbf{n}}\right) \Delta l$, where $\vec{\mathbf{n}}$ is a unit vector that is normal to both $ABCD$ and the element dl. We therefore conclude that the current going across dl in the direction of $\vec{\mathbf{n}}$ is:

$$dI = \left(\vec{\mathbf{K}} \cdot \vec{\mathbf{n}}\right) dl \qquad [3.14]$$

Figure 3.3. *Surface current density* $\vec{\mathbf{K}}$

This equation can be obtained directly from [3.7] by substituting $\vec{\mathbf{J}}$ by $\vec{\mathbf{K}}$ and dS by dl. By analogy with [3.8], we also write:

$$\vec{\mathbf{K}} = \sum_{\alpha} \sigma_\alpha \vec{\mathbf{v}}_\alpha \qquad [3.15]$$

where σ_α represents the *surface charge of type α carriers*. Parallel to [3.9], the total current flowing through a curve Γ is:

$$I = \int_\Gamma \left(\vec{\mathbf{K}} \cdot \vec{\mathbf{n}}\right) dl \qquad [3.16]$$

For example, if a ribbon of no thickness and width L has a surface current $\vec{\mathbf{K}}$ flowing through it along its length, the current intensity measured (on the ribbon) is $I = \int_L \left(\vec{\mathbf{K}} \cdot \vec{\mathbf{n}}\right) dL$, where $\vec{\mathbf{n}}$ is a unit vector perpendicular to L.

When two of the three dimensions of the medium (of the current) are much smaller than the third (filiform or curved medium), a vector current line can be defined, at any point on the line, by multiplying the current I by the vector $\vec{\mathbf{n}}$ tangent to the curve at this point, as depicted in Figure 3.4:

$$\vec{\mathbf{I}} = I\vec{\mathbf{n}} \qquad [3.17]$$

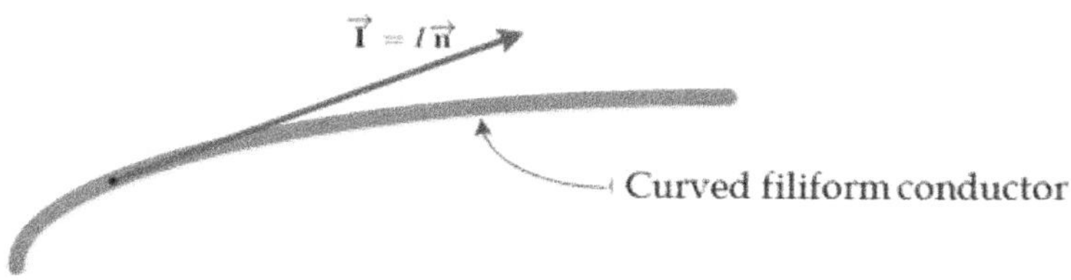

Figure 3.4. *Current traveling over a line (filiform body). For a color version of this figure, see www.iste.co.uk/abdesselam/electrodynamics1.zip*

The current flowing through the wire is then:

$$I = \left|\vec{\mathbf{I}}\right| \qquad [3.18]$$

If the wire is neutral, the current line vector $\vec{\mathbf{I}}$ is always tangent to the wire, regardless of the frame of reference chosen. Analogously to equations [3.8] and [3.15], we also write that:

$$\vec{\mathbf{I}} = \sum_\alpha \lambda_\alpha \vec{\mathbf{v}}_\alpha \qquad [3.19]$$

where λ_α is the *linear charge* of α-type carriers.

A current distribution is *invariant under a transformation* if this transformation leaves the distribution identical to itself. A current distribution can be invariant by translation and/or rotation around an axis. Similarly, there may be symmetry or antisymmetry planes for current distributions:

– *Symmetry plane*: if M and M' are symmetrical points with respect to (P), the current at M' is symmetrical to the current at M, that is, $\vec{\mathbf{J}}_{\parallel}(M') = \vec{\mathbf{J}}_{\parallel}(M')$ and $\vec{\mathbf{J}}_{\perp}(M') = -\vec{\mathbf{J}}_{\perp}(M')$.

– *Antisymmetry plane*: if M and M' are symmetrical points with respect to (P), the current at M' is the anti-symmetrical of the current at M, that is, $\vec{\mathbf{J}}_{\parallel}(M') = -\vec{\mathbf{J}}_{\parallel}(M')$ and $\vec{\mathbf{J}}_{\perp}(M') = \vec{\mathbf{J}}_{\perp}(M')$.

3.3. Conservation of charge

Consider, within a given medium, a fixed closed surface S enclosing a volume τ, as shown in Figure 3.5. Between instants t and $t + \Delta t$, charges in the inner volume close to the surface with velocities directed outward leave the volume, while charges outside the volume with velocities directed toward the surface enter the volume. The variation in the charge Q_{int} contained inside S is therefore (i.e. the charge transported by the current):

$$\begin{aligned} \Delta Q_{\text{int}} &= \int_{\tau} [\rho(t + \Delta t) - \rho(t)]\, d\tau \\ &= -\oint_{S} \left(\sum_{\alpha} n_{\alpha} q_{\alpha} \vec{\mathbf{v}}_{\alpha} \Delta t \right) \cdot d\vec{\mathbf{S}} = -\oint_{S} \vec{\mathbf{J}} \Delta t \cdot d\vec{\mathbf{S}} \end{aligned} \quad [3.20]$$

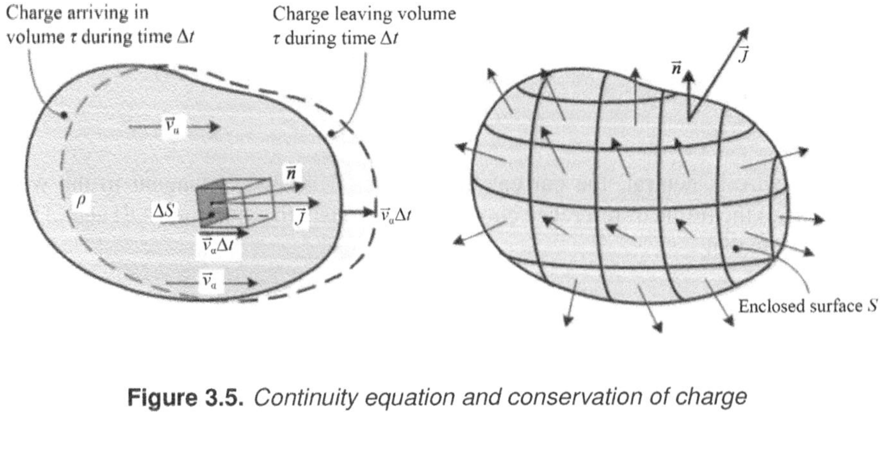

Figure 3.5. *Continuity equation and conservation of charge*

where:

$$\rho(M,t) = \sum_{\alpha} q_\alpha n_\alpha (M,t) \quad [3.21]$$

The $-$ sign on the right is necessary, because when $\vec{\mathbf{v}}_\alpha$ is in the direction of $d\vec{\mathbf{S}}$, the charges leave the volume, that is, the enclosed charge decreases. We divide [3.20] by Δt and then take the limit $\Delta t \to 0$:

$$\oint_S \vec{\mathbf{J}} \cdot d\vec{\mathbf{S}} = -\frac{dQ_{\text{int}}}{dt} = -\int_\tau \frac{\partial \rho}{\partial t} d\tau \quad [3.22]$$

Using the divergence theorem $\oint_S \vec{\mathbf{J}} \cdot d\vec{\mathbf{S}} = \int_\tau \vec{\nabla} \cdot \vec{\mathbf{J}}\, d\tau$, we then obtain: $\int_\tau \frac{\partial \rho}{\partial t} d\tau = -\int_\tau \vec{\nabla} \cdot \vec{\mathbf{J}}\, d\tau$. Since τ is arbitrary, the quantities under the integral signs must be equal at any point in space, that is:

$$\vec{\nabla} \cdot \vec{\mathbf{J}} + \frac{\partial \rho}{\partial t} = 0 \quad [3.23]$$

This is the *continuity equation*. The continuity equation states that an increase in charge density cannot be achieved at a point M, unless the current entering is greater than the current leaving at that point. The equation of continuity expresses the *conservation of charge*.

The continuity equation [3.23] can also take a more appropriate form. Indeed, the charge density is also related to $\vec{\mathbf{E}}$ through the local Gaussian equation $\rho = \varepsilon_0 \vec{\nabla} \cdot \vec{\mathbf{E}}$. Consequently, we can write $\vec{\nabla} \cdot \left(\vec{\mathbf{J}} + \varepsilon_0 \frac{\partial \vec{\mathbf{E}}}{\partial t} \right) = 0$, or alternatively:

$$\vec{\nabla} \cdot \vec{\mathbf{J}}_{\text{total}} = 0 \quad [3.24]$$

with:

$$\vec{\mathbf{J}}_{\text{total}} = \vec{\mathbf{J}} + \vec{\mathbf{J}}_{\text{d}} = \vec{\mathbf{J}} + \varepsilon_0 \frac{\partial \vec{\mathbf{E}}}{\partial t} \quad [3.25]$$

$\vec{\mathbf{J}}_{\text{d}} = \varepsilon_0 \frac{\partial \vec{\mathbf{E}}}{\partial t}$ is the *displacement current*. Equation [3.24] expresses the conservation of the *total current* $\vec{\mathbf{J}}_{\text{total}}$ at any point in space.

3.4. Kirchhoff's first law and approximation of quasi-stationary regimes

In a *stationary* or *permanent* regime, that is, when $\frac{\partial \rho}{\partial t} = 0$ (no creation or loss of charge at any point), the continuity equation [3.23] reduces to $\vec{\nabla} \cdot \vec{\mathbf{J}} = 0$, which means that:

$$\oint_S \vec{\mathbf{J}} \cdot d\vec{\mathbf{S}} = 0 \qquad [3.26]$$

for any closed surface S within the medium. We therefore conclude that in a permanent regime, the current vector has a conservative flow. In other words, the current entering this closed surface is equal to the current leaving it. This is Kirchhoff's first law.

Now consider a portion of a conductor (in the shape of a tube, for simplification), delimited by a surface S_1 on one side, a surface S_2 on the other side and by the external lateral surface S_L (located between S_1 and S_2), as depicted in Figure 3.6. As the total surface $S = S_1 \cup S_L \cup S_2$ is closed, it is possible to write that $0 = \oint_S \vec{\mathbf{J}} \cdot d\vec{\mathbf{S}} = \int_{S_1} \vec{\mathbf{J}} \cdot d\vec{\mathbf{S}}_1 + \int_{S_L} \vec{\mathbf{J}} \cdot d\vec{\mathbf{S}} + \int_{S_2} \vec{\mathbf{J}} \cdot d\vec{\mathbf{S}}_2$. The integral $\int_{S_L} \vec{\mathbf{J}} \cdot d\vec{\mathbf{S}}$ is zero, because the charges cannot leave the conductor across S_L. Since $I_1 = \oint_{S_1} \vec{\mathbf{J}} \cdot \left(-d\vec{\mathbf{S}}_1\right)$ and $I_2 = \int_{S_2} \vec{\mathbf{J}} \cdot d\vec{\mathbf{S}}_2$, it can be derived that:

$$I_1 = I_2 \qquad [3.27]$$

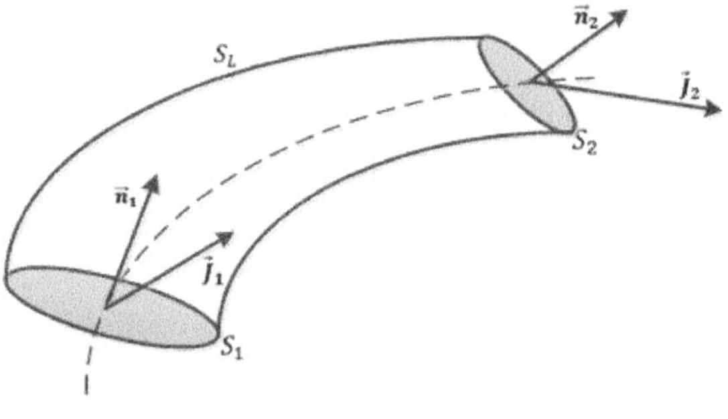

Figure 3.6. *Illustration of Kirchhoff's first law for a conductor in the shape of a tube*

The current flowing through any section of a filiform conductor is the same everywhere, regardless of its position. The current intensity in a conductor (in a permanent regime) is the current intensity through any section.

The same reasoning can be applied to a node, that is a point where several filiform conductors are joined. The sum of the intensities of the currents entering a node is equal to the sum of the intensities of the currents leaving the same node.

This is the *Kirchhoff's current law* or *Kirchhoff's first law*. This law follows directly from the conservation of electric charge, taking into account the fact that in a permanent regime, the charges cannot accumulate at any point in the circuit. Charges arriving at a node compensate for those leaving it. This law can be illustrated as follows. Let us write, by convention, that $s_k = -1$ if the current is oriented toward the node and $s_k = +1$ if the current is in the opposite direction, that is, leaving the node. We therefore obtain:

$$\sum_k s_k I_k = 0 \quad \Longleftrightarrow \quad \sum_{\text{incoming current}} I_k = \sum_{\text{outgoing current}} I_k \qquad [3.28]$$

If the conductor or circuit is sufficiently small, the time taken for the propagation of the electromagnetic field from one end of the circuit to the other is negligible compared to the characteristic time of the evolution of the field (e.g. the period T, in the case of a sinusoidal field). In other words, if the circuit size is very small compared to the wavelength of the signal, that is, $L \ll \lambda$, the following approximation is then always possible:

$$\vec{\nabla} \cdot \vec{\mathbf{J}} \simeq 0 \quad \Longleftrightarrow \quad \frac{\partial \rho}{\partial t} \simeq 0 \qquad [3.29]$$

This is the approximation of quasi-stationary regimes. In this approximation, any variation in the source is immediately felt at all points on the circuit. It immediately follows that (i) the intensity of the current is the same at any point on the conductor and (ii) a circuit remains therefore neutral at any point in the quasi-stationary regime. It must be specified that this approximation does not apply to capacitors. Kirchhoff's law, stated for permanent regime, remains valid also for quasi-stationary regimes.

3.5. Conduction model for charged gas

An electric current in a conductor is a movement of all conduction electrons within the conductor, most often produced due to an electric field. Indeed, each charge q placed in an electric field is subject to the force $\vec{\mathbf{F}} = q\vec{\mathbf{E}}$, which causes its acceleration. A conductor may be said to be at equilibrium, if there is no displacement (of charges) observed on the macroscopic level within the conductor. The macroscopic electric field is therefore zero and the electric potential (macroscopic) is the same everywhere. If a disequilibrium in electric potential appears between the ends of a conductor, an electric field $\vec{\mathbf{E}}$ is created and causes the displacement of conduction electrons. This displacement is carried out in the direction that will reduce the disequilibrium. Indeed, the electric field is created by a disequilibrium of electric charges. The direction of the field is from the region of accumulation of positive charges toward the region of accumulation of negative charges. A negative charge is subject to a force that is opposite to the field and will therefore be directed toward the positive charges with the aim of annihilating these and thereby attenuating the field.

If conduction electrons are subject only to the influence of the electric field, they would be constantly accelerated and their velocity would keep increasing. The conductivity would then be infinite, which contradicts the experimental observations. In reality, the electrons are also subject to braking forces (we can name them as the actions of positive ions in the crystalline lattice, the repulsive action between conduction electrons, thermal agitation forces, actions due to impurities, etc.), which prevent them from being infinitely accelerated in this manner.

As a result, a conduction electron in a conductive medium undergoes multiple collisions that cause it to lose energy. In a first-order approximation, the action of these shocks can be modeled by a braking force of the form $-\mu\vec{v}$ (similar to friction force), which will allow the conduction electron to take a limiting velocity. This phenomenon will be studied in detail in the following sections.

3.5.1. *The drift–diffusion equation*

In many materials, including good conductors (metals, ionized gases, etc.) and bad conductors (such as insulators and semiconductors), charge carriers can be modeled as an ideal or perfect gas in a medium, called a *plasma*. For simplicity's sake, we assume that there are two types of charge carriers of equal size but opposite sign, $\pm q$, with masses $m_\pm$ and densities $n_\pm$ (number per unit volume), respectively. These charges can be holes and electrons in a semiconductor, ions of opposite charge in an electrolyte solution, or electrons and nuclei in a metal.

When an electric field is applied, positive charges move in the direction of the field, while negative charges move in opposite directions. These charge carriers interact with the medium through frictional forces (reflecting the collision phenomena observed in the medium). In addition to electrical and frictional forces, particles exert a force on each other through a pressure force, caused by thermal agitation. In fact, this term exists even for uncharged particles. For an ideal or perfect gas, the pressure, P, measured at each point, has the value:

$$P = nk_BT \qquad [3.30]$$

where n is the charge density, T is the absolute temperature and k_B is the Boltzmann constant.

The pressure force exerted on an elementary volume element $\Delta x \Delta y \Delta z$, shown in Figure 3.7, is:

$$\vec{\mathbf{F}}_P \simeq \left[\frac{P\left(x - \frac{\Delta x}{2}, y, z\right) - P\left(x + \frac{\Delta x}{2}, y, z\right)}{\Delta x} \vec{\mathbf{i}} + \frac{P\left(x, y - \frac{\Delta y}{2}, z\right) - P\left(x, y + \frac{\Delta y}{2}, z\right)}{\Delta y} \vec{\mathbf{j}} + \frac{P\left(x, y, z - \frac{\Delta z}{2}\right) - P\left(x, y, z + \frac{\Delta z}{2}\right)}{\Delta z} \vec{\mathbf{k}} \right] \Delta x \Delta y \Delta z \quad [3.31]$$

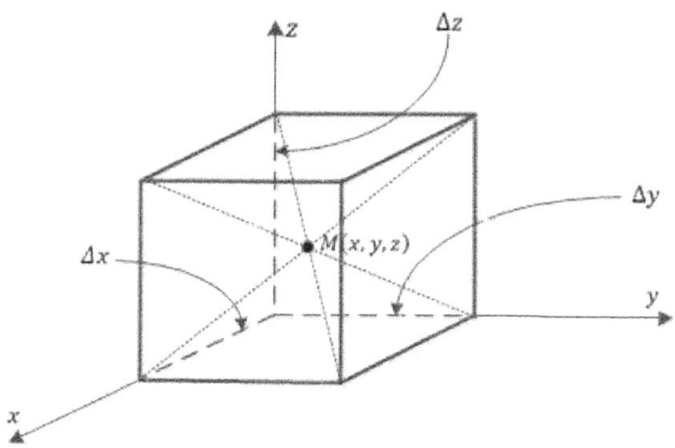

Figure 3.7. *Pressure force acting on an infinitesimal rectangular volume element*

In particular, if $\Delta x, \Delta y, \Delta z \longrightarrow 0$ (i.e. the volume reduces to an infinitesimal size around a single point), we obtain:

$$\vec{\mathbf{F}}_P = -\vec{\nabla} P dx dy dz \quad [3.32]$$

Applying Newton's second law, for the various charge carriers contained in the volume element $dxdydz$:

$$m_\pm n_\pm dxdydz \frac{\partial \vec{\mathbf{v}}_\pm}{\partial t} = \pm q n_\pm dxdydz \vec{\mathbf{E}} - \frac{m_\pm n_\pm dxdydz}{\tau_\pm} \vec{\mathbf{v}}_\pm - \vec{\nabla} P dxdydz \quad [3.33]$$

Upon simplifying by $dxdydz$ on both sides, we get:

$$m_\pm \frac{\partial \vec{\mathbf{v}}_\pm}{\partial t} = \pm q \vec{\mathbf{E}} - \frac{m_\pm}{\tau_\pm} \vec{\mathbf{v}}_\pm - \frac{1}{n_\pm} \vec{\nabla} \left(n_\pm k_B T\right) \quad [3.34]$$

This is the charge carrier *governing equation*. This is a first-order vector differential equation with constant coefficients. It admits the particular solution $\overrightarrow{\mathbf{v}}_{\pm}^{(p)} = \frac{\tau_{\pm}}{m_{\pm}}\left[\pm q\overrightarrow{\mathbf{E}} - \frac{1}{n_{\pm}}\overrightarrow{\nabla}\left(n_{\pm}k_B T\right)\right]$, while the homogeneous part (i.e. without the right-hand side) has the solution $\overrightarrow{\mathbf{A}}_{\pm}e^{-\frac{t}{\tau_{\pm}}}$, where $\overrightarrow{\mathbf{A}}_{\pm}$ is a constant vector that must be determined based on the initial conditions. Therefore, the general solution is $\overrightarrow{\mathbf{v}}_{\pm}(t) = \overrightarrow{\mathbf{A}}_{\pm}e^{-\frac{t}{\tau_{\pm}}} + \frac{\tau_{\pm}}{m_{\pm}}\left[\pm q\overrightarrow{\mathbf{E}} - \frac{1}{n_{\pm}}\overrightarrow{\nabla}\left(n_{\pm}k_B T\right)\right]$. Assuming that $\overrightarrow{\mathbf{v}}_{\pm}(0) = 0$, we obtain $\overrightarrow{\mathbf{A}}_{\pm} = -\frac{\tau_{\pm}}{m_{\pm}}\left[\pm q\overrightarrow{\mathbf{E}} - \frac{1}{n_{\pm}}\overrightarrow{\nabla}\left(n_{\pm}k_B T\right)\right]$ and consequently:

$$\overrightarrow{\mathbf{v}}_{\pm}(t) = \frac{\tau_{\pm}}{m_{\pm}}\left[\pm q\overrightarrow{\mathbf{E}} - \frac{1}{n_{\pm}}\overrightarrow{\nabla}\left(n_{\pm}k_B T\right)\right]\left(1 - e^{-\frac{t}{\tau_{\pm}}}\right) \quad [3.35]$$

As the term $\overrightarrow{\mathbf{A}}_{\pm}e^{-\frac{t}{\tau_{\pm}}}$ tends to zero in some $\tau_{\pm}$ (i.e. when $t \longrightarrow \infty$), we reach the limiting velocities (for carriers):

$$\langle\overrightarrow{\mathbf{v}}_{\pm}\rangle = \frac{\tau_{\pm}}{m_{\pm}}\left[\pm q\overrightarrow{\mathbf{E}} - \frac{1}{n_{\pm}}\overrightarrow{\nabla}\left(n_{\pm}k_B T\right)\right] \quad [3.36]$$

These average velocities of the carriers in the medium are called *drift velocities*. They are very low. Equation [3.36] is called the *drift–diffusion equation*. Therefore, the current vectors for each carrier are given by:

$$\overrightarrow{\mathbf{J}}_{\pm} = \rho_{\pm}\langle\overrightarrow{\mathbf{v}}_{\pm}\rangle = \pm q n_{\pm}\langle\overrightarrow{\mathbf{v}}_{\pm}\rangle = \pm\rho_{\pm}\mu_{\pm}\overrightarrow{\mathbf{E}} - D_{\pm}\overrightarrow{\nabla}\rho_{\pm} \quad [3.37]$$

where $\mu_{\pm}$ are called *mobility coefficients of the particles* and $D_{\pm}$ are their *diffusion coefficients*:

$$\begin{aligned} \mu_{\pm} &= \frac{q\tau_{\pm}}{m_{\pm}} \\ D_{\pm} &= \frac{k_B T\tau_{\pm}}{m_{\pm}} \end{aligned} \quad [3.38]$$

The system is assumed to be at constant temperature. We find that the ratio $\frac{D_{\pm}}{\mu_{\pm}}$ is the same for both carriers and has the dimension of a voltage, called *thermal voltage*:

$$\frac{D_{\pm}}{\mu_{\pm}} = \frac{k_B T}{q} \quad [3.39]$$

This is called the *Einstein equation*.

At equilibrium (i.e. when the net current of each carrier is zero), equations [3.37] can be rewritten in terms of the potential V as follows: $\overrightarrow{\mathbf{J}}_{\pm} = 0 = \mp\rho_{\pm}\mu_{\pm}\overrightarrow{\nabla}V - D_{\pm}\overrightarrow{\nabla}\rho_{\pm}$, which gives:

$$\overrightarrow{\nabla}\left[\pm\frac{\mu_{\pm}}{D_{\pm}}V + \ln\rho_{\pm}\right] = \overrightarrow{\nabla}\left[\pm\frac{q}{k_BT}V + \ln\rho_{\pm}\right] = 0 \qquad [3.40]$$

The term in square brackets is a constant. It therefore follows that the densities $\rho_{\pm}$ are related to the potential V by the *Boltzmann distribution*:

$$\rho_{\pm} = \pm\rho_0 e^{\mp\frac{qV}{k_BT}} \qquad [3.41]$$

where $\pm\rho_0$ are the carrier densities at equilibrium, that is, at $V = 0$. They are of the same magnitude, since the system is initially neutral. To find the spatial dependence of V and, consequently, ρ, we simply use Poisson's equation:

$$\Delta V = -\frac{\rho_+ + \rho_-}{\varepsilon_0} = -\frac{\rho_0}{\varepsilon_0}\left(e^{-\frac{qV}{k_BT}} - e^{\frac{qV}{k_BT}}\right) = \frac{2\rho_0}{\varepsilon_0}\sinh\left(\frac{qV}{k_BT}\right) \qquad [3.42]$$

This equation is known as the *Poisson–Boltzmann equation*, because the charge densities obey the Boltzmann distributions.

Note that when the potential is very low (so that $\frac{qV}{k_BT} \ll 1$), we can take the approximation $\sinh\left(\frac{qV}{k_BT}\right) \sim \frac{qV}{k_BT}$ and, consequently, equation [3.42] reduces to:

$$\Delta V - \frac{V}{l_d^2} = 0 \qquad [3.43]$$

where $l_d = \sqrt{\frac{\varepsilon_0 k_BT}{2\rho_0 q}}$ is called the *Debye length*.

EXAMPLE 3.2. (Electrode at the Potential V_0).– Consider an infinite plane (called an electrode here) placed at $x = 0$ and taken to the potential V_0, as shown in Figure 3.8(a). Since the plane is infinite, the potential only varies with the x coordinate. Therefore, equation [3.42] is reduced to:

$$\frac{d^2V}{dx^2} - \frac{1}{l_d^2}\sinh\widetilde{V} = 0 \qquad [3.44]$$

where $\widetilde{V} = \frac{qV}{k_B T}$. By multiplying [3.44] by $\frac{d\widetilde{V}}{dx}$, we find $\frac{d}{dx}\left[\frac{1}{2}\left(\frac{d\widetilde{V}}{dx}\right)^2 - \frac{\cosh \widetilde{V}}{l_d^2}\right] = 0$, which implies that:

$$\frac{1}{2}\left(\frac{d\widetilde{V}}{dx}\right)^2 - \frac{\cosh \widetilde{V}}{l_d^2} = C_1 \qquad [3.45]$$

where C_1 is an integration constant. As the potential V and the associated field $E_x = -\frac{dV}{dx}$ must be zero at infinity (i.e. for $x \longrightarrow \infty$), we must have $C_1 = -\frac{1}{l_d^2}$. It therefore follows that:

$$\frac{d\widetilde{V}}{dx} = -\widetilde{E}_x = \left[\frac{2}{l_d^2}\left(\cosh \widetilde{V} - 1\right)\right]^{1/2} = \begin{cases} -\dfrac{2}{l_d}\sinh\dfrac{\widetilde{V}}{2} & \text{for } x > 0 \\ \dfrac{2}{l_d}\sinh\dfrac{\widetilde{V}}{2} & \text{for } x < 0 \end{cases} \qquad [3.46]$$

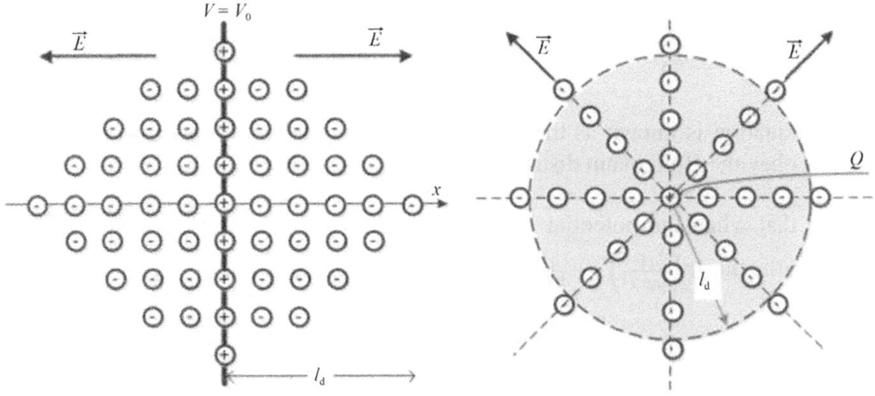

Case (a): Electrode placed at $x = 0$ and held at potential V_0 **Case (b)**: Point-like charge Q inserted in plasma

Figure 3.8. *Accumulation of moving charges of opposite polarity around any point-like charge inserted in a conductor (described by the drift–diffusion equation) to protect its field over distances greater than the Debye length*

The different signs that appear in the preceding expression are required, since the potential V must diminish as we move away from either side of the electrode (the electric field is directed outward from each side). A direct integration of the preceding equation immediately yields:

$$\frac{\tanh \frac{\widetilde{V}}{4}}{\tanh \frac{\widetilde{V_0}}{4}} = \begin{cases} e^{-\frac{x}{l_d}} & \text{for } x > 0 \\ e^{\frac{x}{l_d}} & \text{for } x < 0 \end{cases} \qquad [3.47]$$

The Debye length is therefore the characteristic distance over which the applied potential exerts an influence (i.e. action length of applied potential).

EXAMPLE 3.3. (Point-like Charge Inside a Plasma).– Let us insert a point-like charge inside a plasma. This charge will induce a weak potential that depends only on the distance r for symmetry reasons, as shown in Figure 3.8(b). In the approximation of weak potentials, equation [3.43] is written in spherical coordinates as follows: $\frac{1}{r^2}\frac{d}{dr}\left(r^2\frac{dV}{dr}\right) - \frac{V}{l_d^2} = 0$, or again:

$$\frac{d\,(rV)}{dr^2} - \frac{(rV)}{l_d^2} = 0 \qquad [3.48]$$

This is a linear differential equation with constant coefficients with the solutions $rV = A_1 e^{-\frac{r}{l_d}} + A_2 e^{\frac{r}{l_d}}$, where A_1 and A_2 are integration constants. Since the potential must decrease far from the charge (i.e. tends to zero), we must take $A_2 = 0$, while in the neighborhood of the charge Q (i.e. when $r \longrightarrow 0$), the potential must approach that of an isolated point-like charge, therefore $A_1 = \frac{Q}{4\pi\varepsilon_0}$. We therefore obtain:

$$V\,(r) = \frac{Q}{4\pi\varepsilon_0 r} e^{-\frac{r}{l_d}} \qquad [3.49]$$

It must be noted that in the neighborhood of Q (i.e. with r being very small), the potential becomes very large and the approximation of small potentials is violated.

3.5.2. *Local Ohm's law*

It has been seen in Example 3.2 that when a charge Q is introduced into a plasma, the mobile charges with opposite polarity accumulate all around it. Beyond the Debye length, these mobile charges tend to mask the effect of this introduced charge. Since l_d is generally very small compared to the system's dimensions, most of the regions in the space outside the Debye zone are electrically neutral, that is, the positive charge density ρ_0 is equal to the negative charge density $-\rho_0$. In these regions, the diffusion term in [3.37] is negligible, since there is no charge density gradient. The total current density is therefore proportional to the electric field:

$$\vec{\mathbf{J}} = \vec{\mathbf{J}}_+ + \vec{\mathbf{J}}_- = \rho_0\left(\vec{\mathbf{v}}_+ - \vec{\mathbf{v}}_-\right) = qn_0\left(\mu_+ + \mu_-\right)\vec{\mathbf{E}} = \gamma\vec{\mathbf{E}} \qquad [3.50]$$

where:

$$\gamma = q^2 n_0 \left(\frac{\tau_+}{m_+} + \frac{\tau_-}{m_-} \right) \qquad [3.51]$$

The quantity γ is called the ohmic conductivity of the medium. Relation [3.50] is called the *local Ohm's law*. It is sometimes more practical to use the inverse of the conductivity X, called, *resistivity* of the medium. Therefore, we write:

$$\vec{\mathbf{E}} = \eta \vec{\mathbf{J}} \qquad [3.52]$$

This is the *local Joule's law*. Let us note that the local Ohm's law is not valid in the Debye zones in proximity of charges. When the Ohm's law is valid, the net charge is zero. This does not contribute to Gauss' law.

The ohmic conductivity of a *perfect* conductor is very high and can also be considered to be infinite. Since all physical currents in [3.50] must be finite, the electric field within the conductor is zero, that is, the potential V is constant (equipotential) everywhere in the conductor:

$$\lim_{\gamma \longrightarrow \infty} \vec{\mathbf{J}} = \gamma \vec{\mathbf{E}} \qquad \Longleftrightarrow \qquad \begin{cases} \vec{\mathbf{E}} = 0 \\ V = \text{const} \\ \vec{\mathbf{J}} = \text{finite} \end{cases} \qquad [3.53]$$

When this is a conductor, the coefficient of mobility of positive ions μ_+ (assumed here to be fixed) is negligible compared to the coefficient of mobility of electrons μ_-. Consequently, the current density $\vec{\mathbf{J}}$ is chiefly produced by the displacements of electrons, that is, $\vec{\mathbf{J}} \approx en_-\mu_-\vec{\mathbf{E}}$. In the case of insulators, the two mobility coefficients, μ_+ and μ_-, are very weak, which results in $\vec{\mathbf{J}} \approx 0$. In semiconductors, atoms are bound to each other by covalent bonds. The binding energies of the electrons involved in these bonds are low. A small influx of energy (for example, small thermal agitation) is enough to release a few electrons from these bonds and they thus become free to move under the action of an electric field, as in a conductor, except that there are fewer of these than in conductors. When a covalent bond is broken, the free electron leaves behind a void called a *hole*. An electron from a neighboring bond may come fill this hole, thus leaving behind another hole in turn, which may be filled by the next electron, etc. Thermal agitation thus creates electron–hole pairs through the rupture of covalent bonds. Electrons and holes move independent of each other. A hole may be filled not only by a valence electron, but also any passing free electron. This is called *electron–hole recombination*. Electrons are carriers of negative charges, while holes are considered to be carriers of positive charges. They are called free carriers. In a semiconductor, there is an equal concentration of negative and positive carriers. As with a conductor, if an electric field is applied inside a semiconductor, holes move in the direction of the field, while the electrons move in the opposite direction. This double displacement will produce an electric current in the semiconductor. Analogous to the preceding results, we will have $\vec{\mathbf{v}}_- = -\mu_- \vec{\mathbf{E}}$ for electrons and $\vec{\mathbf{v}}_+ = \mu_+ \vec{\mathbf{E}}$ for the holes. The corresponding conductivities are $\gamma_- = en_-\mu_-$ and $\gamma_+ = en_+\mu_+$, where n_- and n_+ are the number of negative and positive charger carriers per unit volume. The total conductivity is the sum $\gamma = \gamma_- + \gamma_+$ and the resistivity of the semiconductor is $\eta = \frac{1}{\gamma}$. The resistivity of a semiconductor decreases with temperature. It is approximately of the form $\eta \approx Ae^{\frac{W}{kT}}$, where A, W and k are positive quantities and T is the absolute temperature. Indeed, when the temperature increases (i.e. for high thermal agitation), there will be an increase in the rupture of bonds and, consequently, the resistivity will decrease.

3.5.3. *Superconductors*

Superconductivity is a phenomenon or effect characterized by the complete absence of electric resistance within certain materials, called *superconductors*. Superconductivity manifests at very low temperatures close to absolute zero. Superconductor materials are an exception to Ohm's law. With negligible collisions ($\tau_\pm = \infty$) and an absolutely low temperature ($T \approx 0$), the governing equation [3.34] is reduced to:

$$\frac{\partial \vec{\mathbf{v}}_\pm}{\partial t} = \pm \frac{q}{m_\pm} \vec{\mathbf{E}} \tag{3.54}$$

Let us multiply [3.54] by the charge carrier densities, which are assumed to be constant:

$$\frac{\partial}{\partial t}\left(\pm q n_\pm \vec{\mathbf{v}}_\pm\right) = \frac{\partial \vec{\mathbf{J}}_\pm}{\partial t} = \frac{q^2 n_\pm}{m_\pm} \vec{\mathbf{E}} = \omega_{p,\pm}^2 \varepsilon_0 \vec{\mathbf{E}} \tag{3.55}$$

where $\omega_{p,\pm}^2 = \frac{q^2 n_\pm}{m_\pm \varepsilon_0}$. The quantities $\omega_{p,\pm}$ are called the *plasma frequencies of each carrier*.

If a superconductor material is placed between electrodes with parallel plates that are very large compared to the spacing so as to neglect edge effects, the electric field and current density $\vec{\mathbf{J}} = \vec{\mathbf{J}}_+ + \vec{\mathbf{J}}_-$ must be perpendicular to the electrodes, taken as the direction Ox. In other words, at a point far from the extremities, the field and current density must be directed in the direction Ox and depend only on the x variable.

Let us derive the equation for conservation of charge with respect to time, $\frac{\partial^2 \rho}{\partial t^2} + \frac{\partial^2}{\partial t \partial x}(J_+ + J_-) = 0$. This implies that:

$$\frac{\partial}{\partial x}\left[\frac{\partial}{\partial t}(J_+ + J_-) + \varepsilon_0 \frac{\partial^2 E}{\partial t^2}\right] = 0 \qquad [3.56]$$

The term between square brackets is therefore independent of the position x. In addition, it is zero according to the boundary conditions. By inserting [3.55] into [3.56], we find:

$$\frac{\partial^2 E}{\partial t^2} + \omega_p^2 E = 0 \qquad [3.57]$$

where $\omega_p^2 = \omega_{p,+}^2 + \omega_{p,-}^2$. The differential equation [3.57] admits the solution:

$$E = A_1 \cos \omega_p t + A_2 \sin \omega_p t \qquad [3.58]$$

where A_1 and A_2 are integration constants. Any initial perturbation causes an oscillating electric field with frequency ω_p. The charges then execute a simple harmonic movement around their equilibrium position.

3.6. Resistance

3.6.1. *Resistance between two parallel electrodes*

The electrodes used here are generally assumed to be perfectly conducting and, therefore, at constant potential. In the immediate neighborhood of the electrodes, the field must be perpendicular to the electrodes. Two perfectly conducting, parallel electrodes, with any arbitrary shape (surface S) and separated by a distance L, enclose a conducting material with conductivity γ, as shown in Figure 3.9(a). The electrodes are stuck to the bases of the conductor (with surface S) and maintained at a potential difference V. The current must circulate tangential to the external surface (the lateral surface of the conductor), since the external medium (here, the free space) has zero conductivity. Therefore, no current can go across the interface. Since the tangential component of the electric field is continuous, there is indeed an electric field outside the conductor, which diminishes as one moves away from the conductor. This three-dimensional field is difficult to calculate, as it depends on three coordinates. It must just be noted that for at longer distances, the field approaches the field of a point-like dipole, since the electrodes have opposite polarities. These systems will be studied later. The electric field inside the conductor is much simpler to calculate, as it is perpendicular to the electrodes, that is, in the direction Ox.

Since the charge density is zero inside the conductor (no accumulation of charges), the local form of Gauss' theorem makes it possible to write $\overrightarrow{\nabla} \cdot \overrightarrow{\mathbf{E}} = \frac{dE_x}{dx} = 0$. This shows that $E_x = E_0$. On the other hand, the integral for the field $\overrightarrow{\mathbf{E}}$ between the two electrodes must be equal to the applied tension V, $\int_0^L E_x dx = V$ and therefore, $E_0 = \frac{V}{L}$. Therefore, the current density is:

$$\overrightarrow{\mathbf{J}} = \sigma E_0 \overrightarrow{\mathbf{i}} = \frac{\gamma V}{L} \overrightarrow{\mathbf{i}} \qquad [3.59]$$

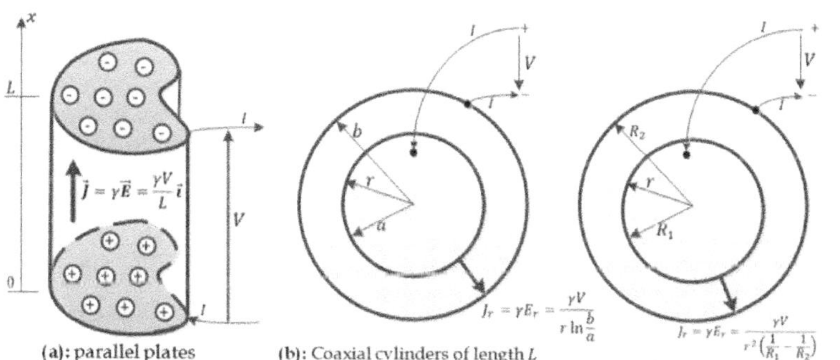

Figure 3.9. *(a) Resistance between two parallel electrodes: Resistance between (b) two coaxial cylindrical electrodes and (c) two concentric spherical electrodes*

The intensity of the total current passing through the electrodes is:

$$I = \oint_S \overrightarrow{\mathbf{J}} \cdot d\overrightarrow{\mathbf{S}} = \frac{\gamma V S}{L} \qquad [3.60]$$

where the surface integral is reduced to a simple product, because $\overrightarrow{\mathbf{J}}$ and $d\overrightarrow{\mathbf{S}}$ are in the same direction. The quantity:

$$R = \frac{V}{I} = \frac{L}{\gamma S} = \frac{\text{(Spacing)}}{\text{(Conductivity)} \times \text{(Surfaceareaoftheelectrode)}} \qquad [3.61]$$

is called the *resistance of the conductor*; it depends only on the geometry of the system and on its conductivity. This is the *global Ohm's law*. We sometimes also use the *conductance* $G = 1/R$. Opposite charges will accumulate on the internal surfaces of the electrodes, with the density $\sigma(x = 0) = \varepsilon_0 E_0 = \frac{\varepsilon_0 V}{L}$ and $\sigma(x = L) = -\sigma(x = 0) = -\varepsilon_0 E_0 = -\frac{\varepsilon_0 V}{L}$, respectively.

EXAMPLE 3.4. (Junction Charge).– Let us calculate the load at the junction of two conductor cables with the same cross-section S, conductivities γ_1 and γ_2, respectively ($\gamma_1 > \gamma_2$), and the same current I flowing through them, as shown in Figure 3.10. Since I and S are the same for the two conductors, we must have the same current density throughout. Consequently, we will have $\gamma_1 E_1 = \gamma_2 E_2$, since $\overrightarrow{\mathbf{J}} = \gamma \overrightarrow{\mathbf{E}}$. From this, it can be derived that $E_2 = \frac{\gamma_1}{\gamma_2} E_1$. Imagine a closed Gaussian surface Σ enclosing the surface of the junction. By applying Gauss' theorem to Σ and with Q denoting the total charge at the junction, we obtain $(E_2 - E_1)\, S = \frac{Q}{\varepsilon_0}$, which gives $Q = \varepsilon_0 S E_1 \left(\frac{\gamma_1}{\gamma_2} - 1\right) = \varepsilon_0 S \gamma_1 E_1 \left(\frac{1}{\gamma_2} - \frac{1}{\gamma_1}\right)$. Since $I = JS = S\gamma_1 E_1 = S\gamma_2 E_2$, it can finally be derived that:

$$Q = \varepsilon_0 I \left(\frac{1}{\gamma_2} - \frac{1}{\gamma_1}\right) \qquad [3.62]$$

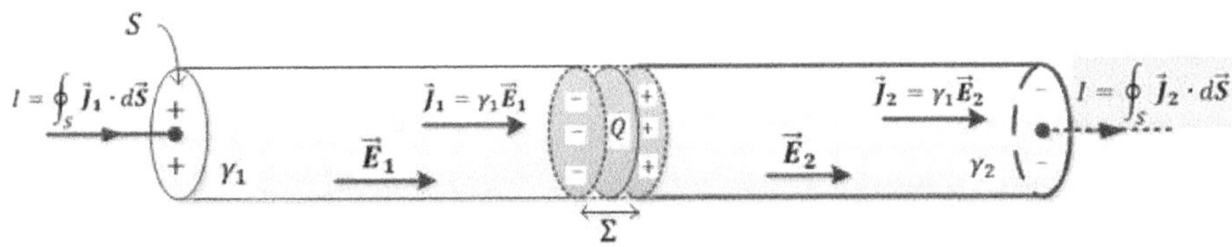

Figure 3.10. *When $\gamma_1 > \gamma_2$ (respectively, $\gamma_2 > \gamma_1$), the charge accumulated on the surface of the junction is positive (respectively, negative)*

EXAMPLE 3.5. (Resistance between Two Cylindrical Electrodes).– Two coaxial cylindrical electrodes, of length L, with internal radius a, and external radius b are maintained at a potential difference V. The two electrodes enclose a conducting material, as seen in Figure 3.9(b). The electric field must be perpendicular to the electrodes, that is, it must be radial. Since the charge density inside the conductor is zero, we must have $\overrightarrow{\nabla} \cdot \overrightarrow{\mathbf{E}} = \frac{1}{r}\frac{d}{dr}(rE_r) = 0$; this gives $E_r = \frac{C}{r}$, where C is an integration constant that can be calculated from the voltage applied between the two electrodes, $\int_a^b E_r dr = C \ln r|_a^b = V$, that is, $C = \frac{V}{\ln\left(\frac{b}{a}\right)}$. Therefore, it is found that $E_r = \frac{V}{r\ln\left(\frac{b}{a}\right)}$. The current density is equal to $J_r = \gamma E_r = \frac{\gamma V}{r\ln\left(\frac{b}{a}\right)}$ and, consequently, the electric current measured between the electrodes is:

$$I = \int_{z=0}^{L} \int_{\varphi=0}^{2\pi} J_r r d\varphi dz = \frac{\gamma V 2\pi L}{\ln\left(\frac{b}{a}\right)} \qquad [3.63]$$

It is clearly seen that I is constant. It therefore follows that:

$$R = \frac{V}{I} = \frac{\ln\left(\frac{b}{a}\right)}{2\pi\gamma L} \qquad [3.64]$$

EXAMPLE 3.6. (Resistance between Two Concentric, Spherical Electrodes).– We proceed in the same way for these two concentric, spherical electrodes maintained at a potential difference V. The radii of the internal and external electrodes are, respectively, R_1 and R_2, as shown in Figure 3.9(c). Due to symmetry, the electric field is purely radial. Since there is no charge density inside the conductor, $\overrightarrow{\nabla} \cdot \overrightarrow{\mathbf{E}} = \frac{1}{r^2}\frac{d}{dr}\left(r^2 E_r\right) = 0$. This yields $E_r = \frac{C}{r^2}$, where C is a constant that can be determined from the applied voltage $\int_{R_1}^{R_2} E_r dr = -\frac{C}{r}\Big|_{R_1}^{R_2} = V$, that is, $C = \frac{R_1 R_2 V}{R_2 - R_1}$. It therefore comes that $E_r = \frac{R_1 R_2 V}{r^2(R_2 - R_1)}$. It can therefore be derived that $J_r = \gamma E_r = \frac{\gamma R_1 R_2 V}{r^2(R_2 - R_1)}$ and consequently:

$$I = \int_{\theta=0}^{\pi} \int_{\varphi=0}^{2\pi} J_r r^2 \sin\theta d\theta d\varphi = \frac{4\pi\gamma R_1 R_2 V}{R_2 - R_1} \qquad [3.65]$$

The current intensity I is constant everywhere between the two electrodes. It is therefore found that:

$$R = \frac{V}{I} = \frac{R_2 - R_1}{4\pi\gamma R_1 R_2} \qquad [3.66]$$

■

For any two electrodes (of arbitrary geometry), the resistance of an ohmic conductor lying between them can be expressed in terms of the $\overrightarrow{\mathbf{E}}$ field as follows:

$$R = \frac{V}{I} = \frac{\int_L \overrightarrow{\mathbf{E}} \cdot d\overrightarrow{\mathbf{l}}}{\oint_S \overrightarrow{\mathbf{J}} \cdot d\overrightarrow{\mathbf{S}}} = \frac{\int_L \overrightarrow{\mathbf{E}} \cdot d\overrightarrow{\mathbf{l}}}{\oint_S \gamma\overrightarrow{\mathbf{E}} \cdot d\overrightarrow{\mathbf{S}}} \qquad [3.67]$$

where S and L are, respectively, any surface that completely encloses one electrode and any path connecting the two electrodes, as shown in Figure 3.11. It must be noted that:

1) The path L must be directed from the electrode with the highest potential to the electrode with the lowest potential, so that the value of the integral of the circulation is equal to the potential difference V. Starting from formula [3.67], it is clear that the resistance depends only on the geometry and conductivity γ and not on the amplitude of the electric field itself. Indeed, if the voltage is multiplied by a factor λ, the field will also increase by the same factor everywhere and, consequently, this factor will disappear in equation [3.67].

2) The conductivity γ may also be a function of the position.

3) The integrals in [3.67] are similar to those in equation [2.59] for the capacitance of a conductor between two electrodes. For the same conductor, the product of

resistance and capacitance is a constant independent of conductor geometry:

$$RC = \frac{\int_L \overrightarrow{\mathbf{E}} \cdot d\,\overrightarrow{\mathbf{l}}}{\gamma \oint_S \overrightarrow{\mathbf{E}} \cdot d\overrightarrow{\mathbf{S}}} \cdot \frac{\varepsilon_0 \oint_S \overrightarrow{\mathbf{E}} \cdot d\overrightarrow{\mathbf{S}}}{\int_L \overrightarrow{\mathbf{E}} \cdot d\,\overrightarrow{\mathbf{l}}} = \frac{\varepsilon_0}{\gamma} \qquad [3.68]$$

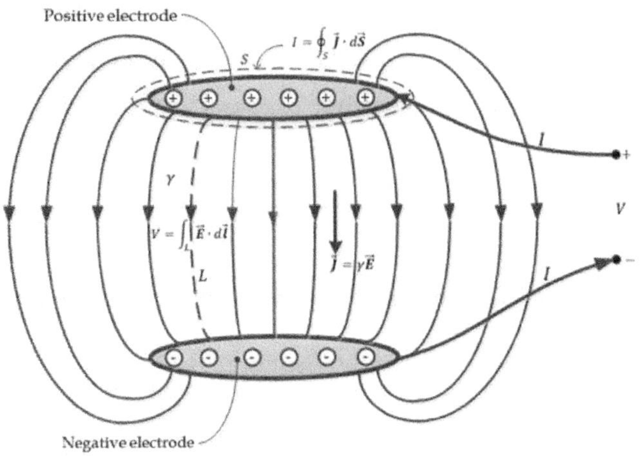

Figure 3.11. *Resistance of any geometry. The ratio $R = \frac{V}{I}$ depends only on the geometry and conductivity of the medium*

Therefore, for a given geometry, if the resistance or capacitance is known, the other quantity is immediately known from [3.68]. For example, it can be noted:

$$\begin{array}{ll} \text{parallelplanes:} & R = \dfrac{l}{\gamma S} \Longrightarrow C = \dfrac{\varepsilon_0 S}{l} \\ \text{coaxialcylinders:} & R = \dfrac{\ln(b/a)}{2\pi\gamma l} \Longrightarrow C = \dfrac{2\pi\varepsilon_0 l}{\ln(b/a)} \\ \text{concentricspheres:} & R = \dfrac{R_2 - R_1}{4\pi\gamma R_1 R_2} \Longrightarrow C = \dfrac{4\pi\varepsilon_0 R_1 R_2}{R_2 - R_1} \end{array} \qquad [3.69]$$

3.6.2. *Association of resistors*

As with capacitors, there are two ways of combining resistors within a circuit.

(a) *Series association*: consider a circuit element with resistors connected one after the other. This is called a *series association*. In the configuration shown in Figure 3.12(a), let $V_1 - V_2$ (respectively, $V_2 - V_3$) denote the voltage at the terminals of R_1 (respectively, R_2). The intensity of the current passing through R_1 (respectively, R_2)

is denoted by I_1 (respectively, I_2). In a stationary regime, the law of conservation of charges imposes that the current intensity be conserved along the whole chain of conductors in series, that is: $I_1 = I_2 = I$. According to Ohm's law, for each of the resistors, we have $V_1 - V_2 = R_1 I_1$ and $V_2 - V_3 = R_2 I_2$. It therefore follows that $V_1 - V_3 = (V_1 - V_2) + (V_2 - V_3) = R_1 I_1 + R_2 I_2 = (R_1 + R_2) I$; this implies that $\frac{V_1 - V_3}{I} = R_1 + R_2$. This therefore behaves as a single resistance R equal to the sum of the resistances in series: $R = R_1 + R_2$. Through recurrence, the preceding result can easily be generalized to any number of resistances $R_1, \ldots, R_n$ in series. The association will behave like a single resistance equal to:

$$R = \sum_{i=1}^{n} R_i \qquad [3.70]$$

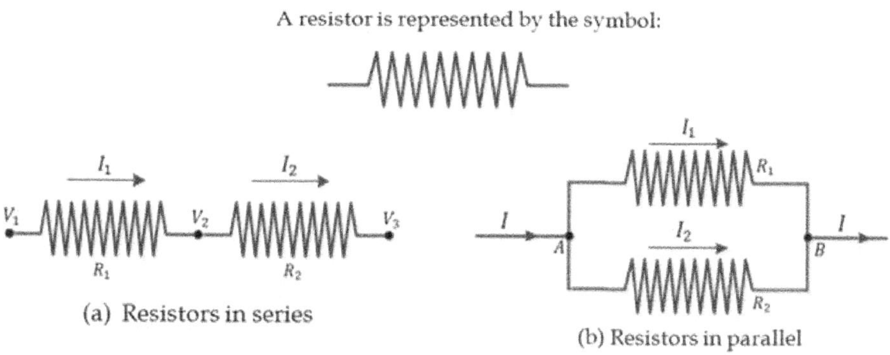

Figure 3.12. *Association of resistances*

It can clearly be seen that the equivalent resistance is larger than the largest of the resistances in the association.

(b) *Parallel association*: if resistors R_1 and R_2 are now arranged in parallel, as in Figure 3.12(b), the current, I, flowing through the combination is equal to the sum of the currents I_1 and I_2 flowing through R_1 and R_2, respectively: $I = I_1 + I_2$. According to Ohm's law, we can write that $V_A - V_B = R_1 I_1 = R_2 I_2$ and, consequently, $I = (V_1 - V_2)\left(\frac{1}{R_1} + \frac{1}{R_2}\right)$. The configuration therefore behaves like a single resistor R equal to $\frac{1}{R} = \frac{1}{R_1} + \frac{1}{R_2}$. This result can be generalized to several resistors, $R_1, \ldots, R_n$, in parallel. The equivalent resistance is equal to:

$$\frac{1}{R} = \sum_{i=1}^{n} \frac{1}{R_i} \qquad [3.71]$$

This relationship shows that the value of R is less than the smallest of the resistances involved in the association.

EXAMPLE 3.7. (Conical Resistor).– A current of intensity I flows through a conductor in the form of a truncated cone of length L, conductivity γ and with the radii of the lower and upper sections a and b, respectively, as shown in Figure 3.13. Consider the disk of thickness dx and radius r located at position x. This disk has an elementary resistance $dR = \frac{dx}{\gamma\pi r^2}$ involved in a series connection. To obtain r as a function of x, it suffices to see that $r = \left(\frac{b-a}{L}\right)x + a$. Therefore, to find the total resistance R, we need to integrate over the whole chain, that is, from 0 and L:

$$R = \int_0^L \frac{dx}{\gamma\pi r^2} = \int_0^L \frac{dx}{\gamma\pi\left[\left(\frac{b-a}{L}\right)x + a\right]^2} = \frac{L}{\gamma\pi ab} \qquad [3.72]$$

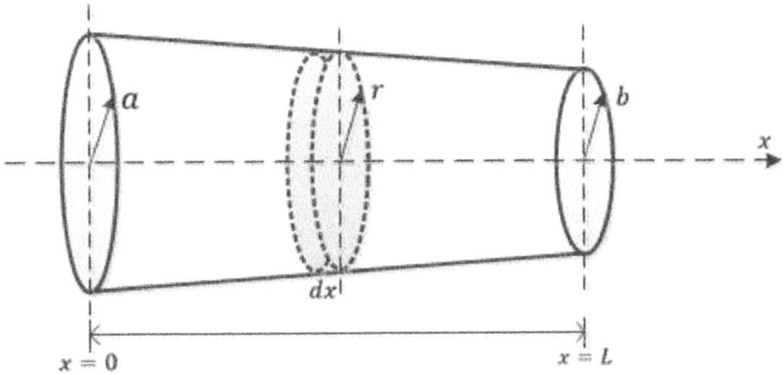

Figure 3.13. *The total resistance of the truncated conical conductor is analogous to elementary resistances* $dR = \frac{dx}{\gamma\pi r^2}$ *connected in series*

3.6.3. *Joule effect*

The *Joule effect* is a thermal effect caused by the displacement of conduction electrons in a conductor. It is therefore a thermal effect that occurs when an electric current flows through a conductor. It is manifested by an increase in the conductor's internal energy and, in general, by a rise in the conductor's temperature. Through frictional forces within the conductor (i.e. electron collisions with the crystal lattice), electrons transmit the electrical energy they have received to the crystal lattice (in calorific form). Each conduction electron of charge $-e$ placed in an electric field $\overrightarrow{\mathbf{E}}$ is subjected to the force $\overrightarrow{\mathbf{F}} = -e\overrightarrow{\mathbf{E}}$. The work done by the electron during an elementary displacement $d\overrightarrow{\mathbf{l}}$ is given by $dW = -e\overrightarrow{\mathbf{E}} \cdot d\overrightarrow{\mathbf{l}} = -e\overrightarrow{\mathbf{E}} \cdot d\overrightarrow{\mathbf{v}}_{-}dt$, where $\overrightarrow{\mathbf{v}}_{-}$ is the velocity of the electrons inside the conductor. In other words, each electron will receive electrical power equal to $-e\vec{\mathbf{E}} \cdot \vec{\mathbf{v}}_-$. If the conductor contains n_- electrons per unit volume, the total power received by all electrons contained in a volume τ will be:

$$\mathsf{P} = -\int_\tau n_- e\vec{\mathbf{E}} \cdot \vec{\mathbf{v}} d\tau = \int_\tau \vec{\mathbf{E}} \cdot \vec{\mathbf{J}} d\tau \qquad [3.73]$$

Since electrons have the same average velocity, their kinetic energy does not vary. The power developed by the electric field is therefore entirely dissipated by the successive collisions the electrons undergo. This power is transmitted throughout the material and converted into heat. We can therefore deduce from [3.73] that the power dissipated per unit volume is equal to $\vec{\mathbf{E}} \cdot \vec{\mathbf{J}}$. Let us now consider an ohmic conductor, through which a current I flows, located between two plane-parallel electrodes whose surface area is S and separated by space L. According to [3.74], the calorific power received by the entire ohmic conductor is:

$$\mathsf{P} = \int_0^L \vec{\mathbf{J}} \cdot \vec{\mathbf{E}} S dl = \frac{J^2 SL}{\gamma} = \frac{I^2 L}{\gamma S} \qquad \Longleftrightarrow \qquad \mathsf{P} = RI^2 = \frac{V^2}{R} = VI \quad [3.74]$$

This is the *Joule's law of heating*. The relationship $sfP = RI^2 = \frac{V^2}{R} = VI$ remains valid for any electrodes.

EXAMPLE 3.8.(Cylindrical and Spherical Resistances).–

(1) Let us calculate the power transmitted to a conductor between two cylindrical electrodes, with inner radius a, outer radius b and length L (Example 3.5):

$$\mathsf{P} = \int_a^b \frac{J_r^2}{\gamma} S dr = \int_a^b \frac{J_r^2}{\gamma} 2\pi r L dr = \frac{\gamma V^2 2\pi L}{\ln^2\left(\frac{b}{a}\right)} \int_a^b \frac{dr}{r} = \frac{\gamma V^2 2\pi L}{\ln\left(\frac{b}{a}\right)} = \frac{V^2}{R} \qquad [3.75]$$

(2) Similarly, for a conductor confined between two spherical electrodes of inner radius R_1 and outer radius R_2, we have (Example 3.6):

$$\begin{aligned}\mathsf{P} &= \int_{R_1}^{R_2} \frac{J_r^2}{\gamma} S dr = \int_{R_1}^{R_2} \frac{J_r^2}{\gamma} 4\pi r^2 dr = \frac{4\pi\gamma R_1^2 R_2^2 V^2}{(R_2 - R_1)^2} \int_{R_1}^{R_2} \frac{dr}{r^2} \\ &= \frac{4\pi\gamma R_1 R_2 V^2}{(R_2 - R_1)} = \frac{V^2}{R}\end{aligned} \qquad [3.76]$$

3.7. Direct current circuits

Before studying the behavior of electric circuits, let us first review a few useful definitions:

1) *Circuit element*: a *circuit element* (such as resistor, inductor, capacitor, voltage source and current source) is an elementary part of an electric circuit that is connected to other elements by wires of negligible resistance, called *electrodes*, as shown in Figure 3.14(a).

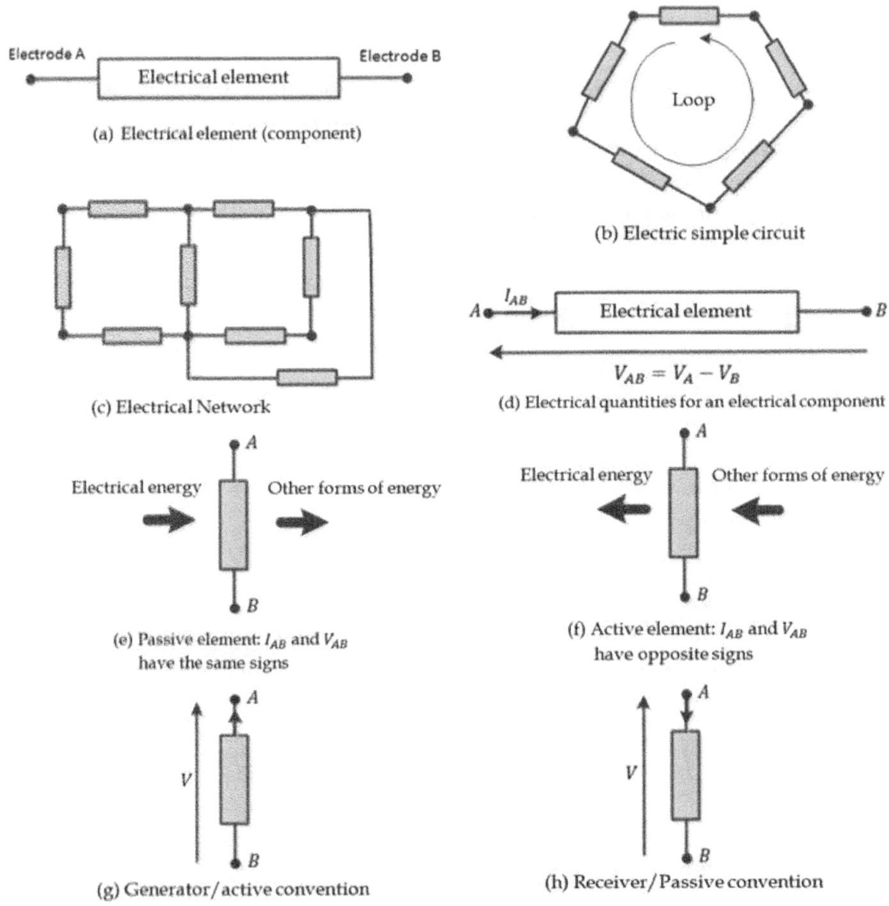

Figure 3.14. *Circuit elements in a simple or complex electric circuit in passive or active mode*

2) *Electric circuit*: an *electric circuit* is an interconnection of various elements in which there is at least one closed path in which current can flow. A circuit is said to be *simple* if all the two-terminal elements are mounted in series (Figure 3.14(b)). It is called a *network* if it is more complex (Figure 3.14(c)).

3) *Node*: a *node* is a point of interconnection that is connected to at least three circuit elements.

4) *Branch*: a *branch* is a circuit section between two nodes.

5) *Mesh*: a *mesh* uses a set if branches that form a closed loop that passes through a given node at least once.

6) In a permanent regime, electrical currents and potentials are independent of time. Charges cannot accumulate: the current flowing into and out of a circuit element is the same. In other words, the current is the same at every point in the circuit.

7) If A and B are the two terminals of a circuit element, I_{AB} is the current traveling through the circuit element from terminal A toward B. This is represented by an arrow, as shown in Figure 3.14(d). I_{AB} is an algebraic value that can be positive or negative, depending on whether the current is indeed traveling in this direction or is traveling in the opposite direction: $I_{AB} = -I_{BA}$. The *voltage* at the terminals of the circuit element is the potential difference $V_{AB} = V_A - V_B$. This is also represented by an arrow (Figure 3.14(d)). If $V_{AB} > 0$ (respectively, $V_{AB} < 0$), then $V_A > V_B$ (respectively, $V_A < V_B$). We clearly have $V_{AB} = -V_{BA}$. The electric state of the circuit element is characterized by the voltage at its terminals, V_{AB}, and the current I_{AB} traveling through it.

8) *Electric power*: an electric current passing through a circuit element is accompanied by an energy exchange with the rest of the circuit. When it passes through a circuit element, an electron receives the power $-e\vec{\mathbf{v}}_- \cdot \vec{\mathbf{E}}$ from the electric force. The power supplied to all the charges contained in the element is $\mathsf{P} = \left(-n_- e\vec{\mathbf{v}}_- \cdot \vec{\mathbf{E}}\right)(SL) = I_{AB}V_{AB}$. This energy is transferred to the circuit element. P is called *electrokinetic power*: this is the power supplied to the circuit element by the electric current.

9) *Passive circuit element*: in a permanent regime, energy cannot accumulate. The power P received by the circuit element must therefore be removed in another form. If $\mathsf{P} > 0$, the circuit element receives energy in electric form and transforms it into mechanical energy (motor), calorific form (heating, etc.), chemical form, radiation, etc. In this case, the circuit element is said to be a *receiver* of electric energy. If $\mathsf{P} > 0$ $\Longrightarrow I_{AB}$ and V_{AB} have the same sign (Figure 3.14(e)). If $V_A > V_B$ (respectively, $V_A < V_B$) $\Longrightarrow V_{AB} > 0$ (respectively, $V_{AB} < 0$) $\Longrightarrow I_{AB} > 0$ (respectively, $I_{AB} < 0$): the current travels within the circuit element from the highest potential toward the lowest potential.

10) *Active circuit element*: if $\mathsf{P} < 0$, it means that the circuit element is supplying the electrical energy. In this case, it is called a *generator*. This energy was supplied to the circuit element itself in some other form (mechanical, chemical, etc.). V_{AB} and I_{AB} have opposite signs: I travels inside the circuit element in the direction of decreasing potentials, as shown in Figure 3.14(f).

11) *Active and passive convention/mode*: when studying electrical circuits, it is best to choose directional conventions for voltages and currents that correspond to reality, so as to have positive quantities. Unfortunately, it is usually difficult to know the positive direction of the current beforehand. We say that we have adopted a passive or active convention based on the direction of the arrows of I and V adopted for a circuit element. Since current flows in a generator in the direction of increasing potential, the voltage arrow and the current arrow will point in the same direction. We will then say we have adopted the generator mode, as shown in Figure 3.14(g). The electrokinetic power received is negative, that is, $\mathrm{P} = -VI$. In a receiver, the current flows through the circuit element in the direction of decreasing potential, that is, V and I will have opposite directions: $\mathrm{P} = VI$. So, if we take arrows of V and I in opposite directions, we will say that we have adopted a passive mode, as in Figure 3.14(h).

3.7.1. *Characteristics of a circuit element*

In a circuit element, the current I and voltage V are related by the reciprocal relationships $V = f(I)$ and $I = g(V)$. The associated graphs in the (V, I) and (I, V) planes are the characteristics of the circuit element. In some cases, these characteristics are nonlinear. Let us look at a few examples:

– *Resistor*: the characteristic of a resistor is linear, as shown in Figure 3.15. Indeed, voltage V and current I (in passive mode) are related by Ohm's law $V = RI \Longleftrightarrow I = GV$. The electrokinetic power is $\mathrm{P} = VI = RI^2 = \frac{V^2}{R}$. We can see that sfP is always positive, that is, the resistor can only receive electrical energy. This energy is dissipated in calorific form: the Joule effect.

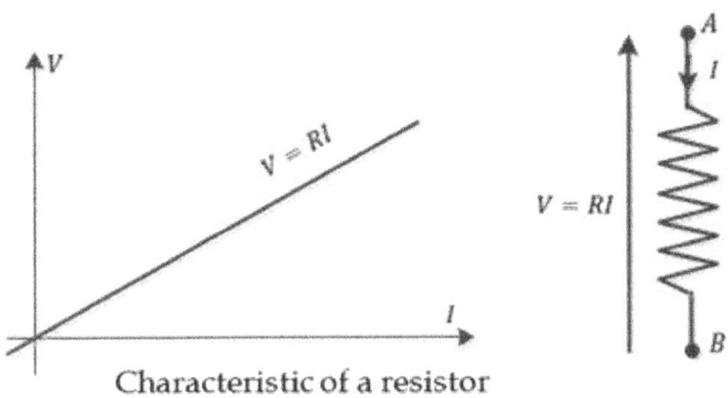

Figure 3.15. *Through Ohm's law, the voltage across a resistor is linearly to the current passing through it*

– *Real voltage generator/Thevenin generator, Pouillet's law*: to operate a closed circuit containing, for example, a resistor R, as shown in Figure 3.16(a), we must necessarily place, between the two points A and B, an electromotive system whose function is to maintain the potential imbalance in order to make a current flow through the circuit. Indeed, if A and B are connected by a conductive wire only, the two terminals will be at the same potential and, consequently, no current will be able to flow through the resistor. *The electromotive force* (EMF) $\mathcal{E}$ of the voltage generator constituted by the electromotor is defined from the power P_G it develops as the ratio:

$$\mathcal{E} = \frac{\mathsf{P}_G}{I} \quad [3.77]$$

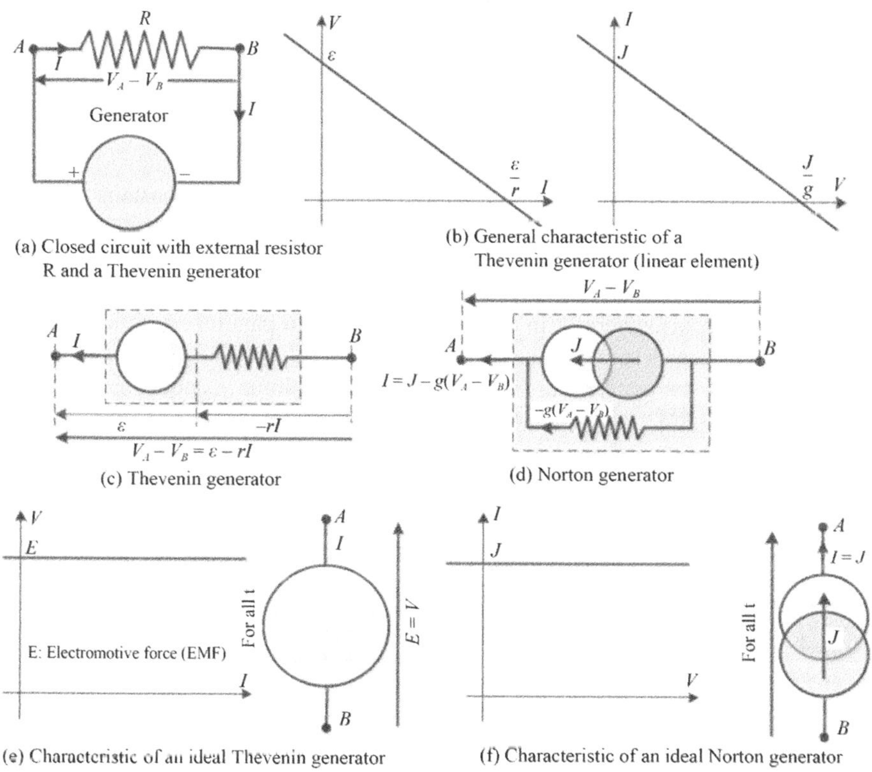

Figure 3.16. *Characteristics of Thevenin and Norton generators (real or ideal)*

The power P_G is used, on the one hand, to provide the power $I\,(V_A - V_B)$ to circulate the charges from B to A ($V_A > V_B$) and is dissipated, on the other hand, in part by a possible Joule effect with power rI^2 between B and A, where r is the internal resistance of the generator. The balance equation is therefore written as:

$$\mathrm{P}_G = \mathcal{E}I = I\left(V_A - V_B\right) + rI^2 \implies V_A - V_B = \mathcal{E} - rI \iff I = J - g\left(V_A - V_B\right) \quad [3.78]$$

This is *Pouillet's law*. It gives the potential difference across a voltage generator as a function of its EMF and internal resistance. Since $V_A > V_B$, we will say that A (respectively, B) is the *positive* (respectively, *negative*) *terminal* of the generator. The characteristic of a generator is shown in Figure 3.16(b). If the internal resistance is zero, it is an *ideal voltage generator/Thevenin generator*. An ideal voltage/Thevenin generator is therefore an electrical element whose voltage across its terminals remains constant, regardless of the current supplied to the rest of the circuit. The characteristic of an ideal generator is a straight line parallel to the current axis, as shown in Figure 3.16(e). Similarly, if the internal conductance g is zero, we will call it an *ideal current/Norton generator*. An ideal current/Norton generator is therefore an electrical element through which a current flows that remains constant whatever the voltage across its terminals. The characteristic is a straight line parallel to the voltage axis, as shown in Figure 3.16(f). A real generator is modeled in the Thevenin (respectively, Norton) representation by the association of an ideal voltage (respectively, current) generator in series (respectively, in parallel), with a resistor r, as shown in Figure 3.16(c) (respectively, Figure 3.16(d)). The form of equation [3.78] forces us to generalize Ohm's law. In an ohmic conductor, we have $\vec{\mathbf{J}} = -\gamma\vec{\mathbf{E}} = -\gamma\vec{\nabla}V$. Clearly, this relationship does not reproduce equation [3.78] by integrating over the whole element. To do this, we need to take into account in the expression of the local Ohm's law, the other causes that allow charges to move (chemical reactions, for example, in batteries, etc.). All these terms can then be combined to form an electromotive field $\vec{\mathbf{f}}_m$. Ohm's law then takes the form:

$$\vec{\mathbf{J}} = \gamma\left(-\vec{\nabla}V + \vec{\mathbf{f}}_m\right) \iff -\vec{\nabla}V = \frac{1}{\gamma}\vec{\mathbf{J}} - \vec{\mathbf{f}}_m \quad [3.79]$$

This is *generalized local Ohm's law*. After integration between the two terminals of the electrical element, we obtain Pouillet's law:

$$V_1 - V_2 = RI - \mathcal{E} \quad [3.80]$$

where $\mathcal{E} = \int_1^2 \vec{\mathbf{f}}_m \cdot d\vec{\mathbf{l}}$ is the *electromotive force*. The expression [3.80] is written in the passive mode;

– *Other passive electrical element*: in general, a passive element, with a voltage V imposed on its terminals, receives power VI, where I is the current flowing through it.

For some passive elements, part rI^2 of this power, where r is the element's internal resistance, is dissipated by the Joule effect, while the remaining part $\mathsf{P}_R = VI - rI^2$ is transformed into mechanical power in the case of a motor, or stored as chemical energy in the case of an electrolyzer. In these cases, the *back-electromotive force* $\mathcal{E}'$ of the receiver is defined from the power P_R actually available to perform its function by the relationship:

$$\mathcal{E}' = \frac{\mathsf{P}_G}{I} \qquad [3.81]$$

Consequently, for these passive elements, we obtain a relationship analogous to Pouillet's law:

$$V = \mathcal{E}' + rI \qquad [3.82]$$

In particular, if the electrical element is an opposing generator of EMF $\mathcal{E}$, we simply have $\mathcal{E}' = -\mathcal{E}$. Just note that the motor characteristic is not rectilinear, as $\mathcal{E}'$ depends on I.

3.7.2. *Association of electrical elements*

There are two types of association. Electrical elements can be connected in series, that is, they all carry the same current, as shown in Figure 3.17(a). They can also be connected in parallel, that is, they are all subject to the same potential difference, as shown in Figure 3.17(b).

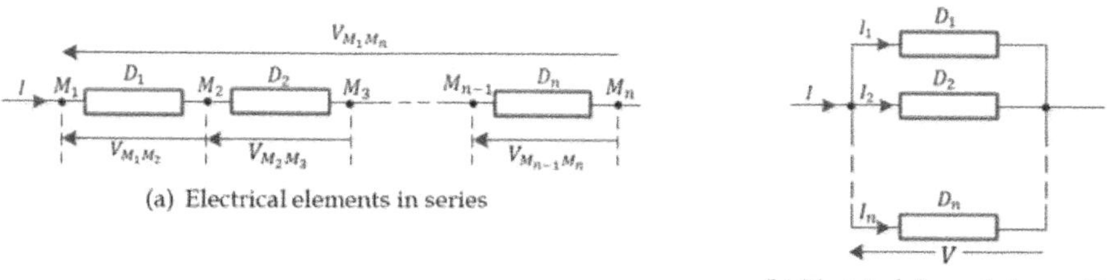

(a) Electrical elements in series

(b) Electrical elements in parallel

Figure 3.17. *Different electrical element associations*

– *Association in series*: the voltage across the equivalent element is equal to the sum of the partial voltages:

$$V_{M_1M_n} = \sum_{k=1}^{n-1} V_{M_kM_{k+1}} \qquad [3.83]$$

– *Association in parallel*: the total current flowing through all of the elements is equal to the sum of the individual currents:

$$I = \sum_{k=1}^{n} I_k \qquad [3.84]$$

3.7.3. Solve DC circuits problems

Solving an electrical network involves determining the currents passing through the various branches of the network based on the characteristics of the elements making up the network. Here are the main results:

1) *Kirchhoff's laws*: Kirchhoff's laws are of two types. The first concerns nodes, while the second applies to meshes. However, before writing these equations, the following rules must be observed: (1) fix the direction of current in each branch of the network (according to the elements involved). Current intensities are to be considered as algebraic quantities. If the current I is positive (respectively, negative), then the physical current is in the indicated direction (respectively, in the opposite direction). (2) Set a direction of travel for each mesh in the network, independently of the choice of current directions in the branches.

a) *Law of nodes*: this law expresses the conservation of charge flow at a node, that is, the fact that there can be no accumulation of charges at a node. For a node with N branches, the law of nodes is written as follows:

$$\sum_{k=1}^{N} \epsilon_k I_k = 0 \qquad [3.85]$$

with $epsilon_k = +1$ (respectively, –1) if the current arrives at the node (respectively, leaves the node). In other words, the algebraic sum of currents is zero at any node of a network.

b) *Mesh law*: this law is based on the property of continuity of electric potential and states that along any mesh of an electrical network, the algebraic sum of voltages is zero. In other words, on a mesh, that is, a portion of circuit whose ends M_1 and M_n are identical, we write that:

$$\sum_{k=1}^{n-1} U_{M_k M_{k+1}} = 0 \qquad [3.86]$$

2) *Simple linear circuits*: in continuous regime, linear electrical circuits comprise only resistors, voltage sources and current sources. The electric current is the same everywhere in a simple circuit. If, in addition, the circuit includes a current source J, the current intensity is fixed in the circuit. Note that it is not possible to connect two current sources in series, unless they are identical. If the circuit has no current source, the value of the current I is calculated by applying Kirchhoff's law to the mesh making up the circuit. For a simple circuit comprising, for example, several voltage sources $E_1, \ldots, E_m$, connected in the same direction, and resistors $R_1, \ldots, R_n$, the current intensity in the circuit is:

$$I = \frac{\sum_{i=1}^{m} E_i}{\sum_{i=1}^{n} R_i} \qquad [3.87]$$

EXAMPLE 3.9.– A motor with back electromotive force $calE'$ and internal resistance r is connected across a voltage source E. Applying the mesh law to this simple circuit, we obtain $E = \mathcal{E}' + rI$ and, therefore, $\mathcal{E}' = E - rI$. Consequently, the power dissipated by the motor (in mechanical form) is $P = \mathcal{E}'I = EI - rI^2$. This power is highest for $I_c = \frac{E}{2r}$. This is because $\frac{dP}{dI}$ cancels out at this value. The power is therefore maximum for this current and equal to $P_{\max} = \frac{E^2}{4r}$. For the motor's internal resistance, the power dissipated by the Joule effect for this current intensity is also $P_J = rI_c^2 = \frac{E^2}{4r} = P_{\max}$. We can see that the resistor dissipates as much power as the motor. ■

3) *Complex circuit analysis*: first, let us consider the case where the circuit contains only voltage sources and resistors, and no current sources. If the electrical circuit contains more than one mesh, and we are looking for the currents in each of the branches, Ohm's law is insufficient. We therefore need to use another method, known as *Kirchhoff's method*, which calculates the current intensity in each branch using the node and mesh laws.

EXAMPLE 3.10.– To illustrate Kirchhoff's method, let us study the circuit shown in Figure 3.18. The circuit contains two nodes (A and B) and three branches (i.e. three unknown currents). If we count positively one current entering the node, the law of nodes indicates that $I_1 - I_2 - I_3 = 0$. This leaves two unknown currents, and the third is obtained from the other two. Similarly, the circuit has three meshes: (1) (B, E_1, R_1, A, R_3, B); (2) (B, R_3, A, R_2, E_2, B); (3) $(B, E_1, R_1, A, R_2, E_2, B)$. The law of meshes states that the algebraic sum of voltages encountered in a mesh is zero. Consider the mesh (1). Initially, the direction of the electromotive forces is known, but not that of the currents. To do this, let us (a) indicate in the circuit the direction of the currents in the various branches and the arrows of the voltages across the resistors (in passive mode); (b) then arbitrarily choose a starting point for the mesh (here point B) and a direction of travel (for example, clockwise); and (c) assign the sign + to voltages whose arrows point in the direction of travel, and the sign – to voltages whose arrows point in the opposite direction. The law of meshes gives for the first mesh:

$$E_1 - V_1 - V_3 = 0 \quad \Longleftrightarrow \quad E_1 - R_1 I_1 - R_3 I_3 = 0 \qquad [3.88]$$

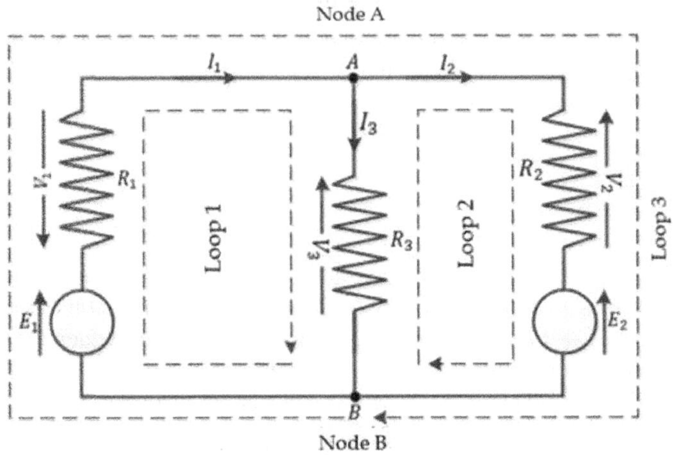

Figure 3.18. *Circuit with two nodes ($n = 2$) and three branches ($b = 3$): this circuit admits two independent currents, which can be determined from two independent meshes*

If the calculation gives a positive value for the current, we must conclude that the chosen direction is correct. Otherwise, the actual current is in the opposite direction to the one chosen at the start. Let us repeat the same procedure for the (3) mesh, that is, $(B, E_1, R_1, A, R_2, E_2, B)$.

In the same way, by choosing B as the point of departure from the mesh, with the direction of travel clockwise, and by assigning the sign $+$ to voltages whose arrows go in the direction of travel, and the sign $-$ to voltages whose arrows go in the opposite direction, the law of meshes allows us to write:

$$E_1 - V_1 - V_2 - E_2 = 0 \quad \Longleftrightarrow \quad E_1 - R_1 I_1 - R_2 I_2 - E_2 = 0 \qquad [3.89]$$

For mesh (2), we also get:

$$V_3 - V_2 - E_2 = 0 \quad \Longleftrightarrow \quad R_3 I_3 - R_2 I_2 - E_2 = 0 \qquad [3.90]$$

These equations are not independent. Equation [3.89] is simply the sum of the other two, that is, [3.80] and [3.90]. To find the current intensities, we need only two equations:

$$\begin{aligned} & E_1 - (R_1 + R_3)\, I_1 - R_3 I_2 = 0 \\ & E_1 - R_1 I_1 - R_2 I_2 - E_2 = 0 \end{aligned} \quad [3.91]$$

The solutions are:

$$\begin{aligned} I_1 &= \frac{R_2 E_1 - R_3 (E_1 - E_3)}{R_1 R_2 + R_2 R_3 - R_3 R_1} \\ I_2 &= \frac{-R_1 E_2 + R_3 (E_1 - E_3)}{R_1 R_2 + R_2 R_3 - R_3 R_1} \\ I_3 &= \frac{R_2 E_1 + R_1 E_2 - 2R_3 (E_1 - E_3)}{R_1 R_2 + R_2 R_3 - R_3 R_1} \end{aligned} \quad [3.92]$$

■

Let us give the method for the most general case of a circuit containing n nodes and b branches. The problem is to find the current intensities in the b branches. Applying Kirchhoff's law to the nodes gives n equations that are not independent. Eliminating one equation yields $(n-1)$ independent equations. So, from the b branch currents and these $(n-1)$ equations, we are left with $k = b - (n-1)$ unknowns. To find these unknowns, we need to search for $k = b - (n-1)$ independent meshes, to which we apply Kirchhoff's law to the meshes. The result is a system containing as many equations as unknowns. The node equations are used progressively to express the various equations of the system solely in terms of the intensities $I_1, \ldots, I_k$. For each independent mesh, i, we obtain an equation of the form $A_{i1}I_1 + A_{i2}I_2 + \cdots + A_{ik}I_k = B_i$, where A_{ij} are homogeneous to resistances and B_i are homogeneous to voltages. Together, these k independent equations form a matrix system of the form:

$$(R_{ij})\,(I_j) = (E_i) \quad [3.93]$$

whose resolution gives us the unique solution, that is, the value of all currents. Now, let us look at what happens when one of the circuit's branches contains a current source. In this case, the value of the current in this branch is known, that is, one less unknown on one side. However, on the other hand, the voltage across the current source is unknown. So, we also have one less equation. The system remains coherent. By writing Kirchhoff's laws for independent meshes (not including the branch containing the current source), we obtain a more general matrix system of the form:

$$(R_{ij})\,(I_j) = (E_j) + \left(R'_{ij}\right)(J_j) \quad [3.94]$$

EXAMPLE 3.11.– Let us determine the current intensities in the circuit shown in Figure 3.19. There are four nodes and six branches, which implies that

$k = 6 - 3 = 3$ are unknowns. However, since the current is fixed in one branch because of the current source, there will only be two unknowns, I_1 and I_2, as shown in Figure 3.19. We therefore need to find two independent meshes that do not contain the current source. Applying Kirchhoff's law to meshes 1 and 2, we obtain:

$$\begin{aligned} &E_1 - R_1 I_1 - R_2 I_2 = 0 \\ &E_2 - R_3 (I_1 - I_2 + J) + R_2 I_2 = 0 \end{aligned} \quad [3.95]$$

or, in matrix form:

$$\begin{pmatrix} R_1 & R_2 \\ R_3 & -(R_2 + R_3) \end{pmatrix} \begin{pmatrix} I_1 \\ I_2 \end{pmatrix} = \begin{pmatrix} E_1 \\ E_2 \end{pmatrix} - \begin{pmatrix} 0 \\ R_3 J \end{pmatrix} \quad [3.96]$$

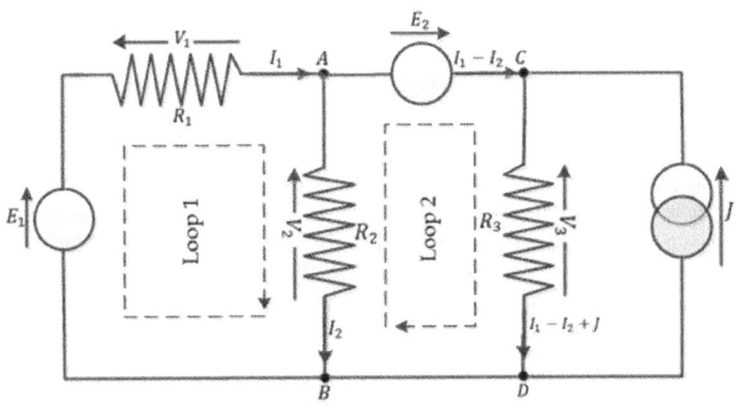

Figure 3.19. *Four-node, six-branch circuit with a current source in one branch*

The solutions are as follows:

$$\begin{aligned} I_1 &= \frac{E_1 (R_2 + R_3) + E_2 R_2 - R_2 R_3 J}{R_1 R_2 + R_2 R_3 + R_3 R_1} \\ I_2 &= \frac{R_3 E_1 - R_1 E_2 + R_1 R_3 J}{R_1 R_2 + R_2 R_3 + R_3 R_1} \end{aligned} \quad [3.97]$$

4) *Superposition theorem*: we have just seen that solving a complex linear circuit reduces us to a system of linear equations relating the currents to the voltage and/or current sources present in the network. Due to linearity, we can state that the current passing in a branch of a circuit comprising voltage and/or current sources is the sum of the currents calculated by successively canceling all but one of the sources: this is the *superposition theorem*. Note that a zero-voltage (respectively, zero-current) source is

a short-circuit (respectively, open circuit). In other words, the superposition theorem states that in a linear circuit with several sources, the current (respectively, voltage) for any element of the circuit is the sum of the currents (respectively, voltages) produced by each source acting independently. To use the superposition theorem, we need to ensure that all circuit elements are linear. Note that the superposition theorem is not applicable to power, as it is not a linear quantity.

EXAMPLE 3.12.– Let us return to the circuit shown in Figure 3.19 and apply the superposition theorem.

a) By canceling all sources except E_1, we obtain the reduced circuit shown in Figure 3.20(a). The current flowing in R_1 is:

$$I_1^{(a)} = \frac{E_1}{R_1 + \frac{R_2 R_3}{R_2 + R_3}} = \frac{E_1 (R_2 + R_3)}{R_1 R_2 + R_2 R_3 + R_3 R_1} \quad [3.98]$$

To obtain the current in R_2, we must use the fact that $R_2 I_2^{(a)} = R_3 \left(I_1^{(a)} - I_2^{(a)} \right)$, which gives:

$$I_2^{(a)} = \frac{E_1 R_3}{R_1 R_2 + R_2 R_3 + R_3 R_1} \quad [3.99]$$

b) By canceling all sources except E_2, the circuit is reduced to that shown in Figure 3.20(b) and, consequently, the current flowing in R_3 is:

$$I_1^{(b)} - I_2^{(b)} = \frac{E_2}{R_3 + \frac{R_1 R_2}{R_1 + R_2}} = \frac{E_2 (R_1 + R_2)}{R_1 R_2 + R_2 R_3 + R_3 R_1} \quad [3.100]$$

where the intensities $I_1^{(b)}$ and $I_2^{(b)}$ are related by the constraint $R_1 I_1^{(b)} = -R_2 I_2^{(b)}$. By simultaneously solving the equations obtained, we find:

$$I_1^{(b)} = \frac{E_2 R_2}{R_1 R_2 + R_2 R_3 + R_3 R_1} \qquad I_2^{(b)} = -\frac{E_3 R_1}{R_1 R_2 + R_2 R_3 + R_3 R_1} \quad [3.101]$$

c) Keeping only the current source J, the circuit reduces to that shown in Figure 3.20(c). The currents $I_1^{(c)}$ and $I_2^{(c)}$ are connected by the relations

$-R_1 I_1^{(c)} = -R_2 I_2^{(c)} = R_3 \left(I_1^{(c)} - I_2^{(c)} + J \right)$. It therefore follows that:

$$I_1^{(c)} = -\frac{J R_2 R_3}{R_1 R_2 + R_2 R_3 + R_3 R_1} \qquad I_2^{(b)} = \frac{J R_1 R_3}{R_1 R_2 + R_2 R_3 + R_3 R_1} \qquad [3.102]$$

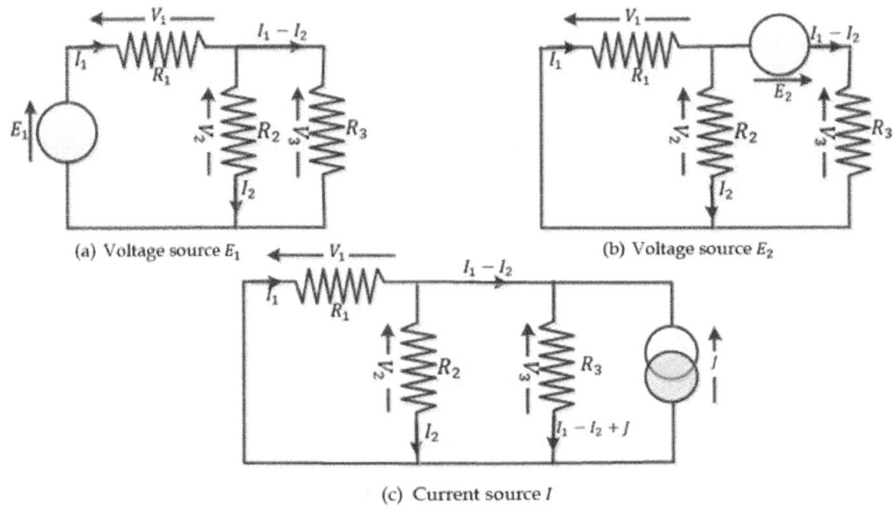

Figure 3.20. *Applying the superposition theorem to the circuit shown in Figure 3.19 gives the same current intensities already obtained*

According to the superposition theorem, we have a total of:

$$\begin{aligned} I_1 &= I_1^{(a)} + I_1^{(b)} + I_1^{(c)} = \frac{E_1 (R_2 + R_3) + E_2 R_2 - R_2 R_3 J}{R_1 R_2 + R_2 R_3 + R_3 R_1} \\ I_2 &= I_2^{(a)} + I_2^{(b)} + I_2^{(c)} = \frac{R_3 E_1 - R_1 E_2 + R_1 R_3 J}{R_1 R_2 + R_2 R_3 + R_3 R_1} \end{aligned} \qquad [3.103]$$

These results are in agreement with those already obtained.

5) *Thevenin and Norton circuits*: a *Thevenin* (respectively, *Norton*) *circuit* is made up of a voltage (respectively, current) source E (respectively, J) in series (respectively, in parallel) with a resistor. Together, they form a linear electrical element. When we connect a Thevenin (respectively, Norton) circuit to another branch, a current I flows through it, and the voltage across it is equal to $V = E - RI$ (respectively, $V = RJ - RI$). These latter relationships are referred to as *Thevenin and Norton element characteristics*.

6) *Thevenin's theorem*: this theorem allows us to replace part of a network with a real voltage generator (a voltage source and a resistor in series) that is electrically equivalent to it, so as to simplify calculations. In other words, if we add a new branch between two nodes A and B of a network, the same current will flow through it as if we were replacing the network with a real voltage generator: (1) whose EMF would be equal to the no-load voltage between the two nodes; (2) whose internal resistance would be the resistance of the network as seen from these nodes, canceling out all sources.

EXAMPLE 3.13.– To calculate the current through resistor R_3 in the network shown in Figure 3.21(a), we can proceed in two steps: (1) calculate the characteristics of the Thevenin generator equivalent to the network as seen from A and B; (2) replace the network with the Thevenin generator to calculate the current through resistor R_3. To do this, let us calculate the no-load voltage, as shown in Figure 3.21(b). The current in the circuit shown in Figure 3.21(b) is $I = \frac{E_1 - E_2}{R_1 + R_2}$ and, consequently, $V_A - V_B = E = E_1 - R_1 I = \frac{E_1 R_2 + E_2 R_1}{R_1 + R_2}$. To calculate the resistance of the equivalent Thevenin circuit, we need to cancel the sources, that is, reduce the circuit to that shown in Figure 3.21(c). Seen from nodes A and B, the network consists of two branches in parallel; this gives a resistance equal to $R_{AB} = \frac{R_1 R_2}{R_1 + R_2}$. To find the current flowing through resistor R_3, we replace the circuit between A and B with its equivalent circuit, as shown in Figure 3.21(d). The current is then:

$$I = \frac{E}{R_3 + R_{AB}} = \frac{R_2 E_1 + R_1 E_2}{R_1 R_2 + R_2 R_3 + R_3 R_1} \qquad [3.104]$$

7) *Norton's theorem*: in the same way, this theorem also allows us to substitute a real current generator (a current source and a resistor in parallel) for part of a network, which is electrically equivalent to it. Indeed, the current flowing in a given branch between A and B is regarded as that produced by a real current generator, that is, $I = J - \frac{V_{AB}}{R_{AB}}$, where J is equal to the current flowing through a short-circuit placed between the two nodes and R_{AB} is the resistance of the network seen from these terminals, canceling out all sources.

EXAMPLE 3.14. Let us return to circuit in Figure 3.22(a) and apply Norton's theorem to calculate the current through R_3. To calculate J, we need to replace R_3 with a short circuit, as shown in Figure 3.22(b). Since nodes A and B are at the same potential, we can write that $0 = V_A - V_B = E_1 - R_1 I_1 = E_2 - R_2 I_2$; this gives $I_1 = \frac{E_1}{R_1}$ and $I_2 = \frac{E_2}{R_2}$ and, therefore, $J = I_1 + I_2 = \frac{E_1 R_2 + E_2 R_1}{R_1 R_2}$. canceling all sources, the resistance seen between A and B is $R_{AB} = \frac{R_1 R_2}{R_1 + R_2}$, as shown in Figure 3.22(c). By replacing the network between A and B with its equivalent circuit, as shown in Figure 3.22(d), we can finally obtain the current in R_3 by writing that $R_3 I_3 = R_{AB} I =$

$R_{AB}(J - I_3)$. Thus, we find:

$$I_3 = \frac{R_{AB}J}{R_3 + R_{AB}} = \frac{R_2E_1 + R_1E_2}{R_1R_2 + R_2R_3 + R_3R_1} \qquad [3.105]$$

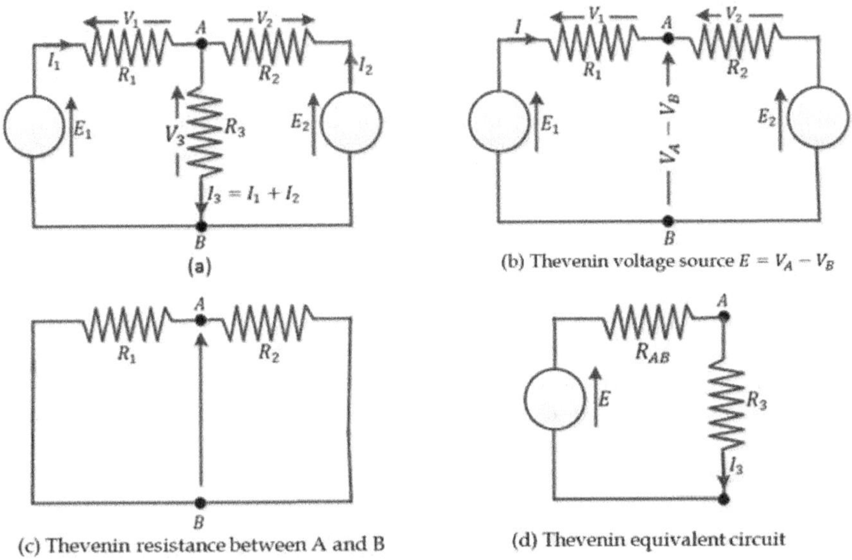

Figure 3.21. *Application of Thevenin's theorem to an electric circuit*

8) *Nonlinear electrical element*: if the circuit contains a nonlinear electrical element, Pouillet's law does not apply. This can be, for example resolved graphically. If a nonlinear element is connected in series with another linear circuit, as shown in Figure 3.23(a), the circuit imposes a linear relationship of the form $V = E - RI$ on the outside of the element. On the other hand, inside the element, there is another nonlinear characteristic to be respected. The relationships must be verified simultaneously. The operating values I and V of the circuit correspond to the point of intersection of the two internal and external characteristics, as shown in Figure 3.23(b).

9) *Charge and discharge of a capacitor*: we will now add capacitors to the circuits. To understand how capacitors work in electrical circuits, let us start with a simple circuit comprising a voltage generator E, a resistor R and a capacitor C, as shown in Figure 3.24(a). When the plates of a capacitor are brought to a potential difference V, we have already seen that the capacitor will gradually charge until it reaches its maximum charge $Q = C|V|$: this is the *transient charging phase* or

transient charging regime. During this transient phase, the load Q will be a function of time, which must be determined. Let us write the law of meshes for a circuit shown in Figure 3.24(a):

$$E - \frac{Q(t)}{C} - RI(t) = 0 \tag{3.106}$$

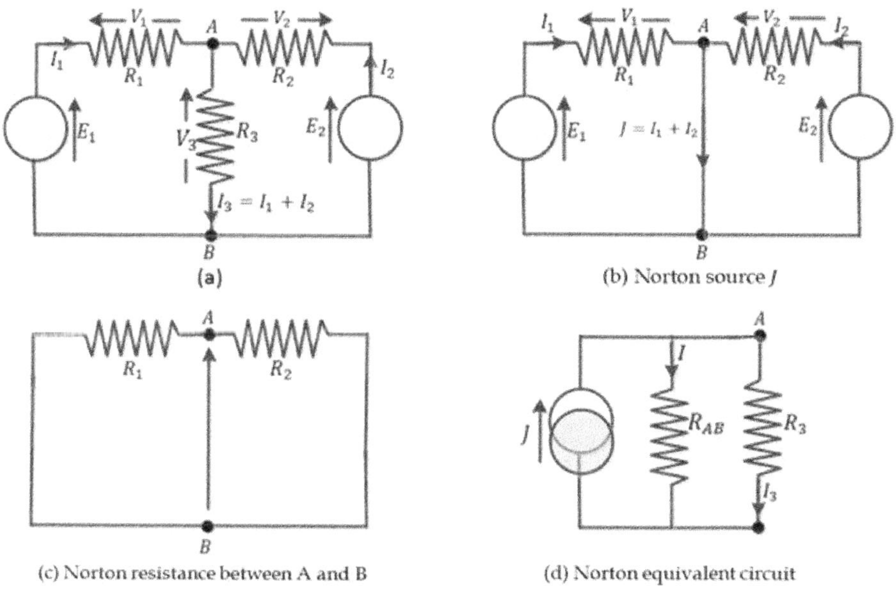

Figure 3.22. *Application of Norton's theorem to an electric circuit*

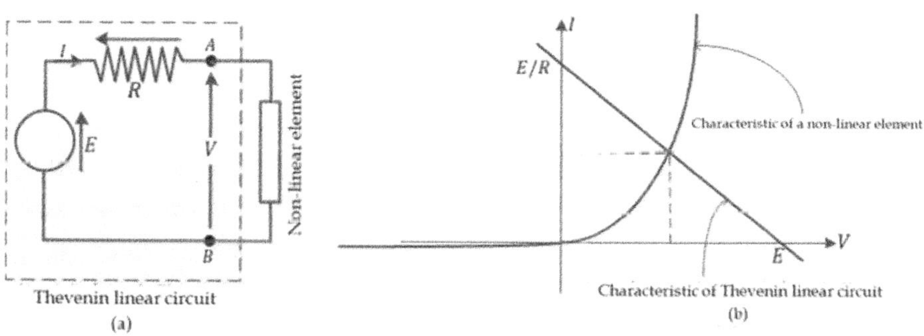

Figure 3.23. *Operating point of a nonlinear element connected in series with a linear Thevenin circuit: this point corresponds to the intersection of the characteristic of the nonlinear element with the linear Thevenin characteristic*

Current intensity is also a function of time, since $I = \frac{dQ}{dt}$. This leads to the differential equation $\frac{dQ}{dt} + \frac{Q}{RC} = \frac{E}{R}$, whose solution is:

$$Q(t) = EC\left(1 - e^{-\frac{t}{RC}}\right) \qquad [3.107]$$

When fully charged, the capacitor charge changes to $Q_{\max} = EC$, as shown in Figure 3.24(b). $\tau = RC$ is called the *characteristic* or *relaxation time*. The electric current is then given as:

$$I(t) = \frac{E}{R}e^{-\frac{t}{RC}} \qquad [3.108]$$

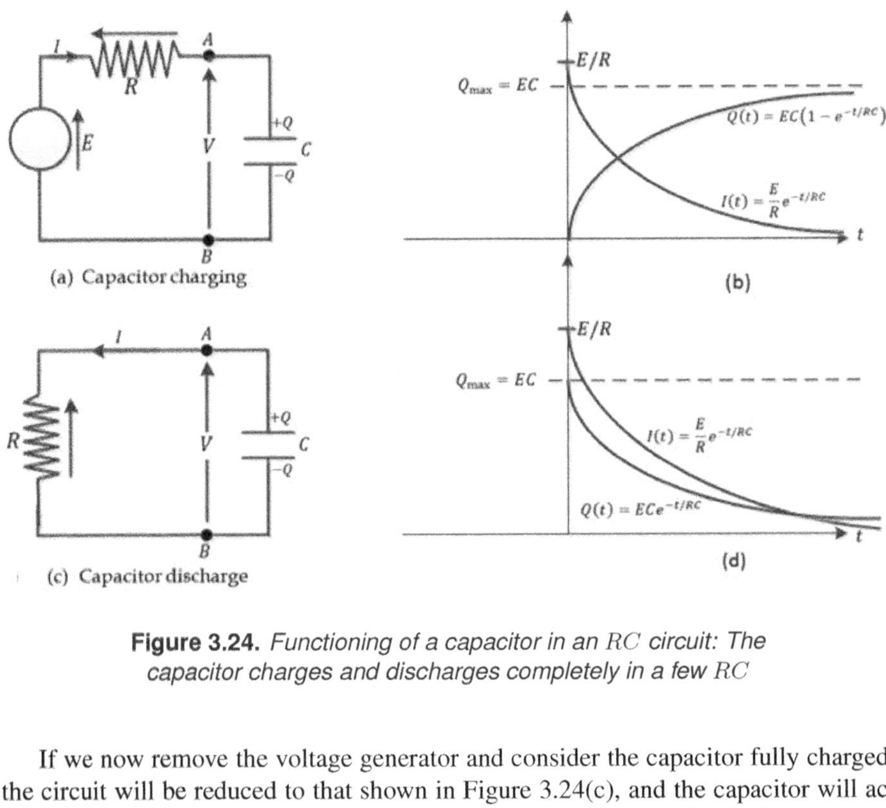

Figure 3.24. *Functioning of a capacitor in an RC circuit: The capacitor charges and discharges completely in a few RC*

If we now remove the voltage generator and consider the capacitor fully charged, the circuit will be reduced to that shown in Figure 3.24(c), and the capacitor will act as a generator, allowing its charge to flow toward the resistor: this is the *capacitor discharge phase*. During this time, the capacitor's charge will decrease and, consequently, the current intensity will be $I = -\frac{dQ}{dt}$ (positive definite). Applying the law of meshes, we find $\frac{Q}{C} - RI = R\frac{dQ}{dt} + \frac{Q}{C} = 0$. The solution to this last differential equation is:

$$Q(t) = Q_{\max} e^{-\frac{t}{RC}} = ECe^{-\frac{t}{RC}} \quad [3.109]$$

as shown in Figure 3.24(d). In this case, the current intensity is given as:

$$I(t) = -\frac{dQ(t)}{dt} = \frac{E}{R}e^{-\frac{t}{RC}} \quad [3.110]$$

3.8. Relaxation time

For a medium of conductivity γ, local Ohm's law gives the value of the current density as a function of the electric field present, that is, $\overrightarrow{\mathbf{J}} = \gamma \overrightarrow{\mathbf{E}}$. Furthermore, the charge density ρ, inside the conductor, is related, on the one hand, to $\overrightarrow{\mathbf{E}}$ by the local Gauss equation, $\overrightarrow{\nabla} \cdot \overrightarrow{\mathbf{E}} = \frac{\rho}{\varepsilon_0}$ and, on the other hand, to $\overrightarrow{\mathbf{J}}$ by the charge conservation equation, $\overrightarrow{\mathbf{J}} + \frac{\partial \rho}{\partial t} = 0$. Combining these equations, we obtain $\frac{\partial \rho}{\partial t} + \frac{\gamma}{\varepsilon_0}\rho = 0$, which gives:

$$\rho(t) = \rho_0 e^{-\frac{t}{\tau}} \quad [3.111]$$

where $\tau = \frac{\varepsilon_0}{\gamma}$ is the *relaxation time*. The solution [3.111] shows that at any point M inside the conductor, the initial charge density decays exponentially with time. If the medium does not have a volume charge at $t = 0$, it will remain uncharged. Charge is therefore carried by ohmic currents to the surface of the conductor or to the interfaces between conductors.

3.8.1. *Discharge of a uniformly charged sphere*

Consider a sphere of radius R_2, permittivity ε_0 (for simplicity) and conductivity γ. The sphere is uniformly charged up to radius $R_1 < R_2$, with charge density ρ_0 at time $t = 0$, as shown in Figure 3.25. Between R_1 and R_2, the sphere contains no charge, that is, it will remain uncharged in this region. The sphere is immersed in a medium of zero conductivity. From [3.111], we can immediately write the expression for the charge density at any instant t:

$$\rho = \begin{cases} \rho_0 e^{-\frac{t}{\tau}} & r < R_1 \\ 0 & r > R_1 \end{cases} \quad [3.112]$$

The total charge of the sphere remains constant at $Q_0 = \frac{4}{3}\pi R_1^3 \rho_0$. However, the volume charge is carried by the ohmic current to the $r = R_2$ interface, where it becomes a surface charge. If we surround the entire sphere with a Gaussian surface of radius $r > R_2$, we can see that the field $\overrightarrow{\mathbf{E}}$ is independent of time, because the charge inside is fixed:

$$E_r\left(r > R_2\right) = \frac{Q_0}{4\pi\varepsilon_0 r^2} \tag{3.113}$$

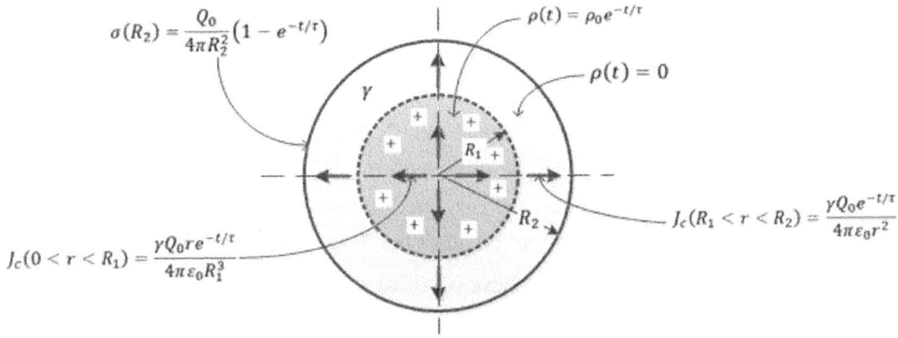

Figure 3.25. *Charge density decreases exponentially with time inside a conductor and disappears after a few minutes. This charge appears at discontinuity interfaces. Initially uncharged regions always remain uncharged, but ohmic currents flow through them*

If the Gauss surface is now inside the sphere, there are two possibilities depending on its radius:

$$\begin{aligned} E_r\left(0 < r < R_1\right) &= \frac{\frac{4}{3}\pi r^3 \rho_0 e^{-\frac{t}{\tau}}}{4\pi\varepsilon_0 r^2} = \frac{\rho_0 r e^{-\frac{t}{\tau}}}{3\varepsilon_0} = \frac{Q_0 r e^{-\frac{t}{\tau}}}{4\pi\varepsilon_0 R_1^3} \\ E_r\left(R_1 < r < R_2\right) &= \frac{\frac{4}{3}\pi R_1^3 \rho_0 e^{-\frac{t}{\tau}}}{4\pi\varepsilon_0 r^2} = \frac{Q_0 e^{-\frac{t}{\tau}}}{4\pi\varepsilon_0 r^2} \end{aligned} \tag{3.114}$$

Clearly, charge will accumulate exponentially over the $r = R_2$ interface, with a surface charge density equal to:

$$\sigma\left(R_2\right) = \varepsilon_0 E_r\left(R_2 + 0\right) - \varepsilon_0 E_r\left(R_2 - 0\right) = \frac{Q_0}{4\pi R_2^2}\left(1 - e^{-\frac{t}{\tau}}\right) \tag{3.115}$$

The charge is transported from the charged region ($r < R_1$) to the surface $r = R_2$ by the conduction current:

$$J_c\,(0 < r < R_1) = \frac{\gamma Q_0 r e^{-\frac{t}{\tau}}}{4\pi\varepsilon_0 R_1^3}$$
$$J_c\,(R_1 < r < R_2) = \frac{\gamma Q_0 e^{-\frac{t}{\tau}}}{4\pi\varepsilon_0 r^2} \qquad [3.116]$$
$$J_c\,(R_2 < r < \infty) = 0$$

Note that (i) the conduction current in the region $R_1 < r < R_2$ is not zero, although the charge density is zero in this region; (ii) the total current, conduction plus displacement, is zero everywhere:

$$-J_c\,(0 < r < R_1) = J_d\,(0 < r < R_1) = \varepsilon_0\frac{\partial E_r}{\partial t} = -\frac{\gamma Q_0 r e^{-\frac{t}{\tau}}}{4\pi\varepsilon_0 R_1^3}$$
$$-J_c\,(R_1 < r < R_2) = J_d\,(R_1 < r < R_2) = \varepsilon_0\frac{\partial E_r}{\partial t} = -\frac{\gamma Q_0 e^{-\frac{t}{\tau}}}{4\pi\varepsilon_0 r^2} \qquad [3.117]$$
$$-J_c\,(R_2 < r < \infty) = J_d\,(R_2 < r < \infty) = \varepsilon_0\frac{\partial E_r}{\partial t} = 0$$

3.8.2. Charging and discharging at the interface between two materials

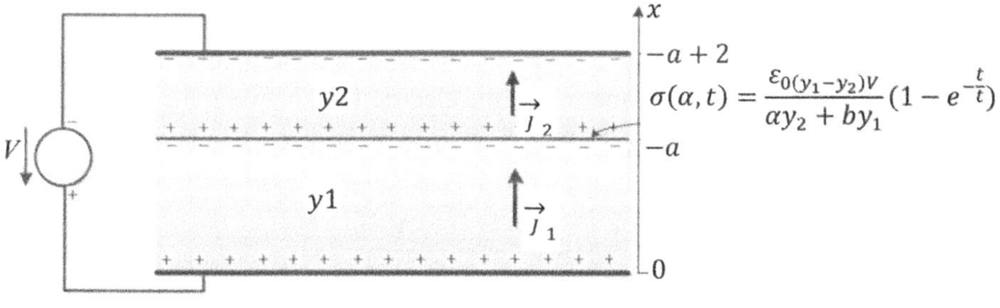

Figure 3.26. *Charge accumulation at the interface of two conductors: in the initial phase, the two conductors behave like capacitors, then like resistors when the permanent regime is reached*

Two parallelepiped conductors of thicknesses a and b, respectively, conductivities γ_1 and γ_2, are placed one on top of the other between two parallel-plate electrodes, as shown in Figure 3.26. By applying a voltage between the two electrodes at $t = 0$, an electric current is observed to flow through the conductors from one electrode to the other. Initially, the current is discontinuous at the boundary between the two conductors due to the temporal variation of the electric field. As a result, there will be an accumulation of charge on their interface, that is, the interface will become charged. During this initial phase, the system will behave like two capacitors in series. In a second phase, that is, after a long period of voltage, the fields approach their permanent regime values. Since there is no longer any temporal variation in $\vec{\mathbf{E}}$, the current density becomes continuous across the interface. In this second phase, the system behaves like two resistors in series. The voltage applied to the electrodes requires that:

$$V = \int_{x=0}^{a+b} E_x dx = E_1 a + E_2 b \qquad [3.118]$$

where E_1 and E_2 are, respectively, the electric fields inside conductors (1) and (2). To establish the expressions for E_1 and E_2, we need to use the equation for conservation of total current, $J_1 + \varepsilon_0 \frac{dE_1}{dt} = J_2 + \varepsilon_0 \frac{dE_2}{dt}$. According to Ohm's law, this last equation is also written as:

$$\gamma_1 E_1 + \varepsilon_0 \frac{dE_1}{dt} = \gamma_2 E_2 + \varepsilon_0 \frac{dE_2}{dt} \qquad [3.119]$$

Using [3.118], we obtain the differential equation governing E_1:

$$\frac{dE_1}{dt} + \frac{1}{\tau} E_1 = \frac{\gamma_2 V}{\varepsilon_0 (a+b)} \qquad [3.120]$$

where $\tau = \frac{\varepsilon_0 (a+b)}{a\gamma_2 + b\gamma_1}$ is the relaxation time of the ensemble and appears here as a weighted average between the relaxation times of each conductor. Equation [3.120] admits the solution $E_1(t) = Ae^{-\frac{t}{\tau}} + \frac{\gamma_2 V}{a\gamma_2 + b\gamma_1}$, where A is a constant to be determined from the initial conditions. At time $t = 0$, the electric field between the electrodes is $E_1(0) = E_2(0) = \frac{V}{a+b}$, which gives $A = \frac{V}{a+b} - \frac{\gamma_2 V}{a\gamma_2 + b\gamma_1} = \frac{b(\gamma_1 - \gamma_2) V}{(a+b)(a\gamma_2 + b\gamma_1)}$. It therefore follows that:

$$\begin{aligned} E_1(t) &= \frac{b(\gamma_1 - \gamma_2) V}{(a+b)(a\gamma_2 + b\gamma_1)} e^{-\frac{t}{\tau}} + \frac{\gamma_2 V}{a\gamma_2 + b\gamma_1} \\ E_2(t) &= \frac{a(\gamma_2 - \gamma_1) V}{(a+b)(a\gamma_2 + b\gamma_1)} e^{-\frac{t}{\tau}} + \frac{\gamma_1 V}{a\gamma_2 + b\gamma_1} \end{aligned} \qquad [3.121]$$

When permanent regime is reached, that is, when $t \longrightarrow \infty$, we deduce that:

$$E_1(\infty) = \frac{\gamma_2 V}{a\gamma_2 + b\gamma_1} \qquad E_2(\infty) = \frac{\gamma_1 V}{a\gamma_2 + b\gamma_1} \qquad [3.122]$$

At the interface between the two conductors, the accumulated surface charge density is given as:

$$\sigma(a,t) = \varepsilon_0 E_2(t) - \varepsilon_0 E_1(t) = \frac{\varepsilon_0 (\gamma_1 - \gamma_2) V}{a\gamma_2 + b\gamma_1} \left(1 - e^{-\frac{t}{\tau}}\right) \qquad [3.123]$$

It is zero at $t = 0$ and tends, when $t \longrightarrow \infty$, toward:

$$\sigma(a, \infty) = \frac{\varepsilon_0 (\gamma_1 - \gamma_2) V}{a\gamma_2 + b\gamma_1} \qquad [3.124]$$

This is the *maximum charge attained at the interface*. The total current supplied by the voltage source is:

$$\begin{aligned} I(t) &= \left(\gamma_1 E_1 + \varepsilon_0 \frac{dE_1}{dt}\right) S = \left(\gamma_2 E_2 + \varepsilon_0 \frac{dE_2}{dt}\right) S \\ &= \left[\frac{\gamma_1\gamma_2}{a\gamma_2 + b\gamma_1} + \frac{ab(\gamma_1 - \gamma_2)^2}{(a+b)^2 (a\gamma_2 + b\gamma_1)} e^{-\frac{t}{\tau}} + \frac{\varepsilon_0}{a+b}\delta(t)\right] VS \end{aligned} \qquad [3.125]$$

where S is the surface area of the electrodes. The last term is the pulse current which instantaneously places a charge on each electrode at $t = 0$. This term is derived from the time derivative of *Heaviside's function*. When permanent regime is reached, that is, when $t \longrightarrow \infty$, we derive that:

$$I(\infty) = \frac{\gamma_1\gamma_2 VS}{a\gamma_2 + b\gamma_1} \qquad [3.126]$$

We therefore conclude that in the early stages, capacitors dominate, while in the steady state, resistors dominate, with relaxation times depending on the relaxation times and the geometry of each region. The system is therefore equivalent to a series combination of a resistor–capacitor in parallel for each region, as shown in Figure 3.27. To observe the discharge phenomenon at the interface once the system reaches permanent regime, we can proceed in two different ways:

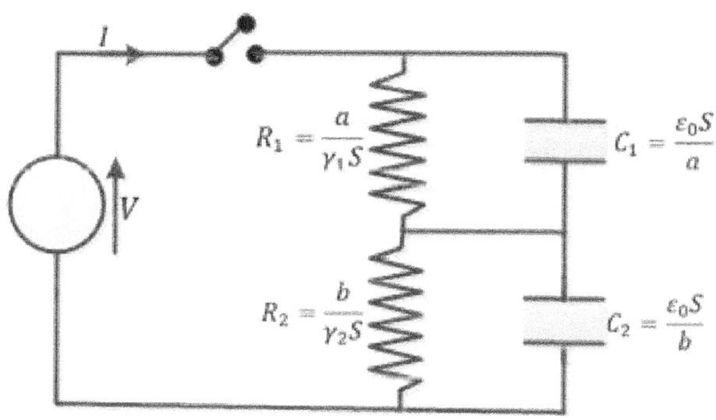

Figure 3.27. *Each region is equivalent to a resistor and a capacitor in parallel (equivalent diagram)*

1) We can instantly open the circuit so that the terminal current is zero, that is, $I = 0$. Using [3.122], we can see in this case that the fields will extinguish in each region with their own relaxation time:

$$E_1(t) = \frac{\gamma_2 V}{\gamma_2 a + \gamma_1 b} e^{-\frac{t}{\tau_1}} \qquad E_2(t) = \frac{\gamma_1 V}{\gamma_2 a + \gamma_1 b} e^{-\frac{t}{\tau_2}} \tag{3.127}$$

where $\tau_1 = \frac{\varepsilon_0}{\gamma_1}$ and $\tau_2 = \frac{\varepsilon_0}{\gamma_2}$. The voltage across the two electrodes and the surface charge density will therefore also decrease in this way:

$$\begin{aligned} V_{\text{cir.open}}(t) &= E_1(t)\,a + E_2(t)\,b = \frac{V}{\gamma_2 a + \gamma_1 b}\left[\gamma_2 a e^{-\frac{t}{\tau_1}} + \gamma_1 b e^{-\frac{t}{\tau_2}}\right] \\ \sigma(a,t) &= \varepsilon_0 E_2(t) - \varepsilon_0 E_1(t) = \frac{\varepsilon_0 V}{\gamma_2 a + \gamma_1 b}\left[\gamma_1 e^{-\frac{t}{\tau_2}} - \gamma_2 e^{-\frac{t}{\tau_1}}\right] \end{aligned} \tag{3.128}$$

2) The two electrodes can be connected by a short-circuit, that is, fixed at the same potential. In this case, equations [3.118] and [3.120] reduce to:

$$E_1 a + E_2 b = 0 \qquad \frac{dE_1}{dt} + \frac{E_1}{\tau} = 0 \tag{3.129}$$

Since at $t = 0$, the surface charge present on the interface cannot change instantaneously, the initial fields $E_1(0)$ and $E_2(0)$ must verify the relation: $\sigma(a,0) = \varepsilon_0 E_2(0) - \varepsilon_0 E_1(0) = -\varepsilon_0\left(\frac{a}{b}+1\right)E_1(0) = \frac{\varepsilon_0(\gamma_1-\gamma_2)V}{\gamma_2 a+\gamma_1 b}$, which gives $E_1(0) = \frac{(\gamma_2-\gamma_1)bV}{(a+b)(\gamma_2 a+\gamma_1 b)}$ and $E_2(0) = \frac{(\gamma_1-\gamma_2)aV}{(a+b)(\gamma_2 a+\gamma_1 b)}$. Consequently, the system [3.129] admits the solutions:

$$E_1(t) = \frac{(\gamma_2-\gamma_1)\,bV}{(a+b)(\gamma_2 a+\gamma_1 b)} e^{-\frac{t}{\tau}} \qquad E_2(t) = \frac{(\gamma_1-\gamma_2)\,aV}{(a+b)(\gamma_2 a+\gamma_1 b)} e^{-\frac{t}{\tau}} \tag{3.130}$$

The short-circuit current and surface charge density are then given by:

$$\begin{aligned} I(t) &= -\frac{(\gamma_1-\gamma_2)^2\,abV}{(a+b)(\gamma_2 a+\gamma_1 b)} e^{-\frac{t}{\tau}} \\ \sigma(a,t) &= \frac{\varepsilon_0(\gamma_1-\gamma_2)\,V}{\gamma_2 a+\gamma_1 b} e^{-\frac{t}{\tau}} \end{aligned} \tag{3.131}$$

3.8.3. *Modeling of an ohmic electrode*

In all of the above, it has been assumed that the electrodes are perfect conductors, that is, they have no resistance and the electric field ends perpendicular to them. Now let us assume that the electrode has a fairly large but finite γ_c conductivity. When current flows along the electrode, there will be an ohmic voltage drop.

Between the two electrodes lies a conductor of permittivity ε_0 and conductivity γ, which can be modeled by a distribution of capacitor–resistors, as shown in Figures 3.28(a) and 4.28(b). The resistance of an electrode element located between $z - \Delta z$ and z is:

$$R\Delta z = \frac{\Delta z}{\gamma_c a d} \qquad [3.132]$$

where $R = frac1\gamma_c ad$ is the resistance per unit length. In section 3.8.2, we saw that a conductor between two electrodes is modeled as a series of parallel resistor–capacitors:

$$C\Delta z = \frac{\varepsilon_0 d\Delta z}{s} \qquad \frac{1}{G\Delta z} = \frac{s}{\gamma d\Delta z} \qquad [3.133]$$

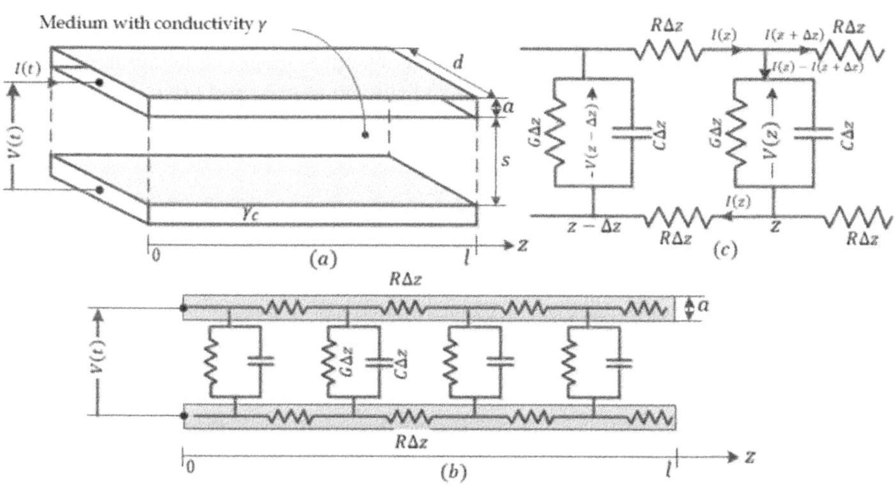

Figure 3.28. *(1) Two electrodes, with parallel plates of finite conductivity γ_c, enclose a conductor of permittivity ε_0 and conductivity γ. (2) This system can be modeled by a resistance–capacitor distribution. (3) Kirchhoff's laws applied to a mesh allow the system to be described by two partial differential equations*

C and G are, respectively, capacitance and conductance per unit length. Let us apply Kirchhoff's laws of voltage and current to the mesh between $z - \Delta z$ and z, as shown in Figure 3.28:

$$\begin{aligned} V(z-\Delta z) - V(z) &= I(z) R\Delta z + I(z) R\Delta z = 2I(z) R\Delta z \\ I(z) - I(z+\Delta z) &= C\Delta z \frac{dV(z)}{dt} + G\Delta z V(z) \end{aligned} \quad [3.134]$$

Dividing by Δz and then passing to the limit $\Delta z \longrightarrow 0$, these equations reduce:

$$\begin{aligned} -\frac{\partial V}{\partial z} &= 2IR \\ -\frac{\partial I}{\partial z} &= C\frac{\partial V}{\partial t} + GV \end{aligned} \quad [3.135]$$

Eliminating I in the first equation gives:

$$\begin{aligned} \frac{\partial^2 V}{\partial z^2} &= 2RC\frac{\partial V}{\partial t} + 2RGV \\ I &= -\frac{1}{2R}\frac{\partial V}{\partial z} \end{aligned} \quad [3.136]$$

This is the *transient diffusion equation.*

a) *Stationary or permanent regime*: if a constant voltage V_0 is applied to the terminals, the solution corresponding to the steady state (reached after a sufficiently long time) must satisfy $\frac{\partial^2 V}{\partial z^2} = 2RGV$, which gives:

$$\begin{aligned} V(z) &= A_1 \sinh\sqrt{2RG}z + A_2 \cosh\sqrt{2RG}z \\ i(z) &= -\frac{1}{2R}\frac{dV}{dz} = -\sqrt{\frac{G}{2R}}\left(A_1 \cosh\sqrt{2RG}z + A_2 \sinh\sqrt{2RG}z\right) \end{aligned} \quad [3.137]$$

where A_1 and A_2 are obtained by the boundary conditions at $z = 0$ and $z = l$: $V(0) = V_0 = A_2$ and $I(l) = 0 = -\sqrt{\frac{G}{2R}}\left(A_1 \cosh\sqrt{2RG}l + A_2 \sinh\sqrt{2RG}l\right)$. It follows that $A_2 = V_0$ and $A_1 = -V_0\frac{\sinh\sqrt{2RG}l}{\cosh\sqrt{2RG}l}$. The solutions obtained are:

$$\begin{aligned} V(z) &= V_0\frac{\sinh\left[\sqrt{2RG}(z-l)\right]}{\cosh\sqrt{2RG}l} \\ I(z) &= -\frac{1}{2R}\frac{dV}{dz} = V_0\sqrt{\frac{G}{2R}}\frac{\cosh\left[\sqrt{2RG}(z-l)\right]}{\cosh\sqrt{2RG}l} \end{aligned} \quad [3.138]$$

b) *Transient regime*: if voltage V_0 is applied at time $t = 0$, the system must take a certain amount of time to reach steady state, and must pass through a so-called *transient regime*. Since [3.136] is linear, we can conclude that the solution governing the system must appear as the sum of the permanent regime solution [3.13] and a transient solution that will vanish with time:

$$V(z) = V_0 \frac{\sinh\left[\sqrt{2RG}\,(z-l)\right]}{\cosh\sqrt{2RG}l} + \hat{V}(z)\,e^{-\alpha t} \tag{3.139}$$

where $\hat{V}(z)$ and α are unknowns for now. Substituting [3.139] into [3.136], we obtain the differential equation:

$$\frac{d^2\hat{V}}{dz^2} + p^2\hat{V} = 0 \tag{3.140}$$

where $p^2 = 2RC\alpha - 2RG$. It admits the trigonometric solution $\hat{V}(z) = a_1 \sin pz + a_2 \cos pz$. As the permanent part in [3.139] already satisfies the boundary condition at $z = 0$, the transient part there must be zero, that is, we must take $a_2 = 0$. The transient solution will contribute to the expression of the current I as follows:

$$I(z,t) = -\frac{1}{2R}\frac{\partial V(z,t)}{\partial z} = V_0\sqrt{\frac{G}{2R}}\frac{\sinh\left[\sqrt{2RG}\,(z-l)\right]}{\cosh\sqrt{2RG}l} + \hat{I}(z)\,e^{-\alpha t} \tag{3.141}$$

where $\hat{I}(z) = -\frac{1}{2R}\frac{\partial\hat{V}(z)}{\partial z} = -\frac{pa_1}{2R}\cos pz$. $\hat{I}(z)$ must be zero at $z = l$, which means that p must satisfy the constraint $pl = (2n+1)\frac{\pi}{2}$ and, consequently, $\alpha_n = \frac{1}{2RC}\left[(2n+1)\frac{\pi}{2l}\right]^2 + \frac{G}{C}$, où $n \in \mathbb{N}$. It follows that the general solution is a superposition of all possible solutions:

$$V(z,t) = V_0 \frac{\cosh\left[\sqrt{2RG}\,(z-l)\right]}{\cosh\sqrt{2RG}l} + \sum_{n=0}^{\infty} A_n \sin\left[\frac{(2n+1)\pi z}{2l}\right] e^{-\alpha_n t} \tag{3.142}$$

This solution satisfies the boundary conditions, but not the initial condition at $t = 0$ when the voltage is applied. Indeed, the voltage distribution must be zero everywhere at $t = 0$, $V(z,0) = 0 = V_0\frac{\cosh\left[\sqrt{2RG}(z-l)\right]}{\cosh\sqrt{2RG}l} + \sum_{n=0}^{\infty} A_n \sin\left[\frac{(2n+1)\pi z}{2l}\right]$. To calculate A_m, simply multiply this equation by $\sin\left[\frac{(2m+1)\pi z}{2l}\right]$ and then integrate z between 0 and l, that is, $0 = \frac{V_0}{\cosh\sqrt{2RG}l}\int_0^l \cosh\left[\sqrt{2RG}\,(z-l)\right] \sin\left[\frac{(2m+1)\pi z}{2l}\right] dz + \sum_{n=0}^{\infty} A_n \int_0^l \sin\left[\frac{(2m+1)\pi z}{2l}\right] \sin\left[\frac{(2n+1)\pi z}{2l}\right] dz$. The second integral is always zero except when $m = n$, when it is $\frac{l}{2}$. The first integral is calculated using the following formula: $\int \cosh a\,(z-l) \sin bz dz = \frac{1}{a^2+b^2}$

$[a \sin bz \sinh a (z-l) - b \cos bz \cosh a (z-l)])$. We therefore deduce that $A_n = -\frac{1}{l^2} \frac{\pi V_0 (2n+1)}{2RG + \left[\frac{(2n+1)\pi}{2l}\right]^2}$. The general solutions are then:

$$
\begin{aligned}
V &= V_0 \frac{\left[\cosh \sqrt{2RG}\,(z-l)\right]}{\cosh \sqrt{2RG} l} - \frac{\pi V_0}{l^2} \sum_{n=0}^{\infty} \frac{(2n+1) \sin\left[\frac{(2n+1)\pi z}{2l}\right] e^{-\alpha_n t}}{2RG + \left[\frac{(2n+1)\pi}{2l}\right]^2} \\
I &= -V_0 \sqrt{\frac{G}{2R}} \frac{\sinh\left[\sqrt{2RG}\,(z-l)\right]}{\cosh \sqrt{2RG} l} + \frac{\pi^2 V_0}{4 l^3 R} \sum_{n=0}^{\infty} \frac{(2n+1)^2 \cos\left[\frac{(2n+1)\pi z}{2l}\right] e^{-\alpha_n t}}{2RG + \left[\frac{(2n+1)\pi}{2l}\right]^2}
\end{aligned}
\quad [3.143]
$$

These series can often be approximated by the first term:

$$
\begin{aligned}
V(z,t) &\simeq V_0 \frac{\left[\cosh \sqrt{2RG}\,(z-l)\right]}{\cosh \sqrt{2RG} l} - \frac{\pi V_0}{l^2} \frac{\sin\left[\frac{\pi z}{2l}\right] e^{-\alpha_0 t}}{2RG + \left[\frac{\pi}{2l}\right]^2} \\
I(z,t) &\simeq -V_0 \sqrt{\frac{G}{2R}} \frac{\sinh\left[\sqrt{2RG}\,(z-l)\right]}{\cosh \sqrt{2RG} l} + \frac{\pi^2 V_0}{4 l^3 R} \frac{\cos\left[\frac{\pi z}{2l}\right] e^{-\alpha_0 t}}{2RG + \left[\frac{\pi}{2l}\right]^2}
\end{aligned}
\quad [3.144]
$$

where $\alpha_0 = \frac{2RG + \left[\frac{\pi}{2l}\right]^2}{2RC}$. The value $\tau_0 = \frac{2RC}{2RG + \left[\frac{\pi}{2l}\right]^2}$ can be taken as the relaxation time for the system.

3.9. Earth atmosphere as a spherical capacitor

Under favorable climatic conditions (good weather), a radial electric field of intensity $-E_0$ directed toward the center of the Earth is measured experimentally at the Earth's surface. The amplitude of the electric field decreases with height above the Earth's surface due to the non-uniform electrical conductivity $\gamma(r)$ of the atmosphere, which is approximately:

$$\gamma(r) = \gamma_0 + a(r-R)^2 \quad [3.145]$$

where γ_0 is the conductivity at ground level (i.e. for $r = R$) and a is a positive constant. R here represents the Earth's radius. Conductivity increases with height, so that the atmosphere acts as a perfect conductor at very great heights. In the permanent regime, the current density, due to spherical symmetry, must satisfy the equation, $\vec{\nabla} \cdot \vec{J} = \frac{1}{r^2} \frac{\partial}{\partial r} \left(r^2 J_r\right) = 0$, which gives the solution

$J_r(r) = \gamma(r) E_r(r) = \frac{C}{r^2}$, where C is a constant to be determined from the boundary condition $E_r(R) = -E_0$. It follows that:

$$J_r(r) = -\frac{\gamma(R) E_0 R^2}{r^2} = -\frac{\gamma_0 E_0 R^2}{r^2} \qquad [3.146]$$

At the Earth's surface, the current density is:

$$J_r(R) = -\gamma_0 E_0 \qquad [3.147]$$

and, consequently, the total current directed radially inward over the whole Earth is:

$$I = \left| J_r(R) 4\pi R^2 \right| = 4\pi \gamma_0 E_0 R^2 \qquad [3.148]$$

The electric field distribution in the atmosphere is obtained from Ohm's law and [3.146]:

$$E_r(r) = \frac{J_r(r)}{\gamma(r)} = -\frac{\gamma_0 E_0 R^2}{\gamma(r) r^2} \qquad [3.149]$$

It therefore follows that the charge density at the Earth's surface is equal to:

$$\sigma(R) = \varepsilon_0 E_r(R) = -\varepsilon_0 E_0 \qquad [3.150]$$

This charge distribution at the Earth's surface is balanced by a positive volume charge distribution in the atmosphere:

$$\begin{aligned} \rho(r) &= \frac{\varepsilon_0}{r^2} \frac{\partial}{\partial r}\left(r^2 E_r\right) = -\frac{\varepsilon_0 \gamma_0 E_0 R^2}{r^2} \frac{\partial}{\partial r}\left(\frac{1}{\gamma(r)}\right) \\ &= \frac{2a\varepsilon_0 \gamma_0 E_0 R^2 (r-R)}{r^2 \left[\gamma_0 + a(r-R)^2\right]^2} \end{aligned} \qquad [3.151]$$

The potential difference between the upper atmosphere and the Earth's surface is:

$$\begin{aligned} V &= -\int_{r=0}^{\infty} E_r(r)\, dr = \gamma_0 E_0 R^2 \int_{r=0}^{\infty} \frac{1}{r^2 \left[\gamma_0 + a(r-R)^2\right]} dr \\ &= \frac{\gamma_0 E_0 R^2}{a\left(R^2 + \frac{\gamma_0}{a}\right)^2} \left[R \ln \frac{\gamma_0}{aR^2} + \frac{R^2 + \frac{\gamma_0}{a}}{R} + \frac{\frac{\pi}{2}\left(R^2 - \frac{\gamma_0}{a}\right)}{\sqrt{\frac{\gamma_0}{a}}} \right] \\ &\simeq \frac{\gamma_0 E_0}{aR^2} \left[R\left(\ln \frac{\gamma_0}{aR^2} + 1\right) + \frac{\pi R^2}{2\sqrt{\frac{\gamma_0}{a}}} \right] \end{aligned} \qquad [3.152]$$

The electric charge on the ground is neutralized by the current flow in a time of the order of $\frac{\varepsilon_0}{\gamma_0}$ (a few minutes). This terrestrial charge is maintained in nature by lightning strikes around the world.

3.10. Space charge

We have already seen that in the permanent regime, the volume charge density in an ohmic conductor with constant conductivity decays over time, that is, it tends toward zero when t becomes sufficiently large. This is because a classical ohmic conductor contains two species of charge carriers of opposite polarities, so charge neutrality is maintained. However, if a single species of charge is injected into a medium, a volume charge distribution, known as *space charge*, may result in permanent regime. Due to the electric force, this ρ volume charge distribution contributes to the electric field, and in turn depends on it. It then becomes necessary to simultaneously satisfy the coupled electrical and mechanical equations. The space charge is therefore a cloud of electrons, located in a space that can be a vacuum, a gas, a liquid or a solid:

– In a vacuum: when a metal object positioned in a vacuum is heated to incandescence, the energy is sufficient for the agitated electrons to leave the surface of the metal, forming a cloud of electrons. This cloud is negatively charged, and can be attracted by any surrounding positively charged object. In this case, the cloud is in fact an electric current flowing in a vacuum. This phenomenon is known as the *Edison effect*. The space charge generates a negative electric field, which is sometimes used to polarize the tube.

– In other media: space charges can also occur in liquids, solids and gases. For example, when a gas is between two electrodes with a large potential difference, electric charges disperse around the electrodes, forming charge zones.

3.10.1. *Vacuum diode, non-ohmic conductor*

Two electrodes A and C (surface S) of a diode are enclosed in a vacuum bulb. By *thermoelectric effect*, the cathode (negative electrode) C emits electrons with no initial velocity, which are attracted to the anode (positive electrode) A maintained at a constant positive potential V_0, as shown in Figure 3.29. The equipotentials correspond to the x = const planes. At the cathode output, electron motion is governed by the differential equation:

$$m\frac{dv}{dt} = -eE = e\frac{dV}{dx} \qquad [3.153]$$

In the permanent regime, the electron's velocity depends only on the position x. It therefore follows that $m\frac{dv}{dt} = m\frac{dv}{dx}\frac{dx}{dt} = mv\frac{dv}{dx}$, which shows that $\frac{d}{dx}\left(\frac{1}{2}mv^2\right) = \frac{d}{dx}\left(eV\right)$ or $\frac{d}{dx}\left[\frac{1}{2}mv^2 - eV\right] = 0$.

The latter equation corresponds to the law of conservation of mechanical energy:

$$\frac{1}{2}mv^2 - eV = \text{const} \qquad [3.154]$$

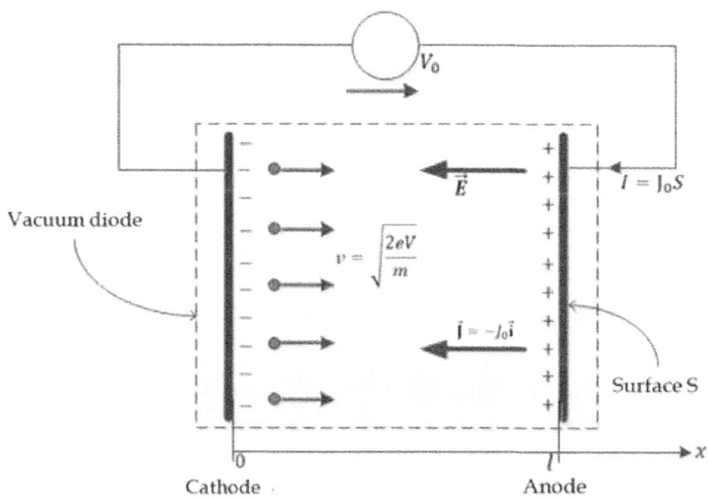

Figure 3.29. *Electrons are injected from the cathode by thermoelectric effect, then attracted by the positive potential of the anode (vacuum diode)*

Since the electron starts with zero velocity and the potential is chosen to be zero at the cathode, this constant is zero. It therefore follows that:

$$v = \left(\frac{2e}{m}V\right)^{1/2} \qquad [3.155]$$

In permanent regime, the current density vector is constant. In fact, we have: $0 = \vec{\mathbf{J}} = \frac{dJ_x}{dx} = 0$ and, therefore, $\vec{\mathbf{J}} = -J_0\,\vec{\mathbf{i}}$. The electric current flows from the anode to the cathode, that is, in the direction of negative x. The sign $-$ is already incorporated in the expression of $\vec{\mathbf{J}}$. J_0 is positive here and related to the charge density by the relation $J_0 = -\rho v$. It follows that:

$$\rho = -J_0\left(\frac{m}{2e}\right)^{1/2} V^{-1/2} \qquad [3.156]$$

To find $V(x)$, we will use the Poisson equation:

$$\frac{d^2V}{dx^2} = -\frac{\rho}{\varepsilon_0} = \frac{J_0}{\varepsilon_0}\left(\frac{m}{2e}\right)^{1/2} V^{-1/2} \qquad [3.157]$$

Multiplying by $2\frac{dV}{dx}$, equation [3.157] is also written as: $\frac{d}{dx}\left(\frac{dV}{dx}\right)^2 = 4\frac{J_0}{\varepsilon_0}\left(\frac{m}{2e}\right)^{1/2} \frac{d}{dx}\left(V^{1/2}\right)$, which shows (after integration) that $\left(\frac{dV}{dx}\right)^2 = 4\frac{J_0}{\varepsilon_0}\left(\frac{m}{2e}\right)^{1/2} V^{1/2}$. Taking the square root of each side, we obtain $\frac{dV}{dx} = 2\sqrt{\frac{J_0}{\varepsilon_0}}\left(\frac{m}{2e}\right)^{1/4} V^{1/4}$. By separating the variables, that is, by writing that $\frac{dV}{V^{1/4}} = 2\sqrt{\frac{J_0}{\varepsilon_0}}\left(\frac{m}{2e}\right)^{1/4} dx$, then (after a second integration) $\frac{4}{3}V^{3/4} = 2\sqrt{\frac{J_0}{\varepsilon_0}}\left(\frac{m}{2e}\right)^{1/4} x + \text{const}$. Since $V(0) = 0$, we find $V^{3/4} = \frac{3}{2}\sqrt{\frac{J_0}{\varepsilon_0}}\left(\frac{m}{2e}\right)^{1/4} x$ and, therefore:

$$V(x) = \left[\frac{9}{4}\frac{J_0}{\varepsilon_0}\left(\frac{m}{2e}\right)^{1/2}\right]^{2/3} x^{4/3} \qquad [3.158]$$

At the anode, we must require that $V(l) = V_0 = \left[\frac{9}{4}\frac{J_0}{\varepsilon_0}\left(\frac{m}{2e}\right)^{1/2}\right]^{2/3} l^{4/3}$, which allows us to find:

$$J_0 = \frac{4}{9}\frac{\varepsilon_0}{l^2}\left(\frac{2e}{m}\right)^{1/2} V_0^{2/3} \qquad [3.159]$$

This is the *Langmuir–Child equation*. The electric potential, electric field and charge density are then given by:

$$\begin{aligned} V(x) &= V_0\left(\frac{x}{l}\right)^{4/3} \\ E(x) &= -\frac{dV}{dx} = -\frac{4}{3}\frac{V_0}{l}\left(\frac{x}{l}\right)^{1/3} \\ \rho(x) &= \varepsilon_0\frac{dE}{dx} = -\frac{4}{9}\varepsilon_0\frac{V_0}{l^2}\left(\frac{x}{l}\right)^{-2/3} \end{aligned} \qquad [3.160]$$

We can see that the charge density at the cathode is infinite, but the total charge between the two electrodes is finite:

$$Q_T = \int_0^l \rho(x)\, S dx = -\frac{4}{3}\varepsilon_0\frac{V_0}{l}S \qquad [3.161]$$

The surface charge density at the anode is equal to $\sigma(l) = -\varepsilon_0 E(l) = \frac{4}{3}\varepsilon_0 \frac{V_0}{l}$. It therefore follows that (Q_A is the charge at the anode):

$$Q_A = \sigma(l) S = \frac{4}{3}\varepsilon_0 \frac{V_0}{l} S \qquad [3.162]$$

The charges Q_T and Q_A are equal, but with opposite polarities. At the cathode, the surface charge density is zero, because the field is zero in its neighborhood.

The trajectory of the electron is determined by writing that $v = \frac{dx}{dt} = \left(\frac{2eV_0}{m}\right)^{1/2} \left(\frac{x}{l}\right)^{2/3}$, which, after separation of the variables, gives $\frac{dx}{x^{2/3}} = \left(\frac{2eV_0}{ml^{4/3}}\right)^{1/2} dt$. By integrating, we find:

$$x = \frac{1}{27}\left(\frac{2eV_0}{ml^{4/3}}\right)^{3/2} t^3 \qquad [3.163]$$

We conclude that the time required for the electron to travel from cathode to anode is $\tau = 3l\left(\frac{m}{2eV_0}\right)^{1/2}$.

3.10.2. *Space charges in materials*

The conductive properties of materials are often studied by charge injection. If a material is placed between the two electrodes instead of in a vacuum, as shown in Figure 3.30, the electrons injected at the cathode collide with the medium so that the velocity becomes proportional to the electric field, due to the mobility of the μ electrons, that is, $\vec{v} = -\mu\vec{\mathbf{E}}$. When electrons penetrate the material, the space charge front located at distance s from the interface obeys the differential equation:

$$\frac{ds}{dt} = -\mu E(s) \qquad [3.164]$$

Although the charge density is non-uniformly distributed behind the wavefront or wave surface, the total charge Q in the material behind the wavefront at time t is related to the current density via the equation $J_0 S = \rho\mu E_x S = -\frac{dQ}{dt} = -\frac{Q}{t}$, giving $Q = -J_0 St$. Gauss' theorem applied to the hypothetical rectangular surface enclosing the charge Q in the material, then allows us to relate the fields of the interface and the charge front, on the one hand, and this charge, on the other hand, through the equation:

$$\oint_S \vec{\mathbf{E}} \cdot d\mathbf{S} = [\varepsilon_0 E(s) - \varepsilon_0 E(0)] S = Q = -J_0 St \qquad [3.165]$$

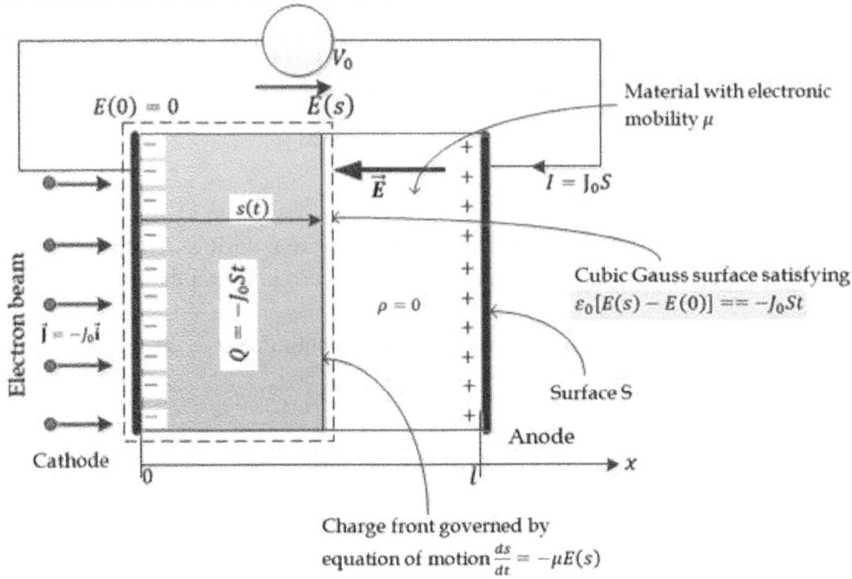

Figure 3.30. *Space charge produced in a material between two electrodes*

The maximum current occurs when $E(0) = 0$, known as *space-charge limited conduction*. To obtain the time dependence of the charge front, simply use [3.165] in [3.164]:

$$\frac{ds}{dt} = \frac{\mu J_0 t}{\varepsilon_0} \quad [3.166]$$

This equation has the solution:

$$s(t) = \frac{\mu J_0 t^2}{2\varepsilon_0} \quad [3.167]$$

Behind the wavefront, Gauss' law allows us to write that $\frac{dE_x}{dx} = \frac{\rho}{\varepsilon_0} = \frac{J_0}{\varepsilon_0 \mu E_x}$ or again:

$$E_x \frac{dE_x}{dx} = \frac{J_0}{\varepsilon_0 \mu} \quad [3.168]$$

Note also that ahead of the wavefront, the charge density is zero and, consequently, the electric field is constant. It follows that the spatial distribution of the electric field is then obtained by integrating [3.168]:

$$E_x = \begin{cases} -\sqrt{\dfrac{2J_0x}{\varepsilon_0\mu}} & \text{for } 0 \leq x \leq s(t) \\ -\sqrt{\dfrac{2J_0s}{\varepsilon_0\mu}} & \text{for } s(t) \leq x \leq l \end{cases} \qquad [3.169]$$

whereas the load distribution is:

$$\rho = \varepsilon_0 \frac{dE_x}{dx} = \begin{cases} -\sqrt{\dfrac{\varepsilon_0 J_0}{2\mu x}} & \text{for } 0 \leq x \leq s(t) \\ 0 & \text{for } s(t) \leq x \leq l \end{cases} \qquad [3.170]$$

The time dependence of the voltage across the material is then:

$$V(t) = -\int_0^l E_x dx = \int_0^{s(t)} \sqrt{\frac{2J_0x}{c_0\mu}} dx + \int_{s(t)}^l \sqrt{\frac{2J_0s}{c_0\mu}} dx = \frac{J_0lt}{c_0} - \frac{\mu J_0^2t^3}{6c_0^2} \qquad [3.171]$$

with the condition that $s(t) \leq l$. These transient solutions are valid until the charge front, given by [3.167], reaches the anode in $s = l$ at the instant $\tau = \sqrt{\frac{2\varepsilon_0 l}{\mu J_0}}$. After this instant, the system is in DC permanent regime, with terminal voltage related to current density as:

$$J_0 = \frac{9}{8}\frac{\varepsilon_0\mu}{l^3}V_0^2 \qquad [3.172]$$

which is the analogue of the Langmuir–Child equation for matrices. Electric field and space charge density in permanent regime are then of the form:

$$E_x = -\frac{3}{2}\frac{V_0}{l}\sqrt{\frac{x}{l}}$$

$$\rho = \varepsilon_0 \frac{dE_x}{dx} = -\frac{3}{4}\frac{\varepsilon_0 V_0}{l}\sqrt{\frac{l}{x}} \qquad [3.173]$$

Chapter 4

Magnetic Field in Vacuum

4.1. Force on a mobile charge

Magnets have been known since Antiquity. The Chinese were the first to make use of the properties of magnets to make compasses. In the 18th century, Benjamin Franklin discovered the electrical nature of lightning. At this time, it was already known that (1) storms interfered with compasses and (2) metallic objects struck by lightning were magnetized. So there is a close connection between electrical and magnetic phenomena. However, a complete theory of electromagnetism did not emerge until the late 19th century. It all started with Ørsted's experiment. He placed a current-carrying conducting wire above a compass and observed that the needle deviated around its axis. He also observed that (1) if the direction of current was reversed, the needle was deflected to the other side, and (2) the force acting upon the needle was not radial. Biot and Savart then demonstrated that the force acting at any point is perpendicular to the direction connecting that point to the conductor and that this force varies inversely with the square of the distance between the two. This is the *Biot and Savart law*. Subsequently, Davy proved experimentally that if a current can cause a magnet to deviate, the converse is also true, that is, a magnet can cause a current to deviate. The definitive development of electromagnetism was completed with the formulation of Maxwell's equations and could only be explained within the context of Einstein's special relativity.

4.1.1. *Theoretical origin of Lorentz force*

Up until this point, we have only studied the effects of static charges. What happens when the charges are in motion? Consider two points-like charges q and q' located, at time t, at the points $\overrightarrow{\mathbf{r}}$ and $\overrightarrow{\mathbf{r}}'$. When they are stationary, the charge q' creates an electrostatic field $\overrightarrow{\mathbf{E}}_{q'}\left(\overrightarrow{\mathbf{r}}\right)$ at the point $\overrightarrow{\mathbf{r}}$. Consequently, the charge q will be subject to the Coulomb force $\overrightarrow{\mathbf{F}}_q\left(\overrightarrow{\mathbf{r}}\right) = q\overrightarrow{\mathbf{E}}_{q'}\left(\overrightarrow{\mathbf{r}}\right)$, which, in turn, will modify the momentum $\vec{\mathbf{p}}$, of q, since $\vec{\mathbf{F}}_q(x, y, z) = \frac{d\vec{\mathbf{p}}}{dt}$. In other words, the electrostatic force arising from q' creates a modification $d\vec{\mathbf{p}}$ over the time interval dt. Therefore, a force corresponds to a transfer of information (in this case, from q' to q) over a short time interval dt. In addition, nothing can travel faster than the velocity of light c. Any transfer of information from one point to another must necessarily take a finite amount of time, which introduces a time lag.

Now suppose that charges q and q' are moving at velocities $\vec{\mathbf{v}}$ and $\vec{\mathbf{v}}'$, respectively. If dt is the time required for the electrostatic field created by q' to propagate from q' to q, then during this time, q' and q will have covered the distances $\vec{\mathbf{v}}'dt$ and $\vec{\mathbf{v}}dt$, respectively. In other words, when q undergoes the electrostatic effects generated by q', the field $\vec{\mathbf{E}}(t - dt)$, seen by q, is directed toward the former position of q' and depends on the distance cdt and not on $\left|\vec{\mathbf{r}} - \vec{\mathbf{r}}'\right|$, as shown in Figure 4.1. We can therefore deduce that the Coulomb force, which is usually expressed as a function of the field $\vec{\mathbf{E}}(t)$, must be corrected, since the propagation of the interaction is assumed to be infinite. Electrical effects cannot therefore be explained by the electrostatic field alone.

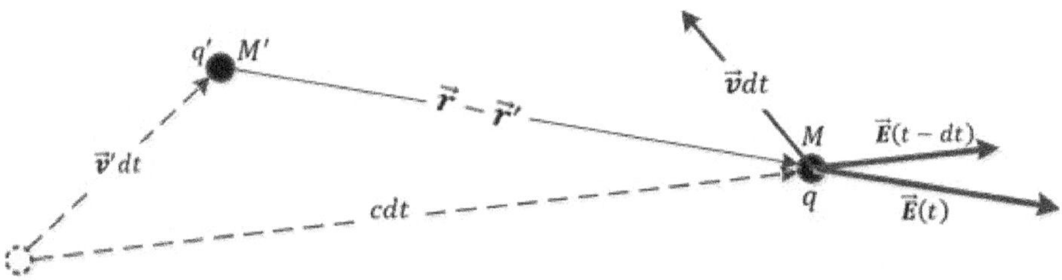

Figure 4.1. *Force applied by a mobile charge q', moving with velocity $\vec{\mathbf{v}}'$, on another mobile charge q, moving at velocity $\vec{\mathbf{v}}$: at time t, the field experienced by the charge q is directed toward the former position of q', measured at the instant $t - dt$*

Experience has shown that the trajectory of the charge q cannot be explained by taking these corrections into account alone, but that Coulomb's law must be modified in this way:

$$\vec{\mathbf{F}}_q(\vec{\mathbf{r}}) \simeq \frac{qq'}{4\pi\varepsilon_0\left|\vec{\mathbf{r}} - \vec{\mathbf{r}}'\right|^3}\Big[(\vec{\mathbf{r}} - \vec{\mathbf{r}}') + \frac{1}{c^2}\left(\left[(\vec{\mathbf{r}} - \vec{\mathbf{r}}')\cdot\vec{\mathbf{v}}\right]\vec{\mathbf{v}}' - \left[(\vec{\mathbf{r}} - \vec{\mathbf{r}}')\cdot\vec{\mathbf{v}}'\right]\vec{\mathbf{v}}\right)\Big]$$
$$= \frac{qq'}{4\pi\varepsilon_0}\left[\frac{\vec{\mathbf{r}} - \vec{\mathbf{r}}'}{\left|\vec{\mathbf{r}} - \vec{\mathbf{r}}'\right|^3} + \frac{\vec{\mathbf{v}}}{c}\times\left(\frac{\vec{\mathbf{v}}'}{c}\times\frac{\vec{\mathbf{r}} - \vec{\mathbf{r}}'}{\left|\vec{\mathbf{r}} - \vec{\mathbf{r}}'\right|^3}\right)\right] \quad [4.1]$$

A second term then appears (called the *magnetic* force), which depends on the speed of the two charges and the speed of light. This second term can be interpreted as the contribution of a magnetic field created by q':

$$\vec{\mathbf{F}}_q\left(\vec{\mathbf{r}}\right) \simeq q\left[\vec{\mathbf{E}}\left(\vec{\mathbf{r}}\right) + \vec{\mathbf{v}} \times \vec{\mathbf{B}}\left(\vec{\mathbf{r}}\right)\right] \qquad [4.2]$$

where $\vec{\mathbf{B}}\left(\vec{\mathbf{r}}\right)$ is a so-called *magnetic field*, created by the charge q':

$$\vec{\mathbf{B}}\left(\vec{\mathbf{r}}\right) = \frac{q'}{4\pi\varepsilon_0 c^2}\frac{\vec{\mathbf{v}}' \times \left(\vec{\mathbf{r}} - \vec{\mathbf{r}}'\right)}{\left|\vec{\mathbf{r}} - \vec{\mathbf{r}}'\right|^3} \equiv \frac{\mu_0 q'}{4\pi}\frac{\vec{\mathbf{v}}' \times \left(\vec{\mathbf{r}} - \vec{\mathbf{r}}'\right)}{\left|\vec{\mathbf{r}} - \vec{\mathbf{r}}'\right|^3} \qquad [4.3]$$

where $\mu_0 = \frac{1}{\varepsilon_0 c^2}$. The factor μ_0 is called the *vacuum magnetic permeability* and it describes the vacuum's ability to allow a magnetic field to pass through. Equation $\mu_0\varepsilon_0 c^2 = 1$ clearly shows that magnetism, electricity and light are closely related. The magnetic field $\vec{\mathbf{B}}$ is perpendicular to both the vector $\vec{\mathbf{r}} - \vec{\mathbf{r}}'$ and the velocity $\vec{\mathbf{v}}'$ of the particle q'. Due to the cross product, the magnetic field is a *pseudo vector*. Equation [4.2] is called the *Lorentz force*. We can easily verify that the ratio between the magnetic force $\vec{\mathbf{F}}_B = q\vec{\mathbf{v}}$ and the Coulomb force $\vec{\mathbf{F}}_E = q\vec{\mathbf{E}}$ is of the order of $(v/c)^2$. This shows that the magnetic force is very weak. This is only true for charges moving at speeds well below the speed of light (magnetostatic approximation).

Since velocity appears in the expression for magnetic force, this force does not obey the principle of action and reaction, unlike electric force. In fact, we have: $\vec{\mathbf{F}}_{B,q} \neq -\vec{\mathbf{F}}_{B,q'}$. Moreover, the magnetic force $\vec{\mathbf{F}}_B$ provides any work. In fact, its power is zero: $\mathsf{P}\left(\vec{\mathbf{F}}_B\right) = \vec{\mathbf{F}}_B \cdot \vec{\mathbf{v}} = q\left(\vec{\mathbf{v}} \times \vec{\mathbf{B}}\right) \cdot \vec{\mathbf{v}} = 0$, because $\vec{\mathbf{v}} \times \vec{\mathbf{B}}$ is perpendicular to $\vec{\mathbf{v}}$. The magnetic force does not modify the kinetic energy and, consequently, there is no modification of the norm of $\vec{\mathbf{v}}$, that is, $\left|\vec{\mathbf{v}}\right| = \text{const}$. On the other hand, this force can modify the direction of $\vec{\mathbf{v}}$.

4.1.2. *Lorentz force*

Consider a charge q moving with velocity $\vec{\mathbf{v}}$ in a region of space where a magnetic field prevails. The charge is subject to the magnetic part of the Lorentz force, empirically determined, perpendicular to both $\vec{\mathbf{v}}$ and $\vec{\mathbf{B}}$ (not to be confused with Laplace's force):

$$\vec{\mathbf{F}}_B = q\left(\vec{\mathbf{v}} \times \vec{\mathbf{B}}\right) \qquad [4.4]$$

The intensity of a magnetic field can also be defined using this force. If dq is a mobile differential elementary charge, then the elementary force experienced will be:

$$d\vec{\mathbf{F}}_B = dq\left(\vec{\mathbf{v}} \times \vec{\mathbf{B}}\right) \quad [4.5]$$

For a distribution over a volume, an area or a line, we write:

$$d\vec{\mathbf{F}}_B = \begin{cases} \rho\left(\vec{\mathbf{v}} \times \vec{\mathbf{B}}\right) d\tau = \vec{\mathbf{J}} \times \vec{\mathbf{B}} d\tau & \left(\vec{\mathbf{J}} = \rho \vec{\mathbf{v}}, \text{ volumedist.}\right) \\ \sigma\left(\vec{\mathbf{v}} \times \vec{\mathbf{B}}\right) dS = \vec{\mathbf{K}} \times \vec{\mathbf{B}} dS & \left(\vec{\mathbf{K}} = \sigma \vec{\mathbf{v}}, \text{ areadist.}\right) \\ \lambda\left(\vec{\mathbf{v}} \times \vec{\mathbf{B}}\right) dl = \vec{\mathbf{I}} \times \vec{\mathbf{B}} dl & \left(\vec{\mathbf{I}} = \lambda \vec{\mathbf{v}}, \text{ linedist.}\right) \end{cases} \quad [4.6]$$

The total force applied to the entire current distribution is obtained by integrating equation [4.6] over the total volume, surface or contour containing the current. That is:

$$\vec{\mathbf{F}}_B = \begin{cases} \int \rho\left(\vec{\mathbf{v}} \times \vec{\mathbf{B}}\right) d\tau = \int \vec{\mathbf{J}} \times \vec{\mathbf{B}} d\tau & \text{(volumedist.)} \\ \int \sigma\left(\vec{\mathbf{v}} \times \vec{\mathbf{B}}\right) dS = \int \vec{\mathbf{K}} \times \vec{\mathbf{B}} dS & \text{(areadist.)} \\ \int \lambda\left(\vec{\mathbf{v}} \times \vec{\mathbf{B}}\right) dl = \int \vec{\mathbf{I}} \times \vec{\mathbf{B}} dl & \text{(linedist.)} \end{cases} \quad [4.7]$$

Furthermore, if there is a net charge with its associated electric field $\vec{\mathbf{E}}$, the force density expression must include the Coulomb term:

$$\begin{cases} \lambda\left(\vec{\mathbf{E}} + \vec{\mathbf{v}} \times \vec{\mathbf{B}}\right) = \lambda\vec{\mathbf{E}} + \vec{\mathbf{I}} \times \vec{\mathbf{B}} & \text{(linedensitydist.)} \\ \sigma\left(\vec{\mathbf{E}} + \vec{\mathbf{v}} \times \vec{\mathbf{B}}\right) = \sigma\vec{\mathbf{E}} + \vec{\mathbf{K}} \times \vec{\mathbf{B}} & \text{(surfacedensitydist.)} \\ \rho\left(\vec{\mathbf{E}} + \vec{\mathbf{v}} \times \vec{\mathbf{B}}\right) = \rho\vec{\mathbf{E}} + \vec{\mathbf{J}} \times \vec{\mathbf{B}} & \text{(volumedensitydist.)} \end{cases} \quad [4.8]$$

In many cases, the net charge is very small. Therefore, it is possible to ignore the Coulomb force. This is often the case for conduction in materials. A net current always flows as a result of the movement of charge carriers (even though the charge density is zero).

4.1.3. *Charge motion in a uniform magnetic field*

4.1.3.1. *Motion in the Serret–Frenet frame*

Let us study the motion of a point-like charge immersed in a uniform static magnetic field. We will assume that the electric field is zero and that all other forces are negligible. The particle's motion is described by the following equation:

$$m\frac{d\vec{\mathbf{v}}}{dt} = q\left(\vec{\mathbf{v}} \times \vec{\mathbf{B}}\right) \qquad [4.9]$$

In the Serret–Frenet frame (attached to the particle), the velocity and acceleration are written as follows:

$$\vec{\mathbf{v}} = v\vec{\mathbf{T}} \qquad \vec{\gamma} = \frac{d\vec{\mathbf{v}}}{dt} = \frac{dv}{dt}\vec{\mathbf{T}} + \frac{v^2}{R}\vec{\mathbf{N}} \qquad [4.10]$$

where R, v, $\vec{\mathbf{T}}$ and $\vec{\mathbf{N}}$ are, respectively, the curvature radius, the norm of velocity, the principal tangent to the trajectory in the direction of motion (defined by $\vec{\mathbf{T}} = \frac{d\vec{\mathbf{r}}}{ds}$, where s is the curvilinear abscissa) and the principal normal directed from the charge to the curvature center (defined by the equation $\frac{d\vec{\mathbf{T}}}{ds} = \frac{d\vec{\mathbf{N}}}{R}$). Clearly, $\vec{\mathbf{N}}$ is perpendicular to $\vec{\mathbf{T}}$. Projecting over $\vec{\mathbf{T}}$, we obtain $m\frac{dv}{dt} = qv\left(\vec{\mathbf{T}} \times \vec{\mathbf{B}}\right) \cdot \vec{\mathbf{T}} = 0$. This shows that $v = v_0$ (v_0 is a constant). Consequently, $\frac{d\vec{\mathbf{v}}}{dt}$ reduces to $\frac{v_0^2}{R}\vec{\mathbf{N}}$ and we can deduce, according to [4.9], that:

$$\vec{\mathbf{N}} = \frac{qR}{mv}\left(\vec{\mathbf{T}} \times \vec{\mathbf{B}}\right) \qquad [4.11]$$

We can see that $\vec{\mathbf{N}}$ is also perpendicular to $\vec{\mathbf{B}}$. Let us write $\vec{\mathbf{v}}$ as $\vec{\mathbf{v}} = \vec{\mathbf{v}}_{\parallel} + \vec{\mathbf{v}}_{\perp}$, where $\vec{\mathbf{v}}_{\parallel}$ (respectively, $\vec{\mathbf{v}}_{\perp}$) is the part of $\vec{\mathbf{v}}$ parallel (respectively, perpendicular) to $\vec{\mathbf{B}}$. Since $\vec{\mathbf{B}}$ is uniform and perpendicular to $\vec{\mathbf{N}}$, we have:

$$0 = m\frac{d\vec{\mathbf{v}}}{dt} \cdot \vec{\mathbf{B}} = m\frac{d}{dt}\left(\vec{\mathbf{v}} \cdot \vec{\mathbf{B}}\right) = mB\frac{dv_{\parallel}}{dt} \qquad [4.12]$$

which implies that $\frac{dv_{\parallel}}{dt} = 0$ and, consequently, $v_{\parallel} = v_{\parallel}^0$ ($v_{\parallel}^0$ is a constant). Since $v^2 = v_{\parallel}^2 + v_{\perp}^2$ and $v = v_0$, we immediately deduce that $v_{\perp} = v_{\perp}^0$. Furthermore, $m\frac{v_0^2}{R}\vec{\mathbf{N}} = q\left(\vec{\mathbf{v}}_{\parallel} + \vec{\mathbf{v}}_{\perp}\right) \times \vec{\mathbf{B}} = q\vec{\mathbf{v}}_{\perp} \times \vec{\mathbf{B}}$. Turning to the norm, we find that $m\frac{v_0^2}{R} = |q|\, v_{\perp}^0 B$. It follows that:

$$R = m\frac{v_0^2}{|q|\, v_{\perp}^0 B} \qquad [4.13]$$

The curvature radius is therefore constant. The result is a trajectory of constant curvature radius, whose curvature center moves with constant velocity $v_{\parallel}^0$ in the direction of $\vec{\mathbf{B}}$. Since the magnetic field is orthogonal to $\vec{\mathbf{N}}$, the trajectory is a helix with axis directed along $\vec{\mathbf{B}}$, as illustrated in Figure 4.2. If $v_{\|} = 0$, the helix is reduced to a circle of radius R, in a plane perpendicular to $\vec{\mathbf{B}}$. When the $\vec{\mathbf{B}}$ field is not uniform, charged particles subjected only to the magnetic field no longer move helically. However, the charges still tend to describe trajectories that wrap around the field lines.

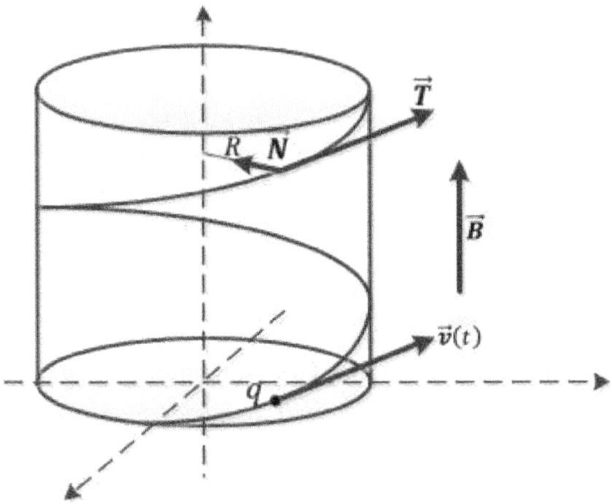

Figure 4.2. *Helicoidal motion of a charged particle along the Oz axis*

4.1.3.2. *Motion in Cartesian coordinates*

The problem can also be resolved using Cartesian coordinates. Let us assume that $\vec{\mathbf{B}} = B\vec{\mathbf{k}}$, $\vec{\mathbf{r}}(0) = 0$ (initial position) and $\vec{\mathbf{v}}(0) = (v_x)_0 \vec{\mathbf{i}} + (v_z)_0 \vec{\mathbf{k}}$ (initial velocity), with $(v_x)_0 > 0$. From [4.9], it can be derived that:

$$m\frac{dv_x}{dt} = qv_yB \qquad m\frac{dv_y}{dt} = -qv_xB \qquad m\frac{dv_z}{dt} = 0 \qquad [4.14]$$

The final equation yields $v_z = (v_z)_0$, that is, the velocity along Oz remains constant. By writing $w = v_x + \mathrm{i}v_y$, the first two equations can be combined in the form $m\frac{dw}{dt} + \mathrm{i}qBw = 0$. This equation admits the solution $w(t) = \lambda e^{-\mathrm{i}\omega_0 t}$, where λ is a constant that can be determined based on the initial conditions. The quantity $\omega_0 = \frac{qB}{m}$ is called the *Larmor frequency*. At the initial instant $t = 0$, we have $w(0) = (v_x)_0$. This makes it possible to obtain $\lambda = (v_x)_0$. We therefore get:

$$v_x(t) = (v_x)_0 \cos\omega_0 t \qquad v_y(t) = -(v_x)_0 \sin\omega_0 t \qquad [4.15]$$

To obtain the expressions $x(t)$, $y(t)$ and $z(t)$, let us integrate [4.15] with respect to time using the initial condition $\overrightarrow{\mathbf{r}}(0) = 0$:

$$x(t) = \frac{(v_x)_0}{\omega_0} \sin \omega_0 t \qquad y(t) = \frac{(v_x)_0}{\omega_0} (\cos \omega_0 t - 1) \qquad z(t) = (v_z)_0 t \qquad [4.16]$$

By eliminating t between the variables x and y, we find:

$$x^2 + \left[y + \frac{(v_x)_0}{\omega_0} \right]^2 = \left[\frac{(v_x)_0}{\omega_0} \right]^2 \qquad [4.17]$$

The trajectory is a helix wrapped around the cylinder with radius $R = \frac{(v_x)_0}{\omega_0}$ and axis $O'z$ passing through the point $O'\left(0, -\frac{(v_x)_0}{\omega_0}, 0\right)$. The rotation is clockwise around $O'z$ when $\omega_0 > 0$ (i.e. $q > 0$) and counterclockwise when $q < 0$.

4.1.4. *Transformation of $\overrightarrow{\mathrm{E}}$ and $\overrightarrow{\mathrm{B}}$ in a change of reference frames*

Consider two inertial reference frames $Oxyz$ and $O'x'y'z'$, one in translation relative to the other. $\overrightarrow{\mathbf{V}}_0$ is the translational velocity of $O'x'y'z'$ relative to $Oxyz$. In classical mechanics, the Lorentz force experienced by an electric charge is the same in both reference frames $q\left(\overrightarrow{\mathbf{E}} + \overrightarrow{\mathbf{v}} \times \overrightarrow{\mathbf{B}}\right) = q\left(\overrightarrow{\mathbf{E}}' + \overrightarrow{\mathbf{v}}' \times \overrightarrow{\mathbf{B}}'\right)$. The charge q is an invariant physical quantity. Since $\overrightarrow{\mathbf{v}} = \overrightarrow{\mathbf{V}}_0 + \overrightarrow{\mathbf{v}}'$, we deduce that $q\left(\overrightarrow{\mathbf{E}} + \overrightarrow{\mathbf{V}}_0 \times \overrightarrow{\mathbf{B}} + \overrightarrow{\mathbf{v}}' \times \overrightarrow{\mathbf{B}}\right) = q\left(\overrightarrow{\mathbf{E}}' + \overrightarrow{\mathbf{v}}' \times \overrightarrow{\mathbf{B}}'\right)$. This last expression must be true for any velocity $\overrightarrow{\mathbf{v}}'$. It follows that:

$$\overrightarrow{\mathbf{E}}' = \overrightarrow{\mathbf{E}} + \overrightarrow{\mathbf{V}}_0 \times \overrightarrow{\mathbf{B}} \qquad \overrightarrow{\mathbf{B}}' = \overrightarrow{\mathbf{B}} \qquad [4.18]$$

These equations show that the electric or magnetic character depends on the reference frame. The fields $\overrightarrow{\mathbf{E}}$ and $\overrightarrow{\mathbf{B}}$ are not independent, but form a whole: the electromagnetic field. According to [4.18], $\overrightarrow{\mathbf{B}}$ does not appear to depend on $\overrightarrow{\mathbf{E}}$. However, these calculations were performed in the context of Galilean relativity. In the context of special relativity, we can show that at low speeds, the second equation of system [4.18] must be replaced by:

$$\overrightarrow{\mathbf{B}}' \simeq \overrightarrow{\mathbf{B}} - \frac{\overrightarrow{\mathbf{V}}_0 \times \overrightarrow{\mathbf{E}}}{c^2} \qquad [4.19]$$

4.1.5. *Hall effect*

The Hall effect, observed in 1880, shows that a wire carrying an electric current I, immersed in a magnetic field perpendicular to it, creates an electric field also perpendicular to the wire. The Hall effect can be used to determine the sign of charge carriers in a conductor. To see this, let us consider a plate through which a current I is flowing and observe the movement of the charge carriers. They are subject to the magnetic part of the Lorentz force, which deflects negatively charged particles (respectively, positively charged particles) toward the top (respectively, bottom) of the plate, as illustrated in Figure 4.3. As a result, the upper side is charged negatively (respectively, positively), while the lower side is charged positively (respectively, negatively). This asymmetry creates an electric field called the *Hall field*, which acts on the moving particles via the Coulomb force, which is opposed to the magnetic force. The Hall field is directed from bottom to top (respectively, from top to bottom) if the charge carriers are charged negatively (respectively, positively). This makes it possible to determine the signs of the charge carriers by examining the sign of *the Hall voltage*.

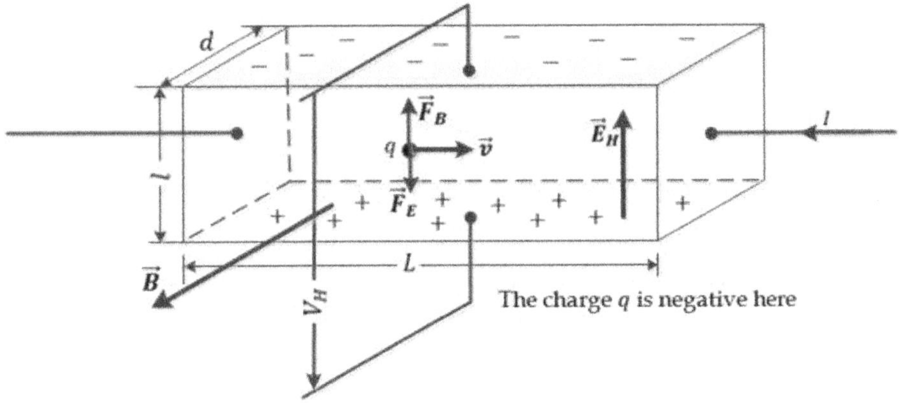

Figure 4.3. *Hall effect for a metal conductor. The magnetic field perpendicular to the plate deflects the charges in a transverse manner, giving rise to an electric field and Hall voltage. The polarity of the voltage is the same as the sign of the charge carriers*

In a transient regime, the magnetic part of the Lorentz force is larger than the Coulomb force produced by the Hall field (which goes from 0 to its maximum value E_H), and the charge carriers are deflected. The permanent regime is reached when $q\vec{\mathbf{E}}_H + q\vec{\mathbf{v}} \times \vec{\mathbf{B}} = \vec{\mathbf{0}}$, that is:

$$E_H = vB \qquad [4.20]$$

and the charge carriers are no longer deflected. In a permanent regime, the Hall field gives rise to Hall voltage:

$$V_H = E_H l = vBl \quad [4.21]$$

where l is the width of the conductive plate. The Hall voltage therefore depends on the velocity of the charge carriers v and the magnetic field strength B. It is possible to relate v to the current strength I through the current density J, $I = JS = nqvS = nqvdl$. This yields:

$$v = \frac{I}{nqdl} \quad [4.22]$$

where d and n are the plate thickness and the number of charge carriers per unit volume, respectively. Consequently, the Hall voltage is:

$$V_H = \frac{IB}{nqd} = R_H \frac{IB}{d} \quad [4.23]$$

$R_H = \frac{1}{nq}$ is called the *Hall constant.*

4.2. Magnetic field, Biot and Savart law

According to [4.3], the magnetic field, created at position $\vec{\mathbf{r}}$ by a point-like charge q' located at position $\vec{\mathbf{r}}'$ and animated by a velocity $\vec{\mathbf{v}}'$, is equal to:

$$\vec{\mathbf{B}}(\vec{\mathbf{r}}) = \frac{\mu_0}{4\pi} \frac{q'\vec{\mathbf{v}'} \times (\vec{\mathbf{r}} - \vec{\mathbf{r}}')}{|\vec{\mathbf{r}} - \vec{\mathbf{r}}'|^3} \quad [4.24]$$

The magnetic field satisfies the superposition principle. If $\vec{\mathbf{B}}_1$ and $\vec{\mathbf{B}}_2$ are two magnetic fields created at point $\vec{\mathbf{r}}$, by charges q_1' and q_2', respectively, then the total magnetic field, measured at the point $\vec{\mathbf{r}}$, will be $\vec{\mathbf{B}} = \vec{\mathbf{B}}_1 + \vec{\mathbf{B}}_2$. Now consider several point charges $q_1', \ldots, q_N'$ moving with the velocities $\vec{\mathbf{v}}'_1, \ldots, \vec{\mathbf{v}}'_N$. The magnetic field produced at position $\vec{\mathbf{r}}$ is the vector sum of the individual fields:

$$\vec{\mathbf{B}}(\vec{\mathbf{r}}) = \frac{\mu_0}{4\pi} \sum_{i=1}^{N} \frac{q_i' \vec{\mathbf{v}}'_i \times (\vec{\mathbf{r}} - \vec{\mathbf{r}}'_i)}{|\vec{\mathbf{r}} - \vec{\mathbf{r}}'_i|^3} \quad [4.25]$$

In the case of a continuous charge distribution, the magnetic field is written as follows:

$$\vec{\mathbf{B}}\left(\vec{\mathbf{r}}\right)=\frac{\mu_0}{4\pi}\int_{\text{dist.}}\frac{dq'\vec{\mathbf{v}}'\times\left(\vec{\mathbf{r}}-\vec{\mathbf{r}}'\right)}{\left|\vec{\mathbf{r}}-\vec{\mathbf{r}}'\right|^3}\qquad [4.26]$$

where $\vec{\mathbf{v}}'$ is the average velocity of the differential charge element dq'. If the charge is distributed over a line with a linear current density $\vec{\mathbf{I}}'$, or a surface with a surface current density $\vec{\mathbf{K}}'$, or a volume with a volume current density $\vec{\mathbf{J}}'$, the differential current elements are used: $\vec{\mathbf{I}}'\left(\vec{\mathbf{r}}'\right)dl'$ (line current), $\vec{\mathbf{K}}'\left(\vec{\mathbf{r}}'\right)dS'$ (surface current) or $\vec{\mathbf{J}}'\left(\vec{\mathbf{r}}'\right)d\tau'$ (volume current). If there is only one type of charge carrier, we write: $\vec{\mathbf{I}}'\left(\vec{\mathbf{r}}'\right)=n_L'\left(\vec{\mathbf{r}}'\right)q'\vec{\mathbf{v}}'$ (n_L' is the density of charge carriers per unit length), $\vec{\mathbf{K}}'\left(\vec{\mathbf{r}}'\right)=n_S'\left(\vec{\mathbf{r}}'\right)q'\vec{\mathbf{v}}'$ (n_S' is the density of charge carriers per unit area) or $\vec{\mathbf{J}}'\left(\vec{\mathbf{r}}'\right)=n_\tau'\left(\vec{\mathbf{r}}'\right)q'\vec{\mathbf{v}}'$ (n_τ' is the density of charge carriers per unit volume), where q' and $\vec{\mathbf{v}}'$ are, respectively, the charge and the average velocity of the charge carriers. If, on the other hand, there are α types of charge carriers (electrons, ions, etc.), with charge q_α' and density n_α', each moving with an average velocity $\vec{\mathbf{v}}'_\alpha$, we must sum over all types $\vec{\mathbf{I}}'\left(\vec{\mathbf{r}}'\right)=\sum_\alpha n_{L,\alpha}'\left(\vec{\mathbf{r}}'\right)q_\alpha'\vec{\mathbf{v}}'_\alpha$, $\vec{\mathbf{K}}'\left(\vec{\mathbf{r}}'\right)=\sum_\alpha n_{S,\alpha}'\left(\vec{\mathbf{r}}'\right)q_\alpha'\vec{\mathbf{v}}'_\alpha$ or $\vec{\mathbf{J}}'\left(\vec{\mathbf{r}}'\right)=\sum_\alpha n_{\tau,\alpha}'\left(\vec{\mathbf{r}}'\right)q_\alpha'\vec{\mathbf{v}}'_\alpha$. Thus, the total magnetic field of a current distribution is obtained by integrating all of the contributions from the differential elements:

$$\vec{\mathbf{B}}\left(\vec{\mathbf{r}}\right)=\begin{cases}\dfrac{\mu_0}{4\pi}\displaystyle\int_{L'}\frac{\vec{\mathbf{I}}'\left(\vec{\mathbf{r}}'\right)\times\left(\vec{\mathbf{r}}-\vec{\mathbf{r}}'\right)}{\left|\vec{\mathbf{r}}-\vec{\mathbf{r}}'\right|^3}dl' & \text{(line current)}\\ \dfrac{\mu_0}{4\pi}\displaystyle\int_{S'}\frac{\vec{\mathbf{K}}'\left(\vec{\mathbf{r}}'\right)\times\left(\vec{\mathbf{r}}-\vec{\mathbf{r}}'\right)}{\left|\vec{\mathbf{r}}-\vec{\mathbf{r}}'\right|^3}dS' & \text{(surface current)}\\ \dfrac{\mu_0}{4\pi}\displaystyle\int_{\tau'}\frac{\vec{\mathbf{J}}'\left(\vec{\mathbf{r}}'\right)\times\left(\vec{\mathbf{r}}-\vec{\mathbf{r}}'\right)}{\left|\vec{\mathbf{r}}-\vec{\mathbf{r}}'\right|^3}d\tau' & \text{(volume current)}\end{cases}\qquad [4.27]$$

This is *Biot and Savart law*. The direction of the magnetic field created by a current element is determined by the right-hand rule. If the index finger points in the direction of the current and the middle finger points toward the position of the (magnetic) field, then the thumb points in the direction of the magnetic field. This is the *three-finger rule*. This magnetic field $\vec{\mathbf{B}}$ can then exert a force on other currents.

For a neutral wire conductor, the current is always tangent to the wire in any reference frame. Therefore, it is possible to replace $\vec{\mathbf{I}}'\left(\vec{\mathbf{r}}'\right)dl'$ by $I'\left(\vec{\mathbf{r}}'\right)d\vec{\mathbf{l}}'$, where $d\vec{\mathbf{l}}'$ is a length element of the wire (since $d\vec{\mathbf{I}}'\left(\vec{\mathbf{r}}'\right)$ and $d\vec{\mathbf{l}}'$ are parallel and point in the same direction). Since the current is the same everywhere along the wire, it is possible to take I' out of the integral. For a filiform conductor, the Biot and Savart law reduces to:

$$\vec{\mathbf{B}}\left(\vec{\mathbf{r}}\right) = \frac{\mu_0 I'}{4\pi}\int_{L'} \frac{d\,\vec{\mathbf{l}}\,' \times \left(\vec{\mathbf{r}} - \vec{\mathbf{r}}\,'\right)}{\left|\vec{\mathbf{r}} - \vec{\mathbf{r}}\,'\right|^3} \qquad [4.28]$$

For a closed circuit, we get:

$$\vec{\mathbf{B}}\left(\vec{\mathbf{r}}\right) = \frac{\mu_0 I'}{4\pi}\oint_{\text{circuit}} \frac{d\,\vec{\mathbf{l}}\,' \times \left(\vec{\mathbf{r}} - \vec{\mathbf{r}}\,'\right)}{\left|\vec{\mathbf{r}} - \vec{\mathbf{r}}\,'\right|^3} \qquad [4.29]$$

Equations [4.28] and [4.29] can be obtained using Biot and Savart law for a volume distribution. Consider a tube forming a circuit (closed or open) of infinitely small cross-section S', that is, tending toward 0. A permanent current I' flows through the tube, as shown in Figure 4.4. The elementary volume $d\tau'$ is written as: $d\tau' = dS'dl'$, where dS' is a transverse elementary surface located at M'' and dl' is an element of wire length. Since $d'\vec{\mathbf{J}}\,'\left(\vec{\mathbf{r}}\,''\right)$ is parallel to the vectors $d'\,\vec{\mathbf{l}}\,'$ and $d'\,\vec{\mathbf{S}}\,'$, it follows that:

$$\begin{aligned}
\vec{\mathbf{B}}\left(\vec{\mathbf{r}}\right) &= \lim_{S'\longrightarrow 0} \frac{\mu_0}{4\pi}\int_{\text{circuit}} dl' \int_{S'} \frac{\vec{\mathbf{J}}\,'\left(\vec{\mathbf{r}}\,''\right) \times \left(\vec{\mathbf{r}} - \vec{\mathbf{r}}\,''\right)}{\left|\vec{\mathbf{r}} - \vec{\mathbf{r}}\,''\right|^3} dS' \\
&= \frac{\mu_0}{4\pi}\int_{\text{circuit}} \frac{\left[\lim_{S'\longrightarrow 0}\int_{S'} \vec{\mathbf{J}}\,'\left(\vec{\mathbf{r}}\,'\right) dS'\right] dl' \times \left(\vec{\mathbf{r}} - \vec{\mathbf{r}}\,'\right)}{\left|\vec{\mathbf{r}} - \vec{\mathbf{r}}\,'\right|^3} \\
&= \frac{\mu_0}{4\pi}\int_{\text{circuit}} \frac{\left[\lim_{S'\longrightarrow 0}\int_{S'} \vec{\mathbf{J}}\,'\left(\vec{\mathbf{r}}\,'\right) \cdot d\vec{\mathbf{S}}\,'\right] d\,\vec{\mathbf{l}}\,' \times \left(\vec{\mathbf{r}} - \vec{\mathbf{r}}\,'\right)}{\left|\vec{\mathbf{r}} - \vec{\mathbf{r}}\,'\right|^3} \\
&= \frac{\mu_0}{4\pi}\int_{\text{circuit}} \frac{I' d\,\vec{\mathbf{l}}\,' \times \left(\vec{\mathbf{r}} - \vec{\mathbf{r}}\,'\right)}{\left|\vec{\mathbf{r}} - \vec{\mathbf{r}}\,'\right|^3}
\end{aligned} \qquad [4.30]$$

4.2.1. *Symmetry properties of the magnetic field*

Symmetry properties are fundamental as they simplify magnetic field calculations. Indeed, these symmetry properties of the field are related to the symmetry and invariance properties of the current sources that give rise to the field. So, if we know the symmetry properties of $\vec{\mathbf{J}}$, we can arrive at the symmetry properties of the magnetic field. Unlike the electric field, the magnetic field is an axial vector or *pseudovector*, i.e., a vector whose direction is defined by the orientation convention of space and which depends on this convention. The vector product of two vectors (respectively, pseudovectors) is a pseudovector (respectively, a vector), while the product of a vector and a pseudo-vector is a pseudovector.

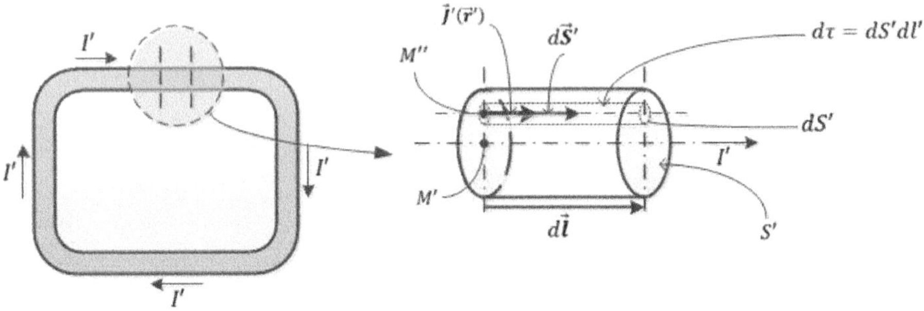

Figure 4.4. *Biot and Savart law for a closed circuit made up of a tube of very small cross-section, tending to zero, with a permanent current I' traveling through it*

As with the electric field, we observe a few simple rules:

– If $\vec{\mathbf{J}}$ is translation-invariant along the Oz axis, the magnetic field $\vec{\mathbf{B}}$ does not depend on z.

– If $\vec{\mathbf{J}}$ is rotation-invariant around the Oz axis (axial symmetry), the magnetic field $\vec{\mathbf{B}}$ is also rotation-invariant.

– If $\vec{\mathbf{J}}$ is translation-invariant along Oz and rotation-invariant around the same axis (cylindrical symmetry), then this is also true of $\vec{\mathbf{B}}$.

– If $\vec{\mathbf{J}}$ is poloidal (i.e. carried by $\vec{\mathbf{e}}_r$ or $\vec{\mathbf{e}}_z \equiv \vec{\mathbf{k}}$), then $\vec{\mathbf{B}}$ is toroidal (i.e. carried by $\vec{\mathbf{e}}_\varphi$).

– If $\vec{\mathbf{J}}$ is toroidal, then $\vec{\mathbf{B}}$ is poloidal.

– If $\vec{\mathbf{J}}$ admits a symmetry plane (P), then the magnetic field $\vec{\mathbf{B}}(M)$ is transformed to the opposite of the symmetric with respect to this plane, as shown in Figure 4.5(a):

$$\vec{\mathbf{B}}(M') = \vec{\mathbf{B}}_{\parallel}(M') + \vec{\mathbf{B}}_{\perp}(M') = -\vec{\mathbf{B}}_{\parallel}(M) + \vec{\mathbf{B}}_{\perp}(M) \qquad [4.31]$$

where M' is the symmetrical point of M with respect to the plane. If $M' \in (P)$, $\vec{\mathbf{B}}(M)$ is perpendicular to the plane at this point.

– If $\vec{\mathbf{J}}$ admits an antisymmetry plane (P'), then the magnetic field $\vec{\mathbf{B}}(M)$ is transformed to:

$$\vec{\mathbf{B}}(M') = \vec{\mathbf{B}}_{\parallel}(M') + \vec{\mathbf{B}}_{\perp}(M') = \vec{\mathbf{B}}_{\parallel}(M) - \vec{\mathbf{B}}_{\perp}(M) \quad [4.32]$$

as seen in Figure 4.5(b). If $M \in (P')$, $\vec{\mathbf{B}}(M)$ is contained in this plane.

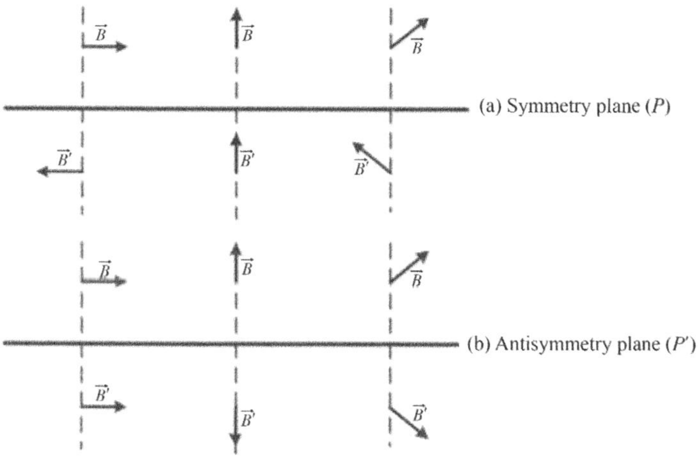

Figure 4.5. *Transformation of a magnetic field $\vec{\mathbf{B}}$ with respect to (a) a symmetry plane; (b) an antisymmetry plane*

4.2.2. *Examples of magnetic field calculation*

Let us look at a few examples of magnetic field calculations to illustrate these concepts.

EXAMPLE 4.1. (Magnetic Field Created by an Infinite Rectilinear Wire).– Consider an infinite wire along the axis Oz carrying a permanent current of intensity I. We wish to calculate the magnetic field at a point M located at a distance r from the wire. For reasons of symmetry, the point M is chosen in the Oxy plane, as shown in Figure 4.6(a). Since $\vec{\mathbf{r}} - \vec{\mathbf{r}}' = r\cos\varphi\,\vec{\mathbf{i}} + r\sin\varphi\,\vec{\mathbf{j}} - z'\vec{\mathbf{k}} = r\vec{\mathbf{e}}_r - z'\vec{\mathbf{k}}$, $|\vec{\mathbf{r}} - \vec{\mathbf{r}}'| = \sqrt{r^2 + z'^2}$ and $d\vec{\mathbf{l}}' = dz'\vec{\mathbf{k}}$, according to the Biot and Savart law [4.28], it is found that:

$$\begin{aligned}\vec{\mathbf{B}}(r,\varphi) &= \frac{\mu_0 I}{4\pi}\int_{-\infty}^{\infty}\frac{\vec{\mathbf{k}}\times\left(r\vec{\mathbf{e}}_r - z'\vec{\mathbf{k}}\right)}{\left(r^2+z'^2\right)^{3/2}}dz' \\ &= \frac{\mu_0 I}{4\pi}\int_{-\infty}^{\infty}\frac{r\vec{\mathbf{e}}_\varphi}{\left(r^2+z'^2\right)^{3/2}}dz' = \frac{\mu_0 I\vec{\mathbf{e}}_\varphi}{2\pi r} = \frac{\mu_0\left(\vec{\mathbf{I}}\times\vec{\mathbf{e}}_r\right)}{2\pi r}\end{aligned} \quad [4.33]$$

where $\vec{\mathbf{e}}_r = \cos\varphi\,\vec{\mathbf{i}} + \sin\varphi\,\vec{\mathbf{j}}$, $\vec{\mathbf{e}}_\varphi = -\sin\varphi\,\vec{\mathbf{i}} + \cos\varphi\,\vec{\mathbf{j}}$ and $\vec{\mathbf{I}} = I\vec{\mathbf{k}}$. Equation [4.33] can also be obtained from the Biot and Savart law for a volume current distribution by taking $\vec{\mathbf{J}}' = \frac{I}{r'}\delta(r')\,\delta(\varphi')\,\vec{\mathbf{k}}$ and, consequently, $\vec{\mathbf{J}}'\times\left(\vec{\mathbf{r}}-\vec{\mathbf{r}}'\right) = \frac{rI}{r'}\delta(r')\,\delta(\varphi')\,\vec{\mathbf{e}}_\varphi$. Therefore, we get:

$$\begin{aligned}\vec{\mathbf{B}} &= \frac{\mu_0 r I}{4\pi}\int_0^{\infty}\int_0^{2\pi}\int_{-\infty}^{\infty}\frac{\delta(r')\,\delta(\varphi')\,\vec{\mathbf{e}}_\varphi}{r'\left(r^2+z'^2\right)^{3/2}}r'dr'd\varphi'dz' \\ &= \frac{\mu_0 r I}{4\pi}\int_{-\infty}^{\infty}\frac{\vec{\mathbf{e}}_\varphi}{\left(r^2+z'^2\right)^{3/2}}dz' = \frac{\mu_0 I\vec{\mathbf{e}}_\varphi}{2\pi r}\end{aligned} \quad [4.34]$$

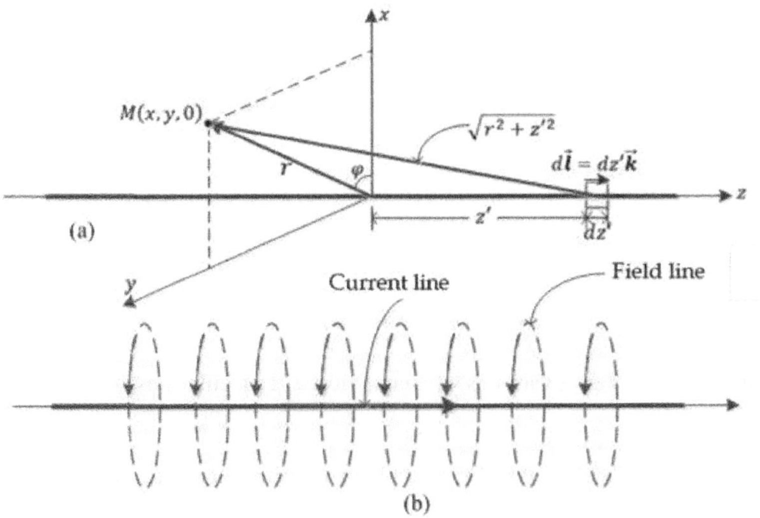

Figure 4.6. *The magnetic field created by an infinite wire, with a permanent current I traveling through it, wraps around the wire and is inversely proportional to the distance r (distance between the point M and the wire). For a color version of this figure, see www.iste.co.uk/abdesselam/electrodynamics1.zip*

Since the field is orthoradial, its field lines are circles located in planes perpendicular to the wire and centered on it. Moreover, since the direction of the current is given (here, in the Oz direction), the orientation of the field on these lines follows Maxwell's corkscrew rule. The field lines also rotate around the wire, as shown in Figure 4.6(b). This result follows from a general property of magnetic fields: $\overrightarrow{\mathbf{B}}$ *field lines are always closed curves that wrap around current lines in the direction indicated by the corkscrew rule.*

EXAMPLE 4.2. (Magnetic Field Created by a Rectilinear Wire of Finite Length).– Suppose the wire now has a finite length L. To simplify the problem, the origin O is chosen to be in the middle of the wire. Consequently, the magnetic field, measured at point $M = (x, y, 0)$, is given as:

$$\overrightarrow{\mathbf{B}}(x,y,0) = \frac{\mu_0 I \overrightarrow{\mathbf{e}}_\varphi}{4\pi r} \left. \frac{z'}{\sqrt{r^2 + z'^2}} \right|_{-\frac{L}{2}}^{\frac{L}{2}} = \frac{\mu_0 I L \overrightarrow{\mathbf{e}}_\varphi}{4\pi r \sqrt{r^2 + \frac{L^2}{4}}} \qquad [4.35]$$

EXAMPLE 4.3. (Magnetic Field Created by a Current Sheet without Thickness).– A constant surface current, $K_0 \overrightarrow{\mathbf{k}}$, flows along a conductive sheet located at $y = 0$. To calculate the magnetic field at $(0, y, 0)$, the sheet is divided into an infinite number of linear currents $K_0 dx'$, as shown in Figure 4.7. Each elementary current $K_0 dx'$ produces the infinitesimal magnetic field:

$$d\overrightarrow{\mathbf{B}} = \frac{\mu_0 K_0 dx'}{4\pi} \int_{-\infty}^{\infty} \frac{dz' \overrightarrow{\mathbf{k}} \times \left[y \overrightarrow{\mathbf{j}} - x' \overrightarrow{\mathbf{i}} - z' \overrightarrow{\mathbf{k}} \right]}{\left(x'^2 + y^2 + z'^2\right)^{3/2}} = -\frac{\mu_0 K_0 \left(y \overrightarrow{\mathbf{i}} + x' \overrightarrow{\mathbf{j}} \right) dx'}{2\pi \left(x'^2 + y^2\right)} \qquad [4.36]$$

Consequently, the total magnetic field is obtained by integrating over the whole sheet:

$$\overrightarrow{\mathbf{B}} = -\frac{\mu_0 K_0 y \overrightarrow{\mathbf{i}}}{2\pi} \int_{-\infty}^{\infty} \frac{dx'}{x'^2 + y^2} = \begin{cases} -\frac{1}{2}\mu_0 K_0 \overrightarrow{\mathbf{i}} & y > 0 \\ \frac{1}{2}\mu_0 K_0 \overrightarrow{\mathbf{i}} & y < 0 \end{cases} \qquad [4.37]$$

Using the *Heaviside function*, it can also be written as:

$$\overrightarrow{\mathbf{B}}(0,y,0) = -\frac{1}{2}\mu_0 K_0 \left(2H(y) - 1\right) \overrightarrow{\mathbf{i}} \qquad [4.38]$$

where $H(y)$ takes the value 1 for $y \geq 0$ and 0 otherwise.

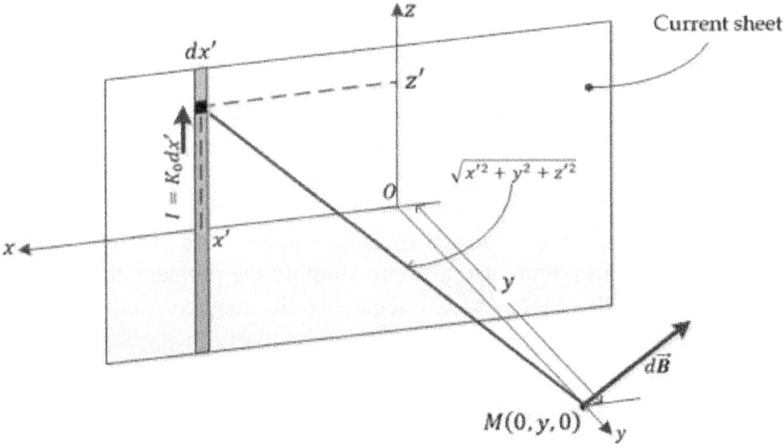

Figure 4.7. *A uniform surface current that is distributed over an infinite planar sheet generates a magnetic field in the opposite direction on each side of the sheet (the sheet is a symmetry plane here). The magnetic field is perpendicular to the surface current, but parallel to the plane of the sheet*

EXAMPLE 4.4. (Plate of Thickness d).– If the current $J_0 \overrightarrow{\mathbf{k}}$ (directed along Oz) is uniform on a plate of thickness d, as shown in Figure 4.8, it is enough to divide the plate into sheets with current $J_0 dy' \overrightarrow{\mathbf{k}}$. The magnetic field of each current sheet is given by $-\frac{1}{2}\mu_0 J_0 dy' \left(2H\left(y-y'\right)-1\right) \overrightarrow{\mathbf{i}}$. Summing the contributions of all of the sheets (the field is directed to $-\overrightarrow{bfi}$ on the left of the sheet and to $\overrightarrow{\mathbf{i}}$ on the right), we obtain:

$$\overrightarrow{\mathbf{B}} = \begin{cases} -\displaystyle\int_{-d/2}^{d/2} \frac{1}{2}\mu_0 J_0 dy' \overrightarrow{\mathbf{i}} = -\frac{1}{2}\mu_0 J_0 d\, \overrightarrow{\mathbf{i}} & y > \frac{d}{2} \\ -\displaystyle\int_{-d/2}^{y} \frac{1}{2}\mu_0 J_0 dy' \overrightarrow{\mathbf{i}} + \int_{y}^{d/2} \frac{1}{2}\mu_0 J_0 dy' \overrightarrow{\mathbf{i}} = -\mu_0 J_0 y\, \overrightarrow{\mathbf{i}} & |y| \leq \frac{d}{2} \\ \displaystyle\int_{-d/2}^{d/2} \frac{1}{2}\mu_0 J_0 dy' \overrightarrow{\mathbf{i}} = \frac{1}{2}\mu_0 J_0 d\, \overrightarrow{\mathbf{i}} & y < -\frac{d}{2} \end{cases} \quad [4.39]$$

EXAMPLE 4.5. (Parallel Sheets without Thickness).– If a second sheet, with a surface current flowing in the opposite direction $-K_0 \overrightarrow{\mathbf{k}}$, is placed at $y = a$ parallel to the first current sheet $K_0 \overrightarrow{\mathbf{k}}$ located at $y = 0$, as shown in Figure 4.9, the respective magnetic fields generated by the two planes are given by:

$$\vec{\mathbf{B}}_1 = \begin{cases} -\frac{1}{2}\mu_0 K_0 \vec{\mathbf{i}} & y > 0 \\ \frac{1}{2}\mu_0 K_0 \vec{\mathbf{i}} & y < 0 \end{cases} \qquad \vec{\mathbf{B}}_2 = \begin{cases} \frac{1}{2}\mu_0 K_0 \vec{\mathbf{i}} & y > a \\ -\frac{1}{2}\mu_0 K_0 \vec{\mathbf{i}} & y < a \end{cases}$$

[4.40]

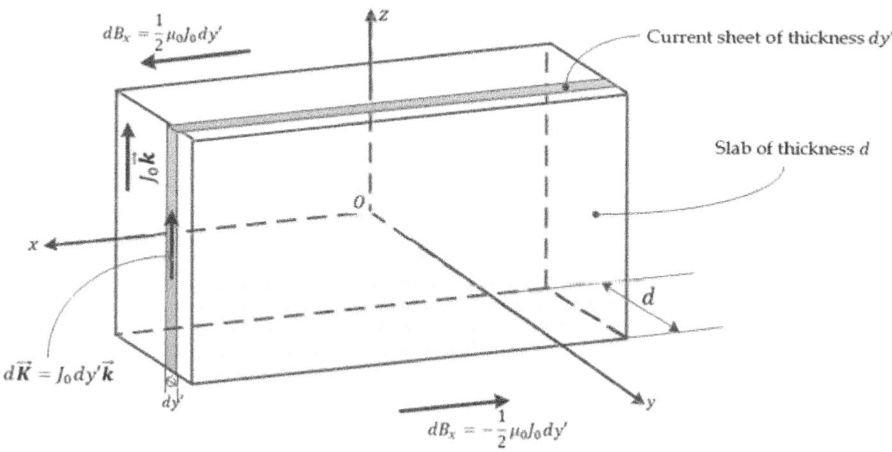

Figure 4.8. *The magnetic field generated by a plate is calculated by superposing the fields created by the current sheets stacked along its thickness*

Therefore, the total magnetic field produced by the entire system is:

$$\vec{\mathbf{B}} = \vec{\mathbf{B}}_1 + \vec{\mathbf{B}}_2 = \begin{cases} -\mu_0 K_0 \vec{\mathbf{i}} & 0 < y < a \\ 0 & \text{elsewhere} \end{cases} \qquad [4.41]$$

It is seen that within the region in between the two planes, the fields are added together, while they are canceled out outside of this region.

EXAMPLE 4.6. (Magnetic Field Created by a Circular Loop).– Let us now consider a circular loop centered on O, of radius R, located in the Oxy plane and traversed by a permanent current I.

1) *On the Oz axis*: we propose to calculate the magnetic field at a point M located on the Oz axis, as shown in Figure 4.10. In cylindrical coordinates, we have:

$d\overrightarrow{\mathbf{l}}' = Rd\varphi'\overrightarrow{\mathbf{e}}_{\varphi}$ and $\overrightarrow{\mathbf{r}} - \overrightarrow{\mathbf{r}}' = z\overrightarrow{\mathbf{k}} - R\overrightarrow{\mathbf{e}}_r$. Biot and Savart law yields:

$$\begin{aligned}\overrightarrow{\mathbf{B}}(0,0,z) &= \frac{\mu_0 I}{4\pi}\int_0^{2\pi}\frac{R\overrightarrow{\mathbf{e}}_{\varphi}\times\left(z\overrightarrow{\mathbf{k}} - R\overrightarrow{\mathbf{e}}_r\right)}{\left(R^2+z^2\right)^{3/2}}d\varphi' \\ &= \frac{\mu_0 IR}{4\pi}\int_0^{2\pi}\frac{z\overrightarrow{\mathbf{e}}_r + R\overrightarrow{\mathbf{k}}}{\left(R^2+z^2\right)^{3/2}}d\varphi'\end{aligned} \qquad [4.42]$$

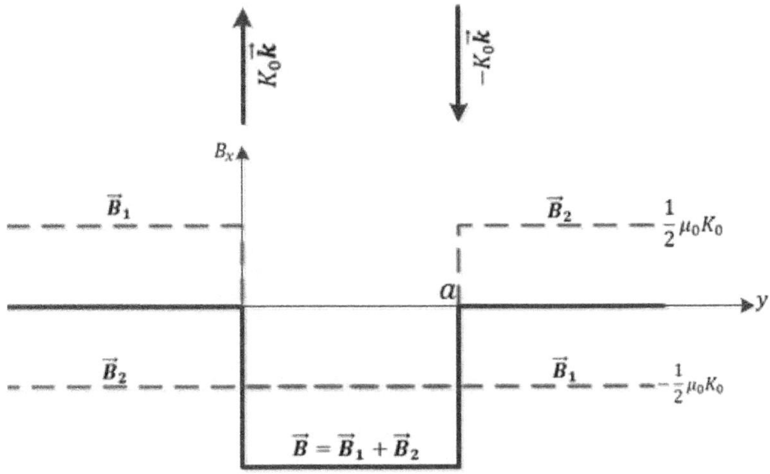

Figure 4.9. *Two sheets with parallel but opposite surface currents have fields that are added together in the region between the sheets and canceled out elsewhere. For a color version of this figure, see www.iste.co.uk/abdesselam/electrodynamics1.zip*

The first integral is clearly zero. Therefore:

$$\overrightarrow{\mathbf{B}}(0,0,z) = \frac{\mu_0 IR^2}{2\left(R^2+z^2\right)^{3/2}}\overrightarrow{\mathbf{k}} = \frac{\mu_0}{2\pi}\frac{\overrightarrow{\mathbf{m}}}{\left(R^2+z^2\right)^{3/2}} \qquad [4.43]$$

where $\overrightarrow{\mathbf{m}} = I\pi R^2\overrightarrow{\mathbf{k}}$ is the *magnetic moment of the loop*. This result can also be obtained by taking $\overrightarrow{\mathbf{J}}' = I\delta(r'-R)\,\delta(z')\,\overrightarrow{\mathbf{e}}_{\varphi}$. This gives $\overrightarrow{\mathbf{J}}'\times\left(\overrightarrow{\mathbf{r}} - \overrightarrow{\mathbf{r}}'\right) = I\delta(r'-R)\,\delta(z')\left(z\overrightarrow{\mathbf{e}}_r + R\overrightarrow{\mathbf{k}}\right)$ and, consequently:

$$\overrightarrow{\mathbf{B}}(0,0,z) = \frac{\mu_0 R^2 I\overrightarrow{\mathbf{k}}}{2\left(R^2+z^2\right)^{3/2}} \qquad [4.44]$$

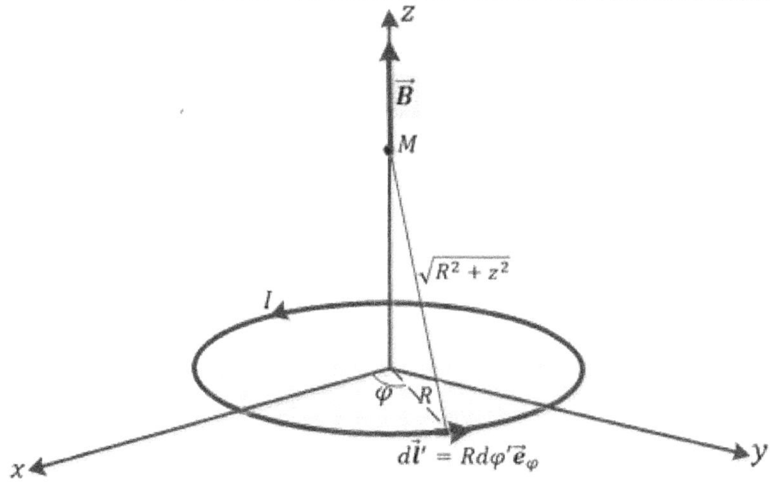

Figure 4.10. *The magnetic field generated by a circular current loop along its axis*

2) *Any point* $(x, 0, z)$: let us now calculate the magnetic field at another point M, which we will take in the Oxz plane to simplify the problem. This point in the plane has polar coordinates (r, φ), as shown in Figure 4.11. Since Oxz is an antisymmetry plane of the current distribution, the field at M lies in this plane, that is, it has no component along Oy. Since $\overrightarrow{\mathbf{r}} - \overrightarrow{\mathbf{r}}' = (r\sin\varphi - R\cos\varphi')\overrightarrow{\mathbf{i}} - R\sin\varphi'\overrightarrow{\mathbf{j}} + r\cos\varphi\overrightarrow{\mathbf{k}}$ and $d\overrightarrow{\mathbf{l}} = -R\sin\varphi' d\varphi'\overrightarrow{\mathbf{i}} + R\cos\varphi' d\varphi'\overrightarrow{\mathbf{j}}$, we find $d\overrightarrow{\mathbf{l}} \times (\overrightarrow{\mathbf{r}} - \overrightarrow{\mathbf{r}}') = R\left[r\cos\varphi\cos\varphi'\overrightarrow{\mathbf{i}} + r\cos\varphi\sin\varphi'\overrightarrow{\mathbf{j}} + (R - r\sin\varphi\cos\varphi')\overrightarrow{\mathbf{k}}\right]d\varphi'$ and $\left|\overrightarrow{\mathbf{r}} - \overrightarrow{\mathbf{r}}'\right| = \sqrt{r^2 + R^2 - 2Rr\sin\varphi\cos\varphi'}$. Biot and Savart law gives:

$$B_x = \frac{\mu_0 I}{4\pi}\int_0^{2\pi}\frac{Rr\cos\varphi\cos\varphi' d\varphi'}{\left(r^2 + R^2 - 2Rr\sin\varphi\cos\varphi'\right)^{3/2}} \qquad [4.45]$$

$$B_y = \frac{\mu_0 I}{4\pi}\int_0^{2\pi}\frac{Rr\cos\varphi\sin\varphi' d\varphi'}{\left(r^2 + R^2 - 2Rr\sin\varphi\cos\varphi'\right)^{3/2}} \qquad [4.46]$$

$$B_z = \frac{\mu_0 I}{4\pi}\int_0^{2\pi}\frac{\left(R^2 - rR\sin\varphi\cos\varphi'\right)d\varphi'}{\left(r^2 + R^2 - 2Rr\sin\varphi\cos\varphi'\right)^{3/2}} \qquad [4.47]$$

It is easy to see that the integral [4.46] is zero. Let us try to write integrals [4.45] and [4.47] in terms of elliptic integrals:

$$K\left(k\right)=\int_{0}^{\frac{\pi}{2}}\sqrt{1-k^{2}\sin^{2}\gamma}d\gamma=\frac{\pi}{2}\left[1+\left(\frac{1}{2}\right)^{2}k^{2}\right.$$
$$\left.+\left(\frac{1\cdot 3}{2\cdot 4}\right)^{2}k^{4}+\left(\frac{1\cdot 3\cdot 5}{2\cdot 4\cdot 6}\right)^{2}k^{6}+\cdots\right] \qquad [4.48]$$

$$E\left(k\right)=\int_{0}^{\frac{\pi}{2}}\frac{d\gamma}{\sqrt{1-k^{2}\sin^{2}\gamma}}=\frac{\pi}{2}\left[1-\left(\frac{1}{2}\right)^{2}k^{2}\right.$$
$$\left.-\left(\frac{1\cdot 3}{2\cdot 4}\right)^{2}\frac{k^{4}}{3}-\left(\frac{1\cdot 3\cdot 5}{2\cdot 4\cdot 6}\right)^{2}\frac{k^{6}}{5}-\cdots\right] \qquad [4.49]$$

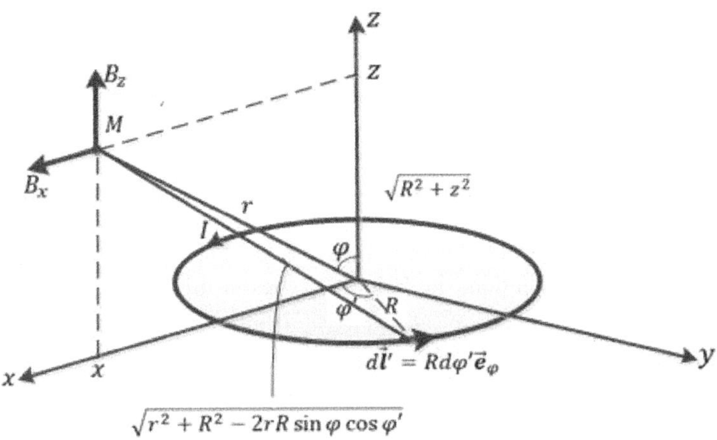

Figure 4.11. *The field $\vec{B}$ admits, outside the Oz axis, another radial component in the plane of the loop*

To do this, note that $r^2+R^2-2Rr\sin\varphi\cos\varphi' = \left(r^2+R^2+2R\sin\varphi\right)\left(1-\ k^2\cos^2\frac{\varphi'}{2}\right)$, where $k^2 = \frac{4Rr\sin\varphi}{r^2+R^2+2Rr\sin\varphi}$. Integrals [4.45] and [4.47] can therefore be rewritten as:

$$B_x=\frac{\mu_0 I}{4\pi}\frac{Rr\cos\varphi}{\left(r^2+R^2+2Rr\sin\varphi\right)^{3/2}}\int_0^{2\pi}\frac{\cos\varphi' d\varphi'}{\left(1-k^2\cos^2\frac{\varphi'}{2}\right)^{3/2}} \qquad [4.50]$$

$$B_z=\frac{\mu_0 I}{4\pi}\frac{R}{\left(r^2+R^2+2Rr\sin\varphi\right)^{3/2}}\int_0^{2\pi}\frac{\left(R-r\sin\varphi\cos\varphi'\right)d\varphi'}{\left(1-k^2\cos^2\frac{\varphi'}{2}\right)^{3/2}} \qquad [4.51]$$

By also writing that $x = r\sin\varphi$ and $z = r\cos\varphi$, we get:

$$B_x = \frac{\mu_0 I z k^2}{4\pi x\sqrt{(R+x)^2+z^2}}\int_0^{2\pi}\frac{2\cos^2\frac{\varphi'}{2}-1}{\left(1-k^2\cos^2\frac{\varphi'}{2}\right)^{3/2}}d\varphi' \quad [4.52]$$

$$B_z = \frac{\mu_0 I k^2}{16\pi x\sqrt{(R+x)^2+z^2}}\int_0^{2\pi}\frac{R-x\left(2\cos^2\frac{\varphi'}{2}-1\right)}{\left(1-k^2\cos^2\frac{\varphi'}{2}\right)^{3/2}}d\varphi' \quad [4.53]$$

where $k^2 = \frac{4Rx}{(R+x)^2+z^2}$. Because of periodicity, we can integrate between $-\pi$ and π instead of between 0 and 2π. Let us carry out the change of variable $\gamma = \frac{\pi}{2} - \frac{\varphi'}{2}$:

$$B_x = \frac{\mu_0 I z k^2}{4\pi x\sqrt{(R+x)^2+z^2}}\int_0^{\frac{\pi}{2}}\frac{\left(2\sin^2\gamma-1\right)d\gamma}{\left(1-k^2\sin^2\gamma\right)^{3/2}} \quad [4.54]$$

$$B_z = \frac{\mu_0 I k^2}{4\pi x\sqrt{(R+x)^2+z^2}}\left[R\int_0^{\frac{\pi}{2}}\frac{d\gamma}{\left(1-k^2\sin^2\gamma\right)^{3/2}} - x\int_0^{\frac{\pi}{2}}\frac{\left(2\sin^2\gamma-1\right)d\gamma}{\left(1-k^2\sin^2\gamma\right)^{3/2}}\right] \quad [4.55]$$

Let us calculate these integrals in terms of elliptic integrals:

$$\int_0^{\frac{\pi}{2}}\frac{d\gamma}{\left(1-k^2\sin^2\gamma\right)^{3/2}} = \frac{1}{k^2-1}\left.\frac{k^2\sin\gamma\cos\gamma}{\sqrt{1-k^2\sin^2\gamma}}\right|_0^{\frac{\pi}{2}} - \frac{E(k)}{k^2-1} = \frac{E(k)}{1-k^2}$$
$$\int_0^{\frac{\pi}{2}}\frac{\left(2\sin^2\gamma-1\right)d\gamma}{\left(1-k^2\sin^2\gamma\right)^{3/2}} = -\frac{2K(k)}{k^2} + \frac{2-k^2}{k^2}\int_0^{\frac{\pi}{2}}\frac{d\gamma}{\left(1-k^2\sin^2\gamma\right)^{3/2}}$$
$$= -\frac{2}{k^2}K(k) + \frac{2-k^2}{k^2}\frac{E(k)}{1-k^2} \quad [4.56]$$

Thus, we find:

$$B_x = \frac{\mu_0 I z}{2\pi x\sqrt{(R+x)^2+z^2}}\left[-K(k) + \frac{R^2+x^2+z^2}{(R-x)^2+z^2}E(k)\right] \quad [4.57]$$

$$B_z = \frac{\mu_0 I}{2\pi\sqrt{(R+x)^2+z^2}}\left[K(k) + \frac{R^2-x^2-z^2}{(R-x)^2+z^2}E(k)\right] \quad [4.58]$$

In particular, for a point on the Oz axis, that is, when $x = 0$ and therefore $k = 0$, the preceding formulas are reduced to $B_x = 0$ and $B_z = \frac{\mu_0 I}{2} \frac{R^2}{(R^2+z^2)^{3/2}}$.

EXAMPLE 4.7. (Helmholtz Coil).– Let us place another loop, identical to the first, at $z = d$. The total magnetic field measured along the Oz axis is calculated by superposing the fields produced by each loop separately:

$$B_z(0,0,z) = \frac{\mu_0 I R^2}{2} \left[\frac{1}{(R^2+z^2)^{3/2}} + \frac{1}{\left(R^2+(z-d)^2\right)^{3/2}} \right] \qquad [4.59]$$

B_z admits a maximum at the point $z = \frac{d}{2}$, that is, $(B_z)_{\max} = B_z\left(0,0,\frac{d}{2}\right) = \frac{\mu_0 I R^2}{(R^2+d^2/4)^{3/2}}$, and has an axis of symmetry passing through this point. In particular, if we fix $d = R$, it can be seen that at the midpoint between the two loops the maximum field is $(B_z)_{\max} = 0.71\frac{\mu_0 I}{R}$, while at the centers of the two loops $B_z(0,0,0) = B_z(0,0,R) = 0.68\frac{\mu_0 I}{R} = 0.96\,(B_z)_{\max}$. B_z is practically uniform between the two loops, as shown in Figure 4.12. This configuration is called a *Helmholtz coil*.

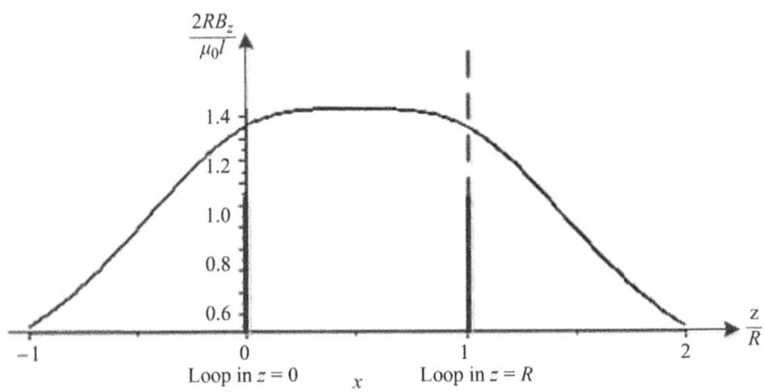

Figure 4.12. *Distribution of magnetic field in the Helmholtz coil: B_z is maximal at $R/2$ and almost uniform in the boundary region between the two loops*

EXAMPLE 4.8. (Field Created by a Square Circuit (or Square Loop)).– A square circuit with side $2a$, center O and located in the Oxy plane, has a permanent current I flowing through it, as shown in Figure 4.13. Let M be a point located on the Oz

axis at a distance z from the center O. We wish to calculate the magnetic field produced by the square circuit at point M. According to Biot and Savart law, we can write:

$$\begin{aligned}\overrightarrow{\mathbf{B}}(0,0,z) &= \frac{\mu_0 I}{4\pi}\left[\int_a^{-a} \frac{dx'\,\overrightarrow{\mathbf{i}} \times \left(z\,\overrightarrow{\mathbf{k}} - x'\,\overrightarrow{\mathbf{i}} - a\,\overrightarrow{\mathbf{j}}\right)}{\left(z^2 + x'^2 + a^2\right)^{3/2}}\right.\\ &+ \int_a^{-a} \frac{dy'\,\overrightarrow{\mathbf{j}} \times \left(z\,\overrightarrow{\mathbf{k}} + a\,\overrightarrow{\mathbf{i}} - y'\,\overrightarrow{\mathbf{j}}\right)}{\left(z^2 + a^2 + y'^2\right)^{3/2}} + \int_{-a}^{a} \frac{dx'\,\overrightarrow{\mathbf{i}} \times \left(z\,\overrightarrow{\mathbf{k}} - x'\,\overrightarrow{\mathbf{i}} + a\,\overrightarrow{\mathbf{j}}\right)}{\left(z^2 + x'^2 + a^2\right)^{3/2}}\\ &+ \left.\int_{-a}^{a} \frac{dy'\,\overrightarrow{\mathbf{j}} \times \left(z\,\overrightarrow{\mathbf{k}} - a\,\overrightarrow{\mathbf{i}} - y'\,\overrightarrow{\mathbf{j}}\right)}{\left(z^2 + a^2 + y'^2\right)^{3/2}}\right] = \frac{\mu_0 I}{\pi}\int_{-a}^{a} \frac{a\,\overrightarrow{\mathbf{k}}\,dx'}{\left(z^2 + x'^2 + a^2\right)^{3/2}}\\ &= \frac{\mu_0 I a^2}{\pi}\frac{2\,\overrightarrow{\mathbf{k}}}{\left(z^2 + a^2\right)\sqrt{z^2 + 2a^2}}\end{aligned} \quad [4.60]$$

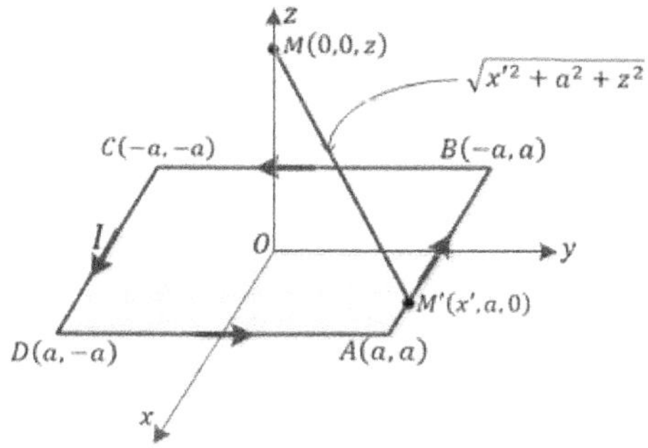

Figure 4.13. *Magnetic field created by a square circuit along the Oz axis*

EXAMPLE 4.9. (Field Created by a Solenoid).– A solenoid is made up of wire wrapped around a cylinder of length L and radius R. It appears as a juxtaposition of N coaxial loops, with $n = N/L$ loops per unit length, as shown in Figure 4.14. The elementary magnetic field $d\overrightarrow{\mathbf{B}}$ created by the ndz' loops, at the point $M\,(0,0,z)$, is:

$$d\overrightarrow{\mathbf{B}} = \frac{\mu_0 R^2 I\,\overrightarrow{\mathbf{k}}\,ndz'}{2\left[R^2 + (z'-z)^2\right]^{3/2}} \quad [4.61]$$

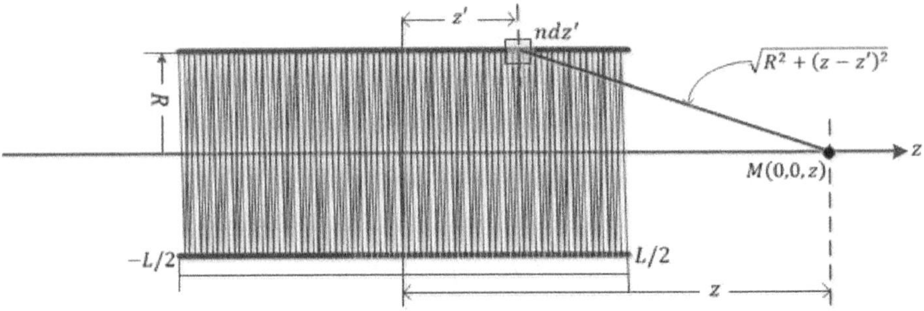

Figure 4.14. *The magnetic field $\vec{\mathbf{B}}$ generated by a solenoid is the superposition of all the magnetic fields created by each loop separately*

The total magnetic field is obtained by integrating along the solenoid, that is, between $-\frac{L}{2}$ and $\frac{L}{2}$:

$$\begin{aligned}\vec{\mathbf{B}}(0,0,z) &= \int_{-\frac{L}{2}}^{\frac{L}{2}} \frac{\mu_0 n R^2 I \vec{\mathbf{k}}\, dz'}{2\left[R^2 + (z'-z)^2\right]^{3/2}} = \frac{\mu_0 n I \vec{\mathbf{k}}}{2} \left.\frac{z'-z}{\sqrt{R^2+(z'-z)^2}}\right|_{-\frac{L}{2}}^{\frac{L}{2}} \\ &= \frac{\mu_0 n I \vec{\mathbf{k}}}{2}\left[\frac{\frac{L}{2}-z}{\sqrt{R^2+\left(\frac{L}{2}-z\right)^2}} + \frac{\frac{L}{2}+z}{\sqrt{R^2+\left(\frac{L}{2}+z\right)^2}}\right]\end{aligned} \quad [4.62]$$

For an infinite solenoid, that is, when $L \longrightarrow \infty$, we get:

$$\vec{\mathbf{B}}(0,0,z) = \mu_0 n I \vec{\mathbf{k}} \quad [4.63]$$

4.3. Conservation of magnetic flux

Biot and Savart law can be used to calculate the magnetic field created by any volume $\vec{\mathbf{J}}$, surface $\vec{\mathbf{K}}$ or line $\vec{\mathbf{I}}$ current distribution. This law is similar to Coulomb's law, which is the basis of the fundamental laws of electrostatics: $\vec{\nabla} \cdot \vec{\mathbf{E}} = \frac{\rho}{\varepsilon_0}$ and $\vec{\nabla} \times \vec{\mathbf{E}} = 0$. The magnetic field $\vec{\mathbf{B}}(\vec{\mathbf{r}})$ created at the point $\vec{\mathbf{r}}$ by a source $\vec{\mathbf{J}}' \equiv \vec{\mathbf{J}}'(\vec{\mathbf{r}}')$ (to make notations more convenient) is written as:

$$\vec{\mathbf{B}}(\vec{\mathbf{r}}) = \frac{\mu_0}{4\pi}\int_{\tau'} \frac{\vec{\mathbf{J}}'(\vec{\mathbf{r}}') \times (\vec{\mathbf{r}} - \vec{\mathbf{r}}')}{|\vec{\mathbf{r}} - \vec{\mathbf{r}}'|^3} d\tau' \quad [4.64]$$

Let us calculate the divergence of $\vec{\mathbf{B}}$:

$$\begin{aligned}\vec{\nabla}\cdot\vec{\mathbf{B}} &= \frac{\mu_0}{4\pi}\int_{\tau'}\vec{\nabla}\cdot\left[\frac{\vec{\mathbf{J}}'\times(\vec{\mathbf{r}}-\vec{\mathbf{r}}')}{|\vec{\mathbf{r}}-\vec{\mathbf{r}}'|^3}\right]d\tau' \\ &= \frac{\mu_0}{4\pi}\int_{\tau'}\left[\frac{\vec{\mathbf{r}}-\vec{\mathbf{r}}'}{|\vec{\mathbf{r}}-\vec{\mathbf{r}}'|^3}\cdot\left(\vec{\nabla}\times\vec{\mathbf{J}}'\right)\right. \\ &\left.+\vec{\mathbf{J}}'\cdot\left(\vec{\nabla}\times\left(\vec{\nabla}\left(\frac{1}{|\vec{\mathbf{r}}-\vec{\mathbf{r}}'|}\right)\right)\right)\right]d\tau'\end{aligned} \qquad [4.65]$$

where identities $\vec{\nabla}\cdot\left(\vec{\mathbf{A}}_1\times\vec{\mathbf{A}}_2\right) = \vec{\mathbf{A}}_2\cdot\left(\vec{\nabla}\times\vec{\mathbf{A}}_1\right) - \vec{\mathbf{A}}_1\cdot\left(\vec{\nabla}\times\vec{\mathbf{A}}_2\right)$ and $\vec{\nabla}\left(\frac{1}{|\vec{\mathbf{r}}-\vec{\mathbf{r}}'|}\right) = -\frac{\vec{\mathbf{r}}-\vec{\mathbf{r}}'}{|\vec{\mathbf{r}}-\vec{\mathbf{r}}'|^3}$ have been used. Since $\vec{\nabla}\times\left[\vec{\nabla}\left(\frac{1}{|\vec{\mathbf{r}}-\vec{\mathbf{r}}'|}\right)\right] = 0$ and $\vec{\nabla}$ acts only on $\vec{\mathbf{r}}$, that is, $\vec{\nabla}\times\vec{\mathbf{J}}' = 0$, we get:

$$\vec{\nabla}\cdot\vec{\mathbf{B}} = 0 \qquad [4.66]$$

This is the *fundamental law of magnetostatics* or the *Maxwell–Thomson equation*. Unlike the Maxwell–Gauss equation, equation [4.66] shows that there is no source term on the right-hand side. There are no magnetic monopole charges. It follows that:

1) If S is a closed surface, the divergence theorem gives $\oint_S \vec{\mathbf{B}}\cdot d\vec{\mathbf{S}} = \int_\tau \vec{\nabla}\cdot\vec{\mathbf{B}}\,d\tau = 0$. This shows that:

$$\Phi_m = \oint_S \vec{\mathbf{B}}\cdot d\vec{\mathbf{S}} = 0 \qquad [4.67]$$

where τ is the volume bounded by S. This is the *conservation law of magnetic flux* Φ_m. The magnetic flux through a closed surface is always zero. In other words, the flux entering a surface is equal to the flux leaving it. Since there are no magnetic charges, the field lines are always closed.

2) The flow through an open surface based on a contour Γ is independent of the choice of surface. If S_1 and S_2 are two distinct surfaces based on Γ, then the surface $S = S_1 + S_2$ forms a closed surface. According to [4.67], it can be written that $0 = \Phi_S - \Phi_{S_1} - \Phi_{S_2}$ and, consequently:

$$\Phi_{S_1} = \Phi_{S_2} \qquad [4.68]$$

4.4. Laplace force

4.4.1. *Definitions*

Consider a neutral conductor traversed by a current density $\overrightarrow{\mathbf{J}}$ and immersed in a magnetic field $\overrightarrow{\mathbf{B}}$. The conductor moves with a velocity $\overrightarrow{\mathbf{v}}_e$ relative to a fixed Galilean frame of reference $Oxyz$. Two types of charge carriers (electrons and ions), indexed by $\pm$, move inside this conductor. Let $q_\pm$ be the charge of these carriers, $n_\pm$ their densities per unit volume, and $\overrightarrow{\mathbf{v}}_\pm$ their average velocities relative to the conductor. The infinitesimal magnetic force exerted on all charge carriers contained in a $d\tau$ elementary volume will be:

$$\begin{aligned} d\overrightarrow{\mathbf{F}} &= n_- d\tau q_- \left(\overrightarrow{\mathbf{v}}_e + \overrightarrow{\mathbf{v}}_-\right) \times \overrightarrow{\mathbf{B}} + n_+ d\tau q_+ \left(\overrightarrow{\mathbf{v}}_e + \overrightarrow{\mathbf{v}}_+\right) \times \overrightarrow{\mathbf{B}} \\ &= \left[n_- q_- \overrightarrow{\mathbf{v}}_- + n_+ q_+ \overrightarrow{\mathbf{v}}_+\right] d\tau \times \overrightarrow{\mathbf{B}} + \left(n_- q_- + n_+ q_+\right) d\tau \overrightarrow{\mathbf{v}}_e \times \overrightarrow{\mathbf{B}} \end{aligned} \qquad [4.69]$$

Since the conductor is neutral, that is, $n_- q_- + n_+ q_+ = 0$, we deduce that:

$$d\overrightarrow{\mathbf{F}} = \left[n_- q_- \overrightarrow{\mathbf{v}}_- + n_+ q_+ \overrightarrow{\mathbf{v}}_+\right] d\tau \times \overrightarrow{\mathbf{B}} = \overrightarrow{\mathbf{J}} \times \overrightarrow{\mathbf{B}} d\tau \qquad [4.70]$$

where $\overrightarrow{\mathbf{J}} = n_- q_- \overrightarrow{\mathbf{v}}_- + n_+ q_+ \overrightarrow{\mathbf{v}}_+$ is the volume current density ($\overrightarrow{\mathbf{v}}_+$ is generally negligible, i.e., positive charges are immobile inside the conductor). The expression [4.70] is called the *Laplace force*. It is denoted by $\overrightarrow{\mathbf{F}}_{\text{(Lap.)}}$. To obtain the total Laplace force exerted on the entire conductor, we must integrate over the entire volume. Note that the Laplace force is electromagnetic in origin and is a reaction of the conductor material to the Lorentz force acting on all of its charged constituents (immobile and mobile). Mobile charges (electrons) are subject to force $n_- q_- \left(\overrightarrow{\mathbf{v}}_e + \overrightarrow{\mathbf{v}}_-\right) \times \overrightarrow{\mathbf{B}}$, while immobile charges (ions) are subject to force $n_+ q_+ \overrightarrow{\mathbf{v}}_e \times \overrightarrow{\mathbf{B}}$. The sum of these two forces makes up the Laplace force.

When the distribution is distributed over a surface or a line, formula [4.70] must be replaced respectively by:

$$d\overrightarrow{\mathbf{F}}_{\text{(Lap.)}} = \begin{cases} \overrightarrow{\mathbf{K}} \times \overrightarrow{\mathbf{B}} dS & \text{(surface current)} \\ \overrightarrow{\mathbf{I}} \times \overrightarrow{\mathbf{B}} dl & \text{(line current)} \end{cases} \qquad [4.71]$$

In the case of a filiform conductor through which a current I flows, the Laplace force exerted on a portion $d\overrightarrow{\mathbf{l}}$ (of the conductor) oriented in the direction of I, is also written as:

$$d\overrightarrow{\mathbf{F}}_{\text{(Lap.)}} = I d\overrightarrow{\mathbf{l}} \times \overrightarrow{\mathbf{B}} \qquad [4.72]$$

because $d\vec{\mathbf{l}}$ and $d\vec{\mathbf{I}}$ are parallel. Consequently, the total force exerted on the entire conductor is calculated by integrating over the entire distribution:

$$\vec{\mathbf{F}}_{(\text{Lap.})} = \begin{cases} \int_\tau \vec{\mathbf{J}} \times \vec{\mathbf{B}} d\tau & \text{(volume distribution)} \\ \int_S \vec{\mathbf{K}} \times \vec{\mathbf{B}} dS & \text{(surface distribution)} \\ \int_L \vec{\mathbf{I}} \times \vec{\mathbf{B}} dl & \text{(line distribution)} \\ \oint_\Gamma I d\vec{\mathbf{l}} \times \vec{\mathbf{B}} & \text{(closed filiform circuit } \Gamma) \end{cases} \qquad [4.73]$$

4.4.2. *Laplace force between two conductors*

When the magnetic force $\vec{\mathbf{B}}$ is created by another conductor (2), the Laplace force $\vec{\mathbf{F}}_1$ exerted by conductor (2) on conductor (1) is:

$$\vec{\mathbf{F}}_1 = \frac{\mu_0}{4\pi} \int_{\tau_1} d\tau_1 \int_{\tau_2} d\tau_2 \frac{\vec{\mathbf{J}}_1(\vec{\mathbf{r}}_1) \times \left[\vec{\mathbf{J}}_2(\vec{\mathbf{r}}_2) \times (\vec{\mathbf{r}}_1 - \vec{\mathbf{r}}_2)\right]}{\left|\vec{\mathbf{r}}_1 - \vec{\mathbf{r}}_2\right|^3} \qquad [4.74]$$

For closed filiform circuits Γ_1 and Γ_2 carrying currents I_1 and I_2, respectively, the Laplace force applied to circuit Γ_1 by the magnetic field created by circuit Γ_2 can be calculated using the following formula:

$$\vec{\mathbf{F}}_1 = \oint_{\Gamma_1} I_1 d\vec{\mathbf{l}}_1 \times \vec{\mathbf{B}}_2(\vec{\mathbf{r}}_1) = \frac{\mu_0}{4\pi} \oint_{\Gamma_1} \oint_{\Gamma_2} I_1 I_2 \frac{d\vec{\mathbf{l}}_1 \times \left[d\vec{\mathbf{l}}_2 \times (\vec{\mathbf{r}}_1 - \vec{\mathbf{r}}_2)\right]}{\left|\vec{\mathbf{r}}_1 - \vec{\mathbf{r}}_2\right|^3} \qquad [4.75]$$

EXAMPLE 4.10. (Laplace Force between Two Infinitely Long, Parallel Wire).– Consider two infinitely long parallel wires (1) and (2), respectively, flowed by currents I_1 and I_2, and separated by a distance d, as shown in Figure 4.15. The magnetic field created by wire (2) and experienced at any point on wire (1) is $\vec{\mathbf{B}}_2 = -\frac{\mu_0 I_2}{4\pi d}\vec{\mathbf{i}}$. It follows that the total force applied to a portion of length L_1 of wire (1) is:

$$\vec{\mathbf{F}}_1 = \int_{-\frac{L_1}{2}}^{\frac{L_1}{2}} I_1 dz_1' \vec{\mathbf{k}} \times \vec{\mathbf{B}}_2 = -\frac{\mu_0 I_1 I_2}{2\pi d} \int_{-\frac{L_1}{2}}^{\frac{L_1}{2}} dz_1' \vec{\mathbf{k}} \times \vec{\mathbf{i}} = -\frac{\mu_0 I_1 I_2 L_1}{2\pi d} \vec{\mathbf{j}} \qquad [4.76]$$

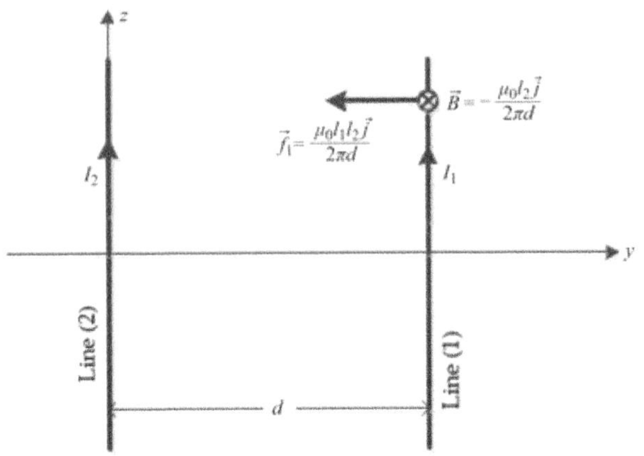

Figure 4.15. *The Laplace force between two wires is inversely proportional to the distance separating the wires. If $I_1 I_2 > 0$ (respectively, $I_1 I_2 < 0$), the two wires are attracted to each other (respectively, repelled by each other)*

The force experienced per unit length is therefore equal to:

$$\vec{\mathbf{f}}_1 = -\frac{\mu_0 I_1 I_2}{2\pi d}\vec{\mathbf{j}} \quad [4.77]$$

If the two currents flow in the same direction $(I_1 I_2 > 0)$, the force is attractive, whereas if they flow in opposite directions $(I_1 I_2 < 0)$, the force is repulsive. This is contrary to the Coulomb force, where opposite charges attract and similar charges repel.

EXAMPLE 4.11. (Plate of Thickness d).– We saw earlier that the magnetic field created by an infinite plate of thickness d, located parallel to the plane Oxz in space $-\frac{d}{2} \leq y \leq \frac{d}{2}$ and carrying the uniform current density $J_0 \vec{\mathbf{k}}$, is governed by equations [4.39]. The force applied by this field on the plate itself (in the Oy direction) per unit area is zero:

$$\vec{\mathbf{f}} = \int_{-\frac{d}{2}}^{\frac{d}{2}} J_0 B_x dy\, \vec{\mathbf{j}} = -\mu_0 J_0^2 \int_{-\frac{d}{2}}^{\frac{d}{2}} y dy\, \vec{\mathbf{j}} = -\mu_0 J_0^2 \frac{y^2}{2}\Bigg|_{-\frac{d}{2}}^{\frac{d}{2}} \vec{\mathbf{j}} = 0 \quad [4.78]$$

A current distribution cannot exert a force on itself. If we take $d \longrightarrow 0$, we always obtain the same thing, that is, a sheet cannot act on itself.

EXAMPLE 4.12. (Laplace Force between Two Parallel Sheets with No Thickness).– The magnetic field created by two sheets traversed by opposing currents $K_0 \vec{\mathbf{k}}$

(sheet (1)) and $-K_0\overrightarrow{\mathbf{k}}$ (sheet (2)), parallel to the Oxz plane, without thickness and separated by a distance d, has the value $-\mu_0 K_0\overrightarrow{\mathbf{i}}$ for $0 \le y \le d$ and 0 elsewhere. As the magnetic field is discontinuous on sheet (2), it is not clear what value to use for the magnetic field. For this reason, the current sheet at $y = d$ must be modeled as a thin volume current with density J_0 and thickness Δ, where $K_0 = J_0\Delta$, as shown in Figure 4.16. To satisfy the magnetic field limit values on both sides of the sheet:

$$\left[\begin{array}{l} B_x(d-\Delta) = -\mu_0 K_0 \\ B_x(d) = 0 \end{array}\right. \qquad [4.79]$$

we take the inside of the Δ thickness sheet:

$$B_x(y) = \frac{\mu_0 K_0}{\Delta}(y-d) \qquad [4.80]$$

Consequently, the force per unit area of the plate is:

$$\overrightarrow{\mathbf{f}} = -\int_{d-\Delta}^{d} \frac{\mu_0 K_0 J_0 (y-d)\overrightarrow{\mathbf{j}}}{\Delta} dy = \frac{\mu_0 K_0 J_0 \Delta}{2}\overrightarrow{\mathbf{j}} = \frac{\mu_0 K_0^2}{2}\overrightarrow{\mathbf{j}} \qquad [4.81]$$

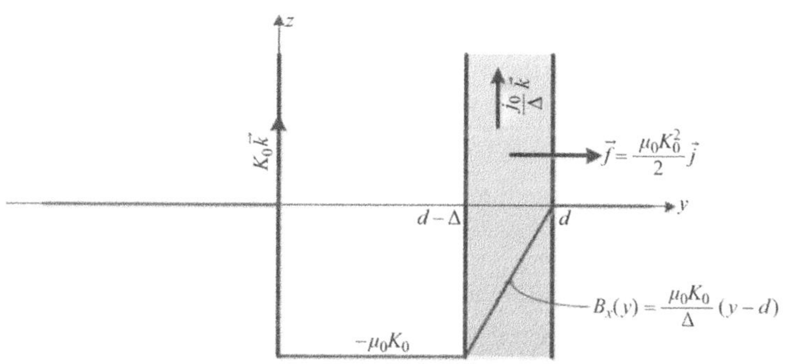

Figure 4.16. *The Laplace force exerted per unit area between two sheets carrying opposite current is constant and repulsive. For a color version of this figure, see www.iste.co.uk/abdesselam/electrodynamics1.zip*

This formula can also be obtained in other ways. Since the magnetic field created by the sheet (2) cannot act on it, the force [4.81] is due solely to the field created by the sheet (1). In fact, we can verify that:

$$\overrightarrow{\mathbf{f}} = \int \overrightarrow{\mathbf{J}}_2 \times \overrightarrow{\mathbf{B}}_1 dy = \int \left[K_0\delta(y-d)\overrightarrow{\mathbf{k}}\right] \times \left[-\frac{1}{2}\mu_0 K_0\overrightarrow{\mathbf{i}}\right] dy = \frac{\mu_0 K_0^2\overrightarrow{\mathbf{j}}}{2}$$

[4.82]

The force acts here to separate the sheets, as the currents are opposite and the sheets repel each other.

4.4.3. *Laplace force work*

Consider a neutral conductor carrying a current density $\overrightarrow{\mathbf{J}}$ and immersed in a magnetic field $\overrightarrow{\mathbf{B}}$. The conductor moves at a drive speed $\overrightarrow{\mathbf{v}}_e$ relative to a fixed Galilean reference frame $Oxyz$. Each volume element $d\tau$ in the conductor undergoes a Laplace force $d\overrightarrow{\mathbf{F}}_{\text{(Lap.)}} = d\overrightarrow{\mathbf{J}} \times \overrightarrow{\mathbf{B}} d\tau$. The work carried out by this force during the displacement of the portion $d\tau$ of $d\overrightarrow{\mathbf{r}} = \overrightarrow{\mathbf{v}}_e dt$ will be:

$$d^2W_{\text{(Lap.)}} = d\overrightarrow{\mathbf{F}}_{\text{(Lap.)}} \cdot d\overrightarrow{\mathbf{r}} = \begin{cases} \left(\overrightarrow{\mathbf{J}} \times \overrightarrow{\mathbf{B}}\right) \cdot d\overrightarrow{\mathbf{r}} d\tau = \left(\overrightarrow{\mathbf{J}} \times \overrightarrow{\mathbf{B}}\right) \cdot \overrightarrow{\mathbf{v}}_e dt d\tau & \text{(volume)} \\ \left(\overrightarrow{\mathbf{K}} \times \overrightarrow{\mathbf{B}}\right) \cdot d\overrightarrow{\mathbf{r}} dS = \left(\overrightarrow{\mathbf{K}} \times \overrightarrow{\mathbf{B}}\right) \cdot \overrightarrow{\mathbf{v}}_e dt dS & \text{(surface)} \\ \left(\overrightarrow{\mathbf{I}} \times \overrightarrow{\mathbf{B}}\right) \cdot d\overrightarrow{\mathbf{r}} dl = I \left(d\overrightarrow{\mathbf{l}} \times \overrightarrow{\mathbf{B}}\right) \cdot \overrightarrow{\mathbf{v}}_e dt & \text{(line)} \end{cases} \quad [4.83]$$

This work done is generally nonzero, as it is not work done by the magnetic force. To clarify this, let us calculate the work of the electromagnetic forces exerted on all of the charge carriers contained in the volume element $d\tau$ of the conductor:

$$d^2W = \sum_{\alpha=\pm} n_\alpha q_\alpha \left[\overrightarrow{\mathbf{E}} + (\overrightarrow{\mathbf{v}}_\alpha + \overrightarrow{\mathbf{v}}_e) \times \overrightarrow{\mathbf{B}}\right] \cdot (\overrightarrow{\mathbf{v}}_\alpha + \overrightarrow{\mathbf{v}}_e)\, d\tau dt \quad [4.84]$$

Note that the work done by the magnetic forces is effectively zero. Since $\left(\overrightarrow{\mathbf{v}}_\pm \times \overrightarrow{\mathbf{B}}\right) \cdot \overrightarrow{\mathbf{v}}_\pm = 0$, $\left(\overrightarrow{\mathbf{v}}_e \times \overrightarrow{\mathbf{B}}\right) \cdot \overrightarrow{\mathbf{v}}_e = 0$ and $\sum_{\alpha=\pm} n_\alpha q_\alpha = 0$ (neutral conductor), we can deduce the following:

$$\begin{aligned} d^2W &= \left(\sum_\alpha n_\alpha q_\alpha \overrightarrow{\mathbf{v}}_\alpha\right) \cdot \left(\overrightarrow{\mathbf{E}} + \overrightarrow{\mathbf{v}}_e \times \overrightarrow{\mathbf{B}}\right) d\tau dt \\ &\quad + \left[\left(\sum_\alpha n_\alpha q_\alpha \overrightarrow{\mathbf{v}}_\alpha\right) \times \overrightarrow{\mathbf{B}}\right] \cdot \overrightarrow{\mathbf{v}}_e d\tau dt \\ &= \overrightarrow{\mathbf{J}} \cdot \left(\overrightarrow{\mathbf{E}} + \overrightarrow{\mathbf{v}}_e \times \overrightarrow{\mathbf{B}}\right) d\tau dt + \left[\overrightarrow{\mathbf{J}} \times \overrightarrow{\mathbf{B}}\right] \cdot \overrightarrow{\mathbf{v}}_e d\tau dt \end{aligned} \quad [4.85]$$

The term:

$$d^2W_{\text{(m)}} = \overrightarrow{\mathbf{J}} \cdot \left(\overrightarrow{\mathbf{E}} + \overrightarrow{\mathbf{v}}_e \times \overrightarrow{\mathbf{B}}\right) d\tau dt \quad [4.86]$$

is called the *electromotive work*, that is, the work supplied to the conductor by a generator to displace free charges with respect to itself. In other words, $d^2W_{(\mathrm{m})}$ is the work required to maintain the current when the conductor moves. Thus, we see that the *electromotive field* is added to the field $\overrightarrow{\mathbf{E}}$:

$$\overrightarrow{\mathbf{f}}_{(\mathrm{m})} = \overrightarrow{\mathbf{v}}_e \times \overrightarrow{\mathbf{B}} \qquad [4.87]$$

resulting from the displacement of the conductor in a magnetic field. The second term $\left(\overrightarrow{\mathbf{J}} \times \overrightarrow{\mathbf{B}}\right) \cdot \overrightarrow{\mathbf{v}}_e d\tau dt$ is the work done by the Laplace force. It corresponds to the work supplied to the conductor to displace it. It is equal to the opposite of the contribution of the field $\overrightarrow{\mathbf{f}}_{(\mathrm{m})}$ to the electromotive work, as the total work done by the magnetic forces is zero.

4.4.4. *Maxwell's theorem*

For a filiform conductor, the work carried out by the Laplace force on a portion, $d\overrightarrow{\mathbf{l}}$, of the conductor is:

$$d^2W_{(\mathrm{Lap.})} = I\left(d\overrightarrow{\mathbf{l}} \times \overrightarrow{\mathbf{B}}\right) \cdot d\overrightarrow{\mathbf{r}} = I\overrightarrow{\mathbf{B}} \cdot \left(d\overrightarrow{\mathbf{r}} \times d\overrightarrow{\mathbf{l}}\right) = I\overrightarrow{\mathbf{B}} \cdot d^2\overrightarrow{\mathbf{S}}_c \qquad [4.88]$$

where $d^2\overrightarrow{\mathbf{S}}_c = d\overrightarrow{\mathbf{r}} \times d\overrightarrow{\mathbf{l}}$ is the area cut (swept) by the $d\overrightarrow{\mathbf{l}}$ element when it is displaced by $d\overrightarrow{\mathbf{r}}$, as illustrated in Figure 4.17. The above equation can also be written in the following form:

$$d^2W_{(\mathrm{Lap.})} = Id^2\Phi_c \qquad [4.89]$$

where $d^2\Phi_c = \overrightarrow{\mathbf{B}} \cdot d^2\overrightarrow{\mathbf{S}}_c$ is the flux of $\overrightarrow{\mathbf{B}}$ cut by $d\overrightarrow{\mathbf{l}}$. Since the intensity I is the same throughout the circuit (quasi-static regime approximation), the work carried out by the Laplace force over the entire conductor in the time period dt is:

$$dW_{(\mathrm{Lap})} = Id\Phi_c \qquad [4.90]$$

where:

$$d\Phi_c = \oint_{\mathrm{circuit}} d^2\Phi_c = \oint_{\mathrm{circuit}} \overrightarrow{\mathbf{B}} \cdot d^2\overrightarrow{\mathbf{S}}_c \qquad [4.91]$$

is the flux cut by the whole circuit. Equation [4.90] is the analytical expression of Maxwell's theorem, which states that the displacement of a closed electric circuit in an external magnetic field generates work performed by the Laplace force, which is equal to the product of the current flowing through the circuit and the flux cut off by the circuit during its displacement. The flux is said to be cut because $\vec{\mathbf{B}}$ is in the form of a field. As it moves, the circuit crosses these field lines, that is, cuts them.

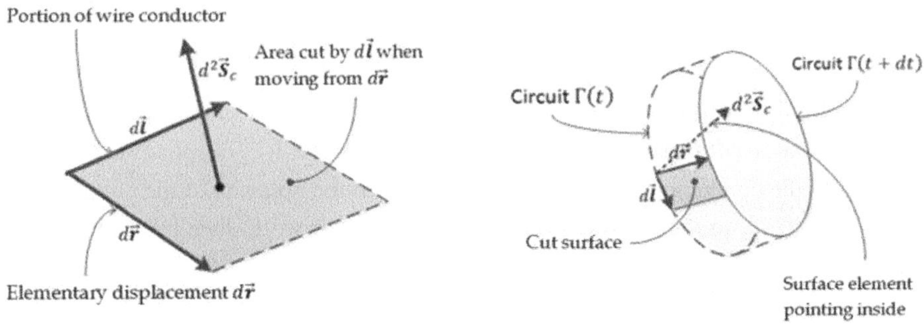

Figure 4.17. *Surface cut by a portion of the filiform conductor during an elementary displacement of a circuit in a magnetic field*

For a static magnetic field, the flux cut $d\Phi_C$ by the circuit as it moves is exactly equal to the variation in total flux $d\Phi = \Phi(t+dt) - \Phi(t)$, where $\Phi(t)$ is the flux measured across any surface supported by the circuit at time t. To demonstrate this, let us consider a deformable circuit $\Gamma(t)$ carrying a current I and moving in a magnetic field $\vec{\mathbf{B}}$ independent of time. At any instant t, this circuit defines a surface $S(t)$ that is supported on $\Gamma(t)$. During its displacement from one instant t to another $t+dt$, the circuit describes a closed surface $S_{(\text{tot.})} = S(t) \cup dS_c \cup S(t+dt)$, where dS_c is the cut lateral surface, that is, swept by the periphery of the circuit, as shown in Figure 4.18. The flux conservation equation $\vec{\nabla} \cdot \vec{\mathbf{B}} = 0$ implies that:

$$\begin{aligned} 0 = \oint_{S_{(\text{tot.})}} \vec{\mathbf{B}} \cdot d\vec{\mathbf{S}}_{(\text{tot.})} &= -\oint_{S(t)} \vec{\mathbf{B}} \cdot d\vec{\mathbf{S}}(t) \\ &- \oint_{dS_c} \vec{\mathbf{B}} \cdot d^2\vec{\mathbf{S}}_c + \oint_{S(t+dt)} \vec{\mathbf{B}} \cdot d\vec{\mathbf{S}}(t+dt) \end{aligned} \quad [4.92]$$

because $d\vec{\mathbf{S}}(t)$ and $d^2\vec{\mathbf{S}}_c$ point inside. Consequently, we obtain $\oint_{dS_c} \vec{\mathbf{B}} \cdot d^2\vec{\mathbf{S}}_c = \oint_{S(t+dt)} \vec{\mathbf{B}} \cdot d\vec{\mathbf{S}}(t+dt) - \oint_{S(t)} \vec{\mathbf{B}} \cdot d\vec{\mathbf{S}}(t)$. It follows that:

$$d\Phi_c = \Phi(t+dt) - \Phi(t) = d\Phi \quad [4.93]$$

where $\Phi(t)$ and $\Phi(t+dt)$ are the flux of $\overrightarrow{\mathbf{B}}$ across the circuit at the instants t and $t+dt$, respectively. From this, it can be derived that:

$$dW_{(\text{Lap.})} = Id\Phi \qquad [4.94]$$

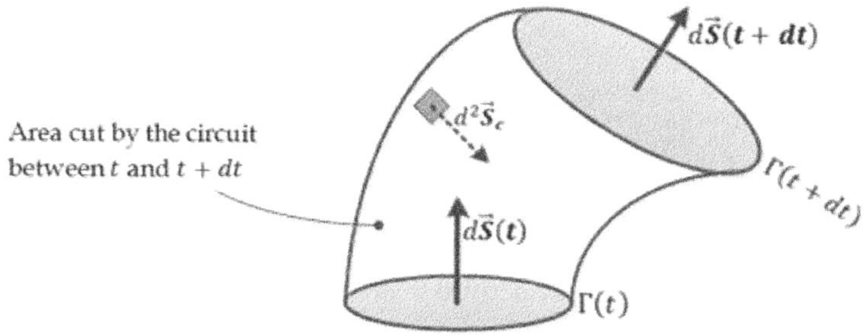

Figure 4.18. *The flux cut Φ_c by the circuit, during its displacement between time t and $t+dt$, is equal to the variation in total flux $\phi(t+dt) - \Phi(t)$*

If I is also stationary, then $dW_{(\text{Lap.})} = d(I\Phi)$ and thus:

$$W_{(\text{Lap.})} = I(\Phi_2 - \Phi_1) \qquad [4.95]$$

4.4.5. *Expression for magnetic potential energy*

Consider a closed filiform circuit in motion, with a current I passing through it and placed in a stationary magnetic field. The variation in the kinetic energy of the circuit, during an infinitesimal displacement, will be:

$$dW_{(\text{cin})} = dW_{(\text{Lap.})} = Id\Phi \qquad [4.96]$$

This means that the circuit has a so-called *magnetic potential energy* $W_{(\text{m})}$, associated with the presence of the external magnetic field. Since the mechanical energy of the circuit is $W_{(\text{mec})} = E_{(\text{kin})} + W_{(\text{m})}$, then:

$$dW_{(\text{m})} = -dW_{(\text{Lap.})} \qquad [4.97]$$

It can thus be concluded that the magnetic potential energy of the circuit is $W_{(\text{m})} = I\Phi + \lambda$, where λ is an additive constant, often chosen to be zero, that is:

$$W_{(\text{m})} = -I\Phi \qquad [4.98]$$

4.5. Ampère's theorem

4.5.1. *Maxwell–Ampère equation*

Let us return to the Biot and Savart law [4.27]. Since $\vec{\nabla}\left(\frac{1}{|\vec{\mathbf{r}}-\vec{\mathbf{r}}'|}\right) = -\frac{\vec{\mathbf{r}}-\vec{\mathbf{r}}'}{|\vec{\mathbf{r}}-\vec{\mathbf{r}}'|^3}$, we write:

$$\vec{\mathbf{B}}\left(\vec{\mathbf{r}}\right) = -\frac{\mu_0}{4\pi}\int_{\tau'}\vec{\mathbf{J}}\left(\vec{\mathbf{r}}'\right)\times\vec{\nabla}\left(\frac{1}{\left|\vec{\mathbf{r}}-\vec{\mathbf{r}}'\right|}\right)d\tau' \quad [4.99]$$

Using the identity $\vec{\mathbf{A}}\times\left(\vec{\nabla}\Phi\right) = \Phi\left(\vec{\nabla}\times\vec{\mathbf{A}}\right) - \vec{\nabla}\times\left(\Phi\vec{\mathbf{A}}\right)$ and noting that $\vec{\nabla}$ acts only on $\vec{\mathbf{r}}$, it can thus be derived that:

$$\begin{aligned}\vec{\mathbf{B}}\left(\vec{\mathbf{r}}\right) &= -\frac{\mu_0}{4\pi}\int_{\tau'}\left[\frac{\vec{\nabla}\times\vec{\mathbf{J}}\left(\vec{\mathbf{r}}'\right)}{\left|\vec{\mathbf{r}}-\vec{\mathbf{r}}'\right|} - \vec{\nabla}\times\left(\frac{\vec{\mathbf{J}}\left(\vec{\mathbf{r}}'\right)}{\left|\vec{\mathbf{r}}-\vec{\mathbf{r}}'\right|}\right)\right]d\tau' \\ &= \frac{\mu_0}{4\pi}\int_{\tau'}\vec{\nabla}\times\left(\frac{\vec{\mathbf{J}}\left(\vec{\mathbf{r}}'\right)}{\left|\vec{\mathbf{r}}-\vec{\mathbf{r}}'\right|}\right)d\tau' = \frac{\mu_0}{4\pi}\vec{\nabla}\times\left(\int_{\tau'}\frac{\vec{\mathbf{J}}\left(\vec{\mathbf{r}}'\right)}{\left|\vec{\mathbf{r}}-\vec{\mathbf{r}}'\right|}d\tau'\right)\end{aligned} \quad [4.100]$$

This last equation clearly shows that $\vec{\nabla}\cdot\vec{\mathbf{B}} = 0$, because $\vec{\nabla}\cdot\left(\vec{\nabla}\times\vec{\mathbf{A}}\right) = 0$, for all $\vec{\mathbf{A}}$.

Using the identity $\vec{\nabla}\times\left(\vec{\nabla}\times\vec{\mathbf{A}}\right) = \vec{\nabla}\left(\vec{\nabla}\cdot\vec{\mathbf{A}}\right) - \Delta\vec{\mathbf{A}}$, let us calculate the curl in expression [4.100]:

$$\begin{aligned}\vec{\nabla}\times\vec{\mathbf{B}} &= \frac{\mu_0}{4\pi}\vec{\nabla}\times\left[\vec{\nabla}\times\left(\int_{\tau'}\frac{\vec{\mathbf{J}}\left(\vec{\mathbf{r}}'\right)}{\left|\vec{\mathbf{r}}-\vec{\mathbf{r}}'\right|}d\tau'\right)\right] \\ &= \frac{\mu_0}{4\pi}\left[\vec{\nabla}\int_{\tau'}\vec{\mathbf{J}}\left(\vec{\mathbf{r}}'\right)\cdot\vec{\nabla}\left(\frac{1}{\left|\vec{\mathbf{r}}-\vec{\mathbf{r}}'\right|}\right)d\tau' - \int_{\tau'}\vec{\mathbf{J}}\left(\vec{\mathbf{r}}'\right)\Delta\left(\frac{1}{\left|\vec{\mathbf{r}}-\vec{\mathbf{r}}'\right|}\right)d\tau'\right]\end{aligned} \quad [4.101]$$

The second integral has the following value:

$$-\frac{\mu_0}{4\pi}\int_{\tau'}\vec{\mathbf{J}}'\Delta\left(\frac{1}{\left|\vec{\mathbf{r}}-\vec{\mathbf{r}}'\right|}\right)d\tau' = \mu_0\int_{\tau'}\vec{\mathbf{J}}'\delta\left(\vec{\mathbf{r}}-\vec{\mathbf{r}}'\right)d\tau' = \mu_0\vec{\mathbf{J}}\left(\vec{\mathbf{r}}\right) \quad [4.102]$$

where $\vec{\mathbf{J}}\left(\vec{\mathbf{r}}'\right) = \vec{\mathbf{J}}'$. To calculate the first term, we simply find that $\vec{\nabla} F\left(\vec{\mathbf{r}} - \vec{\mathbf{r}}'\right) = -\vec{\nabla}' F\left(\vec{\mathbf{r}} - \vec{\mathbf{r}}'\right)$:

$$\begin{aligned} &\frac{\mu_0}{4\pi}\int_{\tau'} \vec{\mathbf{J}}' \cdot \vec{\nabla}\left(\frac{1}{\left|\vec{\mathbf{r}} - \vec{\mathbf{r}}'\right|}\right) d\tau' = -\frac{\mu_0}{4\pi}\int_{\tau'} \vec{\mathbf{J}}' \cdot \vec{\nabla}'\left(\frac{1}{\left|\vec{\mathbf{r}} - \vec{\mathbf{r}}'\right|}\right) d\tau' \\ &= -\frac{\mu_0}{4\pi}\left[\int_{\tau'} \vec{\nabla}' \cdot \left(\frac{\vec{\mathbf{J}}'}{\left|\vec{\mathbf{r}} - \vec{\mathbf{r}}'\right|}\right) d\tau' - \int_{\tau'} \left(\frac{\vec{\nabla}' \cdot \vec{\mathbf{J}}'}{\left|\vec{\mathbf{r}} - \vec{\mathbf{r}}'\right|}\right) d\tau'\right] \end{aligned} \qquad [4.103]$$

The second integral is clearly zero, since $\vec{\nabla}' \cdot \vec{\mathbf{J}}\left(\vec{\mathbf{r}}'\right) = 0$ (quasi-static approximation), and consequently, the previous equation reduces to:

$$\begin{aligned} &\frac{\mu_0}{4\pi}\int_{\tau'} \vec{\mathbf{J}}' \cdot \vec{\nabla}\left(\frac{1}{\left|\vec{\mathbf{r}} - \vec{\mathbf{r}}'\right|}\right) d\tau' = -\frac{\mu_0}{4\pi}\int_{\tau'} \vec{\nabla}' \cdot \left(\frac{\vec{\mathbf{J}}'}{\left|\vec{\mathbf{r}} - \vec{\mathbf{r}}'\right|}\right) d\tau' \\ &= -\frac{\mu_0}{4\pi}\oint_{S'} \frac{\vec{\mathbf{J}}' \cdot d\vec{\mathbf{S}}'}{\left|\vec{\mathbf{r}} - \vec{\mathbf{r}}'\right|} \end{aligned} \qquad [4.104]$$

The surface S' completely encloses the current distribution, that is, S' is found outside, in a region with zero current, $\vec{\mathbf{J}}\Big|_{\vec{\mathbf{r}}' \in S'} = 0$. It therefore follows that this surface integral is zero. Finally, we obtain:

$$\vec{\nabla} \times \vec{\mathbf{B}}\left(\vec{\mathbf{r}}\right) = \mu_0 \vec{\mathbf{J}}\left(\vec{\mathbf{r}}\right) \qquad [4.105]$$

This is the *Maxwell–Ampère equation*. If we take the divergence of both sides in [4.105], the left-hand side is zero because the divergence of the curl of any vector field is always zero. To guarantee this, magnetic field systems must have currents with no divergence so that no charge can accumulate. The currents must always flow in a closed loop.

4.5.2. *Integral form of Maxwell–Ampère equation*

If S is any open surface supported by a contour Γ, then according to Curl's theorem or Stokes's theorem:

$$\oint_{\Gamma} \vec{\mathbf{B}} \cdot d\vec{\mathbf{l}} = \int_S \left(\vec{\nabla} \times \vec{\mathbf{B}}\right) \cdot d\vec{\mathbf{S}} = \int_S \mu_0 \vec{\mathbf{J}} \cdot d\vec{\mathbf{S}} \qquad [4.106]$$

This is *Ampère's theorem*. It can also be written as:

$$\oint_{\Gamma} \vec{\mathbf{B}} \cdot d\vec{\mathbf{l}} = \mu_0 I_{\text{encl.}} \qquad [4.107]$$

The Ampère theorem is stated as follows: *the circulation of* $\overrightarrow{\mathbf{B}}$ *along a closed curve oriented* Γ*, called Ampère's contour, is equal to* μ_0 *times the algebraic sum of the currents entwined by the contour, that is, the current flow through an open surface bounded by* Γ. This result is equivalent to Gauss' theorem for the electrostatic field. The Gauss and Ampère theorems associate $\overrightarrow{\mathbf{E}}$ and $\overrightarrow{\mathbf{B}}$, respectively, with their sources, the charge Q and the current I.

4.5.3. *A few applications of Ampère's theorem*

Ampère's theorem is used to calculate the magnetic field $\overrightarrow{\mathbf{B}}$ when it is constant, for reasons of symmetry, on a certain fictitious closed contour. Here are a few applications.

EXAMPLE 4.13. (Line Current).– Let us find the expression for the magnetic field created by an infinitely long wire, located along the Oz axis and carrying a permanent current I, as shown in Figure 4.19. For reasons of symmetry, the magnetic field $\overrightarrow{\mathbf{B}}$ is independent of the variables φ and z. Let us imagine a circle Γ of radius r surrounding the wire (the wire passes through the center of the circle). Let us apply Ampère's law:

$$\oint_\Gamma \overrightarrow{\mathbf{B}} \cdot d\overrightarrow{\mathbf{l}} = \oint_\Gamma B_\varphi r d\varphi = \mu_0 I_{\text{encl.}} = \mu_0 I \qquad [4.108]$$

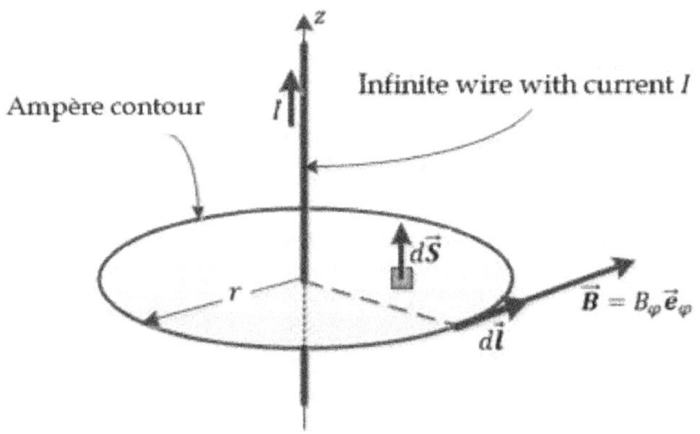

Figure 4.19. *Ampère's theorem applied to an infinitely long wire: the circulation of the magnetic field* $\overrightarrow{\mathbf{B}} = B_\varphi \overrightarrow{\mathbf{e}}_\varphi$ *along the imaginary circle of radius* r *is equal to* μ_0 *times the intensity of the current flowing through the wire*

Due to the cylindrical symmetry around the wire, the B_φ component is constant around the circle. The previous expression therefore reduces to $2\pi r B_\varphi = \mu_0 I$. The result is:

$$B_\varphi = \frac{\mu_0 I}{2\pi r} \quad [4.109]$$

EXAMPLE 4.14. (Surface Current Over the Lateral Surface of a Cylinder).– A surface current $K_0 \overrightarrow{\mathbf{k}}$ flows over the surface of an infinitely long hollow cylinder of radius a. In this case, a direct calculation of the magnetic field is more difficult using the Biot and Savart law. On the other hand, it is much easier to use Ampère's law, since the magnetic field must be in the direction of $\overrightarrow{\mathbf{e}}_\varphi$ and, by symmetry, it can only depend on r and not on φ and z. Let choose a circular contour of radius r as shown in Figure 4.20(a). Since $d\overrightarrow{\mathbf{l}} = rd\varphi\overrightarrow{\mathbf{e}}_\varphi$ is in the same direction as $\overrightarrow{\mathbf{B}}$, the scalar product between $d\overrightarrow{\mathbf{B}}$ and $d\overrightarrow{\mathbf{l}}$ reduces to a simple product, that is, $B_\varphi r d\varphi$. For $r < a$, no current passes through the surface bounded by the contour, while for $r > a$, the current is purely perpendicular to the contour surface:

$$\oint_L \frac{\overrightarrow{\mathbf{B}}}{\mu_0} \cdot d\overrightarrow{\mathbf{l}} = \int_0^{2\pi} \frac{B_\varphi}{\mu_0} r d\varphi = \frac{2\pi r B_\varphi}{\mu_0} = \begin{cases} K_0 2\pi a = I & r > a \\ 0 & r < a \end{cases} \quad [4.110]$$

where I is the total current in the cylinder. The magnetic field is therefore:

$$B_\varphi = \begin{cases} \dfrac{\mu_0 K_0 a}{r} = \dfrac{\mu_0 I}{2\pi r} & r > a \\ 0 & r < a \end{cases} \quad [4.111]$$

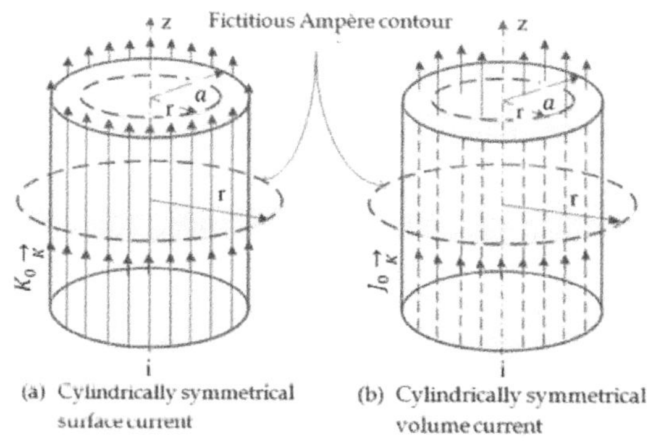

Figure 4.20. *Calculation of the magnetic field by Ampère's law for a cylinder of infinite length traversed by (a) a surface current parallel to its axis and (b) a volume current also parallel to its axis*

Outside the cylinder, the magnetic field is the same as if the current were concentrated along the axis like a line current.

EXAMPLE 4.15. (Volume Flow Inside a Cylinder Along its Axis).– if the current in the cylinder is uniformly distributed over the volume along its axis, that is, like $J_0\overrightarrow{\mathbf{k}}$, then the current surrounding the entire cylinder is crossed by the total current $I = J_0\pi a^2$. If the radius of the contour is smaller than the radius of the cylinder, that is, if $r < a$, the current flowing through the surface bounded by the contour will be $J_0\pi r^2 = Ifracr^2a^2$, as shown in Figure 4.20(b). It follows that:

$$\oint_L \frac{B_\varphi}{\mu_0} r d\varphi = \frac{2\pi r B_\varphi}{\mu_0} = \begin{cases} J_0\pi a^2 = I & r > a \\ J_0\pi r^2 = \dfrac{Ir^2}{a^2} & r < a \end{cases} \quad [4.112]$$

and consequently:

$$B_\varphi = \begin{cases} \dfrac{\mu_0 J_0 a^2}{2r} = \dfrac{\mu_0 I}{2\pi r} & r > a \\ \dfrac{\mu_0 J_0 r}{2} = \dfrac{\mu_0 I r}{2\pi a^2} & r < a \end{cases} \quad [4.113]$$

EXAMPLE 4.16.– (Circular Loop with Radius R).– Let us verify Ampère's law for a circular loop centered on O, with radius R, located in the Oxy plane and carrying a permanent current I. To do this, let us choose the Ampère contour $ABCDA$ given in Figure 4.21. The points A, B, C and D are located very far from the loop, that is, at infinity. Since the magnetic field at infinity is zero, the integral circulation $\oint_{ABCDA} \overrightarrow{\mathbf{B}} \cdot d\overrightarrow{\mathbf{l}}$ reduces to:

$$\oint_{ABCDA} \overrightarrow{\mathbf{B}} \cdot d\overrightarrow{\mathbf{l}} = \int_{AB} \overrightarrow{\mathbf{B}} \cdot d\overrightarrow{\mathbf{l}} = \int_{-\infty}^{\infty} B_z dz \quad [4.114]$$

where B_z is given by formula [4.44]. Inserting this expression into [4.114], we obtain:

$$\int_{-\infty}^{\infty} B_z dz = \int_{-\infty}^{\infty} \frac{\mu_0 R^2 I}{2\left(R^2 + z^2\right)^{3/2}} dz = \frac{\mu_0 R^2 I}{2} \frac{z}{R^2\sqrt{R^2 + z^2}}\Bigg|_{-\infty}^{\infty} = \mu_0 I \quad [4.115]$$

Clearly, the current density flowing through the surface bounded by the Ampère contour is $\overrightarrow{\mathbf{J}} = I\delta\left(y + R\right)\delta\left(z\right)\overrightarrow{\mathbf{i}}$, because the current is tangential to the loop. It follows that:

$$\mu_0 \int_S \overrightarrow{\mathbf{J}} \cdot d\overrightarrow{\mathbf{S}} = \mu_0 \int_{-\infty}^{\infty}\int_{-\infty}^{\infty} I\delta\left(y + R\right)\delta\left(z\right) dydz = \mu_0 I \quad [4.116]$$

Ampère's theorem is therefore satisfied.

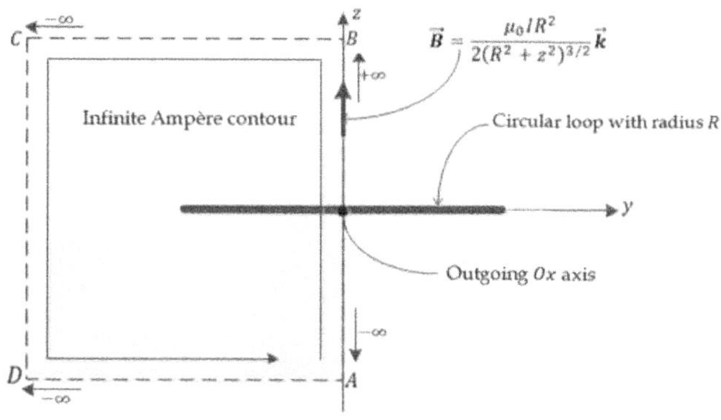

Figure 4.21. *Verification of the Ampère theorem for a circular loop over a rectangular contour $ABCDA$ whose ends are located at infinity: the magnetic field is given by expression [4.44] along the segment AB and is zero on the other sides of the rectangle*

EXAMPLE 4.17. (Infinite Solenoid).– Consider an infinite solenoid containing n loops per unit length, each loop carrying a permanent current I. For reasons of symmetry, the magnetic field is directed along Oz, that is, $\overrightarrow{\mathbf{B}} = B_z\overrightarrow{\mathbf{k}}$. Choose a rectangular Ampère contour with width L and infinite length, as shown in Figure 4.22. By applying Ampère's theorem to this contour, we obtain $LB_z = \mu_0 NI$ and consequently:

$$B_z = \frac{\mu_0 NI}{L} = \mu_0 nI \qquad [4.117]$$

where N is the number of loops that go across the rectangle.

EXAMPLE 4.18. (Circular Solenoid).– A circular solenoid containing N loops caries a permanent current I. This solenoid is made by wrapping a conducting wire over a torus. Assuming that the median plane of the torus coincides with the Oxy plane, it is possible to construct a circle of radius r parallel to the Oxy plane inside the torus, as shown in Figure 4.23. The magnetic field $\overrightarrow{\mathbf{B}}$ within the solenoid is of the form $\overrightarrow{\mathbf{B}} = B_\varphi\overrightarrow{\mathbf{e}}_\varphi$. By integrating the component B_φ of the field around the circle, we obtain $2\pi r B_\varphi = \mu_0 NI$ and therefore:

$$B_\varphi = \frac{\mu_0 NI}{2\pi r} \qquad [4.118]$$

Note that as long as the loop remains inside the torus, moving it up or down makes no difference. The field is independent of z but decreases with $\frac{1}{r}$.

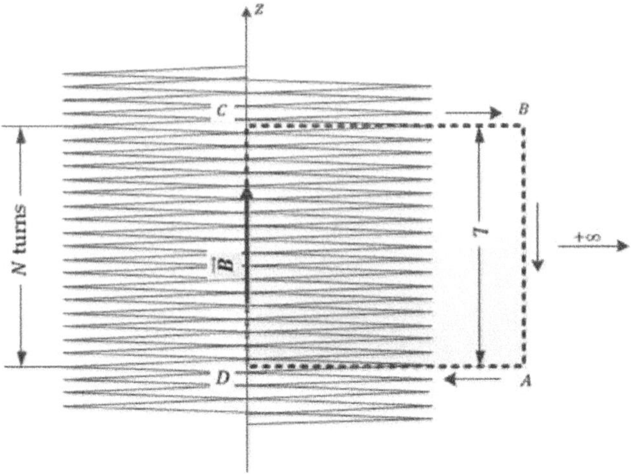

Figure 4.22. *Ampère's theorem applied to a solenoid: the magnetic field* B_z *is constant along the segment* AB *and zero over* BC, CD *and* DA

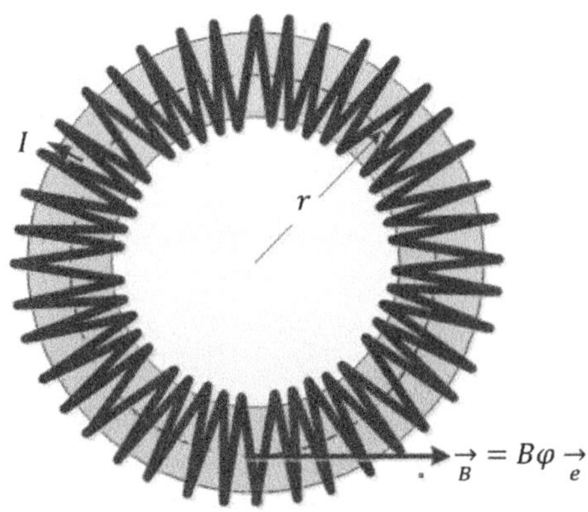

Figure 4.23. *The circle within (respectively, outside) encloses the current* NI *(respectively, no current)*

EXAMPLE 4.19. (Magnetostriction).– Consider a conducting fluid flowing in an infinitely long cylinder of axis Oz and radius R. Assume that the current density vector in the fluid is oriented along Oz and that its expression at a point M with cylindrical coordinates (r, φ, z) is $\overrightarrow{\mathbf{J}}(M) = J(r)\overrightarrow{\mathbf{k}}$ for $r \leq R$ and $\overrightarrow{\mathbf{J}}(M) = 0$ for $r > R$. This current distribution creates an orthoradial magnetic field, and its component B_φ depends only on r. Applying Ampère's theorem to the field line passing through M (circle of axis Oz and radius r), we obtain:

$$\oint \overrightarrow{\mathbf{B}} \cdot d\overrightarrow{\mathbf{l}} = 2\pi r B_{\varphi}(r) = \mu_0 \int \overrightarrow{\mathbf{J}} \cdot d\overrightarrow{\mathbf{S}} = 2\pi\mu_0 \int_0^r r' J(r')\, dr' \qquad [4.119]$$

which means that:

$$B_{\varphi}(r) = \frac{\mu_0}{r} \int_0^r r' J(r')\, dr' \qquad [4.120]$$

The fluid is assumed to be locally neutral. Therefore, the only electromagnetic force acting on a volume element $d\tau = r dr d\varphi dz$ of the fluid is the Laplace force:

$$d\overrightarrow{\mathbf{F}}_{\text{Lap}} = \overrightarrow{\mathbf{J}}(r) \times \overrightarrow{\mathbf{B}}(r)\, d\tau = -J(r)\, B_{\varphi}(r)\, d\tau\, \overrightarrow{\mathbf{e}}_r \qquad [4.121]$$

This force is radial and causes the fluid to tighten around Oz: this is the *magnetostriction effect*. In addition to the Laplace force, there are also pressure forces in the fluid. If $P(M)$ is the pressure at point M, the volume element $d\tau$ is also subject to the pressure force:

$$d\overrightarrow{\mathbf{F}}_P = -\overrightarrow{\nabla} P(M)\, d\tau \qquad [4.122]$$

Assuming that the volume element is in equilibrium, we get:

$$\overrightarrow{\nabla} P(M) = -J(r)\, B_{\varphi}(r)\, \overrightarrow{\mathbf{e}}_r \qquad [4.123]$$

We conclude that pressure depends only on r and that:

$$\frac{dP}{dr} = -J(r)\, B_{\varphi}(r) \qquad [4.124]$$

Furthermore, since $\overrightarrow{\nabla} \times \overrightarrow{\mathbf{B}} = \mu_0 \overrightarrow{\mathbf{J}} = \frac{1}{r}\frac{d(rB_{\varphi})}{dr}\overrightarrow{\mathbf{k}}$, it can be derived that $\mu_0 J(r) = \frac{1}{r}\frac{d(rB_{\varphi})}{dr}$ and consequently:

$$\frac{dP}{dr} = -\frac{1}{\mu_0}\frac{B_{\varphi}}{r}\frac{d(rB_{\varphi})}{dr} = -\frac{1}{2\mu_0 r^2}\frac{d\left(r^2 B_{\varphi}^2\right)}{dr} \qquad [4.125]$$

Integrating this equation yields:

$$P(r) = P(R) + \frac{1}{2\mu_0} \int_r^R \frac{dr'}{r'^2}\frac{d\left(r'^2 B_{\varphi}^2\right)}{dr'} \qquad [4.126]$$

Calculate the average pressure over a fluid cross-section:

$$\begin{aligned}\langle P\rangle &= \frac{1}{\pi R^2}\int_0^R 2\pi r P(r)\,dr = P(R) + \frac{1}{\mu_0 R^2}\int_{r=0}^R r dr \int_{r'=r}^R \frac{dr'}{r'^2}\frac{d\left(r'^2 B_\varphi^2\right)}{dr'} \\ &= P(R) + \frac{1}{\mu_0 R^2}\int_{r'=0}^R \frac{dr'}{r'^2}\frac{d\left(r'^2 B_\varphi^2\right)}{dr'}\int_{r=0}^{r'} r dr \\ &= P(R) + \frac{1}{2\mu_0 R^2}\int_0^R dr' \frac{d\left(r'^2 B_\varphi^2\right)}{dr'} = P(R) + \frac{B_\varphi^2(R)}{2\mu_0}\end{aligned} \quad [4.127]$$

The average overpressure is therefore equal to:

$$\langle \Delta P\rangle = \langle P\rangle - P(R) = \frac{B_\varphi^2(R)}{2\mu_0} \quad [4.128]$$

Since $B_\varphi(R) = \frac{\mu_0 I}{2\pi R}$, where I is the intensity of the current transported by the fluid, it finally comes that:

$$\langle \Delta P\rangle = \frac{\mu_0 I^2}{8\pi^2 R^2} \quad [4.129]$$

Note that the quantity $\frac{B^2}{2\mu}$ has the dimension of a pressure: it is the magnetic pressure. In some applications, the fluid is considered to be a perfect gas, so we can write:

$$P = nk_B T = \frac{B_\varphi^2(R)}{2\mu_0} \quad [4.130]$$

where n, T and k_B are, respectively, the number of constituents per unit volume of the gas, the gas temperature and the Boltzmann constant. This is the *Bennett equation.*

4.6. Magnetic field continuity at a sheet interface

Assume that a surface current distribution $\overrightarrow{\mathbf{K}}$ separates space into two zones, 1 and 2, referred to here as the *current sheet.* Consider a fictitious closed, infinitesimal cylindrical surface enclosing a volume τ and passing through the sheet. The axis of the cylinder is perpendicular to the sheet, as shown in Figure 4.24(a). The conservation of flow through this surface allows us to write:

$$0 = \oint_{\text{cylinder}} \overrightarrow{\mathbf{B}}\cdot d\overrightarrow{\mathbf{S}} = \int_{S_1} \overrightarrow{\mathbf{B}}_1\cdot d\overrightarrow{\mathbf{S}}_1 + \int_{S_2} \overrightarrow{\mathbf{B}}_2\cdot d\overrightarrow{\mathbf{S}}_2 + \int_{S_L} \overrightarrow{\mathbf{B}}\cdot d\overrightarrow{\mathbf{S}}_L \quad [4.131]$$

where S_L is the lateral surface. By taking $\tau \longrightarrow 0$, that is, $S_L \longrightarrow 0$ and $S_1 \longrightarrow S_2$, we obtain $\int_{S_1} \vec{\mathbf{B}}_1 \cdot d\vec{\mathbf{S}}_1 + \int_{S_2} \vec{\mathbf{B}}_2 \cdot d\vec{\mathbf{S}}_2 = 0$ and, consequently:

$$\int_S \left(\vec{\mathbf{B}}_2 - \vec{\mathbf{B}}_1\right) \cdot d\vec{\mathbf{S}} = \int_S \left(\vec{\mathbf{B}}_2 - \vec{\mathbf{B}}_1\right) \cdot \vec{\mathbf{n}}_{12} dS = 0 \quad [4.132]$$

where $\vec{\mathbf{n}}_{12}$ is the normal to the sheet, directed here from medium 1 to medium 2. Since S can be any surface, it can thus be concluded that:

$$\vec{\mathbf{n}}_{12} \cdot \left(\vec{\mathbf{B}}_2 - \vec{\mathbf{B}}_1\right) = 0 \quad \Longleftrightarrow \quad (B_n)_2 = (B_n)_1 \quad [4.133]$$

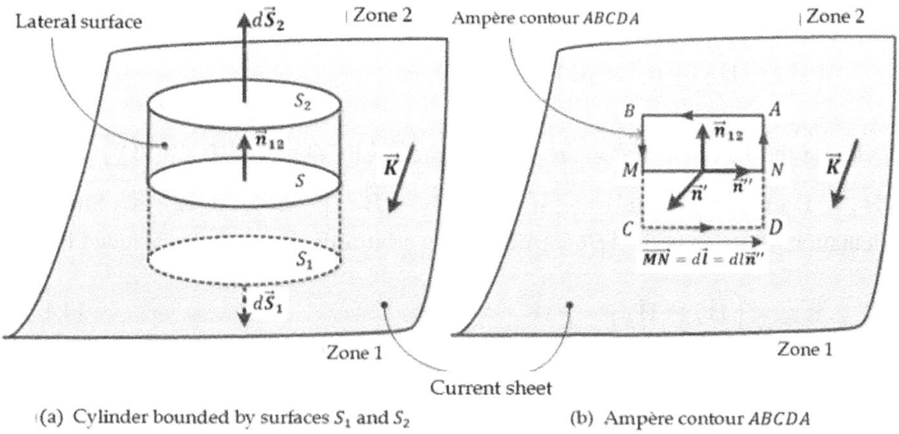

Figure 4.24. *(a) Continuity of the component normal of the magnetic field $\vec{\mathbf{B}}$ at the interface of a current sheet $\vec{\mathbf{K}}$. (b) Discontinuity of the component tangential of the magnetic field $\vec{\mathbf{B}}$ at the interface of a current sheet $\vec{\mathbf{K}}$*

The normal component of the magnetic field $\vec{\mathbf{B}}$ therefore remains continuous across a sheet.

Now imagine the infinitesimal Ampère contour $ABCDA$ shown in Figure 4.24(b) and apply Ampère's theorem:

$$\oint_C \vec{\mathbf{B}}\ d\vec{\mathbf{l}} = \int_{AB} \vec{\mathbf{B}} \cdot d\vec{\mathbf{l}} + \int_{BC} \vec{\mathbf{B}} \cdot d\vec{\mathbf{l}} + \int_{CD} \vec{\mathbf{B}} \cdot d\vec{\mathbf{l}} + \int_{DA} \vec{\mathbf{B}} \cdot d\vec{\mathbf{l}} = \mu_0 I_{\text{encl.}}$$

[4.134]

where $I_{\text{encl.}}$ is the enclosed current, that is, the current flowing through the $ABCD$ surface:

$$I_{\text{encl.}} = \int_{MN} \vec{\mathbf{K}} \cdot \vec{\mathbf{n}}' dl \qquad [4.135]$$

$\vec{\mathbf{n}}'$ is a unit vector tangent to the sheet and perpendicular to the surface $ABCD$. By taking $DA \longrightarrow 0$, equation [4.134] reduces to:

$$\int_{MN} \left(\vec{\mathbf{B}}_1 - \vec{\mathbf{B}}_2\right) \cdot d\vec{\mathbf{n}}'' dl = \mu_0 \int_{MN} \vec{\mathbf{K}} \cdot \vec{\mathbf{n}}' dl \qquad [4.136]$$

where $\vec{\mathbf{n}}''$ is the unit vector carried by $\overrightarrow{\mathbf{MN}}$. Since the vector $\overrightarrow{\mathbf{MN}}$ is arbitrary, we must have:

$$\left(\vec{\mathbf{B}}_1 - \vec{\mathbf{B}}_2\right) \cdot \vec{\mathbf{n}}'' = \mu_0 \vec{\mathbf{K}} \cdot \vec{\mathbf{n}}' \qquad [4.137]$$

Using the fact that $\vec{\mathbf{n}}'' = \vec{\mathbf{n}}_{12} \times \vec{\mathbf{n}}'$, we finally obtain: $\left(\vec{\mathbf{B}}_1 - \vec{\mathbf{B}}_2\right) \cdot \vec{\mathbf{n}}'' = \left(\vec{\mathbf{B}}_1 - \vec{\mathbf{B}}_2\right) \cdot (\vec{\mathbf{n}}_{12} \times \vec{\mathbf{n}}') = \left[\vec{\mathbf{n}}_{12} \times \left(\vec{\mathbf{B}}_2 - \vec{\mathbf{B}}_1\right)\right] \cdot \vec{\mathbf{n}}' = \mu_0 \vec{\mathbf{K}} \cdot \vec{\mathbf{n}}'$. Since the orientation of the contour $ABCDA$ is chosen arbitrarily, it must be concluded that:

$$\vec{\mathbf{n}}_{12} \times \left(\vec{\mathbf{B}}_2 - \vec{\mathbf{B}}_1\right) = \mu_0 \vec{\mathbf{K}} \qquad [4.138]$$

The tangential component of $\vec{\mathbf{B}}$ is therefore discontinuous as it passes through a sheet. Equation [4.138] can be rewritten in another way. Multiplying both sides of [4.138] by $\vec{\mathbf{n}}_{12}$, it is found that $\mu_0 \vec{\mathbf{n}}_{12} \times \vec{\mathbf{K}} = \vec{\mathbf{n}}_{12} \times \left(\vec{\mathbf{n}}_{12} \times \left(\vec{\mathbf{B}}_2 - \vec{\mathbf{B}}_1\right)\right) = \left[\vec{\mathbf{n}}_{12} \cdot \left(\vec{\mathbf{B}}_2 - \vec{\mathbf{B}}_1\right)\right] \vec{\mathbf{n}}_{12} - (\vec{\mathbf{n}}_{12} \cdot \vec{\mathbf{n}}_{12}) \left(\vec{\mathbf{B}}_2 - \vec{\mathbf{B}}_1\right) = -\left(\vec{\mathbf{B}}_2 - \vec{\mathbf{B}}_1\right)$, because $\vec{\mathbf{n}}_{12} \cdot \left(\vec{\mathbf{B}}_2 - \vec{\mathbf{B}}_1\right) = 0$ and $\left|\vec{\mathbf{n}}_{12}\right| = 1$. This finally leads to:

$$\vec{\mathbf{B}}_2 - \vec{\mathbf{B}}_1 = \mu_0 \vec{\mathbf{K}} \times \vec{\mathbf{n}} \qquad [4.139]$$

4.7. Potential vector $\vec{\mathbf{A}}$

4.7.1. *Uniqueness of* $\vec{\mathbf{A}}$

A mathematical consequence of the flux conservation equation $\vec{\nabla} \cdot \vec{\mathbf{B}} = 0$ is that it is always possible to introduce a field $\vec{\mathbf{A}}$ so that:

$$\vec{\mathbf{B}} = \vec{\nabla} \times \vec{\mathbf{A}} \qquad [4.140]$$

as the divergence of the curl of any vector is always zero. The $\vec{\mathbf{A}}$ field is called the *magnetic vector potential*. It is often easier to calculate $\vec{\mathbf{A}}$ and then obtain the magnetic field $\vec{\mathbf{B}}$ from [4.140]. Ampère's theorem allows us to express the vector potential as a function of the current density as follows:

$$\vec{\nabla} \times \vec{\mathbf{B}} = \vec{\nabla} \times \left(\vec{\nabla} \times \vec{\mathbf{A}}\right) = \vec{\nabla}\left(\vec{\nabla} \cdot \vec{\mathbf{A}}\right) - \Delta\vec{\mathbf{A}} = \mu_0 \vec{\mathbf{J}} \qquad [4.141]$$

The field $\vec{\mathbf{A}}$ is not precisely defined because it is always possible to add the gradient of a scalar function Φ to the solution $\vec{\mathbf{A}}$ without changing the value of $\vec{\mathbf{B}}$. In fact, the vector potentials $\vec{\mathbf{A}}$ and $\vec{\mathbf{A}}' = \vec{\mathbf{A}} + \vec{\nabla}\Phi$ produce exactly the same magnetic field $\vec{\mathbf{B}}$, since the curl of the gradient vector of any function is always zero. The Helmholtz–Hodge theorem states that to specify the vector field uniquely, we must give its curl and divergence, and that far from the sources, the fields must tend toward zero. Let us briefly recall how this theorem works. This can be done by setting the curl and divergence of $\vec{\mathbf{A}}$ and then determining $\vec{\mathbf{A}}$. Is there another vector, $\vec{\mathbf{A}}'$, different from $\vec{\mathbf{A}}$, with the same curl and divergence? Let us write $\vec{\mathbf{A}}' = \vec{\mathbf{A}} + \vec{\mathbf{a}}$ and try to prove that $\vec{\mathbf{a}}$ is equal to zero. By definition, $\vec{\nabla} \times \vec{\mathbf{A}}'$ and $\vec{\nabla} \times \vec{\mathbf{A}}$ must be equal: $\vec{\nabla} \times \vec{\mathbf{A}}' = \vec{\nabla} \times \left(\vec{\mathbf{A}} + \vec{\mathbf{a}}\right) = \vec{\nabla} \times \vec{\mathbf{A}}$. This implies that $\vec{\nabla} \times \vec{\mathbf{a}} = 0$, that is, the curl of $\vec{\mathbf{a}}$ must be zero. Consequently, it is possible to write $\vec{\mathbf{a}}$ as the gradient of a scalar function Φ, that is, $\vec{\mathbf{a}} = \vec{\nabla}\Phi$. Similarly, preservation of the divergence value between $\vec{\mathbf{A}}$ and $\vec{\mathbf{A}}'$ requires that the divergence of $\vec{\mathbf{a}}$ is also zero. In fact, we have $\vec{\nabla} \cdot \vec{\mathbf{A}}' = \vec{\nabla} \cdot \left(\vec{\mathbf{A}} + \vec{\mathbf{a}}\right) = \vec{\nabla} \cdot \vec{\mathbf{A}}$ and, consequently, $\vec{\nabla} \cdot \vec{\mathbf{a}} = 0$. Since $\vec{\mathbf{a}} = \vec{\nabla}\Phi$, we can conclude that the Laplacian of Φ is zero:

$$\vec{\nabla} \cdot \vec{\mathbf{a}} = \Delta\phi = 0 \qquad [4.142]$$

This equation is similar to the Poisson equation obtained for electrical potential:

$$\Delta V = -\frac{\rho}{\varepsilon_0} \qquad \Longleftrightarrow \qquad V = \frac{1}{4\pi\varepsilon_0}\int_{\tau'} \frac{\rho' d\tau'}{\left|\vec{\mathbf{r}} - \vec{\mathbf{r}}'\right|} \qquad [4.143]$$

The solution [4.143] shows that the potential V is zero far from the charge distribution. Moreover, if sources are zero elsewhere, the potential is also zero everywhere. Conversely, if the potential is different from zero in one region, it means that there are sources in other regions. Thus, if we compare Φ to V, we deduce that the function Φ is zero everywhere. Thus, equation [4.142] gives $\vec{\mathbf{a}} = 0$ and, consequently, $\vec{\mathbf{A}}' = \vec{\mathbf{A}}$. This proves once again the *Helmholtz–Hodge theorem*.

4.7.2. Potential vector of a current distribution

We saw in the previous section that to fix the choice of the potential vector $\overrightarrow{\mathbf{A}}$, that is, to select a configuration among the equivalent physical solutions, an additional constraint must be imposed on the divergence of $\overrightarrow{\mathbf{A}}$. In other words, the divergence of the potential vector must be specified. In this case, we are talking about fixing a gauge. Consider the simplest case:

$$\overrightarrow{\nabla} \cdot \overrightarrow{\mathbf{A}} = 0 \qquad [4.144]$$

This is called the *Coulomb gauge*. Other choices are also possible. In this case, Ampère's theorem [4.141] is reduced to:

$$\Delta \overrightarrow{\mathbf{A}} = -\mu_0 \overrightarrow{\mathbf{J}} \qquad [4.145]$$

This equation has the same form as the Poisson equation $\Delta V = -\frac{\rho}{\varepsilon_0}$, except that it has three components. Each component of [4.145] admits a solution of the form [4.143] and, therefore, the vector solution of [4.145] is also analogous to [4.143]:

$$\overrightarrow{\mathbf{A}}\left(\overrightarrow{\mathbf{r}}\right) = \frac{\mu_0}{4\pi} \int_{\tau'} \frac{\overrightarrow{\mathbf{J}}\left(\overrightarrow{\mathbf{r}}'\right)}{\left|\overrightarrow{\mathbf{r}} - \overrightarrow{\mathbf{r}}'\right|} d\tau' \qquad [4.146]$$

This expression has already appeared in [4.100]. When it is a surface or linear current distribution, formula [4.146] must be replaced by:

$$\overrightarrow{\mathbf{A}}\left(\overrightarrow{\mathbf{r}}\right) = \begin{cases} \dfrac{\mu_0}{4\pi} \displaystyle\int_{S'} \frac{\overrightarrow{\mathbf{K}}\left(\overrightarrow{\mathbf{r}}'\right)}{\left|\overrightarrow{\mathbf{r}} - \overrightarrow{\mathbf{r}}'\right|} dS' & \text{(surface current)} \\ \dfrac{\mu_0}{4\pi} \displaystyle\int_{L'} \frac{\overrightarrow{\mathbf{I}}\left(\overrightarrow{\mathbf{r}}'\right)}{\left|\overrightarrow{\mathbf{r}} - \overrightarrow{\mathbf{r}}'\right|} dl' & \text{(line current)} \\ \dfrac{\mu_0 I}{4\pi} \displaystyle\oint_{\Gamma} \frac{d\overrightarrow{\mathbf{l}}'}{\left|\overrightarrow{\mathbf{r}} - \overrightarrow{\mathbf{r}}'\right|} & \text{(closed filiform circuit)} \end{cases} \qquad [4.147]$$

For a point charge q with velocity $\overrightarrow{\mathbf{v}}_q$, the vector potential is calculated from the current density $\overrightarrow{\mathbf{J}}\left(\overrightarrow{\mathbf{r}}'\right) = q\delta\left(\overrightarrow{\mathbf{r}}' - \overrightarrow{\mathbf{r}}_q(t)\right)\overrightarrow{\mathbf{v}}_q$, where $\overrightarrow{\mathbf{r}}_q(t)$ describes the particle's trajectory, that is, its position in time. From this, we deduce that:

$$\overrightarrow{\mathbf{A}} = \frac{\mu_0 q \overrightarrow{\mathbf{v}}_q}{4\pi\left|\overrightarrow{\mathbf{r}} - \overrightarrow{\mathbf{r}}_q(t)\right|} \qquad [4.148]$$

If $\vec{\mathbf{B}}$ is uniform, it can be reproduced from the potential vector:

$$\vec{\mathbf{A}}\left(\vec{\mathbf{r}}\right) = \frac{1}{2}\left(\vec{\mathbf{B}} \times \vec{\mathbf{r}}\right) \qquad [4.149]$$

In fact, we verify that:

$$\vec{\nabla} \times \vec{\mathbf{A}} = \frac{1}{2}\vec{\nabla} \times \left(\vec{\mathbf{B}} \times \vec{\mathbf{r}}\right) = \frac{1}{2}\begin{vmatrix} \vec{\mathbf{i}} & \vec{\mathbf{j}} & \vec{\mathbf{k}} \\ \frac{\partial}{\partial x} & \frac{\partial}{\partial y} & \frac{\partial}{\partial z} \\ zB_y - yB_z & xB_z - zB_x & yB_x - xB_y \end{vmatrix} = \vec{\mathbf{B}} \qquad [4.150]$$

The potential vector [4.149] satisfies the Coulomb gauge.

In summary, there are four ways to calculate a magnetic field $\vec{\mathbf{B}}$:

1) using the Biot and Savart law;

2) using Ampère's theorem by calculating the circulation of the field along a specific contour;

3) using flux conservation if the form of the field in a given region is known;

4) using the vector potential, by calculating $\vec{\mathbf{A}}$ and then finding $\vec{\mathbf{B}}$ from equation [4.140].

EXAMPLE 4.20. (Current Line).– Consider a wire of length L, along the axis Oz and carrying a current I. The origin O is placed in the middle of the wire. We wish to calculate the potential vector at any point $M = (r, \varphi, z)$ in space, as shown in Figure 4.25. According to [4.147] (second formula), the potential vector is:

$$\begin{aligned} \vec{\mathbf{A}} &= \frac{\mu_0 I}{4\pi}\int_{L'} \frac{d\vec{\mathbf{l}}'}{\left|\vec{\mathbf{r}} - \vec{\mathbf{r}}'\right|} = \frac{\mu_0 I\vec{\mathbf{k}}}{4\pi}\int_{-\frac{L}{2}}^{\frac{L}{2}} \frac{dz'}{\sqrt{r^2 + (z'-z)^2}} \\ &= \frac{\mu_0 I\vec{\mathbf{k}}}{4\pi}\ln\left(\frac{\sqrt{r^2 + \left(\frac{L}{2} - z\right)^2} - z + \frac{L}{2}}{\sqrt{r^2 + \left(\frac{L}{2} + z\right)^2} - z - \frac{L}{2}}\right) \\ &= \frac{\mu_0 I\vec{\mathbf{k}}}{4\pi}\left(\sinh^{-1}\frac{-z + \frac{L}{2}}{r} + \sinh^{-1}\frac{z + \frac{L}{2}}{r}\right) \end{aligned} \qquad [4.151]$$

We can see that $\vec{\mathbf{A}}$ depends only on the variables r and z. $\vec{\mathbf{A}}$ can also be obtained from [4.146] by choosing a volume current density restricted along Oz, that is, of the form $\overrightarrow{\mathbf{J}}\left(\overrightarrow{\mathbf{r}}'\right) = I\delta\left(x'\right)\delta\left(y'\right)\overrightarrow{\mathbf{k}}$. The magnetic field is calculated using [4.140]:

$$\overrightarrow{\mathbf{B}} = \overrightarrow{\nabla} \times \overrightarrow{\mathbf{A}} = -\frac{\partial A_3}{\partial r}\overrightarrow{\mathbf{e}}_\varphi = \frac{\mu_0 I}{4\pi r}\left[\frac{-z+\frac{L}{2}}{\sqrt{r^2+\left(\frac{L}{2}+z\right)^2}} + \frac{z+\frac{L}{2}}{\sqrt{r^2+\left(\frac{L}{2}-z\right)^2}}\right]\overrightarrow{\mathbf{e}}_\varphi \qquad [4.152]$$

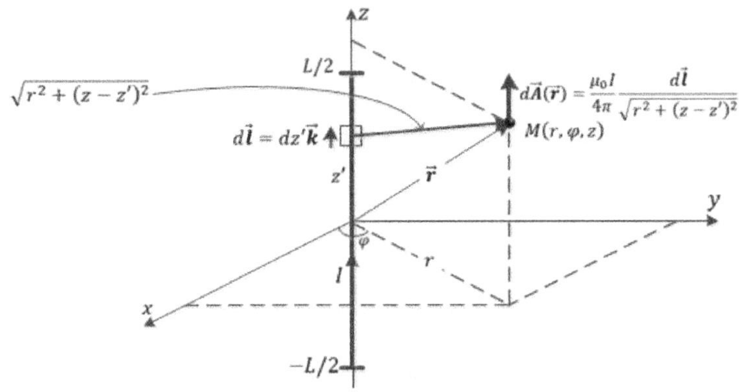

Figure 4.25. *Potential vector of a current line of finite length L*

When L becomes very large, formula [4.152] approaches the expression for the magnetic field created by an infinitely long current line:

$$\overrightarrow{\mathbf{B}} = \frac{\mu_0 I\overrightarrow{\mathbf{e}}_\varphi}{2\pi}\left[\frac{L^2+4z^2}{L^2-4z^2} - \frac{4r^2\left(L^4+24L^2z^2+16z^4\right)}{\left(L^2-4z^2\right)^3} + \frac{6r^4\left(L^6+60L^4z^2+240L^2z^4+64z^6\right)}{\left(L^2-4z^2\right)^5} + \cdots\right] \qquad [4.153]$$

$$\lim_{L\to\infty}\overrightarrow{\mathbf{B}} = \frac{\mu_0 I}{2\pi r}\overrightarrow{\mathbf{e}}_\varphi$$

while the potential vector [4.151] reduces to:

$$\overrightarrow{\mathbf{A}} = \frac{\mu_0 I\overrightarrow{\mathbf{k}}}{4\pi}\left[\ln\left(\frac{L-2z}{r}\right) + \ln\left(\frac{L+2z}{r}\right) + \frac{r^2}{\left(L-2z\right)^2} + \frac{r^2}{\left(L+2z\right)^2} - \frac{3}{2}\frac{r^4}{\left(L-2z\right)^4} - \frac{3}{2}\frac{r^4}{\left(L+2z\right)^4} + \cdots\right] = \frac{\mu_0 I\overrightarrow{\mathbf{k}}}{4\pi}\left[\ln\left(\frac{L^2-4z^2}{r^2}\right) + \frac{2r^2\left(L^2+4z^2\right)}{\left(L^2-4z^2\right)^2} - \frac{3r^4\left(L^4+24L^2z^2+16z^4\right)}{\left(L^2-4z^2\right)^4} + \cdots\right] \qquad [4.154]$$

These developments are valid for $L > 2(r + |z|)$. In particular, if $L \longrightarrow \infty$, the potential vector becomes infinite:

$$\overrightarrow{\mathbf{A}} = -\frac{\mu_0 I \overrightarrow{\mathbf{k}}}{2\pi} \ln r + \lim_{L \longrightarrow \infty} \frac{\mu_0 I \overrightarrow{\mathbf{k}}}{2\pi} \left[\ln L + \frac{r^2}{L^2} - \frac{3}{2} \frac{r^4}{L^4} + \right.$$
$$\left. 2\left(-1 + 6\frac{r^2}{L^2} - 30\frac{r^4}{L^4}\right)\frac{z^2}{L^2} + \cdots \right] \simeq -\frac{\mu_0 I \overrightarrow{\mathbf{k}}}{2\pi} \ln r + \lim_{L \longrightarrow \infty} \frac{\mu_0 I \overrightarrow{\mathbf{k}}}{2\pi} \ln L \quad [4.155]$$

Eliminating the infinite part (which here plays the role of an infinite additive constant), we remain with:

$$\overrightarrow{\mathbf{A}} = -\frac{\mu_0 I \overrightarrow{\mathbf{k}}}{2\pi} \ln r \quad [4.156]$$

The field [4.156] can also be determined in another way. Writing $\overrightarrow{\mathbf{A}} = A_z \overrightarrow{\mathbf{k}}$ and noting that $\overrightarrow{\mathbf{B}}$ must be equal to $\frac{\mu_0 I}{2\pi r} \overrightarrow{\mathbf{e}}_\varphi$, we write $\left(\overrightarrow{\nabla} \times \overrightarrow{\mathbf{A}}\right)_\varphi = -\frac{\partial A_z}{\partial r} = \frac{\mu_0 I}{2\pi r}$. We obtain $\frac{\partial A_z}{\partial r} = -\frac{\mu_0 I}{2\pi r}$. After integration, we find $A_z = -\frac{\mu_0 I}{2\pi} \ln r + \lambda$, where λ is the infinite additive constant (eliminated earlier). ■

Equation [4.156] provides a useful form of the potential vector expression for a current line. If we consider a filament extending to infinity in the Oz direction, located at position $(0, 0)$ (in the Oxy plane) and traversed by an infinitesimal current dI, as shown in Figure 4.26(a), the potential vector created by this current at point $M(x, y, z)$ is written as:

$$d\overrightarrow{\mathbf{A}}(x, y) = -\frac{\mu_0}{2\pi} \ln \sqrt{x^2 + y^2} dI \overrightarrow{\mathbf{k}} = -\frac{\mu_0}{4\pi} \ln\left[x^2 + y^2\right] dI \overrightarrow{\mathbf{k}} \quad [4.157]$$

Furthermore, if dI belongs to a larger current distribution directed along Oz and traversing a surface S' in the Oxy plane, this distribution can still be considered as a superposition of filaments located at points (x', y'), as shown in Figure 4.26(b), by carrying out the substitution $dI = J_z(x', y') dx'dy'$. The total potential vector produced by the distribution is therefore:

$$\overrightarrow{\mathbf{A}}(x, y) = -\frac{\mu_0}{4\pi} \int_{S'} J_z(x', y') \ln\left[(x - x')^2 + (y - y')^2\right] dx'dy' \overrightarrow{\mathbf{k}} \quad [4.158]$$

Integration is performed on the transverse surface S' over which the distribution extends.

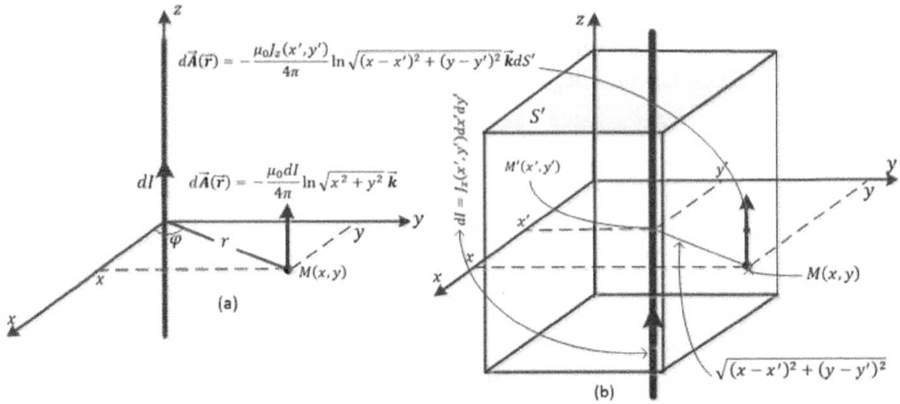

Figure 4.26. *The bar is seen here as a superposition of filaments: the potential vector* $\overrightarrow{\text{A}}$ *is the sum of the elementary potential vectors created by each filament passing through* (x', y')*. For a color version of this figure, see www.iste.co.uk/abdesselam/electrodynamics1.zip*

EXAMPLE 4.21. (Current surface).– A surface current $K_0\overrightarrow{\mathbf{k}}$, of width a, is modeled as a superposition of several current lines, as shown in Figure 4.27(a). The infinitesimal potential vector created by the current line $K_0 dx'$ passing through the point (x, y) will be:

$$dA_z(x,y) = -\frac{\mu_0 K_0 dx'}{4\pi} \ln\left[(x-x')^2 + y^2\right] \qquad [4.159]$$

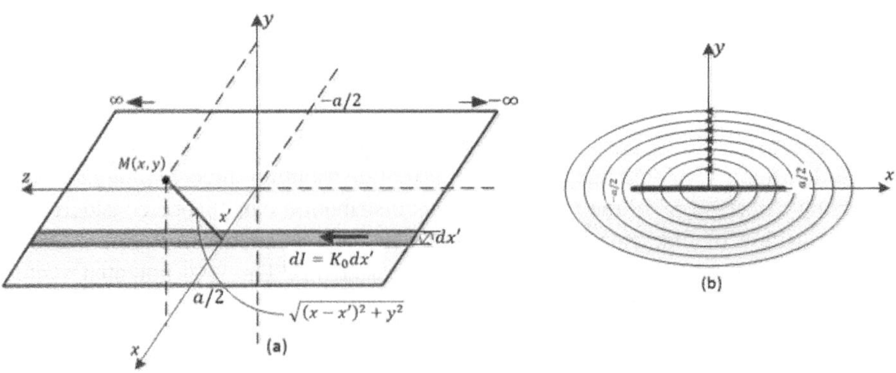

Figure 4.27. *(a) The potential vector of a current surface* $K_0\overrightarrow{\text{k}}$ *of width* a *is obtained by superposing the potential vectors of each line taken separately. (b) The field lines are the lines on which the* A_z *component is constant*

Thus, the total potential vector is obtained by integrating over the width of the distribution:

$$\begin{aligned} A_z &= -\frac{\mu_0 K_0}{4\pi} \int_{-a/2}^{a/2} \ln\left[(x-x')^2+y^2\right] dx' \\ &= -\frac{\mu_0 K_0}{4\pi}\left[\left(\frac{a}{2}-x\right)\ln\left[\left(x-\frac{a}{2}\right)^2+y^2\right]\right. \\ &\left. +\left(\frac{a}{2}+x\right)\ln\left[\left(x+\frac{a}{2}\right)^2+y^2\right]-2a+2y\tan^{-1}\left(\frac{ay}{x^2+y^2-\frac{a^2}{4}}\right)\right] \end{aligned} \quad [4.160]$$

where we have used the identity $\tan^{-1}(a-b)+\tan^{-1}(a+b) = \tan^{-1}\frac{2a}{1-a^2+b^2}$. The magnetic field is calculated from [4.140]:

$$\begin{aligned} \overrightarrow{\mathbf{B}} &= \frac{\partial A_z}{\partial y}\overrightarrow{\mathbf{i}} - \frac{\partial A_z}{\partial x}\overrightarrow{\mathbf{j}} - \frac{\mu_0 K_0}{4\pi}\left[2\tan^{-1}\left(\frac{ay}{y^2+x^2-\frac{a^2}{4}}\right)\overrightarrow{\mathbf{i}}\right. \\ &\left. = +\ln\frac{\left(x+\frac{a}{2}\right)^2+y^2}{\left(x-\frac{a}{2}\right)^2+y^2}\overrightarrow{\mathbf{j}}\right] \end{aligned} \quad [4.161]$$

Since this is a two-dimensional field, it is easier to trace the field lines here. To obtain the field lines, it is enough to write $\frac{dy}{dx} = \frac{B_y}{B_x} = \frac{-\partial A_z/\partial x}{\partial A_z/\partial y}$ or again $\frac{\partial A_z}{\partial x}dx + \frac{\partial A_z}{\partial y}dy = dA_z = 0$. This shows that:

$$A_z = \text{const} \quad [4.162]$$

We can see that the potential vector is constant along a field line. The field lines in Figure 4.27(b) are simply the curves $A_z = \text{const}$.

EXAMPLE 4.22. (Circular Loop).– A permanent current I flows through a circular loop of radius R, located in the Oxy plane. The axis Oz passes through the center of the loop, O. Let us calculate the potential vector $\overrightarrow{\mathbf{A}}$ at the point $M(x, 0, z)$ (located in the Oxz plane), as already seen in Figure 4.11. According to equation [4.147], we write that:

$$\overrightarrow{\mathbf{A}} = \frac{\mu_0 IR}{4\pi}\int_0^{2\pi} \frac{-\sin\varphi'\,\overrightarrow{\mathbf{i}} + \cos\varphi'\,\overrightarrow{\mathbf{j}}}{\sqrt{r^2+R^2-2rR\sin\varphi\cos\varphi'}}d\varphi' \quad [4.163]$$

The first integral is zero. Let us use the fact that $r^2+R^2-2r\sin\varphi\cos\varphi' = \left(r^2+R^2+2r\sin\varphi\right)\left(1-k^2\cos^2\frac{\varphi'}{2}\right)$, where $k^2 = \frac{4rR\sin\varphi}{r^2+R^2+2rR\sin\varphi} = \frac{4Rx}{(R+x)^2+z^2}$.

The second integral becomes:

$$A_y = \frac{\mu_0 I}{4\pi} \frac{R}{\sqrt{r^2 + R^2 + 2rR\sin\varphi}} \int_0^{2\pi} \frac{\left(2\cos^2\frac{\varphi'}{2} - 1\right) d\varphi'}{\sqrt{1 - k^2\cos^2\frac{\varphi'}{2}}} \qquad [4.164]$$

Let us perform the change of variable $\gamma = \frac{\pi}{2} - \frac{\varphi'}{2}$:

$$\begin{aligned} A_y &= \frac{\mu_0 I}{\pi} \frac{R}{\sqrt{r^2 + R^2 + 2rR\sin\varphi}} \int_0^{\frac{\pi}{2}} \frac{\left(2\sin^2\gamma - 1\right) d\gamma}{\sqrt{1 - k^2\sin^2\gamma}} \\ &= \frac{\mu_0 I}{\pi} \frac{R}{\sqrt{r^2 + R^2 + 2rR\sin\varphi}} \left[\left(\frac{2}{k^2} - 1\right) K(k) - \frac{2}{k^2} E(k)\right] \\ &= \frac{\mu_0 I}{2\pi} \sqrt{\frac{R}{x}} \left[\left(\frac{2}{k} - k\right) K(k) - \frac{2}{k} E(k)\right] \end{aligned} \qquad [4.165]$$

where $\frac{R}{\sqrt{r^2+R^2+2rR\sin\varphi}} = \frac{k}{2}\sqrt{\frac{R}{x}}$. To obtain the magnetic field $\overrightarrow{\mathbf{B}}$, let us calculate the curl of $\overrightarrow{\mathbf{A}}$ (with respect to M), noting that the unit vector $\overrightarrow{\mathbf{j}}$ plays here the role of the vector $\overrightarrow{\mathbf{e}}_\varphi$ in cylindrical coordinates:

$$\overrightarrow{\mathbf{B}} = -\frac{\partial A_y}{\partial z}\overrightarrow{\mathbf{i}} + \frac{1}{x}\frac{\partial(xA_y)}{\partial x}\overrightarrow{\mathbf{k}} \qquad [4.166]$$

Using the derivation rules for elliptic integrals $\frac{dK(k)}{dk} = \frac{E(k)}{k(1-k^2)} - \frac{K(k)}{k}$, $\frac{dE(k)}{dk} = \frac{E(k)-K(k)}{k}$ and the fact that $\frac{\partial k}{\partial z} = -\frac{zk^3}{4Rx}$, we get:

$$\begin{aligned} B_x &= -\frac{\partial A_y}{\partial z} = -\frac{\mu_0 I}{2\pi}\sqrt{\frac{R}{x}}\frac{\partial k}{\partial z}\frac{d}{dk}\left[\left(\frac{2}{k} - k\right) K(k) - \frac{2}{k} E(k)\right] \\ &= \frac{\mu_0 I z}{2\pi x\sqrt{(R+x)^2 + z^2}}\left[-K(k) + \frac{2-k^2}{2(1-k^2)}E(k)\right] \\ &= \frac{\mu_0 I z}{2\pi x\sqrt{(R+x)^2 + z^2}}\left[-K(k) + \frac{R^2+x^2+z^2}{(R-x)^2+z^2}E(k)\right] \end{aligned} \qquad [4.167]$$

and:

$$
\begin{aligned}
B_z &= \frac{1}{x}\frac{\partial (xA_y)}{\partial x} = \frac{\mu_0 I}{4\pi x}\sqrt{\frac{R}{x}}\left[\left(\frac{2}{k}-k\right)K(k)-\frac{2}{k}E(k)\right] \\
&+ \frac{\mu_0 I}{2\pi}\sqrt{\frac{R}{x}}\frac{\partial k}{\partial x}\frac{d}{dk}\left[\left(\frac{2}{k}-k\right)K(k)-\frac{2}{k}E(k)\right] \\
&= \frac{\mu_0 I}{2\pi\sqrt{(R+x)^2+z^2}}\left\{\frac{2R}{k}\left[-\frac{2}{k^2}\frac{\partial k}{\partial x}+\frac{1}{2x}\left(\frac{2}{k}-k\right)\right]K(k)\right. \\
&\left.+\frac{2R}{k}\left[-\frac{2-k^2}{k^2(1-k^2)}\frac{\partial k}{\partial x}-\frac{1}{kx}\right]E(k)\right\}
\end{aligned}
\qquad [4.168]
$$

Since $\frac{1}{k^2}=\frac{(R+x)^2+z^2}{4Rx}$, it comes that $\frac{\partial k}{\partial x}=\frac{k^3\left(R^2+z^2-x^2\right)}{8Rx^2}=\frac{k^3}{4Rx}\left(\frac{2R}{k^2}-x-R\right)$. We can therefore deduce that:

$$
B_z = \frac{\mu_0 I}{2\pi\sqrt{(R+x)^2+z^2}}\left[K(k)+\frac{R^2-x^2-z^2}{(R-x)^2+z^2}E(k)\right] \qquad [4.169]
$$

This is in agreement with results [4.57] and [4.58]. As the M point approaches the Oz axis, that is, when $x \longrightarrow 0$, we have the following approximations: $k \simeq 2\sqrt{\frac{R}{R^2+z^2}}\sqrt{x}$, $K(k) \simeq \frac{\pi}{2}\left(1+\frac{k^2}{4}\right) = \frac{\pi}{2}\left(1+\frac{Rx}{R^2+z^2}\right)$ and $E(k)\simeq\frac{\pi}{2}\left(1-\frac{k^2}{4}\right)=\frac{\pi}{2}\left(1-\frac{Rx}{R^2+z^2}\right)$. It therefore follows that:

$$
\overrightarrow{\mathbf{A}}(x,0,z) \simeq -\frac{1}{2}\mu_0 I\left(\frac{R^2}{R^2+z^2}\right)^{3/2}\sqrt{\frac{x}{R}}\overrightarrow{\mathbf{j}} \qquad [4.170]
$$

Clearly, the potential $\overrightarrow{\mathbf{A}}$ is zero for $x=0$. Therefore, it is not possible to obtain the magnetic field $\overrightarrow{\mathbf{B}}$ at this point. This result can be verified directly from formula [4.147]. In fact, we have: $\overrightarrow{\mathbf{A}}=\frac{\mu_0 I}{4\pi}\int_0^{2\pi}\frac{R\overrightarrow{\mathbf{e}}_{\varphi'}d\varphi'}{\sqrt{z^2+R^2}}=0$. Equation [4.170] produces an infinite field in the neighborhood of $x=0$:

$$
\begin{aligned}
\overrightarrow{\mathbf{B}} &= -\frac{\partial A_y}{\partial z}\overrightarrow{\mathbf{i}}+\frac{1}{x}\frac{\partial (xA_y)}{\partial x}\overrightarrow{\mathbf{k}} \\
&= -\frac{3\mu_0 I z}{2R^2}\left(\frac{R^2}{R^2+z^2}\right)^{3/2}\sqrt{\frac{x}{R}}\overrightarrow{\mathbf{i}}-\frac{3\mu_0 I R^{5/2}}{4\sqrt{x}\left(R^2+z^2\right)^{3/2}}\overrightarrow{\mathbf{k}}
\end{aligned}
\qquad [4.171]
$$

We can see that the B_x component is zero for $x=0$, while the B_z component tends toward infinity when x tends toward 0.

4.7.3. *Potential vector and magnetic flux*

Using Stokes's theorem, the *magnetic flux* through a surface can be expressed as a curvilinear integral of the potential vector:

$$\Phi = \int_S \vec{\mathbf{B}} \cdot d\vec{\mathbf{S}} = \int_S \left(\vec{\nabla} \times \vec{\mathbf{A}}\right) \cdot d\vec{\mathbf{S}} = \oint_\Gamma \vec{\mathbf{A}} \cdot d\,\vec{\mathbf{l}} \qquad [4.172]$$

EXAMPLE 4.23. (Square Circuit (or Square Loop) Flux).– A square circuit, with side $2a$, center O and located on the plane Oxy, has a current I flowing through it. We wish to calculate the magnetic field flux $\vec{\mathbf{B}}$ induced by the circuit across the surface S, shown in Figure 4.28. The flux of the $\vec{\mathbf{B}}$ field can be calculated directly by integrating the normal component on the loop surface. This method is simple but cumbersome. Another solution is to superpose the actions of the potential vectors created by each of the four sides of the circuit $ABCD$ on the surface contour $A'B'C'D'$. This gives us eight terms, since each side of the circuit will act on the two sides of the surface parallel to it:

$$\begin{aligned}\Phi = &\int_{a-b}^{-a+b} A_x\left(AB, A'B'\right) dx + \int_{a-b}^{-a+b} A_x\left(AB, D'C'\right) dx \\ &+ \int_{a-b}^{-a+b} A_y\left(BC, B'C'\right) dy + \int_{a-b}^{-a+b} A_y\left(BC, A'D'\right) dy \\ &+ \int_{-a+b}^{a-b} A_x\left(CD, C'D'\right) dx + \int_{-a+b}^{a-b} A_x\left(CD, B'A'\right) dx \\ &+ \int_{-a+b}^{a-b} A_y\left(DA, D'A'\right) dy + \int_{-a+b}^{a-b} A_y\left(DA, C'B'\right) dy\end{aligned} \qquad [4.173]$$

where $A_x\left(AB, A'B'\right)$ is the potential vector created by the line current AB along the path $A'B'$, etc. For reasons of symmetry, we write:

$$\Phi = 4\int_{-a+b}^{a-b} A_x\left(CD, C'D'\right) dx + 4\int_{-a+b}^{a-b} A_x\left(CD, B'A'\right) dx \qquad [4.174]$$

where:

$$\begin{aligned}A_x\left(CD, C'D'\right) &= \frac{\mu_0 I}{4\pi}\left[\sinh^{-1}\left(\frac{-x+a}{b}\right) + \sinh^{-1}\left(\frac{x+a}{b}\right)\right] \\ A_x\left(CD, B'A'\right) &= \frac{\mu_0 I}{4\pi}\left[\sinh^{-1}\left(\frac{-x+a}{2a-b}\right) + \sinh^{-1}\left(\frac{x+a}{2a-b}\right)\right]\end{aligned} \qquad [4.175]$$

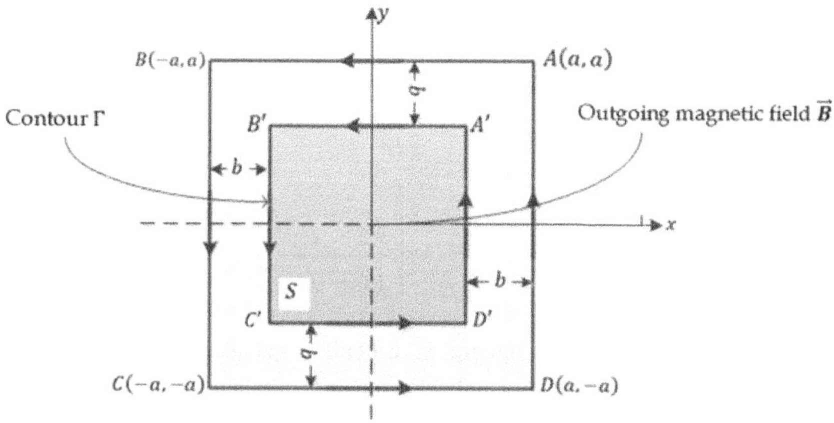

Figure 4.28. *Magnetic field flux created by a square circuit through an internal surface $A'B'C'D'$*

It therefore follows that:

$$
\begin{aligned}
\Phi = & \frac{\mu_0 I}{\pi}\int_{-a+b}^{a-b}\left[\sinh^{-1}\left(\frac{x+a}{b}\right)+\sinh^{-1}\left(\frac{x+a}{2a-b}\right)+\right.\\
& \left.\sinh^{-1}\left(\frac{-x+a}{b}\right)+\sinh^{-1}\left(\frac{-x+a}{2a-b}\right)\right]dx\\
= & \frac{2\mu_0 I}{\pi}\int_{-a+b}^{a-b}\left[\sinh^{-1}\left(\frac{x+a}{b}\right)+\sinh^{-1}\left(\frac{x+a}{2a-b}\right)\right]dx\\
= & \frac{2\mu_0 I}{\pi}\left[(2a-b)\sinh^{-1}\left(\frac{2a-b}{b}\right)-b\sinh^{-1}\left(\frac{b}{2a-b}\right)+\right.\\
& \left.2\left(\sinh^{-1}1-\sqrt{2}\right)(a-b)\right]
\end{aligned}
\qquad [4.176]
$$

If b is very small, we obtain approximately $\Phi \sim \frac{4\mu_0 I a}{\pi}[\sinh^{-1}\left(\frac{2a}{b}\right)+\sinh^{-1}1-\sqrt{2}]$. Φ becomes infinite for $b=0$ (the magnetic field is infinite in the neighborhood of the wires).

EXAMPLE 4.24. (Circular Loop Flux).– Consider a circular loop of radius R, located in the Oxy plane and carrying a current I. We want to calculate the flux of the magnetic field induced by the loop through the circular surface S, of radius a, located inside the loop and bounded by the contour Γ, as shown in Figure 4.29. Noting that here, x plays the role of the variable r in polar coordinates and $\overrightarrow{\mathbf{j}}$ the role of $\overrightarrow{\mathbf{e}}_{\varphi}$, the flux of the field $\overrightarrow{\mathbf{B}}$ across the surface S is:

$$
\Phi = \int_S \overrightarrow{\mathbf{B}}\cdot d\overrightarrow{\mathbf{S}} = \int_0^a B_z 2\pi x dx \qquad [4.177]
$$

where $B_z = \frac{\mu_0 I}{2\pi(R+x)}\left[K(k) + \frac{R^2-x^2}{R^2+x^2}E(k)\right]$ is the component normal to S. Since $k^2 = \frac{4xR}{(x+R)^2}$, we get $x = R\frac{1-\sqrt{1-k^2}}{1+\sqrt{1-k^2}}$. It follows that:

$$\begin{aligned}\Phi &= \mu_0 I \int_0^a \left[K(k) + \frac{R^2-x^2}{R^2+x^2}E(k)\right]\frac{xdx}{R+x} \\ &= \mu_0 IR \int_0^{k_a}\left[K(k) + \frac{2\sqrt{1-k^2}}{2-k^2}E(k)\right]\frac{1-\sqrt{1-k^2}}{\left(1+\sqrt{1-k^2}\right)^2}\frac{kdk}{\sqrt{1-k^2}}\end{aligned} \qquad [4.178]$$

where $k_a = \frac{4Ra}{(R+a)^2} < 1$. This integral is difficult to calculate. However, it is still possible to get around this problem and calculate Φ in another, simple way, using the potential vector $\overrightarrow{\mathbf{A}}$ and the formula [4.172]. In this case, the flux is written as: $\Phi = \oint_\Gamma \overrightarrow{\mathbf{A}} \cdot d\overrightarrow{\mathbf{l}} = \int_0^{2\pi} A_y a dy$, where $A_y = \frac{\mu_0 I}{2\pi}\sqrt{\frac{R}{a}}\left[\left(\frac{2}{k_a} - k_a\right)K(k_a) - \frac{2}{k_a}E(k_a)\right]$. We thus find:

$$\begin{aligned}\Phi &= \frac{\mu_0 I}{2\pi}\sqrt{\frac{R}{a}}\left[\left(\frac{2}{k_a} - k_a\right)K(k_a) - \frac{2}{k_a}E(k_a)\right] \times (2\pi a) \\ &= \frac{\pi}{2}\mu_0 I\sqrt{Ra}\left[-k_a + \sum_{n=1}^{\infty}\left(\frac{1\cdot 3\cdots(2n-1)}{2\cdot 4\cdots(2n)}\right)^2 \frac{2n(2-k_a)+k_a-4}{(2n-1)}k_a^{2n-1}\right]\end{aligned} \qquad [4.179]$$

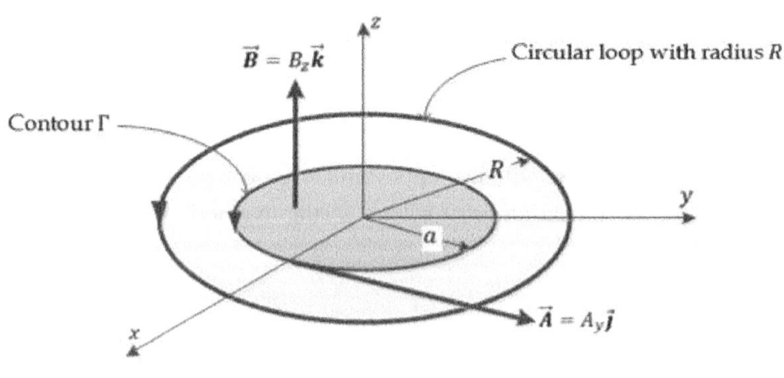

Figure 4.29. *Flux across a surface S with radius $a < R$, bounded by a contour Γ, inside a loop with radius R*

This series diverges for $a = R$. The flux through the loop (of the proper loop's field) is infinite ($\overrightarrow{\mathbf{B}}$ is infinite in the neighborhood of the loop).

4.8. Magnetic scalar potential

Although not as important as the vector potential $\overrightarrow{\mathbf{A}}$, the magnetic scalar potential, referred to here as V_m, is extremely useful for solving problems relating to magnetic fields in areas of space where the current density is zero, that is, $\overrightarrow{\mathbf{B}} = 0$. In this case, the $\overrightarrow{\mathbf{B}}$ field can be expressed as the gradient of a scalar potential V_m, called the *magnetic scalar potential*:

$$\overrightarrow{\mathbf{B}} = -\mu_0 \overrightarrow{\nabla} V_m \qquad [4.180]$$

There is no general prescription for obtaining the expression of the magnetic scalar potential V_m. However, to obtain V_m for a closed filiform circuit Γ carrying a permanent current I, we proceed as follows. According to the Biot and Savart law, we write:

$$dV_m = \overrightarrow{\nabla} V_m \cdot d\overrightarrow{\mathbf{r}} = -\frac{\overrightarrow{\mathbf{B}} \cdot d\overrightarrow{\mathbf{r}}}{\mu_0} = -\frac{I}{4\pi} \oint_\Gamma \frac{d\overrightarrow{\mathbf{r}} \cdot \left[d\overrightarrow{\mathbf{l}}' \times \left(\overrightarrow{\mathbf{r}} - \overrightarrow{\mathbf{r}}' \right) \right]}{\left| \overrightarrow{\mathbf{r}} - \overrightarrow{\mathbf{r}}' \right|^3}$$

$$[4.181]$$

It is interesting to relate this result to the solid angle Ω, under which the surface bounded by the closed circuit Γ is observed by an observer located at $\overrightarrow{\mathbf{r}}$, as shown in Figure 4.30:

$$\Omega = \int_{S'} \frac{d\overrightarrow{\mathbf{S}}' \cdot \left(\overrightarrow{\mathbf{r}}' - \overrightarrow{\mathbf{r}} \right)}{\left| \overrightarrow{\mathbf{r}}' - \overrightarrow{\mathbf{r}} \right|^3} = -\int_{S'} \frac{d\overrightarrow{\mathbf{S}}' \cdot \left(\overrightarrow{\mathbf{r}} - \overrightarrow{\mathbf{r}}' \right)}{\left| \overrightarrow{\mathbf{r}} - \overrightarrow{\mathbf{r}}' \right|^3} \qquad [4.182]$$

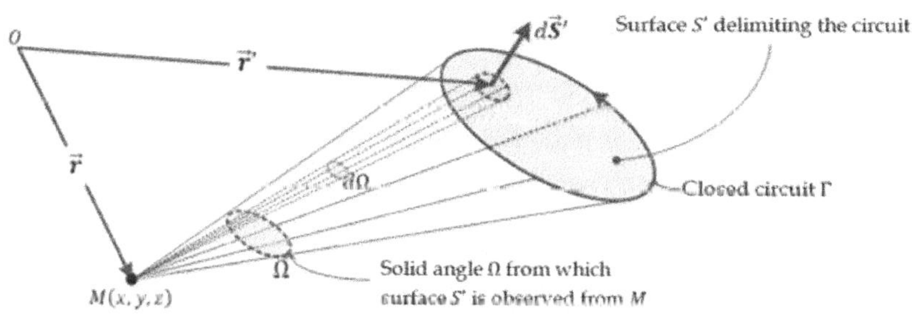

Figure 4.30. *Solid angle Ω under which an observer located at the point $\overrightarrow{\mathbf{r}}$ observes the closed circuit Γ*

Clearly, the form of the surface S' is of no importance. We will choose the simplest form that best suits our calculations. If the observer, located at $d\vec{\mathbf{r}}$, experiences a displacement of the quantity $d\vec{\mathbf{r}}$, then the electric circuit is displaced by $-d\vec{\mathbf{r}}$. This shows that the solid angle bounded by the Γ circuit changes. In particular, the area gained (positive) or lost (negative) by each $d\vec{\mathbf{l}}'$ segment of the circuit as it moves is worth $-d\vec{\mathbf{r}} \times d\vec{\mathbf{l}}'$ and, consequently, the solid angle will change by the quantity:

$$d\Omega = \oint_{\Gamma} \frac{\left[-d\vec{\mathbf{r}} \times d\vec{\mathbf{l}}'\right] \cdot (\vec{\mathbf{r}}' - \vec{\mathbf{r}})}{\left|\vec{\mathbf{r}}' - \vec{\mathbf{r}}\right|^3} = \oint_{\Gamma} \frac{d\vec{\mathbf{r}} \cdot \left[d\vec{\mathbf{l}}' \times (\vec{\mathbf{r}} - \vec{\mathbf{r}}')\right]}{\left|\vec{\mathbf{r}} - \vec{\mathbf{r}}'\right|^3} \qquad [4.183]$$

as shown in Figure 4.31. Taking into account this expression in [4.181], we finally obtain $dV_m = -\frac{I}{4\pi} d\Omega$. It follows that:

$$V_m = -\frac{I\Omega}{4\pi} \qquad [4.184]$$

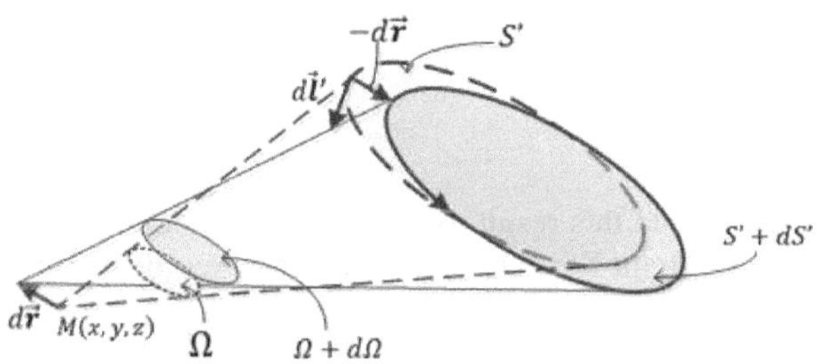

Figure 4.31. *Variation in the solid angle Ω when the observer is displaced by $d\vec{\mathbf{r}}$*

EXAMPLE 4.25.– (Magnetic Scalar Potential of a Circular Loop).– Let us calculate the magnetic scalar potential of a loop of radius R, located in the Oxy plane, center O and flowed by a current I, at a point M located along the Oz axis, as shown in Figure 4.32. Since $\vec{\mathbf{r}} - \vec{\mathbf{r}}' = z\vec{\mathbf{k}} - r'\vec{\mathbf{e}}_r$, $\left|\vec{\mathbf{r}} - \vec{\mathbf{r}}'\right| = \sqrt{z^2 + r'^2}$, $d\vec{\mathbf{S}}' = r'dr'd\varphi'\vec{\mathbf{k}}$ and $(\vec{\mathbf{r}} - \vec{\mathbf{r}}') \cdot d\vec{\mathbf{S}}' = zr'dr'd\varphi'$, we obtain:

$$-\Omega = \int_{S'} \frac{d\vec{\mathbf{S}}' \cdot (\vec{\mathbf{r}} - \vec{\mathbf{r}}')}{\left|\vec{\mathbf{r}} - \vec{\mathbf{r}}'\right|^3} = \int_0^R \int_0^{2\pi} \frac{zr'dr'd\varphi'}{(z^2 + r'^2)^{3/2}} = 2\pi \left(\frac{z}{|z|} - \frac{z}{\sqrt{z^2 + R^2}}\right)$$

[4.185]

and, consequently:

$$V_m = \frac{I}{2}\left(\frac{z}{|z|} - \frac{z}{\sqrt{z^2 + R^2}}\right) \qquad [4.186]$$

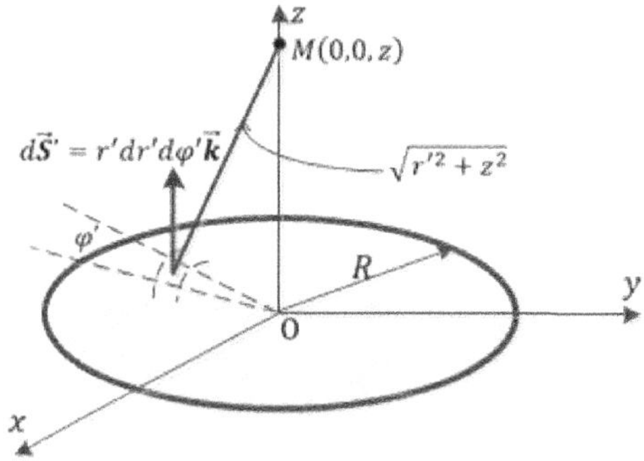

Figure 4.32. *Magnetic scalar potential of a circular loop of radius R*

The magnetic field induced by this potential at the point $(0, 0, z)$ is:

$$B_z(0, 0, z) = -\mu_0 \frac{\partial V_m}{\partial z} = \frac{\mu_0 I}{2} \frac{R^2}{(z^2 + R^2)^{3/2}} \qquad [4.187]$$

EXAMPLE 4.26. (Magnetic Scalar Potential of a Solenoid).– A solenoid of length L, made up of N circular loops, carries a permanent current I. Let us calculate the magnetic scalar potential along the axis of the solenoid, noted here Oz (O is situated in the middle of the solenoid). The elementary scalar potential created by dn loops located at z' is:

$$dV_m(z) = -\frac{I}{2} \frac{z - z'}{\sqrt{(z - z')^2 + R^2}} dn = -\frac{IN}{2L} \frac{z - z'}{\sqrt{(z - z')^2 + R^2}} dz' \qquad [4.188]$$

Consequently, the total potential V_m is obtained after integrating along the solenoid:

$$V_m(z) = -\frac{NI}{2L}\left(\sqrt{\left(z + \frac{L}{2}\right)^2 + R^2} - \sqrt{\left(z - \frac{L}{2}\right)^2 + R^2}\right) \qquad [4.189]$$

The magnetic field induced by V_m is:

$$B_z(0,0,z) = \frac{\mu_0 NIz}{2L}\left(\frac{1}{\sqrt{\left(z+\frac{L}{2}\right)^2+R^2}} - \frac{1}{\sqrt{\left(z-\frac{L}{2}\right)^2+R^2}}\right) \qquad [4.190]$$

In particular, for an infinite solenoid, that is, when $L \to \infty$, the previous formula reduces to:

$$V_m(0,0,z) = -\frac{NIz}{2R} \qquad [4.191]$$

Chapter 5

Electric and Magnetic Moments

5.1. Electric dipole

The presence of matter modifies the electric field, because even if matter is electrically neutral (from the point of view of total charge), the electric field inside matter can cause: a) charge movement, i.e. a *conduction effect* in conductive materials; or b) small charge displacements, called *polarization*, in *dielectrics*. This small imbalance in charge distribution (which then appears as another electric field source added to the first applied field) can have a significant effect on the fields inside and outside these materials. In dielectric mediums, electrons are tightly bound to the nucleus and are not mobile. However, if an electric field is applied, the electron cloud is slightly displaced relative to the positive nucleus (the electron cloud is no longer symmetrical and its center of gravity no longer coincides with that of the nucleus' positive charge), as shown in Figure 5.1. The material is then said to acquire an *electrical polarization*. These slightly separated charges give rise to an *induced dipole*. Note that the ions in a molecule can also undergo small relative displacements, giving rise to *ionic polarization*.

As a first approximation, an electric dipole is modeled by two point charges, $+q$ and $-q$, separated from each other by a distance $\vec{\mathrm{d}}$ (connected, for example, by a fixed bar of length d). Here, $\vec{\mathrm{d}}$ represents the relative position of the positive charge to the negative charge, as shown in Figure 5.2(a). This system is called an *electric or electrostatic dipole*. The *dipole moment* of the charge pair, $\vec{\mathrm{p}}$, is defined as follows:

$$\vec{\mathrm{p}} = q\,\vec{\mathrm{d}} \qquad [5.1]$$

The induced dipoles then become the source of another electric field, which will superpose on the first. In addition, they tend to align themselves in the direction of the applied electric field under the effect of a couple, as shown in Figure 5.2(b). The total electric field is therefore modified. When the external electric field is suppressed, these dipoles may disappear or persist inside the material. Non-induced permanent dipoless are also found in nature. In some cases, they will randomly orient themselves so that the average overall dipole moment is zero, or orient themselves together in any direction.

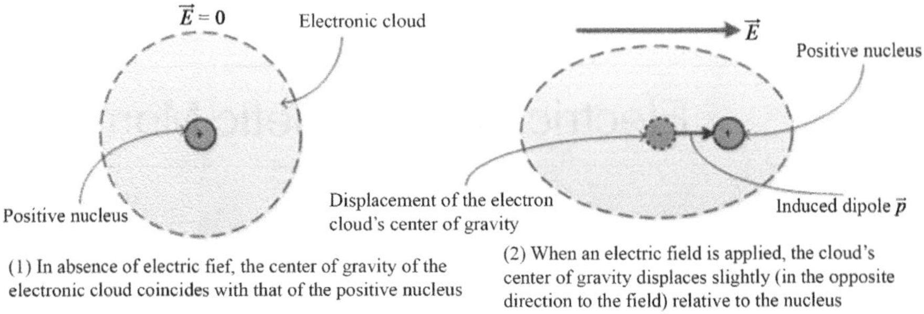

Figure 5.1. *Mechanism for creating an electric dipole inside an atom under the action of an external electric field. For a color version of this figure, see www.iste.co.uk/abdesselam/electrodynamics1.zip*

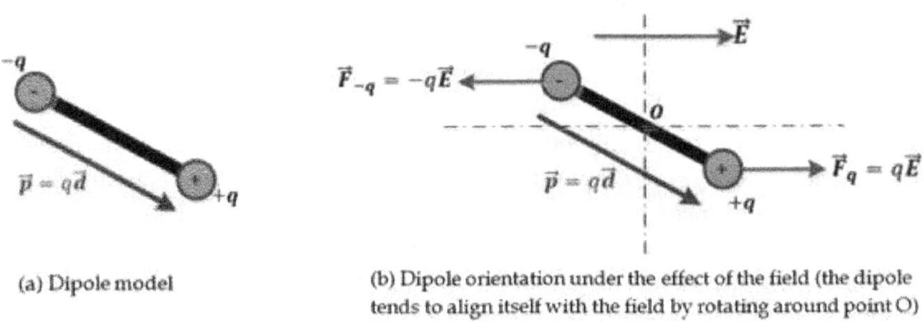

Figure 5.2. *Model of an electric dipole. For a color version of this figure, see www.iste.co.uk/abdesselam/electrodynamics1.zip*

5.1.1. *Field and potential created by an electric dipole*

The simplest model of an electric dipole is a system of two opposite charges, $+q$ and $-q$, located, respectively, at points A and B, separated by a small vector displacement $\vec{d}$ directed from the negative charge to the positive charge. The potential created by this dipole at a point $M\,(x, y, z)$ is

$$\begin{aligned} V\left(\overrightarrow{\mathbf{r}}\right) &= \frac{q}{4\pi\varepsilon_0}\left[\frac{1}{\left|\overrightarrow{\mathbf{AM}}\right|} - \frac{1}{\left|\overrightarrow{\mathbf{BM}}\right|}\right] = \frac{q}{4\pi\varepsilon_0}\left[\frac{1}{\left|\overrightarrow{\mathbf{r}} - \frac{1}{2}\overrightarrow{\mathbf{d}}\right|} - \frac{1}{\left|\overrightarrow{\mathbf{r}} + \frac{1}{2}\overrightarrow{\mathbf{d}}\right|}\right] \\ &= \frac{q}{4\pi\varepsilon_0}\left[\frac{1}{\sqrt{r^2 + \frac{d^2}{4} - \overrightarrow{\mathbf{r}}\cdot\overrightarrow{\mathbf{d}}}} - \frac{1}{\sqrt{r^2 + \frac{d^2}{4} + \overrightarrow{\mathbf{r}}\cdot\overrightarrow{\mathbf{d}}}}\right] \end{aligned} \quad [5.2]$$

as shown in Figure 5.3. Far from the dipole, the potentials of the two charges tend to cancel each other out. In dielectric materials, the separation between the charges is of the order of atomic dimensions and therefore very small compared to the r position away from the dipole, where we wish to calculate the field. To obtain an approximation of $V\left(\overrightarrow{\mathbf{r}}\right)$ when the point M is very far from the dipole, we write:

$$\begin{aligned} &\frac{1}{\sqrt{r^2 + \frac{d^2}{4} \pm \overrightarrow{\mathbf{r}}\cdot\overrightarrow{\mathbf{d}}}} = \frac{1}{r} - \frac{1}{2r}\left(\frac{d^2}{4r^2} \pm \frac{\overrightarrow{\mathbf{r}}\cdot\overrightarrow{\mathbf{d}}}{r^2}\right) + \frac{3}{8r}\left(\frac{d^2}{4r^2} \pm \frac{\overrightarrow{\mathbf{r}}\cdot\overrightarrow{\mathbf{d}}}{r^2}\right)^2 \\ &-\frac{5}{16r}\left(\frac{d^2}{4r^2} \pm \frac{\overrightarrow{\mathbf{r}}\cdot\overrightarrow{\mathbf{d}}}{r^2}\right)^3 + \cdots = \frac{1}{r} \mp \frac{1}{2r}\frac{\overrightarrow{\mathbf{r}}\cdot\overrightarrow{\mathbf{d}}}{r^2} \\ &+\frac{1}{8r}\left[3\left(\frac{\overrightarrow{\mathbf{r}}\cdot\overrightarrow{\mathbf{d}}}{r^2}\right)^2 - \frac{d^2}{r^2}\right] \mp \frac{1}{16r}\left[5\left(\frac{\overrightarrow{\mathbf{r}}\cdot\overrightarrow{\mathbf{d}}}{r^2}\right)^2 - 3\frac{d^2}{r^2}\right]\frac{\overrightarrow{\mathbf{r}}\cdot\overrightarrow{\mathbf{d}}}{r^2} + \cdots \end{aligned} \quad [5.3]$$

and, consequently:

$$\begin{aligned} V\left(\overrightarrow{\mathbf{r}}\right) &= \frac{1}{4\pi\varepsilon_0}\left\{\frac{1}{r} + \frac{1}{8r}\left[5\left(\frac{\overrightarrow{\mathbf{r}}\cdot\overrightarrow{\mathbf{d}}}{r^2}\right)^2 - 3\frac{d^2}{r^2}\right] + \cdots\right\}\left(\frac{\overrightarrow{\mathbf{r}}\cdot\overrightarrow{\mathbf{p}}}{r^2}\right) \\ &= \frac{p}{4\pi\varepsilon_0}\left\{\frac{1}{r^2}\cos\theta + \frac{d^2}{8r^4}\left[5\cos^3\theta - 3\cos\theta\right] + \cdots\right\} \end{aligned} \quad [5.4]$$

where θ is the angle measured between the vectors $\overrightarrow{\mathbf{d}}$ and $\overrightarrow{\mathbf{r}}$. For the first order of approximation in $\frac{p}{r}$, we obtain:

$$V\left(\overrightarrow{\mathbf{r}}\right) = \frac{p\cos\theta}{4\pi\varepsilon_0 r^2} = \frac{\overrightarrow{\mathbf{p}}\cdot\overrightarrow{\mathbf{r}}}{4\pi\varepsilon_0 r^3} \quad [5.5]$$

To calculate the electric field, we use the expression $\overrightarrow{\mathbf{E}} = -\overrightarrow{\nabla} V = -\frac{\partial V}{\partial r}\overrightarrow{\mathbf{e}}_r - \frac{1}{r}\frac{\partial V}{\partial \theta}\overrightarrow{\mathbf{e}}_\theta - \frac{1}{r\sin\theta}\frac{\partial V}{\partial \varphi}\overrightarrow{\mathbf{e}}_\varphi$ (in spherical coordinates):

$$\begin{aligned}\overrightarrow{\mathbf{E}} = \frac{p}{4\pi\varepsilon_0}\left[\frac{2\cos\theta}{r^3} + \frac{d^2\left(5\cos^3\theta - 3\cos\theta\right)}{2r^5} + \cdots\right]\overrightarrow{\mathbf{e}}_r + \\ \frac{p}{4\pi\varepsilon_0}\left[\frac{\sin\theta}{r^3} + \frac{d^2\left(15\cos^2\theta - 3\right)\sin\theta}{8r^5} + \cdots\right]\overrightarrow{\mathbf{e}}_\theta\end{aligned} \quad [5.6]$$

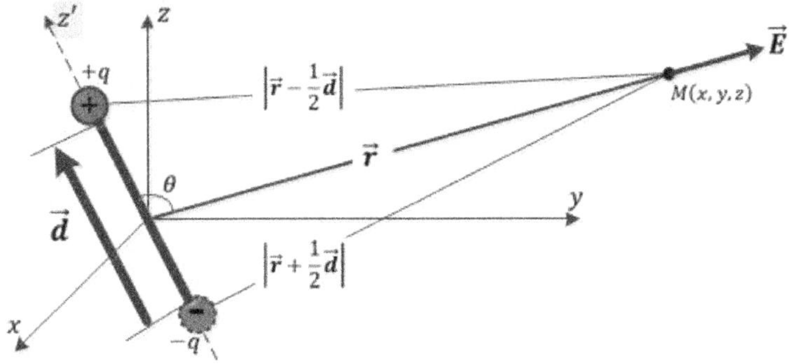

Figure 5.3. *The electric potential (respectively, the field) created by the dipole at point M is equal to the superposition of the individual potentials (respectively, fields) produced by each charge separately*

Since $\overrightarrow{\mathbf{k}}' = \cos\theta\,\overrightarrow{\mathbf{e}}_r - \sin\theta\,\overrightarrow{\mathbf{e}}_\theta = \frac{\cos\theta}{r}\overrightarrow{\mathbf{r}} - \sin\theta\,\overrightarrow{\mathbf{e}}_\theta$ and $\overrightarrow{\mathbf{p}} = p\,\overrightarrow{\mathbf{k}}'$, we deduce that:

$$\begin{aligned}\overrightarrow{\mathbf{E}} = \frac{p\cos\theta\,\overrightarrow{\mathbf{r}}}{4\pi\varepsilon_0 r^4}\left[3 + \frac{5d^2\left(7\cos^2\theta - 1\right)}{8r^2} + \cdots\right] - \frac{\overrightarrow{\mathbf{p}}}{4\pi\varepsilon_0 r^3}[1+ \\ \frac{3d^2\left(5\cos^2\theta - 1\right)}{8r^2} + \cdots\right] = -\frac{\overrightarrow{\mathbf{p}} - 3p\cos\theta\,\overrightarrow{\mathbf{e}}_r}{4\pi\varepsilon_0 r^3} - \frac{d^2}{4\pi\varepsilon_0 r^5}\left[\frac{3\left(5\cos^2\theta - 1\right)\overrightarrow{\mathbf{p}}}{8} - \right. \\ \left.\frac{5p\cos\theta\left(7\cos^2\theta - 1\right)\overrightarrow{\mathbf{r}}}{8r}\right] + \cdots = -\frac{\overrightarrow{\mathbf{p}} - 3\left(\overrightarrow{\mathbf{p}}\cdot\overrightarrow{\mathbf{e}}_r\right)\overrightarrow{\mathbf{e}}_r}{4\pi\varepsilon_0 r^3} - \\ \frac{1}{4\pi\varepsilon_0 r^5}\left[\frac{3}{8}\left(5\frac{\left(\overrightarrow{\mathbf{d}}\cdot\overrightarrow{\mathbf{r}}\right)^2}{r^2} - d^2\right)\overrightarrow{\mathbf{p}} - \frac{5\left(\overrightarrow{\mathbf{p}}\cdot\overrightarrow{\mathbf{r}}\right)}{8r^2}\left(7\frac{\left(\overrightarrow{\mathbf{d}}\cdot\overrightarrow{\mathbf{r}}\right)^2}{r^2} - d^2\right)\overrightarrow{\mathbf{r}}\right] + \cdots\end{aligned}$$

[5.7]

At the first order of approximation in $\frac{p}{r}$, the previous equation is limited to:

$$\vec{\mathbf{E}} = -\frac{\vec{\mathbf{p}} - 3p\cos\theta\,\vec{\mathbf{e}}_r}{4\pi\varepsilon_0 r^3} = -\frac{\vec{\mathbf{p}} - 3\left(\vec{\mathbf{p}}\cdot\vec{\mathbf{e}}_r\right)\vec{\mathbf{e}}_r}{4\pi\varepsilon_0 r^3} = \frac{2p\cos\theta}{4\pi\varepsilon_0 r^3}\vec{\mathbf{e}}_r + \frac{p\sin\theta}{4\pi\varepsilon_0 r^3}\vec{\mathbf{e}}_\theta \qquad [5.8]$$

5.1.2. *Equipotential surfaces and field lines*

To construct equipotential surfaces and field lines of the dipole, simply note that the electric potential and electric field are symmetric by revolution about the axis carrying the electric dipole, namely, Oz'. Equipotential surfaces are defined by $V = V_0$. It follows that:

$$r(\theta) = \sqrt{\frac{p\cos\theta}{4\pi\varepsilon_0 V_0}} \qquad [5.9]$$

The median plane defined by $\theta = \frac{\pi}{2}$ $(z' = 0)$ is the equipotential surface $V = 0$. Equipotential surfaces appear as curves in the plane. The field lines are obtained from equation $\vec{\mathbf{E}} \times d\,\vec{\mathbf{l}} = 0$. Given that the dipole field component E_φ is zero, it follows that $\left(E_r\vec{\mathbf{e}}_r + E_\theta\vec{\mathbf{e}}_\theta\right) \times \left(dr\,\vec{\mathbf{e}}_r + rd\theta\,\vec{\mathbf{e}}_\theta + r\sin\theta d\varphi\,\vec{\mathbf{e}}_\varphi\right) = 0$. This yields $d\varphi = 0$ and $E_r rd\theta = E_\theta dr$. Using the expressions of the components E_r and E_θ from equation [5.8], we get:

$$d\varphi = 0 \qquad \frac{dr}{r} = \frac{E_r d\theta}{E_\theta} = \frac{2\cos\theta d\theta}{\sin\theta} \qquad [5.10]$$

The field lines are obtained after integrating equations [5.10]: $\int d\varphi = K_1$ and $\int\frac{dr}{r} = \int\frac{2d(\sin\theta)}{\sin\theta} + \ln K_2$ or again:

$$\varphi = K_1 \qquad r = K_2\sin^2\theta \qquad [5.11]$$

where K_1 and K_2 are constants. K_2 represents the field line for $\theta = \frac{\pi}{2}$. All field lines start from the positive charge and end at the negative charge. Equipotential surfaces (equation [5.9]) and field lines (equation [5.11]), depicted in Figure 5.4, are valid only for $r \gg d$.

5.1.3. *Dipole moment of a charge distribution*

If there is more than one pair of charges, it is possible to generalize the definition of the dipole moment (equation [5.1]) to all charges present in the system:

$$\vec{\mathbf{p}} = \sum_i q_i\vec{\mathbf{r}}_i \qquad [5.12]$$

where $\vec{\mathbf{r}}_i$ is the position of the charge q_i relative to the origin. When the total charge or net charge of the system is zero ($\sum_i q_i = 0$), the dipole moment is independent of the choice of origin, because if we replace $\vec{\mathbf{r}}_i$ by $\vec{\mathbf{r}}_i + \vec{\mathbf{r}}_0$ (where $\vec{\mathbf{r}}_0$ is the vector separating two origins O and O'), we find $\vec{\mathbf{p}} = \sum_i q_i \left(\vec{\mathbf{r}}_i + \vec{\mathbf{r}}_0\right) = \sum_i q_i \vec{\mathbf{r}}_i + \vec{\mathbf{r}}_0 (\sum_i q_i) = \sum_i q_i \vec{\mathbf{r}}_i$. For a continuous distribution, formula [5.12] can be generalized as follows:

$$\vec{\mathbf{p}} = \int_{\text{for all}_q} \vec{\mathbf{r}}\, dq \qquad [5.13]$$

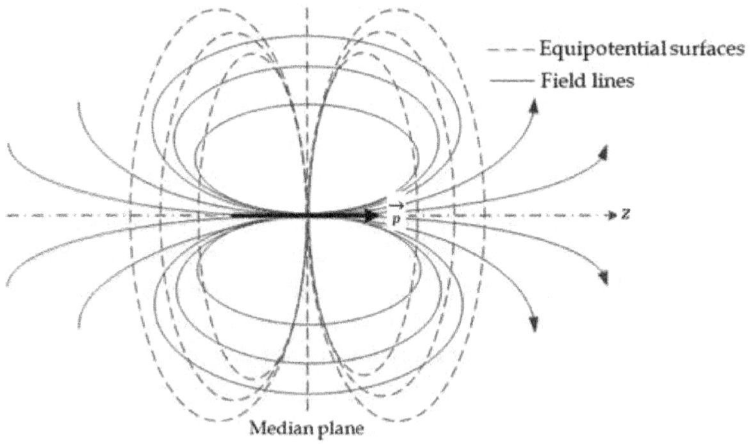

Figure 5.4. *Field lines and equipotential surfaces of an electric dipole. For a color version of this figure, see www.iste.co.uk/abdesselam/electrodynamics1.zip*

Away from any dipole distribution, the electric potential and field are given by [5.5] and [5.8], where the dipole moment $\vec{\mathbf{p}}$ is given by equation [5.12] (for a discontinuous distribution) or equation [5.13] (for a continuous distribution). For linear, surface, or volume distributions, we simply write:

$$\vec{\mathbf{p}} = \begin{cases} \displaystyle\int_L \lambda \vec{\mathbf{r}}\, dl & \text{(line distribution)} \\ \displaystyle\int_S \sigma \vec{\mathbf{r}}\, dS & \text{(surface distribution)} \\ \displaystyle\int_\tau \rho \vec{\mathbf{r}}\, d\tau & \text{(volume distribution)} \end{cases} \qquad [5.14]$$

In the case of a single dipole, we take $\rho\left(\vec{\mathbf{r}}\right) = q\delta\left(\vec{\mathbf{r}} - \frac{1}{2}\vec{\mathbf{d}}\right) - q\delta\left(\vec{\mathbf{r}} + \frac{1}{2}\vec{\mathbf{d}}\right)$.

5.2. Electric multipole moments

5.2.1. *Third-order expansion in* $\frac{1}{r}$

Consider a charge distribution of density $\rho\left(\overrightarrow{\mathbf{r}}'\right)$ occupying a volume τ', which does not extend beyond a distance $r'_{\max}$ from the origin of the coordinates O. Calculate the potential at a point M with coordinates (x, y, z) such that $r \gg r'_{\max}$. The term $\frac{1}{\left|\overrightarrow{\mathbf{r}}-\overrightarrow{\mathbf{r}}'\right|}$ in the potential expression is written as:

$$\begin{aligned}\frac{1}{\left|\overrightarrow{\mathbf{r}}-\overrightarrow{\mathbf{r}}'\right|} &= \frac{1}{r}\frac{1}{\sqrt{1+\frac{r'^2}{r^2}-\frac{2\overrightarrow{\mathbf{r}}'\cdot\overrightarrow{\mathbf{r}}}{r^2}}} = \frac{1}{r}\left[1-\frac{1}{2}\left(\frac{r'^2}{r^2}-\frac{2\overrightarrow{\mathbf{r}}'\cdot\overrightarrow{\mathbf{r}}}{r^2}\right)\right.\\ &\left.+\frac{3}{8}\left(\frac{r'^2}{r^2}-\frac{2\overrightarrow{\mathbf{r}}'\cdot\overrightarrow{\mathbf{r}}}{r^2}\right)^2+\cdots\right] = \frac{1}{r}+\frac{\overrightarrow{\mathbf{r}}'\cdot\overrightarrow{\mathbf{r}}}{r^3}+\frac{3\left(\overrightarrow{\mathbf{r}}'\cdot\overrightarrow{\mathbf{r}}\right)^2-r'^2r^2}{2r^5}+\cdots\\ &= \frac{1}{r}+\frac{\overrightarrow{\mathbf{r}}'\cdot\overrightarrow{\mathbf{r}}}{r^3}+\frac{1}{2r^5}\sum_{i,j-1}^{3}3\left(x_ix'_i\right)\left(x_jx'_j\right)-x_ix_ix'_jx'_j\\ &= \frac{1}{r}+\frac{\overrightarrow{\mathbf{r}}'\cdot\overrightarrow{\mathbf{r}}}{r^3}+\frac{1}{2r^5}\sum_{i,j=1}^{3}x_ix_j\left(3x'_ix'_j-\delta_{ij}r'^2\right)+\cdots\end{aligned} \quad [5.15]$$

where terms of order $\left(\frac{r'}{r}\right)^3$ and above have been neglected. Substituting the previous development into the expression of $V\left(\overrightarrow{\mathbf{r}}\right)=\frac{1}{4\pi\varepsilon_0}\int_{\tau'}\frac{\rho\left(\overrightarrow{\mathbf{r}}'\right)d\tau'}{\left|\overrightarrow{\mathbf{r}}-\overrightarrow{\mathbf{r}}'\right|}$, we find:

$$\begin{aligned}V\left(\overrightarrow{\mathbf{r}}\right) &= \frac{1}{4\pi\varepsilon_0 r}\int_{\tau'}\rho\left(\overrightarrow{\mathbf{r}}'\right)d\tau'+\frac{1}{4\pi\varepsilon_0 r^3}\left(\overrightarrow{\mathbf{r}}\cdot\int_{\tau'}\rho\left(\overrightarrow{\mathbf{r}}'\right)\overrightarrow{\mathbf{r}}'d\tau'\right)\\ &+\frac{1}{2}\sum_{i,j=1}^{3}\frac{x_ix_j}{4\pi\varepsilon_0 r^5}\int_{\tau'}\rho\left(\overrightarrow{\mathbf{r}}'\right)\left[3x'_ix'_j-\delta_{ij}r'^2\right]d\tau'+\cdots\\ &= \frac{q'}{4\pi\varepsilon_0 r}+\frac{\overrightarrow{\mathbf{r}}\cdot\overrightarrow{\mathbf{p}}'}{4\pi\varepsilon_0 r^3}+\frac{1}{2}\sum_{i,j=1}^{3}\frac{x_ix_jQ'_{ij}}{4\pi\varepsilon_0 r^5}+\cdots\end{aligned} \quad [5.16]$$

where:

$$\begin{aligned}q' &= \int_{\tau'}\rho\left(\overrightarrow{\mathbf{r}}'\right)d\tau'\\ \overrightarrow{\mathbf{p}}' &- \int_{\tau'}\rho\left(\overrightarrow{\mathbf{r}}'\right)\overrightarrow{\mathbf{r}}'d\tau'\\ Q'_{ij} &= \int_{\tau'}\rho\left(\overrightarrow{\mathbf{r}}'\right)\left[3x'_ix'_j-\delta_{ij}r'^2\right]d\tau'\end{aligned} \quad [5.17]$$

represent, respectively, the *total charge of the source*, the *dipole moment of the distribution* and the ij components of the *electric quadrupole moment* in Cartesian coordinates. The terms:

$$\begin{aligned} V_{\text{mon.}} &= \frac{q'}{4\pi\varepsilon_0 r} \\ V_{\text{dip.}} &= \frac{\vec{\mathbf{r}} \cdot \vec{\mathbf{p}}'}{4\pi\varepsilon_0 r^3} \\ V_{\text{quad.}} &= \frac{1}{2}\sum_{i,j=1}^{3} \frac{x_i x_j Q_{ij}}{4\pi\varepsilon_0 r^5} \end{aligned} \qquad [5.18]$$

are called, respectively, the *monopole term*, the *dipole term*, the *quadrupole term*, etc., of the potential.

Equation [5.18] is known as *the multipole expansion formula of the potential.* When $q' \neq 0$, the monopole term is dominant at long distances. It is equivalent to the potential created by a point charge of value q' located at the origin. If $q' = 0$, the monopole term is zero.

If, furthermore, $\vec{\mathbf{p}}' \neq 0$, the dipolar term dominates at very long distance, and we find the same expression as for a dipole made up of two opposite charges. If $q' = 0$ and $\vec{\mathbf{p}}' = 0$, then $V_{\text{mon.}} = V_{\text{dip.}} = 0$.

In this case, we must consider the higher order terms.

For a set of point-like charges, equations [5.17] are reduced to:

$$\begin{aligned} q' &= \sum_{\alpha=1}^{N} q'_\alpha \\ \vec{\mathbf{p}}' &= \sum_{\alpha=1}^{N} q'_\alpha \vec{\mathbf{r}}'_{q_\alpha} \\ Q'_{ij} &= \sum_{\alpha=1}^{N} q'_\alpha \left[3x'_{q'_\alpha,i} x'_{q'_\alpha,j} - \delta_{ij} r'^2_{q'_\alpha}\right] \end{aligned} \qquad [5.19]$$

EXAMPLE 5.1. (Distribution Rotation-Invariant around Oz).– If the distribution is rotation-invariant around the Oz axis (axial symmetry relative to Oz), that is, independent of φ ($\rho \equiv \rho(r', \theta')$), then:

$$\begin{aligned}
Q_{11} &= \int_0^{\infty}\int_0^{\pi}\int_0^{2\pi} \rho\left(3r'^2\sin^2\theta'\cos^2\varphi' - r'^2\right) r'^2 dr' \sin\theta' d\theta' d\varphi' \\
&= \pi \int_0^{\infty}\int_0^{\pi} \rho\left[3\sin^2\theta' - 2\right] r'^4 dr' \sin\theta' d\theta' \\
Q_{22} &= Q_{11} \\
Q_{33} &= \int_0^{\infty}\int_0^{\pi}\int_0^{2\pi} \rho\left(3r'^2\cos^2\theta' - r'^2\right) r'^2 dr' \sin\theta' d\theta' d\varphi' \\
&= 2\pi \int_0^{\infty}\int_0^{\pi} \rho\left[2 - 3\sin^2\theta'\right] r'^4 dr' \sin\theta' d\theta' = -2Q_{11}
\end{aligned} \qquad [5.20]$$

The other components are zero, that is, $Q_{ij} = 0$ for $i \neq j$. In fact, we find:

$$\begin{aligned}
Q_{12} = Q_{21} &= \int_0^{\infty}\int_0^{\pi}\int_0^{2\pi} 3\rho r'^4 \sin^3\theta' \sin\varphi' \cos\varphi' dr' d\theta' d\varphi' = 0 \\
Q_{13} = Q_{31} &= \int_0^{\infty}\int_0^{\pi}\int_0^{2\pi} 3\rho r'^4 \sin^2\theta' \cos\theta' \cos\varphi' dr' d\theta' d\varphi' = 0 \\
Q_{23} = Q_{32} &= \int_0^{\infty}\int_0^{\pi}\int_0^{2\pi} 3\rho r'^4 \sin^2\theta' \cos\theta' \sin\varphi' dr' d\theta' d\varphi' = 0
\end{aligned} \qquad [5.21]$$

EXAMPLE 5.2. (Linear Distribution).– Let us calculate the total charge, the dipole moment and the quadrupole moment components of a linear charge density of type $\rho\left(\overrightarrow{\mathbf{r}}'\right) = \lambda(z')\,\delta(x')\,\delta(y')$ for $-\frac{L}{2} \leq z' \leq \frac{L}{2}$ and 0 elsewhere. We obtain:

$$\begin{aligned}
q' &= \int_{-\infty}^{\infty}\int_{-\infty}^{\infty}\int_{-\frac{L}{2}}^{\frac{L}{2}} \lambda(z')\,\delta(x')\,\delta(y')\,dx'dy'dz' = \int_{-\frac{L}{2}}^{\frac{L}{2}} \lambda(z')\,dz' \\
\overrightarrow{\mathbf{p}}' &= \int_{-\infty}^{\infty}\int_{-\infty}^{\infty}\int_{-\frac{L}{2}}^{\frac{L}{2}} \lambda(z')\,\overrightarrow{\mathbf{r}}'\delta(x')\,\delta(y')\,dx'dy'dz' = \int_{-\frac{L}{2}}^{\frac{L}{2}} z'\lambda(z')\,dz'\,\overrightarrow{\mathbf{k}} \\
Q_{11} &= \int_{-\infty}^{\infty}\int_{-\infty}^{\infty}\int_{-\frac{L}{2}}^{\frac{L}{2}} \lambda(z')\left(3x'^2 - r'^2\right)\delta(x')\,\delta(y')\,dx'dy'dz' \\
&= -\int_{-\frac{L}{2}}^{\frac{L}{2}} z'^2\lambda(z')\,dz' \\
Q_{22} &= \int_{-\infty}^{\infty}\int_{-\infty}^{\infty}\int_{-\frac{L}{2}}^{\frac{L}{2}} \lambda(z')\left(3y'^2 - r'^2\right)\delta(x')\,\delta(y')\,dx'dy'dz' = Q_{11} \\
Q_{33} &= -Q_{11} - Q_{22} = 2\int_{-\frac{L}{2}}^{\frac{L}{2}} z'^2\lambda(z')\,dz'
\end{aligned} \qquad [5.22]$$

The other components are zero, that is, $Q_{ij} = 0, \forall i \neq j$.

1) In particular, if $\lambda(z') = \lambda$, then:

$$\begin{array}{ll} q' = \lambda \int_{-\frac{L}{2}}^{\frac{L}{2}} dz' = \lambda L & \overrightarrow{\mathbf{p}}' = \lambda \int_{-\frac{L}{2}}^{\frac{L}{2}} z' dz' \overrightarrow{\mathbf{k}} = 0 \\ Q_{11} = Q_{22} = -\lambda \int_{-\frac{L}{2}}^{\frac{L}{2}} z'^2 dz' = -\frac{\lambda L^3}{12} & \\ Q_{33} = -2Q_{11} = \frac{\lambda L^3}{6} & \end{array} \quad [5.23]$$

2) Similarly, for $\lambda(z') = \mu z'$, we get:

$$\begin{array}{ll} q' = 0 & \overrightarrow{\mathbf{p}}' = \frac{\lambda L^3}{12} \overrightarrow{\mathbf{k}} \\ Q_{11} = Q_{22} = 0 & Q_{33} = -2Q_{11} = 0 \end{array} \quad [5.24]$$

EXAMPLE 5.3. (Distribution of Four Point-like Charges).– Let us calculate the total charge, dipole moment and quadrupole moment components of the charge distribution shown in Figure 5.5. The charge distribution of the system is given by:

$$\rho(\overrightarrow{\mathbf{r}}') = q\left[\delta\left(x - \frac{L}{2}\right)\delta\left(y - \frac{L}{2}\right) - \delta\left(x - \frac{L}{2}\right)\delta\left(y + \frac{L}{2}\right) - \delta\left(x + \frac{L}{2}\right)\delta\left(y - \frac{L}{2}\right) + \delta\left(x + \frac{L}{2}\right)\delta\left(y + \frac{L}{2}\right)\right]\delta(z) \quad [5.25]$$

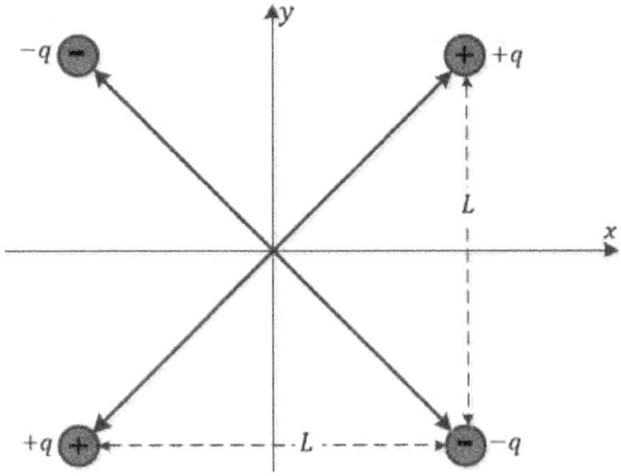

Figure 5.5. *Distribution of four point-like charges arranged on a square of side L*

The formulas in [5.19] can then be applied directly.

1) *Total charge*:

$$q_{\text{tot.}} = (+q) + (+q) + (-q) + (-q) = 0 \qquad [5.26]$$

2) *Dipole moment*:

$$\begin{aligned}\vec{\mathbf{p}} &= (+q)\frac{L}{2}\left(\vec{\mathbf{i}} + \vec{\mathbf{j}}\right) + (+q)\frac{L}{2}\left(-\vec{\mathbf{i}} - \vec{\mathbf{j}}\right) + (-q)\frac{L}{2}\left(\vec{\mathbf{i}} - \vec{\mathbf{j}}\right) \\ &+ (-q)\frac{L}{2}\left(-\vec{\mathbf{i}} + \vec{\mathbf{j}}\right) = 0\end{aligned} \qquad [5.27]$$

3) *Quadrupole moment*: the xx component of the quadrupole moment tensor is:

$$\begin{aligned}Q_{xx} = (+q)&\left[3\left(\frac{L}{2}\right)^2 - \frac{L^2}{2}\right] + (-q)\left[3\left(\frac{-L}{2}\right)^2 - \frac{L^2}{2}\right] + \\ (+q)&\left[3\left(\frac{-L}{2}\right)^2 - \frac{L^2}{2}\right] + (-q)\left[3\left(\frac{L}{2}\right)^2 - \frac{L^2}{2}\right] = 0\end{aligned} \qquad [5.28]$$

In the same way, we find $Q_{yy} = Q_{zz} = 0$. For the other components, we have:

$$\begin{aligned}Q_{xy} = Q_{yx} &= 3q\left(\frac{L}{2}\cdot\frac{L}{2}\right) - 3q\left(\frac{L}{2}\cdot\frac{-L}{2}\right) + 3q\left(\frac{-L}{2}\cdot\frac{-L}{2}\right) \\ &- 3q\left(\frac{-L}{2}\cdot\frac{L}{2}\right) = 3qL^2 \\ Q_{xz} = Q_{zx} &= Q_{yz} = Q_{zy} = 0\end{aligned} \qquad [5.29]$$

It follows that the quadrupole moment is:

$$\begin{pmatrix} 0 & 3qL^2 & 0 \\ 3qL^2 & 0 & 0 \\ 0 & 0 & 0 \end{pmatrix} \qquad [5.30]$$

EXAMPLE 5.4. (Ellipsoid of Revolution).– Let us determine the components of the quadrupole moment of a uniformly charged ellipsoid of revolution and rotation-invariant around Oz. The ellipsoid equation is $\frac{x^2+y^2}{b^2} + \frac{z^2}{a^2} = 1$, where a and b are constants.

1) *Total charge*:

$$\begin{aligned}q' &= \int_{\text{ellipsoid}} \rho_0 d\tau' = \int_{z'=-a}^{a} \int_{s'=0}^{b\sqrt{1-\frac{z'^2}{a^2}}} \int_{\varphi'=0}^{2\pi} \rho_0 s' ds' dz' d\varphi' \\ &= \pi\rho_0 b^2 \int_{-a}^{a} \left(1 - \frac{z'^2}{a^2}\right) dz' = \frac{4}{3}\pi\rho_0 ab^2\end{aligned} \qquad [5.31]$$

2) *Dipole moment*:

$$\overrightarrow{\mathbf{p}}' = \int_{\text{ellipsoid}} \rho_0 \overrightarrow{\mathbf{r}}' d\tau' = \int_{-a}^{a} \int_{0}^{b\sqrt{1-\frac{z'^2}{a^2}}} \int_{0}^{2\pi} \rho_0 \left(s' \cos\varphi' \overrightarrow{\mathbf{i}} + s' \sin\varphi' \overrightarrow{\mathbf{j}} + z' \overrightarrow{\mathbf{k}} \right) s' ds' dz' d\varphi' = \pi\rho_0 b^2 \int_{-a}^{a} z' \left(1 - \frac{z'^2}{a^2}\right) dz' \overrightarrow{\mathbf{k}} = 0 \quad [5.32]$$

3) *Quadrupole moment*:

$$\begin{aligned}
Q_{xx} &= \int_{\text{ellipsoid}} \left(3x'^2 - r'^2\right) \rho_0 d\tau' = \int_{z'=-a}^{a} \int_{s'=0}^{b\sqrt{1-\frac{z'^2}{a^2}}} \int_{\varphi'=0}^{2\pi} \rho_0 \\
&\quad \times \left(3s'^2 \cos^2\varphi' - s'^2 - z'^2\right) s' ds' dz' d\varphi' \\
&= \pi\rho_0 \int_{z'=-a}^{a} \int_{s'=0}^{b\sqrt{1-\frac{z'^2}{a^2}}} \left(s'^2 - 2z'^2\right) s' ds' dz' \\
&= -\frac{4}{15}\pi\rho_0 a b^2 \left(a^2 - b^2\right), \\
Q_{yy} &= \int_{\text{ellipsoid}} \left(3y'^2 - r'^2\right) \rho_0 d\tau' = \int_{z'=-a}^{a} \int_{s'=0}^{b\sqrt{1-\frac{z'^2}{a^2}}} \int_{\varphi'=0}^{2\pi} \rho_0 \\
&\quad \times \left(3s'^2 \sin^2\varphi' - s'^2 - z'^2\right) s' ds' dz' d\varphi' \\
&= \pi\rho_0 \int_{z'=-a}^{a} \int_{s'=0}^{b\sqrt{1-\frac{z'^2}{a^2}}} \left(s'^2 - 2z'^2\right) s' ds' dz' \\
&= -\frac{4}{15}\pi\rho_0 a b^2 \left(a^2 - b^2\right) \\
Q_{zz} &= -Q_{xx} - Q_{yy} = \frac{8}{15}\pi\rho_0 a b^2 \left(a^2 - b^2\right) \\
Q_{xy} &= Q_{yx} = Q_{xz} = Q_{zx} = Q_{yz} = Q_{zy} = 0
\end{aligned} \quad [5.33]$$

5.2.2. *Tensor form of multipole expansion*

Let us denote $x'_1 \equiv x'$, $x'_2 \equiv y'$ and $x'_3 \equiv z'$ as the components of $\overrightarrow{\mathbf{r}}'$ according to $\overrightarrow{\mathbf{i}}$, $\overrightarrow{\mathbf{j}}$ and $\overrightarrow{\mathbf{k}}$. The electrical quadrupole moment of the distribution can be represented by a 3×3 matrix:

$$Q^{(2)} = \begin{pmatrix} Q_{11} & Q_{12} & Q_{13} \\ Q_{21} & Q_{22} & Q_{23} \\ Q_{31} & Q_{32} & Q_{33} \end{pmatrix} \quad [5.34]$$

$$= \begin{pmatrix} \int_{\tau'} \rho\left(\overrightarrow{\mathbf{r}}'\right)\left[3x_1'^2 - r'^2\right] d\tau' & \int_{\tau'} 3\rho\left(\overrightarrow{\mathbf{r}}'\right) x_1' x_2' d\tau' & \int_{\tau'} 3\rho\left(\overrightarrow{\mathbf{r}}'\right) x_1' x_3' d\tau' \\ \int_{\tau'} 3\rho\left(\overrightarrow{\mathbf{r}}'\right) x_2' x_1' d\tau' & \int_{\tau'} \rho\left(\overrightarrow{\mathbf{r}}'\right)\left[3x_2'^2 - r'^2\right] d\tau' & \int_{\tau'} 3\rho\left(\overrightarrow{\mathbf{r}}'\right) x_2' x_3' d\tau' \\ \int_{\tau'} 3\rho\left(\overrightarrow{\mathbf{r}}'\right) x_3' x_1' d\tau' & \int_{\tau'} 3\rho\left(\overrightarrow{\mathbf{r}}'\right) x_3' x_2' d\tau' & \int_{\tau'} \rho\left(\overrightarrow{\mathbf{r}}'\right)\left[3x_3'^2 - r'^2\right] d\tau' \end{pmatrix}$$

This moment is a second-order tensor. It is clearly seen that:

$$Q_{11} + Q_{22} + Q_{33} = \int_{\tau'} \rho\left(\overrightarrow{\mathbf{r}}'\right)\left[3x'^2 + 3y'^2 + 3z'^2 - 3r'^2\right] d\tau' = 0 \quad [5.35]$$

that is, the trace of the quadrupole moment tensor is zero. We can deduce that:

$$Q_{33} = -Q_{11} - Q_{22} \quad [5.36]$$

The dipole moment is a vector, that is, a first-order tensor. In fact, we can verify that:

$$\overrightarrow{\mathbf{p}}' \cdot \overrightarrow{\mathbf{r}}' = p_x' x' + p_y' y' + p_z' z' = \sum_{j=1}^{3} Q_j^{(1)} x_j' \quad [5.37]$$

with:

$$Q^{(1)} = \begin{pmatrix} \int_{\tau'} \rho\left(\overrightarrow{\mathbf{r}}'\right) x_1' d\tau' \\ \int_{\tau'} \rho\left(\overrightarrow{\mathbf{r}}'\right) x_2' d\tau' \\ \int_{\tau'} \rho\left(\overrightarrow{\mathbf{r}}'\right) x_3' d\tau' \end{pmatrix} \quad [5.38]$$

The total charge is a zero-order tensor, that is, a scalar. We use the notation $Q^{(0)} = q'$. In general, it can be written that:

$$V(M) = \frac{1}{4\pi\varepsilon_0 r} \sum_{k=0}^{\infty} \frac{1}{k!} \frac{\sum_{j_1=1}^{3} \cdots \sum_{j_k=1}^{3} Q_{j_1 \cdots j_k}^{(k)} x_{j_1}' \cdots x_{j_k}'}{r^{2k}} \quad [5.39]$$

where $Q^{(k)}$, called the 2^k-pole moment, is a tensor of order k.

As with the dipole moment, the 2^k-pole moment does not depend on the origin if all lower order moments are zero. Let us illustrate this, for example, for the $(1, 1)$

component of the quadrupole moment. In fact, if (x'_i, y'_i, z'_i) are the coordinates of the charge q_i relative to a point O' with coordinates (x_0, y_0, z_0) in the reference frame $\left(O, \overrightarrow{\mathbf{i}}, \overrightarrow{\mathbf{j}}, \overrightarrow{\mathbf{k}}\right)$, then:

$$Q'_{11} = \sum_{i=1}^{N} q_i \left(3x'^2_i - r'^2_i\right) = \sum_{i=1}^{N} q_i \left(2\left[x_i^2 - 2x_i x_0 + x_0^2\right] - \left[y_i^2 - 2y_i y_0 + y_0^2\right] - \right.$$

$$\left.\left[z_i^2 - 2z_i z_0 + z_0^2\right]\right) = \sum_{i=1}^{N} q_i \left(2x_i^2 - y_i^2 - z_i^2\right) - 2x_0 \sum_{i=1}^{N} q_i x_i + y_0 \sum_{i=1}^{N} q_i y_i$$

$$+ z_0 \sum_{i=1}^{N} q_i z_i + \left(x_0^2 + y_0^2 + z_0^2\right) x_0 \sum_{i=1}^{N} q_i = \sum_{i=1}^{N} q_i \left(3x_i^2 - r_i^2\right) = Q_{11} \qquad [5.40]$$

because $Q^{(0)} = Q^{(1)} = 0$, that is:

$$\sum_{i=1}^{N} q_i = \sum_{i=1}^{N} q_i x_i = \sum_{i=1}^{N} q_i y_i = \sum_{i=1}^{N} q_i z_i = 0.$$

5.2.3. *Multipole expansion and spherical functions*

Consider a distribution $\rho\left(\overrightarrow{\mathbf{r}}'\right)$ that occupies a volume τ', which does not extend beyond a distance $r'_{\max}$ from the coordinate origin. Another expansion of the scalar potential $V\left(\overrightarrow{\mathbf{r}}'\right)$ induced by the distribution $\rho\left(\overrightarrow{\mathbf{r}}'\right)$ is also possible using the *generating function of Legendre polynomials*. In fact, we write:

$$\frac{1}{\left|\overrightarrow{\mathbf{r}} - \overrightarrow{\mathbf{r}}'\right|} = \frac{1}{r_> \sqrt{1 - 2\frac{r_<}{r_>}\cos\gamma + \left(\frac{r_<}{r_>}\right)^2}} = \frac{1}{r_>} \sum_{n=0}^{\infty} \left(\frac{r_<}{r_>}\right)^n P_n(\cos\gamma) \qquad [5.41]$$

where γ is the angle measured between $\overrightarrow{\mathbf{r}}$ and $\overrightarrow{\mathbf{r}}'$, $r_< = \min(r, r')$, and $r_> = \max(r, r')$. It therefore follows that:

$$V\left(\overrightarrow{\mathbf{r}}\right) = \frac{1}{4\pi\varepsilon_0} \int_{\tau'} \frac{\rho\left(\overrightarrow{\mathbf{r}}'\right) d\tau'}{\left|\overrightarrow{\mathbf{r}} - \overrightarrow{\mathbf{r}}'\right|} = \frac{1}{4\pi\varepsilon_0} \int_{\tau'} \sum_{n=0}^{\infty} \frac{r_<^n}{r_>^{n+1}} P_n(\cos\gamma)\, \rho\left(\overrightarrow{\mathbf{r}}'\right) d\tau' \qquad [5.42]$$

EXAMPLE 5.5. (Potential at a Point M Located along the Oz-Axis).– When we want to calculate the potential at a point M located along the axis Oz, that is, if we take

$\overrightarrow{\mathbf{r}} = z\overrightarrow{\mathbf{k}}$, then γ here represents the polar angle θ' of the vector $\overrightarrow{\mathbf{r}}'$. It therefore follows that:

$$V(0,0,z) = \frac{1}{4\pi\varepsilon_0}\sum_{n=0}^{\infty}\frac{1}{z^{n+1}}\int_{\tau'} r'^{n+2}\rho\left(\overrightarrow{\mathbf{r}}'\right)P_n\left(\cos\theta'\right)dr'\sin\theta' d\theta' d\varphi' \qquad [5.43]$$

This formula is only valid for $z \geq r'_{\max}$.

EXAMPLE 5.6. (Line Distribution).– If the distribution is along the Oz axis, that is, is of the form $\rho\left(\overrightarrow{\mathbf{r}}'\right) = \lambda\left(z'\right)\delta\left(x'\right)\delta\left(y'\right)$ and occupies the interval between $z'_{\min}$ and $z'_{\min} > 0$, where $z'_{\min} > 0$, then γ represents, in this case, the polar angle of $\overrightarrow{\mathbf{r}}$. Thus, the potential $V\left(\overrightarrow{\mathbf{r}}\right)$ is written as:

$$\begin{aligned} V\left(\overrightarrow{\mathbf{r}}\right) &= \frac{1}{4\pi\varepsilon_0}\int_{-\infty}^{\infty}\int_{-\infty}^{\infty}\int_{z'_{\min}}^{z'_{\max}}\sum_{n=0}^{\infty}\frac{P_n\left(\cos\theta\right)}{r^{n+1}}z'^n \\ &\times\lambda\left(z'\right)\delta\left(x'\right)\delta\left(y'\right)dx'dy'dz' = \frac{1}{4\pi\varepsilon_0}\sum_{n=0}^{\infty}\frac{P_n\left(\cos\theta\right)}{r^{n+1}}\int_{z'_{\min}}^{z'_{\max}} z'^n\lambda\left(z'\right)dz' \end{aligned} \qquad [5.44]$$

for $r \geq z'_{\max}$, and:

$$\begin{aligned} V\left(\overrightarrow{\mathbf{r}}\right) &= \frac{1}{4\pi\varepsilon_0}\int_{-\infty}^{\infty}\int_{-\infty}^{\infty}\int_{z'_{\min}}^{z'_{\max}}\sum_{n=0}^{\infty}r^n P_n\left(\cos\theta\right) \\ &\times\frac{\lambda\left(z'\right)}{z^{n+1}}\delta\left(x'\right)\delta\left(y'\right)dx'dy'dz = \frac{1}{4\pi\varepsilon_0}\sum_{n=0}^{\infty}r^n P_n\left(\cos\theta\right)\int_{z'_{\min}}^{z'_{\max}}\frac{\lambda\left(z'\right)}{z'^{n+1}}dz' \end{aligned} \qquad [5.45]$$

for $r \leq z'_{\min}$. For $z'_{\min} < r < z'_{\max}$, the integration interval $[z'_{\min}, z'_{\max}]$ is divided into two intervals $[z'_{\min}, r]$ and $[r, z'_{\max}]$:

$$V\left(\overrightarrow{\mathbf{r}}\right) = \frac{1}{4\pi\varepsilon_0}\sum_{n=0}^{\infty}P_n\left(\cos\theta\right)\left[\frac{1}{r^{n+1}}\int_{z'_{\min}}^{r} z'^n\lambda\left(z'\right)dz' + r^n\int_{r}^{z'_{\max}}\frac{\lambda\left(z'\right)}{z'^{n+1}}dz'\right] \qquad [5.46]$$

5.2.4. *Expansion on spherical harmonics*

According to the *addition formula* (see Appendix 6 of this book online at: www.iste.co.uk/abdesselam/electrodynamics1.zip):

$$P_l\left(\cos\gamma\right) = \frac{4\pi}{2l+1}\sum_{m=-l}^{l}Y_l^m\left(\theta,\varphi\right)\overline{Y_l^m\left(\theta',\varphi'\right)} \qquad [5.47]$$

where γ is the angle measured between (θ, φ) and (θ', φ'), we write that:

$$V\left(\vec{\mathbf{r}}\right) = \frac{1}{\varepsilon_0} \sum_{l=0}^{\infty} \sum_{m=-l}^{l} \frac{Y_l^m(\theta, \varphi)}{(2l+1)\, r^{l+1}} \int_{\tau'} \rho\left(\vec{\mathbf{r}}'\right) r'^{l+2} \overline{Y_l^m(\theta', \varphi')} dr' \sin\theta' d\theta' d\varphi' \qquad [5.48]$$

valid only when $r > r'_{\max}$. The function $Y_l^m(\theta, \varphi)$ is called the *spherical harmonic function of degree l and order m* and $\overline{Y_l^m(\theta, \varphi)}$ its complex conjugate. It is given by:

$$Y_l^m(\theta, \varphi) = \sqrt{\frac{(2l+1)(l-m)!}{4\pi (l+m)!}} P_l^m(\cos\theta)\, e^{im\varphi} \qquad [5.49]$$

where $-l \leq m \leq l$. Using the notation:

$$Q_{l,m} = \int_{\tau'} \rho\left(\vec{\mathbf{r}}'\right) r'^l \overline{Y_l^m(\theta', \varphi')} d\tau' = \int_{\tau'} \rho\left(\vec{\mathbf{r}}'\right) r'^{l+2} \overline{Y_l^m(\theta', \varphi')} dr' \sin\theta' d\theta' d\varphi' \qquad [5.50]$$

we obtain:

$$V\left(\vec{\mathbf{r}}\right) = \frac{1}{\varepsilon_0} \sum_{l=0}^{\infty} \sum_{m=-l}^{l} \frac{Q_{l,m}}{2l+1} \frac{Y_l^m(\theta, \varphi)}{r^{l+1}} \qquad [5.51]$$

$Q_{l,m}$ is the mth *component of the electric multipole moment of order l* of the distribution.

In spherical and Cartesian bases, the components of the multipole moments, of orders 0, 1 and 2, are related by the following relations:

$$Q_{0,0} = \int_{\tau'} \rho\left(\vec{\mathbf{r}}'\right) \sqrt{\frac{1}{4\pi}} d\tau' = \sqrt{\frac{1}{4\pi}} \int_{\tau'} \rho\left(\vec{\mathbf{r}}'\right) d\tau' = \sqrt{\frac{1}{4\pi}} q'$$

$$Q_{1,0} = \int_{\tau'} \rho\left(\vec{\mathbf{r}}'\right) r' \sqrt{\frac{3}{4\pi}} \cos\theta' d\tau' = \sqrt{\frac{3}{4\pi}} \int_{\tau'} \rho\left(\vec{\mathbf{r}}'\right) z' d\tau' = \sqrt{\frac{3}{4\pi}} p_z$$

$$Q_{1,\pm1} = \int_{\tau'} \rho\left(\vec{\mathbf{r}}'\right) r' \left(\mp\sqrt{\frac{3}{8\pi}}\right) \sin\theta' e^{\mp i\varphi} d\tau'$$

$$= \mp\sqrt{\frac{3}{8\pi}} \int_{\tau'} \rho\left(\vec{\mathbf{r}}'\right) (x' \mp iy')\, d\tau' = \mp\sqrt{\frac{3}{8\pi}} (p_x \mp ip_y)$$

$$Q_{2,0} = \int_{\tau'} \rho\left(\vec{\mathbf{r}}'\right) r'^2 \sqrt{\frac{5}{16\pi}} \left(3\cos^2\theta' - 1\right) d\tau' = \sqrt{\frac{5}{16\pi}} Q_{zz}$$

$$\begin{aligned} Q_{2,\pm1} &= \sqrt{\frac{5}{24\pi}}\left(Q_{xz} \mp \mathrm{i}Q_{yz}\right) \\ Q_{2,\pm2} &= \sqrt{\frac{5}{96\pi}}\left(Q_{xx} - Q_{yy} \mp 2\mathrm{i}Q_{xy}\right) \end{aligned} \quad [5.52]$$

EXAMPLE 5.7. (Spherical Distribution).–

1) Suppose that the density $\rho\left(\overrightarrow{\mathbf{r}}'\right)$ is distributed on a sphere of radius R and that it admits the following expansion in the spherical harmonic basis:

$$\rho\left(\overrightarrow{\mathbf{r}}'\right) = \sum_{l=0}^{\infty}\sum_{m=-l}^{l} \rho_{l,m}\left(r'\right) Y_l^m\left(\theta', \varphi'\right) \quad [5.53]$$

Since the spherical harmonics obey the orthogonality relations:

$$\int_{\theta=0}^{\pi}\int_{\varphi=0}^{2\pi} Y_l^m\left(\theta', \varphi'\right)\overline{Y_{l'}^{m'}\left(\theta', \varphi'\right)} \sin\theta' d\theta' d\varphi' = \delta_{ll'}\delta_{mm'} \quad [5.54]$$

we get:

$$\begin{aligned} Q_{l,m} &= \int_0^R\int_0^{\pi}\int_0^{2\pi}\sum_{l'=0}^{\infty}\sum_{m'=-l'}^{l'} \rho_{l',m'}\left(r'\right) r'^{l+2} Y_{l'}^{m'}\left(\theta', \varphi'\right)\overline{Y_l^m\left(\theta', \varphi'\right)} \\ &\quad \times dr' \sin\theta' d\theta' d\varphi' = \int_0^R \sum_{l'=0}^{\infty}\sum_{m'=-l'}^{l'} r'^{l+2}\rho_{l',m'}\left(r'\right)\delta_{l,l'}\delta_{m,m'} dr' \\ &= \int_0^R r'^{l+2}\rho_{l,m}\left(r'\right) dr' \end{aligned} \quad [5.55]$$

2) Let us find the dipole moment of a sphere of radius R, centered on O and with charge density $\rho\left(\overrightarrow{\mathbf{r}}'\right) = az'$ for $r' \leq R$ and 0 elsewhere, where a is a constant. In terms of the spherical harmonic function $Y_1^0\left(\theta', \varphi'\right)$, we write:

$$\rho\left(\overrightarrow{\mathbf{r}}'\right) = az' = ar'\cos\theta' = \sqrt{\frac{4\pi}{3}} ar' Y_1^0\left(\theta', \varphi'\right) \quad [5.56]$$

It therefore follows that:

$$Q_{1,0} = \int_0^R r'^3\sqrt{\frac{4\pi}{3}} ar' dr' = \frac{2\sqrt{\pi} a R^5}{5\sqrt{3}} \quad [5.57]$$

which yields:

$$p_z = 2\sqrt{\frac{\pi}{3}}Q_{1,0} = \frac{4\pi}{15}aR^5. \qquad [5.58]$$

3) Similarly, when $\rho\left(\vec{\mathbf{r}}'\right) = az'^2$ for $r' \leq R$, we get:

$$\rho\left(\vec{\mathbf{r}}'\right) = az'^2 = ar'^2\cos^2\theta' = \frac{1}{3}ar'^2\left[2\sqrt{\pi}Y_0^0\left(\theta',\varphi'\right) + \frac{4\sqrt{\pi}}{\sqrt{5}}Y_2^0\left(\theta',\varphi'\right)\right] \qquad [5.59]$$

that is, $\rho_{0,0} = \frac{2\sqrt{\pi}}{3}ar'^2$ and $\rho_{2,0} = \frac{4\sqrt{\pi}}{3\sqrt{5}}ar'^2$. By substituting these results in equation [5.55], it becomes:

$$Q_{0,0} = \frac{2\sqrt{\pi}}{15}aR^5 \qquad Q_{2,0} = \frac{4\sqrt{\pi}}{21\sqrt{5}}aR^7 \qquad [5.60]$$

and consequently:

$$q' = \frac{2\pi}{15}aR^5 \qquad Q_{zz} = \frac{16\pi}{105}aR^7 = \frac{8}{7}q'R^2 \qquad [5.61]$$

5.2.5. *Dipole in a uniform external electric field*

Consider an electric dipole in a region of space with a uniform electric field, as shown in Figure 5.6(a). The resultant of the forces applied to the dipole is:

$$\vec{\mathbf{F}}_{\text{ext}} = -q\vec{\mathbf{E}}_{\text{ext}}(N) + q\vec{\mathbf{E}}_{\text{ext}}(P) = q\left[\vec{\mathbf{E}}_{\text{ext}}(P) - \vec{\mathbf{E}}_{\text{ext}}(N)\right] \qquad [5.62]$$

Since the field is uniform, that is, $\vec{\mathbf{E}}_{\text{ext}}(P) = \vec{\mathbf{E}}_{\text{ext}}(N)$, we conclude that $\vec{\mathbf{F}}_{\text{ext}} = 0$. The dipole is therefore subjected to a pair of forces. If the dipole cannot turn, nothing happens. If the dipole can turn, it performs a free oscillation, or more generally, a precessional movement around its stable orientation. This motion is damped, and the dipole eventually aligns itself along $+\vec{\mathbf{E}}_{\text{ext}}$ after a few τ, where τ is the *relaxation time*. The torque is given by:

$$\begin{aligned}\vec{\mathbf{N}}_{\text{ext}/O'} &= \overrightarrow{\mathbf{O'N}} \times \left(-q\vec{\mathbf{E}}_{\text{ext}}\right) + \overrightarrow{\mathbf{O'P}} \times \left(q\vec{\mathbf{E}}_{\text{ext}}\right) \\ &= \left(\overrightarrow{\mathbf{NO'}} + \overrightarrow{\mathbf{O'P}}\right) \times q\vec{\mathbf{E}}_{\text{ext}} = q\overrightarrow{\mathbf{NP}} \times \vec{\mathbf{E}}_{\text{ext}} = \vec{\mathbf{p}} \times \vec{\mathbf{E}}_{\text{ext}}\end{aligned} \qquad [5.63]$$

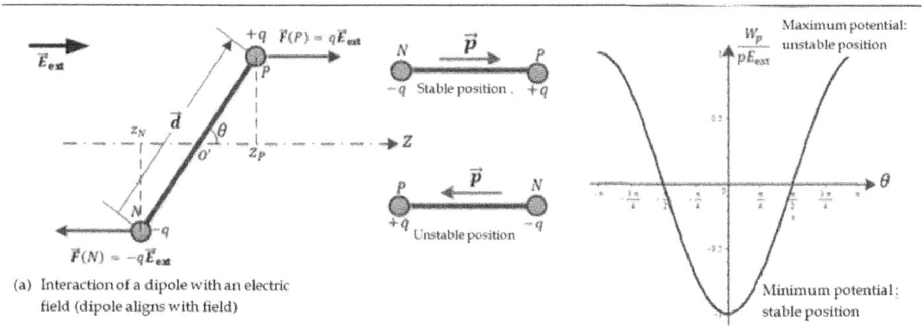

(a) Interaction of a dipole with an electric field (dipole aligns with field)

(b) Allure of potential energy

Figure 5.6. *A system made up of several charges such that $\sum_i q_i = 0$: N, P and q are, respectively, the barycenter of negative charges, the barycenter of positive charges and q the sum of positive charges. For a color version of this figure, see www.iste.co.uk/abdesselam/electrodynamics1.zip*

This torque will influence the dipole by rotating it relative to the point O', so that $\vec{\mathbf{p}}$ is aligned in the direction of $\vec{\mathbf{E}}_{\text{ext}}$. The torque relative to another point O is equal to:

$$\vec{\mathbf{N}}_{\text{ext}/O} = \overrightarrow{\mathbf{OO}}' \times \vec{\mathbf{F}}_{\text{ext}} + \vec{\mathbf{N}}_{\text{ext}/O'} \qquad [5.64]$$

Since $\vec{\mathbf{F}}_{\text{ext}} = 0$, we deduce that $\vec{\mathbf{N}}_{\text{ext}/O} = \vec{\mathbf{N}}_{\text{ext}/O'}$, for all O.

The dipole is in equilibrium when $\vec{\mathbf{N}}_{\text{ext}/O'} = 0$, i.e. if $\vec{\mathbf{p}}$ is parallel to $\vec{\mathbf{E}}_{\text{ext}}$. There are therefore two equilibrium orientations: (1) $\vec{\mathbf{p}}$ is oriented along $+\vec{\mathbf{E}}_{\text{ext}}$; or (2) $\vec{\mathbf{p}}$ oriented along $-\vec{\mathbf{E}}_{\text{ext}}$. Moreover, when the dipole is moved away from its equilibrium position, the torque always brings it back towards $+\vec{\mathbf{E}}_{\text{ext}}$. We can therefore conclude that equilibrium is stable except when $\vec{\mathbf{p}}$ is oriented along $+\vec{\mathbf{E}}_{\text{ext}}$. Let us calculate the dipole's potential energy. Choosing $\vec{\mathbf{E}}_{\text{ext}}$ parallel to Oz and assuming that $\vec{\mathbf{p}}$ makes an angle θ with $\vec{\mathbf{E}}_{\text{ext}}$, we obtain:

$$W_p = W_p(N) + W_p(P) = -qV_{\text{ext}}(N) + qV_{\text{ext}}(P) = q\left[V_{\text{ext}}(P) - V_{\text{ext}}(N)\right] \qquad [5.65]$$

However, at any point, the potential is linked to the electric field by the equation $\vec{\mathbf{E}}_{\text{ext}} = -\vec{\nabla} V_{\text{ext}}$ and therefore, $V_{\text{ext}} = -\int E_{\text{ext}} dz = -E_{\text{ext}} z + \text{const}$. It follows that:

$$W_p = -qE_{\text{ext}}(z_P - z_N) = -qdE_{\text{ext}} \cos\theta = -\vec{\mathbf{d}} \cdot \vec{\mathbf{E}}_{\text{ext}} \qquad [5.66]$$

The graph of the potential function $W_p(\theta)$ is shown in Figure 5.6(b).

5.2.6. *Dipole in a nonuniform external field*

Consider an infinitesimal electric dipole consisting of a negative charge $-q$ located at the point N of position $\vec{\mathbf{r}}$, with respect to a point O chosen here as the origin, and a positive charge $+q$ located at the point P of position $\vec{\mathbf{r}} + \vec{\mathbf{d}}$, where $\left|\vec{\mathbf{r}}\right| gg \left|\vec{\mathbf{d}}\right|$. This dipole is immersed in a nonuniform electric field, that is, a field dependent on the position $\vec{\mathbf{r}}$. The resultant of the forces applied to the dipole is nonzero. As a result, the dipole undergoes rotation and displacement. It will be assumed that the size of the dipole is very small compared to the characteristic distance of variation of the electrostatic field. For simplicity's sake, the calculations will be carried out in Cartesian coordinates. However, the results can be considered valid for other coordinate systems. Let us calculate the component along Ox of the force applied to the dipole:

$$F^x_{\text{ext}} = q\left[E^x_{\text{ext}}(P) - E^x_{\text{ext}}(N)\right] \quad [5.67]$$

Since N and P are close to each other, it is possible to carry out a limited expansion to first order in the neighborhood of the point O' (located at the center of the dipole):

$$\begin{aligned} E^x_{\text{ext}}(P) &\simeq E^x_{\text{ext}}(O') + (x_P - x_{O'}) \left.\frac{\partial E^x_{\text{ext}}}{\partial x}\right|_{O'} + (y_P - y_{O'}) \left.\frac{\partial E^x_{\text{ext}}}{\partial y}\right|_{O'} \\ &\quad + (z_P - z_{O'}) \left.\frac{\partial E^x_{\text{ext}}}{\partial z}\right|_{O'} \\ E^x_{\text{ext}}(N) &\simeq E^x_{\text{ext}}(O') + (x_N - x_{O'}) \left.\frac{\partial E^x_{\text{ext}}}{\partial x}\right|_{O'} + (y_N - y_{O'}) \left.\frac{\partial E^x_{\text{ext}}}{\partial y}\right|_{O'} \\ &\quad + (z_N - z_{O'}) \left.\frac{\partial E^x_{\text{ext}}}{\partial z}\right|_{O'} \end{aligned} \quad [5.68]$$

and, therefore, we write:

$$\begin{aligned} F^x_{\text{ext}} &\simeq q\left[(x_P - x_N) \left.\frac{\partial E^x_{\text{ext}}}{\partial x}\right|_{O'} + (y_P - y_N) \left.\frac{\partial E^x_{\text{ext}}}{\partial y}\right|_{O'} + (z_P - z_N) \left.\frac{\partial E^x_{\text{ext}}}{\partial z}\right|_{O'}\right] \\ &= p_x \left.\frac{\partial E^x_{\text{ext}}}{\partial x}\right|_{O'} + p_y \left.\frac{\partial E^x_{\text{ext}}}{\partial y}\right|_{O'} + p_z \left.\frac{\partial E^x_{\text{ext}}}{\partial z}\right|_{O'} = \left(\vec{\mathbf{p}} \cdot \vec{\nabla}\right) \left.E^x_{\text{ext}}\right|_{O'} \end{aligned} \quad [5.69]$$

This result can be generalized to F^y_{ext} and F^z_{ext}. It follows that:

$$\begin{aligned} \vec{\mathbf{F}}_{\text{ext}} &= \left(\vec{\mathbf{p}} \cdot \vec{\nabla}\right) \left.E^x_{\text{ext}}\right|_{O'} \vec{\mathbf{i}} + \left(\vec{\mathbf{p}} \cdot \vec{\nabla}\right) \left.E^y_{\text{ext}}\right|_{O'} \vec{\mathbf{j}} + \left(\vec{\mathbf{p}} \cdot \vec{\nabla}\right) \left.E^z_{\text{ext}}\right|_{O'} \vec{\mathbf{k}} \\ &= \left(\vec{\mathbf{p}} \cdot \vec{\nabla}\right) \left.\vec{\mathbf{E}}_{\text{ext}}\right|_{O'} \end{aligned} \quad [5.70]$$

It then appears that the force exerted on the dipole depends on:

1) the orientation of the dipole with respect to field variations;

2) field variations: the greater the variations, the greater the force exerted.

Equation [5.70] can be rewritten in a form that is often more practical. According to the identity $\vec{\nabla}\left(\vec{\mathbf{A}}\cdot\vec{\mathbf{B}}\right)=\vec{\mathbf{A}}\times\left(\vec{\nabla}\times\vec{\mathbf{B}}\right)+\left(\vec{\mathbf{A}}\cdot\vec{\nabla}\right)\vec{\mathbf{B}}+\vec{\mathbf{B}}\times\left(\vec{\nabla}\times\vec{\mathbf{A}}\right)+\left(\vec{\mathbf{B}}\cdot\vec{\nabla}\right)\vec{\mathbf{A}}$, we get:

$$\begin{aligned}\vec{\nabla}\left(\vec{\mathbf{p}}\cdot\vec{\mathbf{E}}_{\text{ext}}\right) &= \vec{\mathbf{E}}_{\text{ext}}\times\left(\vec{\nabla}\times\vec{\mathbf{p}}\right)+\left(\vec{\mathbf{E}}_{\text{ext}}\cdot\vec{\nabla}\right)\vec{\mathbf{p}}+\vec{\mathbf{p}}\times\left(\vec{\nabla}\times\vec{\mathbf{E}}_{\text{ext}}\right)\\ &\quad+\left(\vec{\mathbf{p}}\cdot\vec{\nabla}\right)\vec{\mathbf{E}}_{\text{ext}}\end{aligned} \qquad [5.71]$$

The first three terms on the right-hand side are zero: the first two because $\vec{\mathbf{p}}$ does not depend on the position O', and the third term because $\vec{\nabla}\times\vec{\mathbf{E}}_{\text{ext}}=0$ in electrostatics. It can thus be derived that:

$$\vec{\mathbf{F}}_{\text{ext}}=\vec{\nabla}\left[\vec{\mathbf{p}}\cdot\vec{\mathbf{E}}_{\text{ext}}\right]\Big|_{O'} \qquad [5.72]$$

In particular, if $\vec{\mathbf{E}}_{\text{ext}}\left(\vec{\mathbf{r}}\right)$ is a field induced by an electric charge q' located at O, that is $\vec{\mathbf{E}}_{\text{ext}}\left(\vec{\mathbf{r}}\right)=\frac{q'}{4\pi\varepsilon_0}\frac{\vec{\mathbf{r}}}{r^3}$, we obtain:

$$\vec{\mathbf{F}}_{\text{ext}}=\frac{q'}{4\pi\varepsilon_0}\left(\frac{\vec{\mathbf{p}}}{r^3}-\frac{3\left(\vec{\mathbf{p}}\cdot\vec{\mathbf{r}}\right)\vec{\mathbf{r}}}{r^5}\right) \qquad [5.73]$$

The torque acting on the dipole, seen from O', is (first-order expansion in $\vec{\mathbf{d}}$):

$$\begin{aligned}\vec{\mathbf{N}}_{\text{ext}/O'} &= \frac{\vec{\mathbf{d}}}{2}\times\left(q\vec{\mathbf{E}}_{\text{ext}}(P)\right)+\frac{-\vec{\mathbf{d}}}{2}\times\left(-q\vec{\mathbf{E}}_{\text{ext}}(N)\right)\\ &= \frac{\vec{\mathbf{p}}}{2}\times\left[\vec{\mathbf{E}}_{\text{ext}}(P)+\vec{\mathbf{E}}_{\text{ext}}(N)\right]=\vec{\mathbf{p}}\times\vec{\mathbf{E}}_{\text{ext}}\Big|_{O'}\end{aligned} \qquad [5.74]$$

Furthermore, the torque $\vec{\mathbf{N}}_{\text{ext}/O}$, seen from the origin O, will be:

$$\begin{aligned}\vec{\mathbf{N}}_{\text{ext}/O} &= \left(\vec{\mathbf{r}}+\frac{\vec{\mathbf{d}}}{2}\right)\times\left(q\vec{\mathbf{E}}_{\text{ext}}(P)\right)+\left(\vec{\mathbf{r}}-\frac{\vec{\mathbf{d}}}{2}\right)\times\left(-q\vec{\mathbf{E}}_{\text{ext}}(N)\right)\\ &= \vec{\mathbf{r}}\times q\left[\vec{\mathbf{E}}_{\text{ext}}(P)-\vec{\mathbf{E}}_{\text{ext}}(N)\right]+\frac{\vec{\mathbf{p}}}{2}\times\left[\vec{\mathbf{E}}_{\text{ext}}(P)+\vec{\mathbf{E}}_{\text{ext}}(N)\right]\\ &= \vec{\mathbf{r}}\times\vec{\mathbf{F}}_{\text{ext}}+\vec{\mathbf{N}}_{\text{ext}/O'}\end{aligned} \qquad [5.75]$$

5.2.7. *Potential energy*

The potential energy of a dipole placed in a nonuniform external field is:

$$W_p = q\left[V_{\text{ext}}(P) - V_{\text{ext}}(N)\right] \tag{5.76}$$

As $V_{\text{ext}}(P) \simeq V_{\text{ext}}(O') + \overrightarrow{\mathbf{O'P}} \cdot \vec{\nabla} V_{\text{ext}}\big|_{O'}$ and $V_{\text{ext}}(N) \simeq V_{\text{ext}}(O') + \overrightarrow{\mathbf{O'N}} \cdot \vec{\nabla} V_{\text{ext}}\big|_{O'}$, we deduce that:

$$W_p = q\left[\overrightarrow{\mathbf{O'P}} - \overrightarrow{\mathbf{O'N}}\right] \cdot \vec{\nabla} V_{\text{ext}}\Big|_{O'} = q\overrightarrow{\mathbf{NP}} \cdot \vec{\nabla} V_{\text{ext}}\Big|_{O'} = -\vec{\mathbf{p}} \cdot \vec{\mathbf{E}}_{\text{ext}}(O') \tag{5.77}$$

$\vec{\mathbf{E}}_{\text{ext}}(O')$ is the field that prevails at point O'. Once again, we assume that the characteristic distances of the potential variation are very large compared to the dipole size. Equation [5.72] already contains the result in equation [5.77]. The latter clearly shows that the force applied by the external electric field $\vec{\mathbf{E}}_{\text{ext}}$ to an infinitesimal electric dipole is derived from the potential in equation [5.77]. This force is therefore conservative.

This result can be generalized to any distribution. Consider a collection of point-like charges q_α in a region of space where there is an external potential V_{ext}. The sources of potential V_{ext} are located very far from the distribution, that is, it satisfies the Poisson equation $\vec{\nabla} \cdot \vec{\mathbf{E}}_{\text{ext}} = -\Delta V_{\text{ext}} = 0$ in the region of space under consideration. If O' represents the center of gravity (weighted by the charges) of the q_α charges, the potential energy of the system is then given by:

$$\begin{aligned}
W_p &= \sum_\alpha q_\alpha V_{\text{ext}}(\vec{\mathbf{r}}_\alpha) = \sum_\alpha q_\alpha V_{\text{ext}}\left(\vec{\mathbf{R}} + \vec{\mathbf{r}}'_\alpha\right) = \sum_\alpha q_\alpha\, V_{\text{ext}}\big|_{\vec{\mathbf{R}}} \\
&\quad + \sum_\alpha q_\alpha \sum_{i=1}^{3} x'_{\alpha,i} \left.\frac{\partial V_{\text{ext}}}{\partial x'_{\alpha,i}}\right|_{\vec{\mathbf{R}}} + \frac{1}{2}\sum_\alpha q_\alpha \sum_{i,j=1}^{3} x'_{\alpha,i}x'_{\alpha,j} \left.\frac{\partial^2 V_{\text{ext}}}{\partial x'_{\alpha,i}x'_{\alpha,j}}\right|_{\vec{\mathbf{R}}} + \cdots \\
&= \sum_\alpha q_\alpha\, V_{\text{ext}}\big|_{\vec{\mathbf{R}}} + \sum_\alpha q_\alpha \sum_{i=1}^{3} x'_{\alpha,i} \left.\frac{\partial V_{\text{ext}}}{\partial x'_{\alpha,i}}\right|_{\vec{\mathbf{R}}} \\
&\quad + \frac{1}{6}\sum_\alpha q_\alpha \left[\sum_{i,j=1}^{3} 3x'_{\alpha,i}x'_{\alpha,j} \left.\frac{\partial^2 V_{\text{ext}}}{\partial x'_{\alpha,i}x'_{\alpha,j}}\right|_{\vec{\mathbf{R}}} - r'^2_\alpha\, \Delta V_{\text{ext}}\big|_{\vec{\mathbf{R}}}\right] + \cdots
\end{aligned}$$

$$
\begin{aligned}
&= \sum_{\alpha} q_\alpha \left. V_{\text{ext}}\right|_{\vec{\mathbf{R}}} + \sum_{\alpha} q_\alpha \sum_{i=1}^{3} x'_{\alpha,i} \left.\frac{\partial V_{\text{ext}}}{\partial x'_{\alpha,i}}\right|_{\vec{\mathbf{R}}} \\
&+\frac{1}{6}\sum_{\alpha} q_\alpha \sum_{i,j=1}^{3}\left[3x'_{\alpha,i}x'_{\alpha,j}-\delta_{ij}r'^2_{\alpha}\right]\left.\frac{\partial^2 V_{\text{ext}}}{\partial x'_{\alpha,i}x'_{\alpha,j}}\right|_{\vec{\mathbf{R}}}+\cdots \\
&= q' \left. V_{\text{ext}}\right|_{\vec{\mathbf{R}}} - \vec{\mathbf{p}}\cdot\left.\vec{\mathbf{E}}_{\text{ext}}\right|_{\vec{\mathbf{R}}} + \frac{1}{6}\sum_{i,j} Q_{ij}\left.\frac{\partial^2 V_{\text{ext}}}{\partial x'_i x'_j}\right|_{\vec{\mathbf{R}}} + \cdots
\end{aligned}
\quad [5.78]
$$

where $q' = \sum_\alpha q_\alpha$, $\vec{\mathbf{p}} = \sum_\alpha q_\alpha \vec{\mathbf{r}}'_\alpha$ and $Q_{ij} = \sum_\alpha q_\alpha \left[3x'_{\alpha,i}x'_{\alpha,j} - \delta_{ij}r'^2_\alpha\right]$.

EXAMPLE 5.8.– Let us find the potential energy of the quadrupole of Figure 5.5 in the potential $V(x,y) = V_0 xy$. Using formula [5.7], we get:

$$
W_p = \frac{1}{6}\left(Q_{xx}\frac{\partial^2 V}{\partial x^2} + 2Q_{xy}\frac{\partial^2 V}{\partial x \partial y} + Q_{yy}\frac{\partial^2 V}{\partial y^2}\right) = qL^2 V_0 \quad [5.79]
$$

5.3. Magnetic dipole moment

5.3.1. *Expressions of magnetic moment*

Magnetic vector potential $\vec{\mathbf{A}}$ can be expanded in the same way as the scalar potential V. For simplicity's sake, we will perform the calculations for a closed filiform circuit. The results obtained can be generalized to other distributions. For a closed filiform circuit Γ, the vector potential is calculated from the following equation:

$$
\vec{\mathbf{A}}\left(\vec{\mathbf{r}}\right) = \frac{\mu_0 I}{4\pi}\oint_{\Gamma'} \frac{d\vec{\mathbf{l}}'}{\left|\vec{\mathbf{r}} - \vec{\mathbf{r}}'\right|} \quad [5.80]
$$

If the current distribution does not extend beyond a maximum radius r'_{max}, then it is possible to write, for $r \gg r'_{\text{max}}$, that $\frac{1}{\left|\vec{\mathbf{r}} - \vec{\mathbf{r}}'\right|} = \frac{1}{r} + \frac{\vec{\mathbf{r}}'\cdot\vec{\mathbf{r}}}{r^3} + \frac{3}{2}\frac{\left(\vec{\mathbf{r}}'\cdot\vec{\mathbf{r}}\right)^2}{r^5} - \frac{1}{2}\frac{r'^2}{r^3} + \cdots$ and consequently:

$$
\begin{aligned}
\vec{\mathbf{A}}\left(\vec{\mathbf{r}}\right) &= \frac{\mu_0 I}{4\pi r}\oint_{\Gamma'} d\vec{\mathbf{l}}' + \frac{\mu_0 I}{4\pi r^3}\oint_{\Gamma'}\left(\vec{\mathbf{r}}\cdot\vec{\mathbf{r}}'\right)d\vec{\mathbf{l}}' \\
&+ \frac{\mu_0 I}{4\pi r^3}\oint_{\Gamma'}\left[\frac{3}{2}\frac{\left(\vec{\mathbf{r}}\cdot\vec{\mathbf{r}}'\right)^2}{r^2} - \frac{1}{2}r'^2\right]d\vec{\mathbf{l}}' + \cdots
\end{aligned}
\quad [5.81]
$$

To calculate the various terms in the above equation, we proceed as follows: let us write Stokes' theorem for the vector field $f\overrightarrow{\mathbf{B}}$, where $\overrightarrow{\mathbf{B}}$ is constant:

$$\begin{aligned}\oint_{\Gamma'} \left(f\overrightarrow{\mathbf{B}}\right) \cdot d\overrightarrow{\mathbf{l}}' &= \int_{S'} \overrightarrow{\nabla}' \times \left(f\overrightarrow{\mathbf{B}}\right) \cdot d\overrightarrow{\mathbf{S}}' = \int_{S'} \left(\overrightarrow{\nabla}' f \times \overrightarrow{\mathbf{B}}\right) \cdot d\overrightarrow{\mathbf{S}}' \\ &= \left(\int_{S'} d\overrightarrow{\mathbf{S}}' \times \overrightarrow{\nabla}' f\right) \cdot \overrightarrow{\mathbf{B}}\end{aligned} \quad [5.82]$$

Since $\overrightarrow{\mathbf{B}}$ is any field, we deduce that:

$$\oint_{\Gamma'} f d\overrightarrow{\mathbf{l}}' = \int_{S'} d\overrightarrow{\mathbf{S}}' \times \overrightarrow{\nabla}' f \quad [5.83]$$

We therefore obtain, for $f = 1$:

$$\oint_{\Gamma'} d\overrightarrow{\mathbf{l}}' = 0 \quad [5.84]$$

and, for $f = \overrightarrow{\mathbf{r}} \cdot \overrightarrow{\mathbf{r}}'$:

$$\oint_{\Gamma'} \left(\overrightarrow{\mathbf{r}} \cdot \overrightarrow{\mathbf{r}}'\right) d\overrightarrow{\mathbf{l}}' = -\int_{S'} \overrightarrow{\mathbf{r}} \times d\overrightarrow{\mathbf{S}}' \quad [5.85]$$

Similarly, for $f = \frac{3\left(\overrightarrow{\mathbf{r}} \cdot \overrightarrow{\mathbf{r}}'\right)^2 - r^2 r'^2}{2r^2}$, it is found that:

$$\oint_{\Gamma'} \frac{\left[3\left(\overrightarrow{\mathbf{r}} \cdot \overrightarrow{\mathbf{r}}'\right)^2 - r^2 r'^2\right] d\overrightarrow{\mathbf{l}}'}{2r^2} = -\int_{S'} \frac{\left[3\left(\overrightarrow{\mathbf{r}} \cdot \overrightarrow{\mathbf{r}}'\right) \overrightarrow{\mathbf{r}} - r^2 \overrightarrow{\mathbf{r}}'\right] \times d\overrightarrow{\mathbf{S}}'}{r^2} \quad [5.86]$$

We conclude that the first term in [5.81] is identically zero, since the sum of displacement vectors around a closed curve is zero. The second term is the contribution of the magnetic dipole moment of the closed circuit. The third term is the second-order contribution to the vector potential, which will be neglected for simplicity. It follows that:

$$\overrightarrow{\mathbf{A}}\left(\overrightarrow{\mathbf{r}}\right) \simeq \frac{\mu_0 I}{4\pi r^3} \int_{S'} d\overrightarrow{\mathbf{S}}' \times \overrightarrow{\mathbf{r}} = \frac{\mu_0}{4\pi r^3} \left(I \int_{S'} d\overrightarrow{\mathbf{S}}'\right) \times \overrightarrow{\mathbf{r}} \quad [5.87]$$

The term:

$$I \int_{S'} d\overrightarrow{\mathbf{S}}' \equiv \overrightarrow{\mathbf{m}}' \quad [5.88]$$

is called the circuit's *magnetic dipole moment* or simply *magnetic moment*. In the first-order approximation, the magnetic vector potential takes the following form:

$$\vec{\mathbf{A}}\left(\vec{\mathbf{r}}\right) = \frac{\mu_0}{4\pi r^3}\vec{\mathbf{m}}' \times \vec{\mathbf{r}} \quad [5.89]$$

This is the vector potential created by the magnetic dipole $\vec{\mathbf{m}}'$. If the vector potential is created by a distribution τ' of magnetic moments, the previous expression can be generalized as follows:

$$\vec{\mathbf{A}}\left(\vec{\mathbf{r}}\right) = \frac{\mu_0}{4\pi}\int_{\tau'}\frac{d\vec{\mathbf{m}}\left(\vec{\mathbf{r}}'\right) \times \left(\vec{\mathbf{r}} - \vec{\mathbf{r}}'\right)}{\left|\vec{\mathbf{r}} - \vec{\mathbf{r}}'\right|^3} = \frac{\mu_0}{4\pi}\int_{\tau'}\frac{\vec{\mathbf{M}}\left(\vec{\mathbf{r}}'\right) \times \left(\vec{\mathbf{r}} - \vec{\mathbf{r}}'\right)}{\left|\vec{\mathbf{r}} - \vec{\mathbf{r}}'\right|^3}d\tau' \quad [5.90]$$

where $\vec{\mathbf{M}}\left(\vec{\mathbf{r}}'\right)$ represents the *density of magnetic moments*.

The expression for the magnetic moment [5.88] can be generalized to other types of distribution. To see this, let us write Stokes' theorem for the vector $\vec{\mathbf{r}}' \times \vec{\mathbf{B}}$, where $\vec{\mathbf{B}}$ is constant: $\oint \left(\vec{\mathbf{r}}' \times \vec{\mathbf{B}}\right) \cdot d\vec{\mathbf{l}}' = \int \vec{\nabla}' \times \left(\vec{\mathbf{r}}' \times \vec{\mathbf{B}}\right) \cdot d\vec{\mathbf{S}}' = \int \left[\left(\vec{\mathbf{B}} \cdot \vec{\nabla}'\right)\vec{\mathbf{r}}' - \vec{\mathbf{B}}\left(\vec{\nabla}' \cdot \vec{\mathbf{r}}'\right)\right] \cdot d\vec{\mathbf{S}}' = -2\int \vec{\mathbf{B}} \cdot d\vec{\mathbf{S}}' \Longrightarrow \vec{\mathbf{B}} \cdot \left(\oint d\vec{\mathbf{l}}' \times \vec{\mathbf{r}}'\right) = \vec{\mathbf{B}} \cdot \left(-2\int d\vec{\mathbf{S}}'\right)$. Since $\vec{\mathbf{B}}$ is arbitrary, we obtain:

$$-\frac{1}{2}\oint d\vec{\mathbf{l}}' \times \vec{\mathbf{r}}' = \int d\vec{\mathbf{S}}' \quad [5.91]$$

Thus, the magnetic moment is written as:

$$\vec{\mathbf{m}}' = I\int_{S'} d\vec{\mathbf{S}}' = \frac{I}{2}\oint_{\Gamma'} \vec{\mathbf{r}}' \times d\vec{\mathbf{l}}' \quad [5.92]$$

This equation is easily generalized to other types of distribution:

$$\vec{\mathbf{m}}' = \begin{cases} \frac{1}{2}\int_{\tau'} \vec{\mathbf{r}}' \times \vec{\mathbf{J}}\left(\vec{\mathbf{r}}'\right) d\tau' & \text{(volume current)} \\ \frac{1}{2}\int_{S'} \vec{\mathbf{r}}' \times \vec{\mathbf{K}}\left(\vec{\mathbf{r}}'\right) dS' & \text{(surface current)} \\ \frac{1}{2}\int_{L'} \vec{\mathbf{r}}' \times \vec{\mathbf{I}}\left(\vec{\mathbf{r}}'\right) dl' & \text{(line current)} \\ \frac{1}{2}q\,\vec{\mathbf{r}}' \times \vec{\mathbf{v}}' & \text{(charge } q' \text{ moving at velocity } \vec{\mathbf{v}}'\text{)} \end{cases} \quad [5.93]$$

EXAMPLE 5.9. (Magnetic Moment of a Loop).– According to [5.88], the magnetic moment of the circular loop depicted in Figure 5.7 is:

$$\vec{\mathbf{m}}' = I\int_{S'} d\vec{\mathbf{S}}' = I\vec{\mathbf{k}}\int_{S'} dS' = \pi R^2 I\vec{\mathbf{k}} \quad [5.94]$$

This formula can also be derived from [5.92]. Indeed, we obtain:

$$\vec{\mathbf{m}}' = \frac{I}{2}\oint_{\Gamma'} \vec{\mathbf{r}}' \times d\vec{\mathbf{l}}' = \frac{I}{2}\int_0^{2\pi}\left(R\cos\varphi\,\vec{\mathbf{i}} + R\sin\varphi\,\vec{\mathbf{j}}\right) \times \left(-R\sin\varphi\,\vec{\mathbf{i}} + R\cos\varphi\,\vec{\mathbf{j}}\right)d\varphi = \frac{IR^2}{2}\int_0^{2\pi} d\varphi\,\vec{\mathbf{k}} = \pi R^2 I\vec{\mathbf{k}} \quad [5.95]$$

EXAMPLE 5.10. (Magnetic Moment of a Rotating Disk).– We wish to calculate the magnetic moment of the uniformly charged disk, shown in Figure 5.8, located in the Oxy plane and rotating around Oz at angular frequency ω. The rotating motion of the disk induces the surface current density:

$$\vec{\mathbf{K}}\left(\vec{\mathbf{r}}'\right) = \sigma_0\left(\vec{\omega} \times \vec{\mathbf{r}}'\right) \quad [5.96]$$

where σ_0 is the charge density on the disk surface. We can easily verify that:

$$\int_{S'} \vec{\mathbf{K}}\left(\vec{\mathbf{r}}'\right) dS' = \sigma_0\omega\int_0^R\int_0^{2\pi} r'^2\left[-\sin\varphi'\,\vec{\mathbf{i}} + \cos\varphi'\,\vec{\mathbf{j}}\right] dr'd\varphi' = 0 \quad [5.97]$$

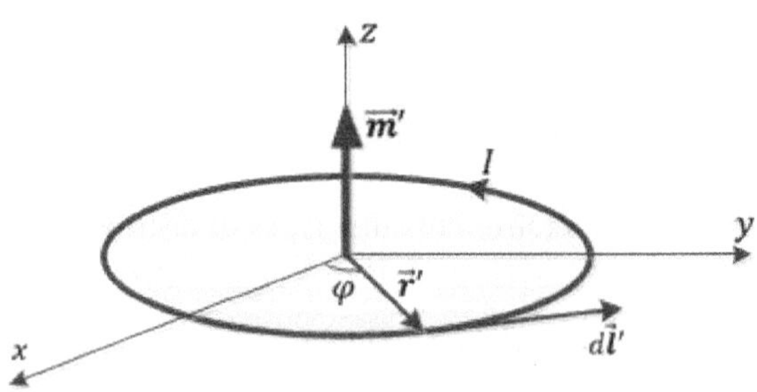

Figure 5.7. *Magnetic moment of a loop with rayon R carrying a current I*

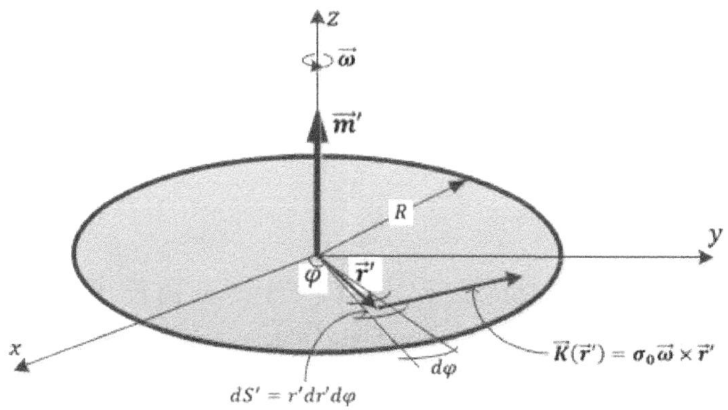

Figure 5.8. *Magnetic moment of a uniformly charged disk, of radius R, rotating around the Oz axis: The moment is equivalent to that of a loop of radius R carrying a current $I = \frac{Q\omega}{4\pi} = \frac{Q}{2T}$, where T is the period of revolution of the disk*

We can see that the zero-order term in the expression for the expansion of $\vec{\mathbf{A}}\left(\vec{\mathbf{r}}'\right)$ is effectively zero. Moreover, the magnetic moment is equal to:

$$\begin{aligned}\vec{\mathbf{m}}' &= \frac{1}{2}\int_{S'}\vec{\mathbf{r}}'\times\vec{\mathbf{K}}\left(\vec{\mathbf{r}}'\right)dS' = \frac{1}{2}\int_{S'}\vec{\mathbf{r}}'\times\left[\sigma_0\left(\vec{\omega}\times\vec{\mathbf{r}}'\right)\right]dS' \\ &= \frac{1}{2}\sigma_0\omega\int_0^R\int_0^{2\pi} r'^3dr'd\varphi'\vec{\mathbf{k}} = \frac{1}{4}\pi\sigma_0\omega R^4\vec{\mathbf{k}} = \frac{1}{4}\omega Q R^2\vec{\mathbf{k}}\end{aligned} \quad [5.98]$$

where $Q = \pi R^2\sigma_0$ is the total disk charge.

The magnetic moment induced by the disk is equivalent to that induced by a loop with current density $I = \frac{Q\omega}{4\pi} = \frac{Q}{2T}$, where T is the period of revolution of the disk.

EXAMPLE 5.11. (Magnetic Moment of a Rotating Sphere).– Let us find the magnetic moment of a uniformly charged sphere of center O, radius R and rotating at angular frequency ω around the axis Oz, as shown in Figure 5.9. Rotation around Oz induces the volume current density:

$$\vec{\mathbf{J}}\left(\vec{\mathbf{r}}'\right) = \rho_0\vec{\mathbf{v}}' = \rho_0\vec{\omega}\times\vec{\mathbf{r}}' = \rho_0\omega r'\sin\theta'\left[-\sin\varphi'\vec{\mathbf{i}} + \cos\varphi'\vec{\mathbf{j}}\right] \quad [5.99]$$

It can be easily verified that:

$$\int_{\text{sphere}} \vec{\mathbf{J}}'d\tau' = \rho_0\omega \int_0^R \int_0^\pi \int_0^{2\pi} r'^3 \sin^2\theta' \left[-\sin\varphi'\vec{\mathbf{i}} + \cos\varphi'\vec{\mathbf{j}}\right] dr'd\theta'd\varphi' = 0 \quad [5.100]$$

where $\vec{\mathbf{J}}' \equiv \vec{\mathbf{J}}\left(\vec{\mathbf{r}}'\right)$. Since $\vec{\mathbf{r}}' \times \vec{\mathbf{J}}\left(\vec{\mathbf{r}}'\right) = \rho_0\vec{\mathbf{r}}' \times \left(\vec{\omega} \times \vec{\mathbf{r}}'\right) = \rho_0\left[r'^2\vec{\omega} - \left(\vec{\mathbf{r}}' \cdot \vec{\omega}\right)\vec{\mathbf{r}}'\right] = \rho_0\omega r'^2\left[\sin^2\theta'\vec{\mathbf{k}} - \sin\theta'\cos\theta'\cos\varphi'\vec{\mathbf{i}} - \sin\theta' \cos\theta' \times \sin\varphi'\vec{\mathbf{j}}\right]$, the magnetic moment will be:

$$\begin{aligned}\vec{\mathbf{m}}' = \frac{1}{2}\int_{\tau'} \vec{\mathbf{r}}' \times \vec{\mathbf{J}}\left(\vec{\mathbf{r}}'\right) d\tau' = \frac{1}{2}\rho_0\omega \int_0^R \int_0^\pi \int_0^{2\pi} r'^2 \Big[\sin^2\theta'\vec{\mathbf{k}} - \\ \sin\theta'\cos\theta'\cos\varphi'\vec{\mathbf{i}} - \sin\theta'\cos\theta'\sin\varphi'\vec{\mathbf{j}}\Big] r'^2dr'\sin\theta'd\theta'd\varphi'\end{aligned} \quad [5.101]$$

The last two integrals are zero. We finally obtain:

$$\begin{aligned}\vec{\mathbf{m}}' &= \frac{\rho_0\omega}{2}\int_0^R \int_0^\pi \int_0^{2\pi} r'^4 \sin^3\theta'dr'd\theta'd\varphi'\vec{\mathbf{k}} \\ &= \frac{\pi\rho_0\omega R^5}{5}\int_0^\pi \sin^3\theta'd\theta'\vec{\mathbf{k}} = \frac{4\pi}{15}\rho_0\omega R^5\vec{\mathbf{k}} = \frac{1}{5}\omega QR^2\vec{\mathbf{k}}\end{aligned} \quad [5.102]$$

where Q is the electric charge carried by the sphere.

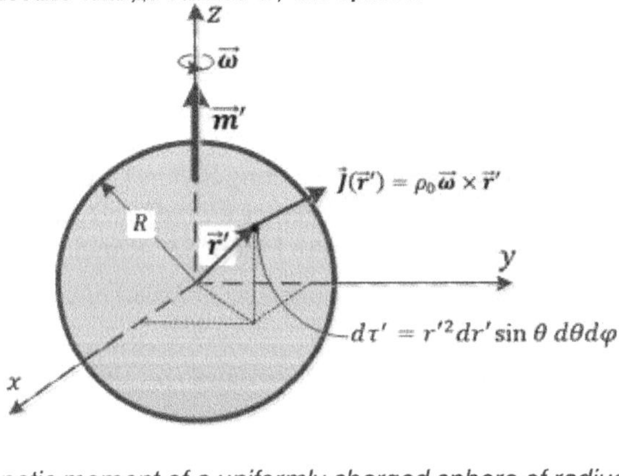

Figure 5.9. *Magnetic moment of a uniformly charged sphere of radius R rotating about the axis Oz: the moment is equivalent to that of a loop of radius R carrying a current $I = \frac{Q\omega}{4\pi} = \frac{2Q}{5T}$ along the equator, where T is the period of rotation of the sphere*

5.3.2. *Magnetic field created by a magnetic moment*

To obtain the magnetic field induced by the magnetic moment $\vec{\mathbf{m}}'$, we must calculate the curl of the potential vector (equation [5.89]):

$$\vec{\mathbf{B}}\left(\vec{\mathbf{r}}\right) = \vec{\nabla} \times \left[\frac{\mu_0\left(\vec{\mathbf{m}}' \times \vec{\mathbf{r}}\right)}{4\pi r^3}\right] = \vec{\nabla}\left(\frac{\mu_0}{4\pi r^3}\right) \times \left(\vec{\mathbf{m}}' \times \vec{\mathbf{r}}\right)$$
$$+\frac{\mu_0}{4\pi r^3}\vec{\nabla} \times \left(\vec{\mathbf{m}}' \times \vec{\mathbf{r}}\right) = -\frac{3\mu_0}{4\pi r^5}\vec{\mathbf{r}} \times \left(\vec{\mathbf{m}}' \times \vec{\mathbf{r}}\right) + \quad [5.103]$$
$$+\frac{\mu_0}{4\pi r^3}\left[\vec{\mathbf{m}}'\left(\vec{\nabla} \cdot \vec{\mathbf{r}}\right) - \left(\vec{\mathbf{m}}' \cdot \vec{\nabla}\right)\vec{\mathbf{r}}\right] = \frac{\mu_0}{4\pi}\left[-\frac{\vec{\mathbf{m}}'}{r^3} + \frac{3\left(\vec{\mathbf{r}} \cdot \vec{\mathbf{m}}'\right)\vec{\mathbf{r}}}{r^5}\right]$$

If it is a continuous distribution of magnetic moments, with a density $\vec{\mathbf{M}}\left(\vec{\mathbf{r}}'\right)$, located in a τ' region of space, as shown in Figure 5.10, the total induced magnetic field will be:

$$\vec{\mathbf{B}} = \frac{\mu_0}{4\pi}\int_{\tau'}\left[-\frac{\vec{\mathbf{M}}\left(\vec{\mathbf{r}}'\right)}{\left|\vec{\mathbf{r}} - \vec{\mathbf{r}}'\right|^3} + \frac{3\left[\left(\vec{\mathbf{r}} - \vec{\mathbf{r}}'\right) \cdot \vec{\mathbf{M}}\left(\vec{\mathbf{r}}\right)\right]\left(\vec{\mathbf{r}} - \vec{\mathbf{r}}'\right)}{\left|\vec{\mathbf{r}} - \vec{\mathbf{r}}'\right|^5}\right] d\tau' \quad [5.104]$$

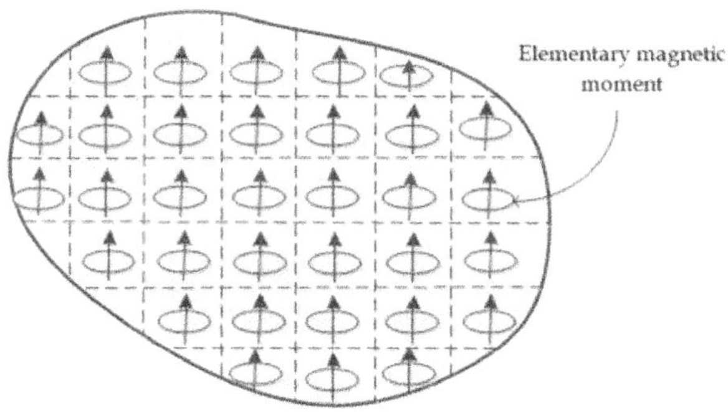

Figure 5.10. *Contribution to the magnetic field by a set of elementary magnetic moments contained in a volume τ'*

Expression [5.103] can also be written as:

$$\vec{\mathbf{B}} = -\mu_0\vec{\nabla}\left(\frac{\vec{\mathbf{m}}' \cdot \vec{\mathbf{r}}}{4\pi r^3}\right) \quad [5.105]$$

Recall that away from sources (of currents), it is always possible to express the magnetic field $\overrightarrow{bfB}$ from a magnetic scalar potential using the relation:

$$\overrightarrow{\mathbf{B}} = -\mu_0 \vec{\nabla} V_m \qquad [5.106]$$

We can therefore conclude that the magnetic scalar potential induced by a small magnetic moment located at the origin is equal to:

$$V_m = \frac{\overrightarrow{\mathbf{m}}' \cdot \overrightarrow{\mathbf{r}}}{4\pi r^3} \qquad [5.107]$$

If several elementary moments are distributed in a volume τ', which does not extend beyond a maximum radius r'_{max}, the total magnetic scalar potential induced by the entire distribution will be:

$$V_m = \int_{\tau'} \frac{\overrightarrow{\mathbf{M}}\left(\overrightarrow{\mathbf{r}}'\right) \cdot \left(\overrightarrow{\mathbf{r}} - \overrightarrow{\mathbf{r}}'\right)}{4\pi \left|\overrightarrow{\mathbf{r}} - \overrightarrow{\mathbf{r}}'\right|^3} d\tau' \qquad [5.108]$$

In particular, a finite Γ circuit can be viewed as a set of microcircuits that join together to form the Γ circuit, as shown in Figure 5.11. Each microcircuit, located at $\overrightarrow{\mathbf{r}}'$, of magnetic moment $Id\overrightarrow{\mathbf{S}}'$ creates an elementary scalar potential $dV_m = \frac{Id\overrightarrow{\mathbf{S}}' \cdot \left(\overrightarrow{\mathbf{r}} - \overrightarrow{\mathbf{r}}'\right)}{4\pi\left|\overrightarrow{\mathbf{r}} - \overrightarrow{\mathbf{r}}'\right|^3}$. Thus, the total magnetic scalar potential measured at the point $\overrightarrow{\mathbf{r}}$ is:

$$V_m = \int dV_m = \int_{S'} \frac{Id\overrightarrow{\mathbf{S}}' \cdot \left(\overrightarrow{\mathbf{r}} - \overrightarrow{\mathbf{r}}'\right)}{4\pi \left|\overrightarrow{\mathbf{r}} - \overrightarrow{\mathbf{r}}'\right|^3} = -\frac{I\Omega}{4\pi} \qquad [5.109]$$

where Ω is the solid angle from which the circuit is observed from position $\overrightarrow{\mathbf{r}}$.

5.3.3. *Multipole expansion of vector potential*

Consider a current distribution $\overrightarrow{\mathbf{J}}\left(\overrightarrow{\mathbf{r}}'\right)$ that occupies a volume τ' that does not extend beyond a maximum radius r'_{max} from the coordinate origin. For $r \gg r'_{\text{max}}$, the potential magnetic vector $\overrightarrow{\mathbf{A}}\left(\overrightarrow{\mathbf{r}}\right)$ is:

$$\overrightarrow{\mathbf{A}}\left(\overrightarrow{\mathbf{r}}\right) = \frac{\mu_0}{4\pi} \int_{\tau'} \frac{\overrightarrow{\mathbf{J}}\left(\overrightarrow{\mathbf{r}}'\right) d\tau'}{\left|\overrightarrow{\mathbf{r}} - \overrightarrow{\mathbf{r}}'\right|} = \frac{\mu_0}{4\pi} \sum_{l=0}^{\infty} \frac{1}{r^{l+1}} \int_{\tau'} r'^l \overrightarrow{\mathbf{J}}\left(\overrightarrow{\mathbf{r}}'\right) P_l\left(\cos\gamma\right) d\tau' \qquad [5.110]$$

where γ is the angle measured between $\vec{\mathbf{r}}$ and $\vec{\mathbf{r}}'$. Furthermore, according to the addition of formula [5.47], it can also be written that:

$$\begin{aligned}\vec{\mathbf{A}}\left(\vec{\mathbf{r}}\right) &= \mu_0 \sum_{l=0}^{\infty}\sum_{m=-l}^{l} \frac{Y_l^m\left(\theta,\varphi\right)}{\left(2l+1\right)r^{l+1}} \int_{\tau'} r'^l \vec{\mathbf{J}}\left(\vec{\mathbf{r}}'\right)\overline{Y_l^m\left(\theta',\varphi'\right)}d\tau' \\ &= \mu_0 \sum_{l=0}^{\infty}\sum_{m=-l}^{l} \frac{Y_l^m\left(\theta,\varphi\right)}{\left(2l+1\right)r^{l+1}} \vec{\mathbf{M}}_{lm}\end{aligned} \quad [5.111]$$

where:

$$\vec{\mathbf{M}}_{lm} = \int_{\tau'} r'^l \vec{\mathbf{J}}\left(\vec{\mathbf{r}}'\right)\overline{Y_l^m\left(\theta',\varphi'\right)}d\tau' \quad [5.112]$$

$\vec{\mathbf{M}}_{lm}$ is the *magnetic moment of degree* l *and order* m in the spherical representation.

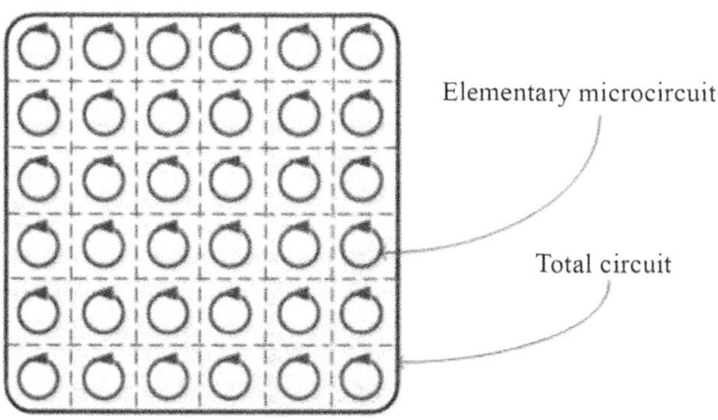

Figure 5.11. *Contribution to the magnetic field of a set of small elementary circuits interconnected to construct a large* Γ' *circuit*

EXAMPLE 5.12. (Rotating Sphere).– For the sphere in Example 5.11, the following results are obtained:

$$\begin{aligned}\vec{\mathbf{M}}_{00} &= \sqrt{\frac{1}{\pi}}\int_{\tau'}\vec{\mathbf{J}}\left(\vec{\mathbf{r}}'\right)d\tau' = 0 \\ \vec{\mathbf{M}}_{10} &= \sqrt{\frac{3}{4\pi}}\int_{\tau'} r'\cos\theta'\vec{\mathbf{J}}\left(\vec{\mathbf{r}}'\right)d\tau' = \sqrt{\frac{3}{4\pi}}\rho_0\omega\int_0^R\int_0^{\pi}\int_0^{2\pi} r'^4 \times \\ &\cos\theta'\sin^2\theta'\left[-\sin\varphi'\vec{\mathbf{i}} + \cos\varphi'\vec{\mathbf{j}}\right]dr'd\theta'd\varphi' = 0\end{aligned} \quad [5.113]$$

and:

$$\begin{aligned}\vec{\mathbf{M}}_{1,\pm 1} &= \mp\sqrt{\frac{3}{8\pi}}\int_{\tau'} r'\sin\theta' e^{\mp i\varphi'}\vec{\mathbf{J}}\left(\vec{\mathbf{r}}'\right)d\tau' \\ &= \mp\sqrt{\frac{3}{8\pi}}\rho_0\omega\int_0^R\int_0^\pi\int_0^{2\pi} r'^4\sin^3\theta'\left[-\sin\varphi'\vec{\mathbf{i}}+\cos\varphi'\vec{\mathbf{j}}\right]e^{\mp i\varphi'}dr'd\theta'd\varphi' \\ &= \mp\frac{4}{5}\sqrt{\frac{2\pi}{3}}\rho_0\omega R^5\left[\mp i\,\vec{\mathbf{i}}+\vec{\mathbf{j}}\right] = \mp\sqrt{\frac{6}{\pi}}\vec{\mathbf{m}}'\times\left[\vec{\mathbf{i}}\pm i\,\vec{\mathbf{j}}\right]\end{aligned} \quad [5.114]$$

5.3.4. *Interaction of the magnetic moment with an external field*

Consider an infinitesimal closed rectangular circuit of dimensions Δx and Δy, located for simplicity in the Oxy plane, carrying a current I and immersed in a nonuniform external magnetic field, as shown in Figure 5.12. The force exerted on the circuit is:

$$\begin{aligned}\vec{\mathbf{F}} &= I\vec{\mathbf{PQ}}\times\vec{\mathbf{B}}\left(x+\frac{\Delta x}{2},y,z\right)+I\vec{\mathbf{QR}}\times\vec{\mathbf{B}}\left(x+\Delta x,y+\frac{\Delta y}{2},z\right) \\ &\quad + I\vec{\mathbf{RS}}\times\vec{\mathbf{B}}\left(x+\frac{\Delta x}{2},y+\Delta y,z\right)+I\vec{\mathbf{SP}}\times\vec{\mathbf{B}}\left(x,y+\frac{\Delta y}{2},z\right)\end{aligned} \quad [5.115]$$

Using Taylor expansions in the neighborhood of the point (x, y, z), to the order $\Delta x\Delta y$ (first order of approximation), we find:

$$\begin{aligned}\vec{\mathbf{F}} &= -I\Delta x\,\vec{\mathbf{i}}\times\frac{\partial\vec{\mathbf{B}}}{\partial y}\Delta y+I\Delta y\,\vec{\mathbf{j}}\times\frac{\partial\vec{\mathbf{B}}}{\partial x}\Delta x \\ &= m_z\left(-\vec{\mathbf{i}}\times\frac{\partial\vec{\mathbf{B}}}{\partial y}+\vec{\mathbf{j}}\times\frac{\partial\vec{\mathbf{B}}}{\partial x}\right) \\ &= m_z\left(\frac{\partial B_z}{\partial x}\vec{\mathbf{i}}+\frac{\partial B_z}{\partial y}\vec{\mathbf{j}}+\frac{\partial B_z}{\partial z}\vec{\mathbf{k}}\right)-m_z\left(\frac{\partial B_x}{\partial x}+\frac{\partial B_y}{\partial y}+\right. \\ &\quad \left.\frac{\partial B_z}{\partial z}\right)\vec{\mathbf{k}} = m_z\vec{\nabla}\left(B_z\right)\end{aligned} \quad [5.116]$$

because $\vec{\nabla}\cdot\vec{\mathbf{B}} = 0$ (the $\vec{\mathbf{B}}$ field has no sources). $m_z = I\Delta x\Delta y$ is the magnetic moment of the circuit. In vector notation, the above equation can also be written as:

$$\vec{\mathbf{F}} = m_z\vec{\nabla}\left(B_z\right) = \vec{\nabla}\left(m_zB_z\right) \quad [5.117]$$

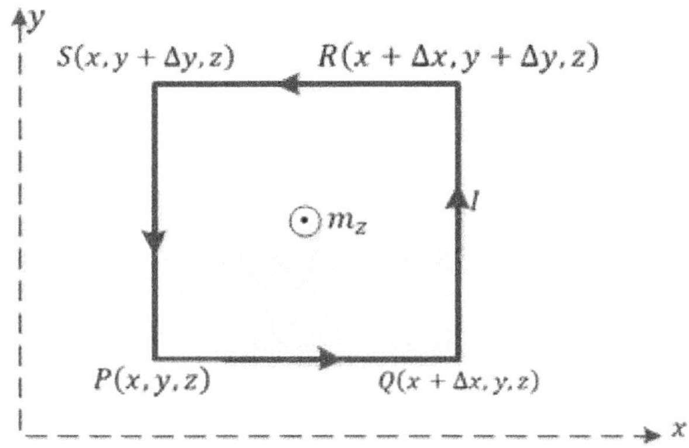

Figure 5.12. *The total force applied to the circuit is equal to the sum of the magnetic forces experienced on each side separately*

If the circuit is oriented arbitrarily in space, we deduce that:

$$\vec{\mathbf{F}} = \vec{\nabla}\left(\vec{\mathbf{m}} \cdot \vec{\mathbf{B}}\right) = \vec{\mathbf{B}} \times \left(\vec{\nabla} \times \vec{\mathbf{m}}\right) + \vec{\mathbf{m}} \times \left(\vec{\nabla} \times \vec{\mathbf{B}}\right) + \left(\vec{\mathbf{B}} \cdot \vec{\nabla}\right)\vec{\mathbf{m}} + \left(\vec{\mathbf{m}} \cdot \vec{\nabla}\right)\vec{\mathbf{B}} = \left(\vec{\mathbf{m}} \cdot \vec{\nabla}\right)\vec{\mathbf{B}} \quad [5.118]$$

because $\vec{\mathbf{m}}$ does not depend on position and the sources of $\vec{\mathbf{B}}$ are far from the magnetic moment. Clearly, if $\vec{\mathbf{B}}$ is uniform, the force $\vec{\mathbf{F}}$ is zero. We conclude from this, according to equation [5.118], that the force applied by an external magnetic field $\vec{\mathbf{B}}$ to a magnetic moment derives from the potential energy:

$$W_p = -\vec{\mathbf{m}} \cdot \vec{\mathbf{B}} \quad [5.119]$$

The force is therefore conservative.

5.3.5. *Torque acting on the circuit*

The torque applied to the circuit, as seen from point P, is equal to:

$$\vec{\mathbf{N}}_{/P} = \frac{\overrightarrow{\mathbf{PQ}}}{2} \times \left[I\overrightarrow{\mathbf{PQ}} \times \vec{\mathbf{B}}\left(x + \frac{\Delta x}{2}, y, z\right)\right] + \frac{\overrightarrow{\mathbf{PS}}}{2} \times \left[I\overrightarrow{\mathbf{SP}} \times \vec{\mathbf{B}}\left(x, y + \frac{\Delta y}{2}, z\right)\right] +$$

$$\left[\overrightarrow{\mathbf{PQ}} + \frac{\overrightarrow{\mathbf{QR}}}{2}\right] \times \left[I\overrightarrow{\mathbf{QR}} \times \overrightarrow{\mathbf{B}}\left(x + \Delta x, y + \frac{\Delta y}{2}, z\right)\right] + \\ \left[\overrightarrow{\mathbf{PS}} - \frac{\overrightarrow{\mathbf{RS}}}{2}\right] \times \left[I\overrightarrow{\mathbf{RS}} \times \overrightarrow{\mathbf{B}}\left(x + \frac{\Delta x}{2}, y + \Delta y, z\right)\right] \tag{5.120}$$

Since $\overrightarrow{\mathbf{RS}} = -\overrightarrow{\mathbf{PQ}}$ and $\overrightarrow{\mathbf{QR}} = \overrightarrow{\mathbf{PS}}$, the previous equation reduces to:

$$\begin{aligned} \overrightarrow{\mathbf{N}}_{/P} \simeq & \frac{\overrightarrow{\mathbf{PQ}}}{2} \times \left[-I\overrightarrow{\mathbf{PQ}} \times \frac{\partial \overrightarrow{\mathbf{B}}}{\partial y}\Delta y + \cdots\right] + \\ & \frac{\overrightarrow{\mathbf{PS}}}{2} \times \left[I\overrightarrow{\mathbf{PS}} \times \frac{\partial \overrightarrow{\mathbf{B}}}{\partial x}\Delta x + \cdots\right] + \\ & \overrightarrow{\mathbf{PQ}} \times \left[I\overrightarrow{\mathbf{PS}} \times \overrightarrow{\mathbf{B}}\left(x + \Delta x, y + \frac{\Delta y}{2}, z\right)\right] + \\ & \overrightarrow{\mathbf{PS}} \times \left[-I\overrightarrow{\mathbf{PQ}} \times \overrightarrow{\mathbf{B}}\left(x + \frac{\Delta x}{2}, y + \Delta y, z\right)\right] \end{aligned} \tag{5.121}$$

At the order of $\Delta x \Delta y$, there remains:

$$\overrightarrow{\mathbf{N}}_{/P} = I\Delta x\,\overrightarrow{\mathbf{i}} \times \left(\Delta y\,\overrightarrow{\mathbf{j}} \times \overrightarrow{\mathbf{B}}\right) - I\Delta y\,\overrightarrow{\mathbf{j}} \times \left(\Delta x\,\overrightarrow{\mathbf{i}} \times \overrightarrow{\mathbf{B}}\right) = m_z\left(B_x\overrightarrow{\mathbf{j}} - B_y\overrightarrow{\mathbf{i}}\right) \tag{5.122}$$

If the circuit lies in the Oyz plane (respectively, the Ozx plane), then $\overrightarrow{\mathbf{N}}_{/P} = m_x\left(B_y\overrightarrow{\mathbf{k}} - B_z\overrightarrow{\mathbf{j}}\right)$ (respectively, $\overrightarrow{\mathbf{N}}_{/P} = m_y\left(B_z\overrightarrow{\mathbf{i}} - B_x\overrightarrow{\mathbf{k}}\right)$). Thus, the total torque acting on a circuit oriented arbitrarily in space is:

$$\begin{aligned} \overrightarrow{\mathbf{N}}_{/P} = & (m_yB_z - m_zB_y)\,\overrightarrow{\mathbf{i}} + (m_zB_x - m_xB_z)\,\overrightarrow{\mathbf{j}} + \\ & (m_xB_y - m_yB_x)\,\overrightarrow{\mathbf{k}} = \overrightarrow{\mathbf{m}} \times \overrightarrow{\mathbf{B}} \end{aligned} \tag{5.123}$$

5.3.6. *Gyromagnetic ratio*

The magnetic moment of a circuit is closely related to the angular momentum associated with the motion of the charge carriers present in the circuit. Consider an infinitesimal volume element $d\tau'$ around a point $\overrightarrow{\mathbf{r}}'$, which contains charge carriers assumed to be identical. The total momentum (relative to the circuit) of the charge carriers contained in $d\tau'$ is given by:

$$d\overrightarrow{\mathbf{p}}' = n\left(\overrightarrow{\mathbf{r}}'\right) m\overrightarrow{\mathbf{v}}' d\tau' \tag{5.124}$$

where m and $n\left(\vec{\mathbf{r}}'\right)$ are, respectively, the mass and concentration of charges. Here, $\vec{\mathbf{v}}'$ is the average velocity of the charge carriers. This same average velocity, also called the *drift velocity*, is used to define the current density vector: $\vec{\mathbf{J}}\left(\vec{\mathbf{r}}'\right) = n\left(\vec{\mathbf{r}}'\right) q\vec{\mathbf{v}}'$. We can therefore write that:

$$d\vec{\mathbf{p}}' = \frac{m}{q}\vec{\mathbf{J}}\left(\vec{\mathbf{r}}'\right) d\tau' \qquad [5.125]$$

If all charge carriers in volume $d\tau'$ are considered as a single particle of mass $mn\left(\vec{\mathbf{r}}'\right)$, the following angular momentum (relative to the circuit) will be assigned to this particle:

$$d\vec{\mathbf{L}}' = m\left(\vec{\mathbf{r}}'\right) d\tau'\vec{\mathbf{r}}' \times \vec{\mathbf{v}} = \frac{m}{q}\vec{\mathbf{r}}' \times \vec{\mathbf{J}}\left(\vec{\mathbf{r}}'\right) d\tau' \qquad [5.126]$$

It therefore follows that we can associate to the entire circuit of volume τ' an angular momentum equal to:

$$\vec{\mathbf{L}}' = m\int_{\tau'}\left(\vec{\mathbf{r}}'\right) d\tau'\vec{\mathbf{r}}' \times \vec{\mathbf{v}} = \frac{m}{q}\int_{\tau'}\vec{\mathbf{r}}' \times \vec{\mathbf{J}}\left(\vec{\mathbf{r}}'\right) d\tau' \qquad [5.127]$$

By comparing this expression with that of the magnetic moment of the circuit, we deduce:

$$\vec{\mathbf{m}}' = \gamma\vec{\mathbf{L}}' \qquad [5.128]$$

where the parameter:

$$\gamma = \frac{q}{2m} \qquad [5.129]$$

is called the *gyromagnetic ratio*.

EXAMPLE 5.13. (Rotating Sphere).– The calculation of the total angular momentum, $\vec{\mathbf{L}}$, of the sphere relative to its center O is almost the same as in Example 5.11. We simply replace Q by $2M$. This yields:

$$\vec{\mathbf{L}}' - \frac{2}{5}\omega MR^2\vec{\mathbf{k}} \qquad [5.130]$$

where M is the total mass of the sphere. This gives a gyromagnetic ratio equal to $\frac{Q}{2M}$.

5.3.7. *Magnetic moment of the hydrogen atom*

We use the following classical model: the hydrogen atom consists of a proton, assumed immobile at the coordinate origin, around which gravitates an electron in a closed trajectory around O. On the one hand, the electron's angular momentum relative to O is given by the expression:

$$\overrightarrow{\mathbf{L}}_e = m_e \overrightarrow{\mathbf{r}} \times \overrightarrow{\mathbf{v}} \qquad [5.131]$$

where $\overrightarrow{\mathbf{r}}$ and $\overrightarrow{\mathbf{v}}$ are, respectively, the position and velocity of the electron. On the other hand, the electron's motion along the closed trajectory will give the hydrogen atom a magnetic moment equal to:

$$\overrightarrow{\mathbf{m}} = -\frac{e}{2} \overrightarrow{\mathbf{r}} \times \overrightarrow{\mathbf{v}} \qquad [5.132]$$

By comparing expressions [5.131] and [5.132], it is found that:

$$\overrightarrow{\mathbf{m}} = -\frac{e}{2m_e} \overrightarrow{\mathbf{L}}_e \qquad [5.133]$$

The magnetism induced by this motion is called *orbital magnetism.* Concerning orbital magnetism, quantum mechanics also predicts this relationship between angular momentum and magnetic moment with the same gyromagnetic ratio as in the classical model. Quantum mechanics also predicts the existence, for almost all particles, of an intrinsic angular momentum, called *spin,* which has no equivalent in classical theory. The spin angular momentum, denoted by $\overrightarrow{\mathbf{s}}$, confers the electron a magnetic moment $\overrightarrow{\mu}$ equal to:

$$\overrightarrow{\mu} = -\frac{e}{m_e} \overrightarrow{\mathbf{s}} \qquad [5.134]$$

The gyromagnetic ratio here is twice that calculated in classical theory. This type of magnetism is called *spin magnetism.* Note that the projection of spin on any axis is quantized and proportional to *Planck's constant,* that is, $s_z = \mathcal{M}\hbar$, where $\mathcal{M}$ can only take one of the values $2s+1$ $\mathcal{M} = -s, -s+1, \ldots, s-1, s$. Here, s is an integer for bosons and a half-integer for fermions. The modulus of spin angular momentum is equal to $\left|\overrightarrow{\mathbf{s}}\right| = \hbar\sqrt{s(s+1)}$. Thus, the projection of the spin magnetic moment is quantized as follows:

$$\mu_z = -\frac{e\hbar}{m_e}\mathcal{M} \qquad [5.135]$$

where $\frac{e\hbar}{m_e}$ is the *Bohr magneton.*

Chapter 6

Boundary Problems for Static Potentials in Vacuum

6.1. Introduction

6.1.1. *Laplace equation*

So far, we have assumed that the distributions of charges and currents are known, allowing us to use the Coulomb or Biot and Savart laws to obtain expressions for fields or potentials. In reality, the distribution of charges is not known a priori in most cases. For example, it is possible to know the total charge on the two plates of a planar capacitor, but the problem of non-uniform charge density remains a mystery. Even when most of the charge density is known, the densities induced on other nearby conductors will generally contribute to the field. In this case, potentials or fields must be deduced from boundary conditions, such as the requirement that the electric field is everywhere perpendicular to the conductor surface or, alternatively, that the conductor surface is equipotential. Finally, the unknown charge density may be calculated from the field at the surface of the conductor.

In this chapter, we will study the static fields created by charges or currents located entirely outside or on the boundaries of a study region, noted here Σ. Since the fields are independent of time, this means that the curl of $\vec{\mathbf{E}}$ is zero, that is, $\vec{\nabla} \times \vec{\mathbf{E}} = 0$. Due to the absence of currents within Σ, the circulation of $\vec{\mathbf{B}}$, along any closed circuit, is zero, that is, $\vec{\nabla} \times \vec{\mathbf{B}} = 0$. The fields $\vec{\mathbf{E}}$ and $\vec{\mathbf{B}}$ can therefore be expressed as the gradient of a potential V. The absence of sources inside Σ also means that the field divergences in this region are zero (the relation $\vec{\nabla} \cdot \vec{\mathbf{B}} = 0$ is always verified). In these conditions, as we already know, the potential V satisfies the Laplace equation:

$$\Delta V = 0 \qquad [6.1]$$

The solutions of [6.1], called the *harmonic functions*, must satisfy the boundary conditions imposed by the problem under study. These functions play an important role in physics. We will consider only continuous and one-to-one functions here, as well as their first and second partial derivatives. Harmonic functions have the following properties:

1) Any linear combination $\lambda_1 V_1 + \cdots + \lambda_n V_n$ of harmonic functions $V_1, \ldots, V_n$, where $\lambda_1, \ldots, \lambda_n$ are constant, is also a harmonic function.

2) Over any closed surface S enclosing a volume τ (within Σ), we write:

$$\oint_S \frac{\partial V}{\partial n} dS = \int_\tau \Delta V d\tau = 0 \qquad [6.2]$$

where $\frac{\partial V}{\partial n}$ is the derivative along the exterior normal to the surface.

3) If the harmonic function V, its first and second derivatives are continuous in Σ, then it cannot have any isolated maximum or minimum in Σ. Let us assume that V admits a maximum at a point M. We surround M with a sphere of very small radius. In this case, the expression $\oint_{\text{sphere}} \frac{\partial V}{\partial n} dS$ will be negative; this contradicts [6.2]. Consequently, V has no isolated maximum or minimum inside Σ. We can conclude that if V represents the electrostatic potential, V has no maximum or minimum outside these charges. Therefore, there is no stable equilibrium position outside these charges.

4) To determine V unambiguously from [6.1], we must know the potential over a particular area S. The boundary problem for equation [6.1] is formulated as follows: let us find a function $V\left(\vec{\mathbf{r}}\right)$ that satisfies equation [6.1] in a region Σ and which takes prescribed values over the region S (limiting Σ).

$$V|_{\vec{\mathbf{r}} \in S} = \Theta\left(\vec{\mathbf{r}}\right) \qquad [6.3]$$

This problem is called the *Dirichlet problem*. In certain cases, equation [6.3] is replaced by:

$$\left.\frac{\partial V}{\partial n}\right|_{\vec{\mathbf{r}} \in S} = \Psi\left(\vec{\mathbf{r}}\right) \qquad [6.4]$$

The problem of finding the solution to equation [6.1] that satisfies boundary condition [6.4] is called the *Neumann problem*. Of course, it is perfectly possible to have mixed boundary conditions with V specified on one part and the normal derivative on another part.

6.1.2. *Uniqueness theorem*

Before we begin the detailed study of the Laplace equation [6.1], let us first prove an important theorem, called the *uniqueness theorem*, which states that the solution of

[6.1] satisfying the Dirichlet or Neumann (or mixed) boundary conditions is unique. This means that regardless of how a solution is obtained, if $V\left(\overrightarrow{\mathbf{r}}\right)$ satisfies [6.1] with the boundary conditions [6.3] and/or [6.4], then it is the correct solution. To demonstrate this, let V_1 and V_2 be two solutions of [6.1]. Each solution satisfies the boundary conditions that are imposed. It is clear that the difference $V = V_1 - V_2$ also satisfies equation $\Delta V = 0$ and additionally satisfies $V|_{\overrightarrow{\mathbf{r}} \in S} = 0$ or $\frac{\partial V}{\partial n}|_{\overrightarrow{\mathbf{r}} \in S} = 0$ (depending on the prescribed boundary conditions). In both cases, the product of V and $\left(\overrightarrow{\nabla} V\right)_n$ is zero on the boundary S. Therefore, using the divergence theorem, it is possible to write:

$$\oint_S V\overrightarrow{\nabla}V \cdot d\overrightarrow{\mathbf{S}} = \int_\tau \overrightarrow{\nabla} \cdot \left(V\overrightarrow{\nabla}V\right) d\tau = \int_\tau \left[\left(\overrightarrow{\nabla}V\right)^2 + V\Delta V\right] d\tau = \int_\tau \left(\overrightarrow{\nabla}V\right)^2 d\tau \tag{6.5}$$

Since $\left(\overrightarrow{\nabla} V\right)^2$ is not negative anywhere, the above integral will not vanish unless $\overrightarrow{\nabla} V = 0$. This implies that V is constant. For the Dirichlet problem, V is zero on the boundary and, consequently, it must be zero everywhere. We can therefore conclude that $V_1 = V_2$. For the Neumann problem, V_1 and V_2 differ only by a constant, because $\frac{\partial V}{\partial n}$ is zero on the boundary. This difference is of no physical significance because the fields obtained by V_1 and V_2 are identical. We deduce that the solution to the Laplace equation is unique.

6.2. 1D Laplace equation

When the system is translation-invariant in all directions except one, Δ is reduced to the ordinary one-dimensional differential operator. Although we are dealing with single-variable differential equations, the problem is still a single-variable 3D space problem. In the most commonly used coordinate systems, it is easy to find the operators and the corresponding solutions:

– *Cartesian coordinates*: if V depends on only one variable, for example x, we write $\frac{d^2V}{dx^2} = 0$ and, consequently:

$$V(x) = Ax + B \tag{6.6}$$

– *Cylindrical coordinates*: if V depends only on r, we obtain $\frac{1}{r}\frac{d}{dr}\left(r\frac{dV}{dr}\right) = 0$ and therefore:

$$V(r) = A\ln r + B \tag{6.7}$$

– *Spherical coordinates*: if V depends only on r, then $\frac{1}{r^2}\frac{d}{dr}\left(r^2\frac{dV}{dr}\right) = 0$ and therefore:

$$V(r) = \frac{A}{r} + B \qquad [6.8]$$

In all three cases, we apply the boundary conditions to find A and B. For $A \neq 0$, the solutions [6.6]–[6.8] diverge, respectively, for $x \longrightarrow \infty$, $r \longrightarrow 0$ and $r \longrightarrow 0$.

EXAMPLE 6.1. (Potential between Two Spherical Electrodes)].– Let us find the potential between two concentric spherical electrodes with radii R_1 and R_2, with the potentials V_1 and V_2. At the boundaries, equation [6.8] gives $\frac{A}{R_1} + B = V_1$ and $\frac{A}{R_2} + B = V_2$. Consequently, we obtain $A = \frac{R_1 R_2 (V_1 - V_2)}{R_2 - R_1}$ and $B = \frac{R_1 V_1 - R_2 V_2}{R_1 - R_2}$.

6.3. 2D Laplace equation

We will introduce a few methods for solving the two-dimensional Laplace equation. Particular attention is paid to the use of boundary conditions. 2D covers, for example:

– Cartesian problems, where there is no dependence on z (problems in the plane);

– cylindrical problems, where again there is no variation with z (polar problems);

– spherical problems, where there is no dependence on the azimuthal angle, etc.

One of the methods used to reduce a partial derivative equation with multiple variables to a set of ordinary differential equations is the *separation of variables*. The task is to find a solution V of [6.1], expressed as a product of functions, each of which depends on just one of the coordinates. Thus, we obtain a differential equation for each of the basic functions. If the set of basic functions for each variable forms a complete basis for the functions defined in the space of this variable, then any solution to [6.1] can be expressed as a linear combination whose constants are determined by the boundary conditions.

6.3.1. *2D Cartesian coordinates*

Let us assume that in a region in space (without any charges inside it), the solutions do not depend on the coordinate z (translation-invariant problem along Oz). In this case, the Laplace equation [6.1] is reduced to:

$$\frac{\partial^2 V}{\partial x^2} + \frac{\partial^2 V}{\partial y^2} = 0 \qquad [6.9]$$

Let us find a solution in the form of the product of a function with only an x-coordinate and another function with only a y-coordinate: $V(x,y) = X(x)\,Y(y)$. It is convenient to use this solution if the system's limits lie in $x =$ const and/or $y =$ const planes. Inserting this solution into [6.9], we obtain $Y\frac{d^2X}{dx^2} + X\frac{d^2Y}{dy^2} = 0$, or again $\frac{1}{X}\frac{d^2X}{dx^2} = -\frac{1}{Y}\frac{d^2Y}{dy^2}$. Partial derivatives thus become total derivatives since each function depends on just one coordinate. As the first ratio depends only on x and the second only on y, the last equation, identically verified for all x and y, can only occur if these ratios do not depend on the corresponding arguments, that is, if they are constant numbers. It therefore splits into two equations: $\frac{1}{X}\frac{d^2X}{dx^2} = -\mathsf{k}^2$ and $\frac{1}{Y}\frac{d^2Y}{dy^2} = \mathsf{k}^2$, where k^2 is called the *separation constant* and can generally be a complex number. These equations can be rewritten in the form:

$$\begin{aligned}\frac{d^2X}{dx^2} + \mathsf{k}^2X = 0\\ \frac{d^2Y}{dy^2} - \mathsf{k}^2Y = 0\end{aligned} \qquad [6.10]$$

6.3.1.1. *Solutions with zero separation constant,* $\mathsf{k}^2 = 0$

When $\mathsf{k}^2 = 0$, the solutions to [6.10] are $X(x) = ax + b$ and $Y(y) = cy + d$, where a, b, c and d are constants. The potential is then given by the product of these solutions, that is:

$$V(x,y) = A + Bx + Cy + Dxy \qquad [6.11]$$

EXAMPLE 6.2. (Hyperbolic Electrode).– An electrode of hyperbolic form, with equation $xy = ab$, is placed above a right-angle earth connection. This electrode is held at potential V_0, as shown in Figure 6.1. The potential $V(x,y)$ must satisfy the boundary conditions: $V(x=0) = 0$, $V(y=0) = 0$ and $V(xy=ab) = V_0$. By imposing these conditions on solution [6.11], it can be derived that:

$$V(x,y) = V_0\frac{xy}{ab} \qquad [6.12]$$

Consequently, the induced electric field will have the value:

$$\vec{\mathbf{E}} = -\vec{\nabla}V = -\frac{V_0}{ab}\left[y\,\vec{\mathbf{i}} + x\,\vec{\mathbf{j}}\right] \qquad [6.13]$$

The field lines depicted in Figure 6.1 are hyperbolas (perpendicular to equipotential lines [6.12]) defined by: $\frac{dy}{dx} = \frac{E_y}{E_x} = \frac{x}{y}$. After integration, this gives:

$$y^2 - x^2 = \text{const} \qquad [6.14]$$

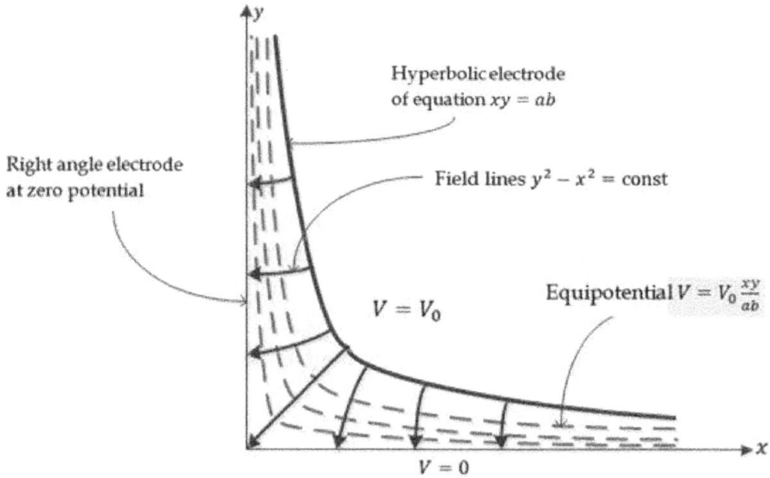

Figure 6.1. *Equipotential and field lines for an electrode with hyperbolic geometry at the potential V_0 above a square corner electrode connected to the earth. For a color version of this figure, see www.iste.co.uk/abdesselam/electrodynamics1.zip*

EXAMPLE 6.3. (Resistance in an Open Box).– A resistive medium is contained between two electrodes, one of which (on the right) extends above and makes a right angle (square corner), as shown in Figure 6.2. Let us try the solutions with a zero separation constant, given by [6.11], in each region delimited by the electrodes:

$$V(x,y) = \begin{cases} A_1 + B_1x + C_1y + D_1xy & 0 \le y \le d \\ A_2 + B_2x + C_2y + D_2xy & d < y \le s \end{cases} \qquad [6.15]$$

Since the potential is constant at the electrodes and continuous across the interface, we obtain:

$$V(x=0, 0 \le y \le d) = A_1 + C_1y = V_0 \Longrightarrow \begin{cases} A_1 = V_0 \\ C_1 = 0 \end{cases}$$

$$V(x=l, 0 \le y \le d) = A_1 + B_1l + C_1y + D_1ly = 0 \Longrightarrow \begin{cases} B_1 = -\dfrac{A_1}{l} = -\dfrac{V_0}{l} \\ D_1 = 0 \end{cases}$$

$$V(x=l, d < y \le s) = A_2 + B_2l + C_2y + D_2ly = 0 \Longrightarrow \begin{cases} B_2 = -\dfrac{A_2}{l} \\ D_2 = -\dfrac{C_2}{l} \end{cases}$$

$$V(x, y = s) = 0 = A_2 + B_2 x + C_2 s + D_2 s x \Longrightarrow \begin{cases} C_2 = -\dfrac{A_2}{s} \\ D_2 = -\dfrac{B_2}{s} = \dfrac{A_2}{sl} \end{cases}$$

$$V\left(x, y = d^+\right) = V\left(x, y = d^-\right) = A_2\left(1 - \frac{x}{l} - \frac{d}{s} + \frac{xd}{sl}\right)$$

$$= V_0 - \frac{V_0}{l}x \Longrightarrow A_2 = \frac{V_0}{1 - \frac{d}{s}} \qquad [6.16]$$

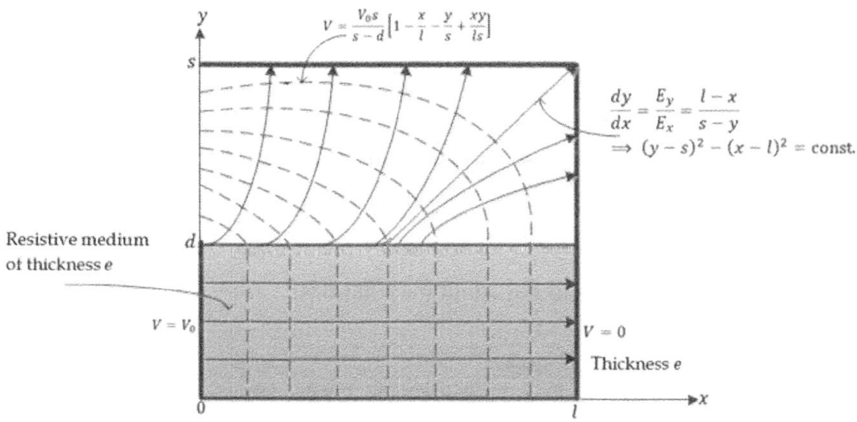

Figure 6.2. *Resistive medium in an open box. For a color version of this figure, see www.iste.co.uk/abdesselam/electrodynamics1.zip*

Therefore, the potential is:

$$V(x, y) = \begin{cases} V_0\left(1 - \dfrac{x}{l}\right) & 0 \leq y \leq d \\ \dfrac{V_0 s}{s - d}\left[1 - \dfrac{x}{l} - \dfrac{y}{s} + \dfrac{xy}{ls}\right] & d < y \leq s \end{cases} \qquad [6.17]$$

with the associated electric field:

$$\vec{\mathbf{E}} = -\vec{\nabla} V = \begin{cases} \dfrac{V_0}{l}\vec{\mathbf{i}} & 0 \leq y \leq d \\ \dfrac{V_0 s}{s - d}\left[\dfrac{\vec{\mathbf{i}}}{l}\left(1 - \dfrac{y}{s}\right) + \dfrac{\vec{\mathbf{j}}}{s}\left(1 - \dfrac{x}{l}\right)\right] & d < y \leq s \end{cases} \qquad [6.18]$$

The electric field is purely tangential at the interfaces $y = 0$ and $y = d^-$. Consequently, no current can pass through these interfaces. The surface charge density over the $y = d$ interface is therefore due only to the normal field above it (below the interface, the normal field is zero):

$$\begin{aligned}\sigma(x, y = d) &= \varepsilon_0 E_y(x, y = d_+) - \varepsilon_0 E_y(x, y = d_-) = \varepsilon_0 E_y(x, y = d_+) \\ &= \frac{\varepsilon_0 V_0}{s-d}\left(1 - \frac{x}{l}\right)\end{aligned} \qquad [6.19]$$

This surface charge is subject to a total shear force equal to:

$$F_x = \int_0^l \sigma E_x(x, y = d)\, e dx = \frac{\varepsilon_0 V_0^2 e}{2(s-d)} \qquad [6.20]$$

where e is the thickness of the resistive medium. If the medium is a liquid, this shear force can be used to pump the fluid.

6.3.1.2. *Solutions with non-zero separation constant,* $\mathrm{k}^2 \neq 0$

When $\mathrm{k}^2 \neq 0$, equations [6.10] admit the solutions:

$$\begin{aligned}X(x) &= A\cos \mathrm{k}x + B\sin \mathrm{k}x \\ Y(y) &= C\cosh \mathrm{k}y + D\sinh \mathrm{k}y = C'e^{\mathrm{k}y} + D'e^{-\mathrm{k}y}\end{aligned} \qquad [6.21]$$

If $\mathrm{k}^2 > 0$ (k is real), the solutions of X are trigonometric and those of Y hyperbolic or exponential. On the other hand, if $\mathrm{k}^2 < 0$ (k is purely imaginary), X becomes hyperbolic (or exponential) and Y trigonometric. To obtain the second solutions, it is sufficient to replace k by ik in [6.21]. The solution of the potential is then given by the product of X and Y:

$$\begin{aligned}V(x, y) &= (A\cos \mathrm{k}x + B\sin \mathrm{k}x)(C\cosh \mathrm{k}y + D\sinh \mathrm{k}y) \\ &= (A\cos \mathrm{k}x + B\sin \mathrm{k}x)\left(C'e^{\mathrm{k}y} + D'e^{-\mathrm{k}y}\right)\end{aligned} \qquad [6.22]$$

For $k^2 > 0$, we must take $V(x, y) = (A\cosh \mathrm{k}x + B\sinh \mathrm{k}x)(C\cos \mathrm{k}y + D\sin \mathrm{k}y) = \left(A'e^{\mathrm{k}x} + B'e^{-\mathrm{k}x}\right)(C\cos \mathrm{k}y + D\sin \mathrm{k}y)$ instead of [6.22]. The solutions of the Laplace equation are thus oscillating in one direction and attenuated in a perpendicular direction. To obtain the most general solution of the Laplace equation, a solution of the type in [6.11] must be added to [6.22], because of linearity. The values of the coefficients A, B, C, D and sfk are determined by the boundary conditions. When the region under study is of infinite extent in the direction of the y coordinates, it is often convenient to use the exponential solutions of [6.22], since these solutions must tend toward zero as y approaches $\pm\infty$. For a finite study region, it is generally more practical to use the hyperbolic expressions of [6.22].

6.3.1.3. *General method of resolution*

The aim here is to find the potential in a box with dimensions $a \times b$, which extends infinitely in the direction Oz. The potentials on faces 1, 2, 3 and 4 are $\Theta_1(y)$, $\Theta_2(x)$, $\Theta_3(y)$ and $\Theta_4(x)$, respectively, as shown in Figure 6.3. Equation $\Delta V = 0$ can be solved as follows:

– We temporarily fix $\Theta_1(y) = \Theta_2(x) = \Theta_3(y) = 0$ and only retain $\Theta_4(x)$.

– We solve the equation $\Delta V_4 = 0$, with the boundary conditions $V_4(x, 0) \equiv \Theta_4(x)$.

– We repeat the procedure for $\Theta_1 = \Theta_2 = \Theta_4 = 0$ and only retain $\Theta_3(y)$. We proceed to solve the equation $\Delta V_3 = 0$, with the boundary conditions $V_3(a, y) \equiv \Theta_3(y)$.

– We continue in the same way to obtain $V_1(x, y)$ et $V_2(x, y)$. The boundary conditions are, respectively, $V_1(0, y) \equiv \Theta_1(y)$ and $V_2(x, b) \equiv \Theta_2(x)$.

– The sum $V = V_1 + V_2 + V_3 + V_4$ clearly satisfies the equation $\Delta V = 0$ and also verifies the four boundary conditions simultaneously.

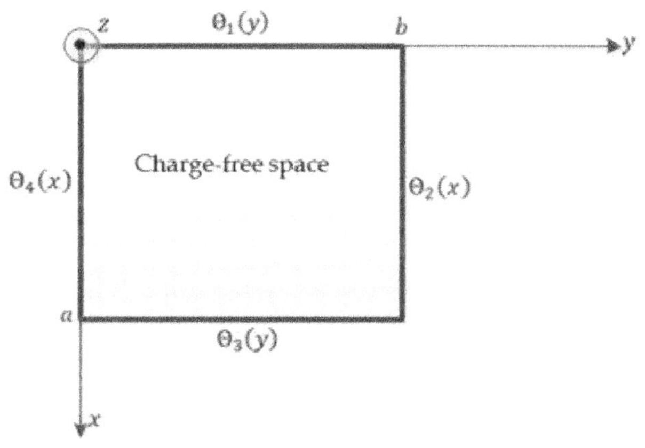

Figure 6.3. *Solve the Laplace equation for an open rectangular box of infinite extent in the z direction and area $a \times b$ in the Oxy plane, with potentials $\Theta_1(y)$, $\Theta_2(x)$, $\Theta_3(y)$ and $\Theta_4(x)$ on all four faces*

Let us try for V_4, $V_4(x, y) = (A \cos kx + B \sin kx)(C \cosh ky + D \sinh ky)$. The boundary conditions $V_4(0, y) = V_4(x, b) = V_4(a, y) = 0$ immediately give:

$$\begin{aligned} &A(C \cosh ky + D \sinh ky) = 0 \\ &(A \cos kx + B \sin kx)(C \cosh kb + D \sinh kb) = 0 \\ &(A \cos ka + B \sin ka)(C \cosh ky + D \sinh ky) = 0 \end{aligned} \qquad [6.23]$$

This means that $A = 0$ and $D = -C\frac{\cosh \mathrm{k}b}{\sinh \mathrm{k}b}$. To have $B \neq 0$ (non-trivial solution), we must choose $\mathrm{k}a = n\pi$, that is, $\mathrm{k} = \frac{n\pi}{a}$ with $n \geq 1$. Consequently, we obtain $V_4(x,y) = \nu_n \sin\frac{n\pi x}{a}\left(\cosh\frac{n\pi y}{a} - \frac{\cosh\frac{n\pi b}{a}}{\sinh\frac{n\pi b}{a}}\sinh\frac{n\pi y}{a}\right) = \nu_n \sin\frac{n\pi x}{a}\frac{\sinh\frac{n\pi(b-y)}{a}}{\sinh\frac{n\pi b}{a}}$. To have the most general solution, we need to sum over all possible values n, $V_4(x,y) = \sum_{n=1}^{\infty}\nu_n \sin\frac{n\pi x}{a}\frac{\sinh\frac{n\pi(b-y)}{a}}{\sinh\frac{n\pi b}{a}}$, because of linearity, where μ_n are constants. Taking $y = 0$, we obtain $V_4(x,0) = \sum_{n=1}^{\infty}\nu_n \sin\frac{n\pi x}{a} \equiv \Theta_4(x)$. This is the Fourier expansion of the function $\Theta_4(x)$. The coefficients ν_n are obtained from Fourier formulas: $\nu_n = \frac{2}{a}\int_0^a \Theta_4(x')\sin\frac{n\pi x'}{a}dx'$. Substituting these coefficients into the expression for $V_4(x,y)$, we find:

$$V_4(x,y) = \sum_{n=1}^{\infty}\frac{2}{a}\left[\int_0^a \Theta_4(x')\sin\frac{n\pi x'}{a}dx'\right]\sin\frac{n\pi x}{a}\frac{\sinh\frac{n\pi(b-y)}{a}}{\sinh\frac{n\pi b}{a}} \qquad [6.24]$$

By applying the same boundary conditions to the second choice of solution, that is, $V_4(x,y) = (A\cosh \mathrm{k}x + B\sinh \mathrm{k}x)(C\cos ky + D\sin \mathrm{k}y)$, the constraints in [6.23] must be replaced by:

$$\begin{aligned}
&A(C\cos \mathrm{k}y + D\sin \mathrm{k}y) = 0\\
&(A\cosh \mathrm{k}x + B\sinh \mathrm{k}x)(C\cos \mathrm{k}b + D\sin \mathrm{k}b) = 0\\
&(A\cosh \mathrm{k}a + B\sinh \mathrm{k}a)(C\cos \mathrm{k}y + D\sin \mathrm{k}y) = 0
\end{aligned} \qquad [6.25]$$

Thus, we find $A = 0$ and $D = -C\frac{\cos \mathrm{k}b}{\sin \mathrm{k}b}$. To have $B \neq 0$, we must take $\mathrm{k} = \frac{in\pi}{a}$. Finally, it can be derived from this that $V_4(x,y) = \nu_n \sin\frac{n\pi x}{a}\left[\cosh\frac{n\pi y}{a} - \frac{\cosh\frac{n\pi b}{a}}{\sinh\frac{n\pi b}{a}}\sinh\frac{n\pi y}{a}\right]$. So, we conclude that this second choice also leads us to the same solution.

By following the same reasoning, we obtain:

$$\begin{aligned}
V_1(x,y) &= \sum_{n=1}^{\infty}\frac{2}{b}\left[\int_0^b \Theta_1(y')\sin\frac{n\pi y'}{b}dy'\right]\sin\frac{n\pi y}{b}\frac{\sinh\frac{n\pi(a-x)}{b}}{\sinh\frac{n\pi a}{b}}\\
V_2(x,y) &= \sum_{n=1}^{\infty}\frac{2}{a}\left[\int_0^a \Theta_2(x')\sin\frac{n\pi x'}{a}dx'\right]\sin\frac{n\pi x}{a}\frac{\sinh\frac{n\pi y}{a}}{\sinh\frac{n\pi b}{a}}\\
V_3(x,y) &= \sum_{n=1}^{\infty}\frac{2}{b}\left[\int_0^b \Theta_3(y')\sin\frac{n\pi y'}{a}dy'\right]\sin\frac{n\pi y}{b}\frac{\sinh\frac{n\pi x}{b}}{\sinh\frac{n\pi a}{b}}
\end{aligned} \qquad [6.26]$$

and consequently:

$$\begin{aligned} V(x,y) = & \sum_{n=1}^{\infty} \frac{2}{b} \left[\int_0^b \Theta_1(y') \sin \frac{n\pi y'}{b} dy' \right] \sin \frac{n\pi y}{b} \frac{\sinh \frac{n\pi(a-x)}{b}}{\sinh \frac{n\pi a}{b}} + \\ & \sum_{n=1}^{\infty} \frac{2}{a} \left[\int_0^a \Theta_2(x') \sin \frac{n\pi x'}{a} dx' \right] \sin \frac{n\pi x}{a} \frac{\sinh \frac{n\pi y}{a}}{\sinh \frac{n\pi b}{a}} + \\ & \sum_{n=1}^{\infty} \frac{2}{b} \left[\int_0^b \Theta_3(y') \sin \frac{n\pi y'}{a} dy' \right] \sin \frac{n\pi y}{b} \frac{\sinh \frac{n\pi x}{b}}{\sinh \frac{n\pi a}{b}} + \\ & \sum_{n=1}^{\infty} \frac{2}{a} \left[\int_0^a \Theta_4(x') \sin \frac{n\pi x'}{a} dx' \right] \sin \frac{n\pi x}{a} \frac{\sinh \frac{n\pi(b-y)}{a}}{\sinh \frac{n\pi b}{a}} \end{aligned} \qquad [6.27]$$

EXAMPLE 6.4.(Spacial Periodic Excitation).– A conductive sheet in the plane $x = 0$ is kept at the periodic potential $V = V_0 \sin ay$, as shown in Figure 6.4. Let us determine the potential $V(x,y)$ at any point in the space (the system is independent of z). To satisfy the prescribed boundary conditions, we use the solution $V(x,y) = \left(A' e^{\mathsf{k}x} + B' e^{-\mathsf{k}x}\right)(C \cos \mathsf{k}y + D \sin \mathsf{k}y)$, with $\mathsf{k} = a$. Since the potential must vanish far from the source, we will write the solution separately for positive and negative values of x, as follows:

$$V(x,y) = \begin{cases} V_0 \sin ay e^{-ax} & x \geq 0 \\ V_0 \sin ay e^{ax} & x < 0 \end{cases} \qquad [6.28]$$

where the coefficients A', B', C and D have been fixed such that the potential is continuous and corresponds to the excitation introduced at $x = 0$. The electric field is therefore:

$$\vec{\mathbf{E}} = \begin{cases} -V_0 a e^{-ax} \left[-\sin ay \, \vec{\mathbf{i}} + \cos ay \, \vec{\mathbf{j}} \right] & x \geq 0 \\ -V_0 a e^{ax} \left[\sin ay \, \vec{\mathbf{i}} + \cos ay \, \vec{\mathbf{j}} \right] & x < 0 \end{cases} \qquad [6.29]$$

The surface charge density accumulated over the sheet is given by (discontinuity of the component normal to $\vec{\mathbf{E}}$ across the sheet):

$$\sigma(x=0) = \varepsilon_0 \left[E_x\left(0^+\right) - E_x\left(0^-\right) \right] = 2\varepsilon_0 V_0 a \sin ay \qquad [6.30]$$

The field lines depicted in Figure 6.4 obey the equation:

$$\frac{dy}{dx} = \frac{E_y}{E_x} = \begin{cases} -\dfrac{1}{\tan ay} & x \geq 0 \\ \frac{1}{\tan ay} & x < 0 \end{cases} \Longrightarrow \begin{cases} \cos ay e^{-ax} = \text{const.} & x \geq 0 \\ \cos ay e^{ax} = \text{const.} & x < 0 \end{cases}$$

[6.31]

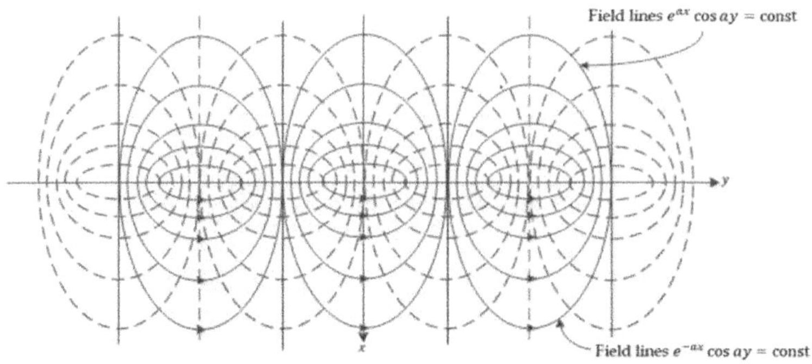

Figure 6.4. *Equipotential and field lines emanating from a conductive sheet with a periodic voltage in space, along the Oy-axis: the field lines start from a positive surface charge and end in a negative surface charge. For a color version of this figure, see www.iste.co.uk/abdesselam/electrodynamics1.zip*

EXAMPLE 6.5. (Rectangular Harmonics).– Let us determine the potential $V(x, y)$ in $2D$ at any point in a region bounded by the lines $y = 0$, $y = b$ and $x = 0$. The potential is zero on the lines $y = 0$ and $y = b$. The $x = 0$ line is held at constant potential V_0, as shown in Figure 6.5. From the above, the general solution is of the form $V(x, y) = \left(Ae^{\mathrm{k}x} + Be^{-\mathrm{k}x}\right)(C\cos \mathrm{k}y + D\sin \mathrm{k}y)$. Since $V(x, y)$ is zero on the lines $y = 0$ and $y = b$, we deduce that $C = 0$ and $sfk = \frac{n\pi}{b}$, since $D \neq 0$ is required to obtain a non-trivial solution. Moreover, $V(x, y)$ must cancel when $x \longrightarrow \infty$; this leads to $A = 0$. Therefore, we finally find, $V(x, y) = \sum_{n=1}^{\infty} B_n e^{-\frac{n\pi x}{b}} \sin\frac{n\pi y}{b}$. For $x = 0$, the previous formula reduces to $V_0 = \sum_{n=1}^{\infty} B_n \sin\frac{n\pi y}{b}$. The B_n coefficients are calculated using Fourier formulas:

$$B_n = \frac{2}{a}\int_0^a V_0 \sin\frac{n\pi y}{b} dy = \begin{cases} 0 & n = 2m \\ \dfrac{4V_0}{\pi(2m+1)} & n = 2m+1 \end{cases} \qquad [6.32]$$

It follows from this that:

$$V(x, y) = \frac{4V_0}{\pi}\sum_{n=0}^{\infty} \frac{e^{-(2n+1)\pi x/b}}{2n+1} \sin\left(\frac{(2n+1)\pi y}{b}\right) \qquad [6.33]$$

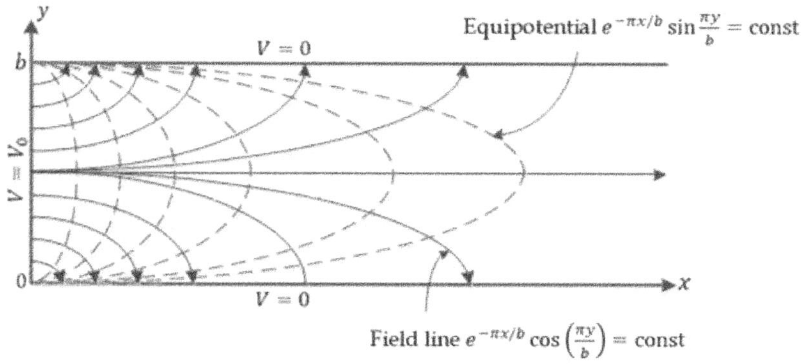

Figure 6.5. *An open box with finite width d in the direction y, stretching infinitely in the directions x and z, has zero potential over all surfaces except the base located in the plane $x = 0$, where $V = V_0$. For a color version of this figure, see www.iste.co.uk/abdesselam/electrodynamics1.zip*

The electric field is therefore:

$$\vec{\mathbf{E}} = -\vec{\nabla}V = -\frac{4V_0}{b}\sum_{n=0}^{\infty}\left[-\sin\left(\frac{(2n+1)\pi y}{b}\right)\vec{\mathbf{i}} + \cos\left(\frac{(2n+1)\pi y}{b}\right)\vec{\mathbf{j}}\right]e^{-(2n+1)\pi x/b} \quad [6.34]$$

Note that for $x \gg b$, the solution is dominated by the first harmonic. The field lines and equipotential lines are depicted in Figure 6.5. The field lines are obtained as follows:

$$\frac{dy}{dx} = \frac{E_y}{E_x} \simeq -\frac{\cos\left(\frac{\pi y}{b}\right)}{\sin\left(\frac{\pi y}{b}\right)} \Longrightarrow e^{-\pi x/b}\cos\left(\frac{\pi y}{b}\right) = \text{const.} \quad [6.35]$$

EXAMPLE 6.6. (Dirichlet Problem in the Plane).– We wish to determine the potential at any point (x, y) in the upper half-plane $\mathcal{P}\,(-\infty < x < \infty, y \geq 0)$. The potential is assumed to be equal to $V_0\,(x)$ along the line $y = 0$, as shown in Figure 6.6(a). Since there is only one boundary condition (on the line $y = 0$) and that $V\,(x, y)$ must disappear when $y \longrightarrow \infty$, we take a solution of the form $e^{-\mathrm{k}y}\,[A\cos \mathrm{k}x + B\sin \mathrm{k}x]$, where A and B are constants. Consequently, the general solution has the value:

$$V\,(x, y) = \int_0^{+\infty} e^{-\mathrm{k}y}\,[A\,(\mathrm{k})\cos \mathrm{k}x + B\,(\mathrm{k})\sin \mathrm{k}x]\,d\mathrm{k} \quad [6.36]$$

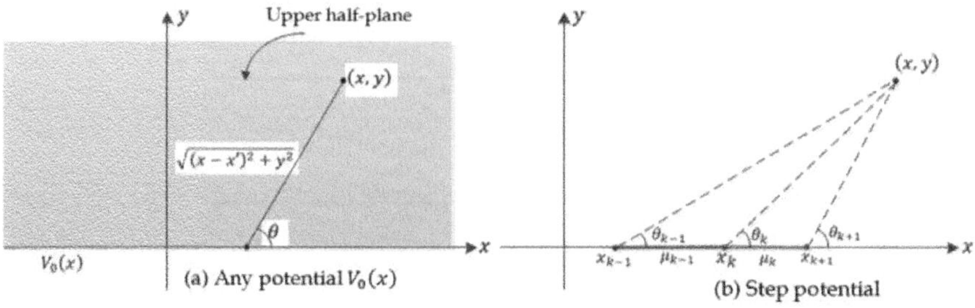

Figure 6.6. *(a) Dirichlet problem for the upper half-plane with any potential along $y=0$. (b) Step potential $V_0(x)=\mu_k$ pour $x_k \leq x \leq x_{k+1}$. For a color version of this figure, see www.iste.co.uk/abdesselam/electrodynamics1.zip*

$A(\mathsf{k})$ and $B(\mathsf{k})$ are chosen so that $V(x,y)$ satisfies the boundary condition $V(x,0)=V_0(x)$, that is, $V_0(x)=\int_0^{+\infty}[A(\mathsf{k})\cos \mathsf{k}x + B(\mathsf{k})\sin \mathsf{k}x]\,d\mathsf{k}$. If $V_0(x)$ is represented by *Fourier integral*, $V_0(x) = \frac{1}{\pi}\int_0^{\infty}\left[\int_{-\infty}^{+\infty} V_0(x')\right.$ $\left.(\cos \mathsf{k}x' \cos \mathsf{k}x + \sin \mathsf{k}x' \times \sin \mathsf{k}x)\,dx'\right]d\mathsf{k}$, it can then be derived from this that $A(\mathsf{k}) = \frac{1}{\pi}\int_{-\infty}^{+\infty} V_0(x')\cos \mathsf{k}x' dx'$ and $B(\mathsf{k}) = \frac{1}{\pi}\int_{-\infty}^{+\infty} V_0(x')\sin \mathsf{k}x' dx'$. By substituting these expressions in [6.36], we obtain:

$$V(x,y)=\frac{1}{\pi}\int_{-\infty}^{+\infty} V_0(x')\left[\int_0^{+\infty} e^{-\mathsf{k}y}\cos \mathsf{k}(x'-x)\,d\mathsf{k}\right]dx' \quad [6.37]$$

To simplify this result, let us calculate the integral appearing between the square brackets:

$$\int_0^{+\infty} e^{-\mathsf{k}y}\cos \mathsf{k}(x'-x)\,d\mathsf{k} = \frac{y}{y^2+(x'-x)^2} \quad [6.38]$$

Therefore, it follows from this that:

$$V(x,y)=\frac{1}{\pi}\int_{-\infty}^{+\infty}\frac{yV_0(x')}{y^2+(x'-x)^2}dx' \quad [6.39]$$

This is the *Poisson formula for a half-plane*. The induced electric field has the value:

$$\vec{\mathbf{E}} = \frac{1}{\pi}\int_{-\infty}^{+\infty}\frac{V_0(x')\left[-2y(x'-x)\,\vec{\mathbf{i}} + \left[y^2-(x'-x)^2\right]\vec{\mathbf{j}}\right]}{\left[y^2+(x'-x)^2\right]^2}dx' \quad [6.40]$$

i) By choosing $V_0(x') = \mu\delta(x')$, we find:

$$V(x,y) = \frac{\mu}{\pi}\frac{y}{x^2+y^2}$$
$$\vec{\mathbf{E}} = \frac{\mu}{\pi}\frac{2xy\,\vec{\mathbf{i}} + (y^2 - x^2)\,\vec{\mathbf{j}}}{\left[x^2+y^2\right]^2} \qquad [6.41]$$

which gives a representation of Dirac distribution $\delta(x) \equiv \lim_{y\longrightarrow 0}\frac{\mu}{\pi}\frac{y}{x^2+y^2}$.

ii) In the case of step potential $V_0(x) = \mu_k$ for $x_k \leq x \leq x_{k+1}$, as in Figure 6.6(b), it comes:

$$V(x,y) = \sum_k \frac{\mu_k y}{\pi}\int_{x_k}^{x_{k+1}} \frac{dx'}{y^2 + (x'-x)^2} = \sum_k \frac{\mu_k}{\pi}\left[\tan^{-1}\frac{x_{k+1}-x}{y} - \tan^{-1}\frac{x_k - x}{y}\right] = \sum_k \frac{\mu_k}{\pi}\left[\tan^{-1}\frac{y}{x_k - x} - \tan^{-1}\frac{y}{x_{k+1}-x}\right]$$
$$= \sum_k \frac{\mu_k}{\pi}\left[\theta_{k+1} - \theta_k\right] \qquad [6.42]$$

6.3.2. *Polar coordinates*

Consider an infinite cylindrical medium which is translation invariant along Oz (i.e. independent of z). Consequently, the Laplace equation takes the following form:

$$\frac{1}{r}\frac{\partial}{\partial r}\left(r\frac{\partial V}{\partial r}\right) + \frac{1}{r^2}\frac{\partial^2 V}{\partial \varphi^2} = 0 \qquad [6.43]$$

This is the *Laplace equation in polar coordinates*. Let us find a solution in the form of the product of two functions: the first depends only on r and the second depends on φ, that is, $V(r,\varphi) = R(r)\,\Phi(\varphi)$. By substituting this expression in [6.43], we obtain $\frac{1}{rR}\frac{d}{dr}\left(r\frac{dR}{dr}\right) + \frac{1}{r^2\Phi}\frac{d^2\Phi}{d\varphi^2} = 0$, or again, $\frac{r}{R}\frac{d}{dr}\left(r\frac{dR}{dr}\right) = -\frac{1}{\Phi}\frac{d^2\Phi}{d\varphi^2} = m^2$, where m is a constant. This can be divided into two differential equations $\frac{d^2\Phi}{d\varphi^2} + m^2\Phi = 0$ and $r\frac{d}{dr}\left(r\frac{dR}{dr}\right) = m^2 R$. The first differential equation admits the solution:

$$\Phi(\varphi) = \begin{cases} A\cos m\varphi + B\sin m\varphi & m \neq 0 \\ A_0\varphi + B_0 & m = 0 \end{cases} \qquad [6.44]$$

where A, A_0 B and B_0 are constants. Since the function $V(r,\varphi)$ must be periodic in φ, because for a value r corresponding to φ and $\varphi + 2\pi$, we must have the same solution V (this is, in fact, the same point on the circle), that is, $\Phi(\varphi + 2\pi) = \Phi(\varphi)$.

It can be concluded that m is an integer. Further, the part $A_0\varphi$ cannot be retained due to the periodicity of Φ. The second differential equation $r\frac{d}{dr}\left(r\frac{dR}{dr}\right) = m^2R$ is easy to resolve. It is clear that its form suggests a monomial solution. Let us try a solution in the form $R(r) \sim r^l$. By injecting this solution into the differential equation, it comes that $l^2R = m^2R$; this gives $l = \pm m$. When $m \neq 0$, $R(r)$ appears as a linear combination of two solutions:

$$R(r) = Cr^m + \frac{D}{r^m} \qquad [6.45]$$

where C and D are constants. For $m = 0$, another solution is obtained. To this end, let us directly integrate the differential equation with $m = 0$, $r\frac{d}{dr}\left(r\frac{dR}{\partial r}\right) = 0$. This implies that $r\frac{dR}{dr} = C_0$, or again $\frac{dR}{dr} = \frac{C_0}{r}$. A second integration gives:

$$R(r) = C_0 \ln r + \frac{D_0}{2} \qquad [6.46]$$

where C_0 and D_0 are constant. Consequently, by virtue of the principle of superposition, the most general solution to the Laplace equation in polar coordinates is:

$$V(r,\varphi) = C_0 \ln r + \frac{D_0}{2} + \sum_{m=1}^{\infty}\left(C_m r^m + \frac{D_m}{r^m}\right)(A_m \cos m\varphi + B_m \sin m\varphi) \quad [6.47]$$

EXAMPLE 6.7. (Inclined Conductive Planes).– Consider two conductive planes (electrodes) that are infinite (in the direction Oz) forming an angle α relative to each other, as shown in Figure 6.7. The upper plane is maintained at the potential V_0, while the lower plane is connected to the ground (i.e. has zero potential). The planes do not touch, but they are close enough that we can neglect the fields at the edge. The electrodes extend from $r = a$ to $r = b$. Let us calculate the approximate capacity per unit length of the structure. To this end, let us try a solution of the type $V(\varphi) = A\varphi + B$ (corresponding to k $= 0$ and with no radial dependence). The boundary conditions imposed on the system require that $V(0) = 0$ and $V(\alpha) = V_0$; this implies that $B = 0$ and $A = \frac{V_0}{\alpha}$. Consequently, we obtain:

$$V(\varphi) = \frac{V_0\varphi}{\alpha} \qquad [6.48]$$

The value of the induced electric field is:

$$E_\varphi(\varphi) = -\frac{1}{r}\frac{dV}{d\varphi} = -\frac{V_0}{r\alpha} \qquad [6.49]$$

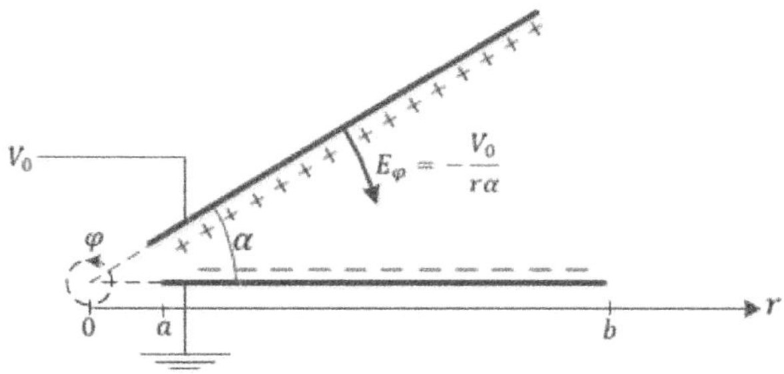

Figure 6.7. *Inclined plane capacitor*

The surface charge density on the upper electrode is $\sigma(\alpha) = -\varepsilon_0 E_\varphi(\alpha) = \frac{\varepsilon_0 V_0}{r\alpha}$ and, consequently, the total charge per unit length, is obtained by integrating over r:

$$\lambda(\alpha) = \int_a^b \sigma(\alpha)\, dr = \frac{\varepsilon_0 V_0}{\alpha} \ln \frac{b}{a} \qquad [6.50]$$

which means that the capacity, per unit length, is:

$$C = \frac{\lambda(\alpha)}{V_0} = \frac{\varepsilon_0}{\alpha} \ln \frac{b}{a} \qquad [6.51]$$

EXAMPLE 6.8. (Dirichlet Problem in a Cylinder).– An infinitely long hollow cylinder, with radius R and axis Oz, is taken to a potential $V_0(\varphi)$ (independent of z). The system is therefore translation-invariant along Oz. We wish to calculate the potential $V(r,\varphi)$ (inside and outside the cylinder) that verifies the Laplace equation [6.43]:

– *Inside the cylinder*: because V cannot be infinite when $r \longrightarrow 0$, the solutions $C_0 \ln r$ and $\frac{D_m}{r^m}$ must not be retained. We must therefore only retain the part $V(r,\varphi) = \frac{D_0}{2} + \sum_{m=1}^{\infty} r^m (A_m \cos m\varphi + B_m \sin m\varphi)$, which is reduced (on the cylinder) to $V(R,\varphi) = \frac{D_0}{2} + \sum_{m=1}^{\infty} R^m (A_m \cos m\varphi + B_m \sin m\varphi) \equiv V_0(\varphi)$. D_0, $R^m A_m$ and $R^m B_m$ therefore appear here as Fourier coefficients associated with $V_0(\varphi)$, that is, $D_0 = \frac{1}{\pi}\int_0^{2\pi} V_0(\phi)\, d\phi$, $A_m = \frac{1}{\pi R^m}\int_0^{2\pi} V_0(\phi) \cos m\phi d\phi$ and $B_m = \frac{1}{\pi R^m}\int_0^{2\pi} V_0(\phi) \sin m\phi d\phi$. By substituting these integrals in $V(r,\varphi)$, we obtain:

$$\begin{aligned} V(r,\varphi) &= \frac{1}{2\pi}\int_0^{2\pi} V_0(\phi) \left[1 + 2\sum_{m=1}^{\infty} \left(\frac{r}{R}\right)^m \cos m(\phi - \varphi)\right] d\phi \\ &= \frac{1}{2\pi}\int_0^{2\pi} V_0(\phi) \frac{R^2 - r^2}{r^2 - 2rR\cos(\phi - \varphi) + R^2} d\phi \end{aligned} \qquad [6.52]$$

and consequently, the induced electric field is:

$$\vec{\mathbf{E}} = \frac{R}{\pi}\int_0^{2\pi} V_0(\phi) \times \frac{\left[2rR - \left(r^2+R^2\right)\cos(\phi-\varphi)\right]\vec{\mathbf{e}}_r + \left(r^2-R^2\right)\sin(\phi-\varphi)\,\vec{\mathbf{e}}_\varphi}{\left[r^2 - 2rR\cos(\phi-\varphi)+R^2\right]^2} d\phi \tag{6.53}$$

– *Outside the cylinder*: since V must cancel when $r \longrightarrow \infty$, the solutions $C_0 \ln r$ and $C_m r^m$ cannot be retained. We will just use the expression $V(r,\varphi) = \frac{D_0}{2} + \sum_{m=1}^{\infty} \frac{1}{r^m}(A_m \cos m\varphi + B_m \sin m\varphi)$.

An analogous calculation gives:

$$V(r,\varphi) = \frac{1}{2\pi}\int_0^{2\pi} V_0(\phi)\frac{r^2-R^2}{r^2-2rR\cos(\phi-\varphi)+R^2}d\phi \tag{6.54}$$

$$\vec{\mathbf{E}} = \frac{R}{\pi}\int_0^{2\pi} V_0(\phi) \times \frac{\left[2rR - \left(r^2+R^2\right)\cos(\phi-\varphi)\right]\vec{\mathbf{e}}_r + \left(R^2-r^2\right)\sin(\phi-\varphi)\,\vec{\mathbf{e}}_\varphi}{\left[r^2 - 2rR\cos(\phi-\varphi)+R^2\right]^2} d\phi \tag{6.55}$$

The solutions [6.52] and [6.54] are called *Poisson integral formulas for the circle.*

We can thus conclude that if we know the expression of the potential on a cylinder of radius R, it is always possible to find its general expression at any point (r,φ) using either [6.52] or [6.54].

EXAMPLE 6.9.– Let us determine the potential inside and outside a cylinder with radius R, assuming that the lateral surface is taken to a potential $V_0(\varphi) = \frac{V_0\varphi}{2\pi}$, $0 \le \varphi < 2\pi$:

– *Within the cylinder* $(r \le R)$: the Fourier coefficients D_0, A_m and B_m associated with $V_0(\varphi)$ are given by:

$$\begin{aligned} D_0 &= \frac{1}{\pi}\int_0^{2\pi}\frac{V_0\varphi d\varphi}{2\pi} = V_0 \\ A_m &= \frac{1}{\pi R^m}\int_0^{2\pi}\frac{V_0\varphi\cos m\varphi}{2\pi}d\varphi = 0 \\ B_m &= \frac{1}{\pi R^m}\int_0^{2\pi}\frac{V_0\varphi\sin m\varphi}{2\pi}d\varphi = -\frac{V_0}{m\pi R^m} \end{aligned} \tag{6.56}$$

Consequently:

$$\begin{aligned} V(r,\varphi) &= V_0\left[\frac{1}{2}-\sum_{m=1}^{\infty}\frac{1}{m\pi}\left(\frac{r}{R}\right)^m \sin m\varphi\right] \\ &= V_0\left[\frac{1}{2}-\frac{1}{\pi}\tan^{-1}\frac{r\sin\varphi}{R-r\cos\varphi}\right] \end{aligned} \qquad [6.57]$$

The radial and angular components of the electric field associated with the potential [6.57] are calculated as follows:

$$\begin{aligned} E_r &= -\frac{\partial V}{\partial r} = \frac{V_0 R}{\pi}\frac{\sin\varphi}{r^2+R^2-2rR\cos\varphi} \\ E_\varphi &= -\frac{1}{r}\frac{\partial V}{\partial \varphi} = \frac{V_0}{\pi}\frac{R\cos\varphi - r}{r^2+R^2-2rR\cos\varphi} \end{aligned} \qquad [6.58]$$

– *Outside the cylinder* $(r > R)$: outside the cylinder, the potential is written $V(r,\varphi) = \frac{D_0}{2} + \sum_{m=1}^{\infty}\frac{1}{r^m}\left(A_m\cos m\varphi + B_m \sin m\varphi\right)$, where the coefficients D_0, A_m and B_m are obtained from the expressions in [6.56] by replacing R with $\frac{1}{R}$, that is, $D_0 = V_0$, $A_m = 0$ and $B_m = -\frac{V_0 R^m}{m\pi}$. We thus obtain:

$$\begin{aligned} V(r,\varphi) &= V_0\left[\frac{1}{2}-\sum_{m=1}^{\infty}\frac{1}{m\pi}\left(\frac{R}{r}\right)^m \sin m\varphi\right] \\ &= V_0\left[\frac{1}{2}-\frac{1}{\pi}\tan^{-1}\frac{R\sin\varphi}{r-R\cos\varphi}\right] \end{aligned} \qquad [6.59]$$

and, consequently:

$$\begin{aligned} E_r &= -\frac{\partial V}{\partial r} = \frac{V_0 r}{\pi}\frac{\sin\varphi}{r^2+R^2-2rR\cos\varphi} \\ E_\varphi &= -\frac{1}{r}\frac{\partial V}{\partial \varphi} = \frac{V_0}{\pi}\frac{r\cos\varphi - R}{r^2+R^2-2rR\cos\varphi} \end{aligned} \qquad [6.60]$$

EXAMPLE 6.10. (Local Pulsation of Potential).– In the case of a potential pulse of the type $\mu\delta(\varphi-\varphi_0)$ localized at the point (R,φ_0), formulas [6.52] and [6.54] immediately give the following results (valid for the inside and outside of the cylinder):

$$\begin{aligned} V(r,\varphi) &= \frac{\left|R^2-r^2\right|}{2\pi}\int_0^{2\pi}\frac{\mu\delta(\phi-\varphi_0)\,d\phi}{r^2-2rR\cos(\phi-\varphi)+R^2} \\ &= \frac{\mu}{2\pi}\frac{\left|r^2-R^2\right|}{r^2-2rR\cos(\varphi-\varphi_0)+R^2} \end{aligned} \qquad [6.61]$$

EXAMPLE 6.11. (Step Potential).– By taking a step potential on the cylinder of the form $V_0\left(\varphi_k \leq \varphi < \varphi_{k+1}\right) = \mu_k$, where $\varphi_0 = 0$ and $\varphi_N = 2\pi$, as illustrated in Figure 6.8, the following is obtained (valid in both cases):

$$
\begin{aligned}
V\left(r,\varphi\right) &= \frac{\left|R^2 - r^2\right|}{2\pi} \sum_{k=0}^{N-1} \int_{\varphi_k}^{\varphi_{k+1}} \frac{\mu_k d\phi}{r^2 + R^2 - 2rR\cos\left(\phi - \varphi\right)} \\
&= \frac{1}{\pi} \sum_{k=0}^{N-1} \mu_k \left[\tan^{-1}\left(\frac{R+r}{|R-r|}\tan\frac{\varphi_{k+1}-\varphi}{2}\right) - \tan^{-1}\left(\frac{R+r}{|R-r|}\tan\frac{\varphi_k-\varphi}{2}\right)\right]
\end{aligned} \quad [6.62]
$$

where we used the identity $\int \frac{d\phi}{p - q\tan\phi} = \frac{2}{\sqrt{p^2-q^2}}\tan^{-1}\left[\sqrt{\frac{p+q}{p-q}}\tan\frac{\phi}{2}\right]$.

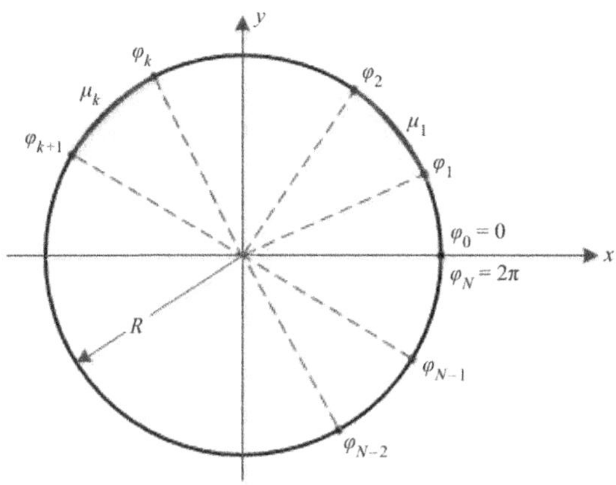

Figure 6.8. *Step potential on a cylinder. For a color version of this figure, see www.iste.co.uk/abdesselam/electrodynamics1.zip*

EXAMPLE 6.12. (Cylinder with a Charged Surface).– Let us assume a hollow cylinder carries a surface charge $\sigma\left(\varphi\right)$. Let us determine the potential inside and outside the cylinder:

– *Insider the cylinder* $r \leq R$: in this zone, the potential is governed by the series $V\left(r,\varphi\right) = \frac{A_0}{2} + \sum_{m=1}^{\infty}\left(\frac{r}{R}\right)^m\left(A_m\cos m\varphi + B_m\sin m\varphi\right)$ and, consequently, the radial and angular components of the induced electric field are:

$$
\begin{aligned}
E_r &= -\frac{\partial V}{\partial r} = -\frac{1}{r}\sum_{m=1}^{\infty} m\left(\frac{r}{R}\right)^m (A_m\cos m\varphi + B_m \sin m\varphi) \\
E_\varphi &= -\frac{1}{r}\frac{\partial V}{\partial \varphi} = -\frac{1}{r}\sum_{m=1}^{\infty} m\left(\frac{r}{R}\right)^m (-A_m\sin m\varphi + B_m \cos m\varphi)
\end{aligned}
\qquad [6.63]
$$

– *Outside the cylinder* $r > R$: in this part of the space, we will use the expression $V(r,\varphi) = C_0 \ln r + \frac{D_0}{2} + \sum_{m=1}^{\infty}\left(\frac{R}{r}\right)^m (C_m \cos m\varphi + D_m \sin m\varphi)$, since the potential of a hollow cylinder with a charge surface is of the order of $\ln r$. Hence, the term $C_0 \ln r$ must be retained. It comes that:

$$
\begin{aligned}
E_r &= -\frac{\partial V}{\partial r} = -\frac{C_0}{r} + \frac{1}{r}\sum_{m=1}^{\infty} m\left(\frac{R}{r}\right)^m (C_m\cos m\varphi + D_m \sin m\varphi) \\
E_\varphi &= -\frac{1}{r}\frac{\partial V}{\partial \varphi} = -\frac{1}{r}\sum_{m=1}^{\infty} m\left(\frac{R}{r}\right)^m (-C_m\sin m\varphi + D_m \cos m\varphi)
\end{aligned}
\qquad [6.64]
$$

At the boundary of the cylinder $r = R$, the $\overrightarrow{\mathbf{E}}$ field must obey the continuity equations: $E_r(R^+,\varphi) = E_r(R^-,\varphi) + \frac{\sigma(\varphi)}{\varepsilon_0}$ and $E_\varphi(R^+,\varphi) = E_\varphi(R^-,\varphi)$, or more precisely:

$$
\begin{aligned}
&-\frac{C_0}{R} + \frac{1}{R}\sum_{m=1}^{\infty} m(A_m + C_m)\cos m\varphi + m(B_m + D_m)\sin m\varphi = \frac{\sigma(\varphi)}{\varepsilon_0} \\
&-\frac{1}{R}\sum_{m=1}^{\infty} m(A_m - C_m)\sin m\varphi + m(-B_m + D_m)\cos m\varphi = 0
\end{aligned}
\qquad [6.65]
$$

Since the functions $\sin m\varphi$ and $\cos m\varphi$ are independent, it can be derived from this that $A_m = C_m$ and $B_m = D_m$. It thus follows that $-\frac{C_0}{R} + \frac{2}{R}\sum_{m=1}^{\infty} m(A_m \cos m\varphi$ $+B_m \sin m\varphi) = \frac{\sigma(\varphi)}{\varepsilon_0}$, which implies that $C_0 = -\frac{R}{2\pi\varepsilon_0}\int_0^{2\pi}\sigma(\varphi)\,d\varphi$, $A_m = \frac{R}{2\pi m\varepsilon_0}$ $\int_0^{2\pi}\sigma(\varphi)\cos m\varphi d\varphi$ and $B_m = \frac{R}{2\pi m\varepsilon_0}\int_0^{2\pi}\sigma(\varphi)\sin m\varphi d\varphi$. By substituting these results into the expressions of internal and external potentials, we obtain:

$$
\begin{aligned}
V(r<R) &= \frac{R}{2\pi\varepsilon_0}\int_0^{2\pi}\sigma'\left[\sum_{m=1}^{\infty}\frac{1}{m}\left(\frac{r}{R}\right)^m \cos m(\varphi-\varphi')\right]d\varphi' \\
&= \frac{R}{4\pi\varepsilon_0}\int_0^{2\pi}\sigma' \ln\left(1+\frac{r^2}{R^2}-\frac{2r}{R}\cos(\varphi-\varphi')\right)d\varphi' \\
V(r>R) &= \frac{R}{2\pi\varepsilon_0}\int_0^{2\pi}\sigma'\left[-\ln r + \sum_{m=1}^{\infty}\frac{1}{m}\left(\frac{R}{r}\right)^m \cos m(\varphi-\varphi')\right]d\varphi' \\
&= \frac{R}{2\pi\varepsilon_0}\int_0^{2\pi}\sigma'\left[-\ln r + \frac{1}{2}\ln\left(1+\frac{R^2}{r^2}-\frac{2R}{r}\cos(\varphi-\varphi')\right)\right]d\varphi'
\end{aligned}
$$

[6.66]

where $\sigma' = \sigma(\varphi')$. The additive constants $\frac{A_0}{2}$ and $\frac{D_0}{2}$ are simply eliminated. These integral equations are equivalent to Poisson formulas [6.52] and [6.54]. These make it possible to calculate the potential at any point (r, φ), depending on the surface charge density distributed over the cylinder. In particular:

i) If $\sigma(\varphi) = \sigma_0$, we deduce that $C_0 = -\frac{R\sigma_0}{\varepsilon_0}$, $A_m = 0$ and $B_m = 0$. Thus, we find $V(r \leq R, \varphi) = 0$ and $V(r > R, \varphi) = -\frac{R\sigma_0}{\varepsilon_0} \ln r$. This leads, of course, to the result already established by Gauss' theorem.

ii) Similarly, for $\sigma(\varphi) = \sigma_0 \cos\varphi$, it comes that $C_0 = 0$, $A_1 = \frac{R\sigma_0}{2\varepsilon_0}$, $A_m = 0$ $(m \geq 2)$ and $B_m = 0$ $(m \geq 1)$. Therefore, it follows from this that:

$$V(r \leq R, \varphi) = \frac{\sigma_0 r \cos\varphi}{2\varepsilon_0}$$
$$V(r > R, \varphi) = \frac{\sigma_0 R^2 \cos\varphi}{2\varepsilon_0 r} \qquad [6.67]$$

EXAMPLE 6.13. (Coaxial Cylinders).– Two coaxial cylinders (hollow) with radii R_1 and R_2 $(R_1 < R_2)$ are at potentials $V_1(\varphi)$ and $V_2(\varphi)$, respectively, as seen in Figure 6.9. Let us determine the potential in three zones: (1) $r \leq R_1$, (2) $r > R_2$ and (3) $R_1 < r \leq R_2$.

1) *The region* $r \leq R_1$: the terms $\ln r$ and $\frac{1}{r^m}$ diverge when $r \longrightarrow 0$. They are therefore eliminated in [6.47]. We take $V(r \leq R_1, \varphi) = \frac{D_0}{2} + \sum_{m=1}^{\infty} r^m$ $(A_m \cos m\varphi + B_m \sin m\varphi)$. For $r = R_1$, the latter equation is reduced to $V_1(\varphi) = \frac{D_0}{2} + \sum_{m=1}^{\infty} R^m \times (A_m \cos m\varphi + B_m \sin m\varphi)$. It follows that: $D_0 = \frac{1}{\pi}\int_0^{2\pi} V_1(\varphi)\, d\varphi$, $A_m = \frac{1}{\pi R_1^m} \times \int_0^{2\pi} V_1(\varphi) \cos m\varphi d\varphi$ and $B_m = \frac{1}{\pi R_1^m}\int_0^{2\pi} V_1(\varphi) \sin m\varphi d\varphi$.

2) *The region* $r > R_2$: the terms r^m and $\ln r$ diverge, this time for $r \longrightarrow \infty$. We therefore retain $V(r > R_2, \varphi) = \frac{D_0'}{2} + \sum_{m=1}^{\infty} \frac{1}{r^m}(A_m' \cos m\varphi + B_m' \sin m\varphi)$. On the boundary $r = R_2$, we obtain $V_2(\varphi) = \frac{D_0'}{2} + \sum_{m=1}^{\infty} \frac{1}{R_2^m}$ $(A_m' \cos m\varphi + B_m' \sin m\varphi)$ and, consequently, $D_0' = \frac{1}{\pi}\int_0^{2\pi} V_2(\varphi)\, d\varphi$, $A_m' = \frac{R_2^m}{\pi}\int_0^{2\pi} V_2(\varphi)\cos m\varphi d\varphi$ and $B_m' = \frac{R_2^m}{\pi}\int_0^{2\pi} V_2(\varphi)\sin m\varphi d\varphi$.

3) *Region* $R_1 < r \leq R_2$: now, there is no reason to remove terms in the expression of V. We will take the complete solution $V(R_1 < r \leq R_2, \varphi) = C_0'' \ln r$ $+\frac{D_0''}{2} + \sum_{m=1}^{\infty}\left(C_m'' r^m + \frac{D_m''}{r^m}\right)(A_m'' \cos m\varphi + B_m'' \sin m\varphi)$. In this region, the potential must obey the restrictions:

$$V_1(\varphi) = C_0'' \ln R_1 + \frac{D_0''}{2} + \sum_{m=1}^{\infty} \left[C_m'' R_1^m + \frac{D_m''}{R_1^m} \right] (A_m'' \cos m\varphi + B_m'' \sin m\varphi)$$
$$V_2(\varphi) = C_0'' \ln R_2 + \frac{D_0''}{2} + \sum_{m=1}^{\infty} \left[C_m'' R_2^m + \frac{D_m''}{R_2^m} \right] (A_m'' \cos m\varphi + B_m'' \sin m\varphi)$$
[6.68]

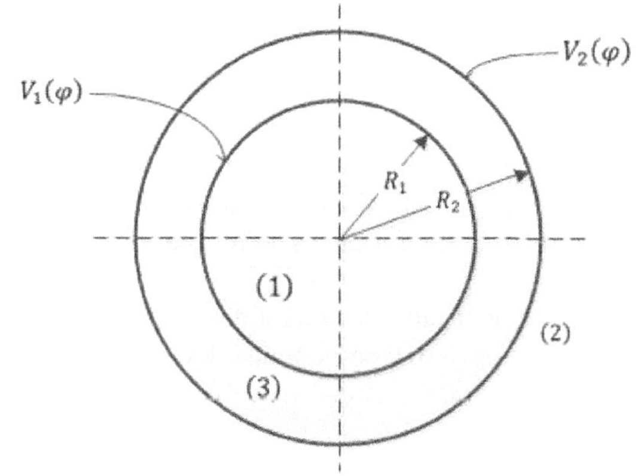

Figure 6.9. *Coaxial cylinders with radii R_1 and R_2 at the potentials $V_1(\varphi)$ and $V_2(\varphi)$*

It follows from this as:

$$C_0'' \begin{Bmatrix} \ln R_1 \\ \ln R_2 \end{Bmatrix} + \frac{D_0''}{2} = \frac{1}{2\pi} \int_0^{2\pi} \begin{Bmatrix} V_1(\varphi) \\ V_2(\varphi) \end{Bmatrix} d\varphi$$
$$\left(C_m'' R_1^m + \frac{D_m''}{R_1^m} \right) \begin{Bmatrix} A_m'' \\ B_m'' \end{Bmatrix} = \frac{1}{\pi} \int_0^{2\pi} V_1(\varphi) \begin{Bmatrix} \cos m\varphi \\ \sin m\varphi \end{Bmatrix} d\varphi \qquad [6.69]$$
$$\left(C_m'' R_2^m + \frac{D_m''}{R_2^m} \right) \begin{Bmatrix} A_m'' \\ B_m'' \end{Bmatrix} = \frac{1}{\pi} \int_0^{2\pi} V_2(\varphi) \begin{Bmatrix} \cos m\varphi \\ \sin m\varphi \end{Bmatrix} d\varphi$$

which are solved to obtain C_0'', D_0'', $C_m'' A_m''$, $D_m'' A_m''$, $C_m'' B_m''$ and $D_m'' B_m''$:

$$C_0'' = \frac{1}{2\pi} \int_0^{2\pi} \frac{V_2(\varphi') - V_1(\varphi')}{\ln R_2 - \ln R_1} d\varphi$$
$$\frac{D_0''}{2} = \frac{1}{2\pi} \int_0^{2\pi} \frac{V_1(\varphi') \ln R_2 - V_2(\varphi') \ln R_1}{\ln R_2 - \ln R_1} d\varphi'$$
$$C_m'' \begin{Bmatrix} A_m'' \\ B_m'' \end{Bmatrix} = \frac{1}{\pi} \int_0^{2\pi} \frac{R_1^m V_1(\varphi') - R_2^m V_2(\varphi')}{R_1^{2m} - R_2^{2m}} \begin{Bmatrix} \cos m\varphi' \\ \sin m\varphi' \end{Bmatrix} d\varphi' \qquad [6.70]$$
$$D_m'' \begin{Bmatrix} A_m'' \\ B_m'' \end{Bmatrix} = \frac{R_1^m R_2^m}{\pi} \int_0^{2\pi} \frac{R_2^m V_1(\varphi') - R_1^m V_2(\varphi')}{R_2^{2m} - R_1^{2m}} \begin{Bmatrix} \cos m\varphi' \\ \sin m\varphi' \end{Bmatrix} d\varphi'$$

and as a result:

$$V(r,\varphi) = \frac{1}{2\pi}\int_0^{2\pi} \frac{\ln(r/R_2)V_1(\varphi') - \ln(r/R_1)V_2(\varphi')}{\ln R_1 - \ln R_2} d\varphi' + \sum_{m=1}^{\infty} \frac{1}{\pi}\int_0^{2\pi} \frac{R_1^m\left(r^{2m} - R_2^{2m}\right)V_1(\varphi') - R_2^m\left(r^{2m} - R_1^{2m}\right)V_2(\varphi')}{\left(R_1^{2m} - R_2^{2m}\right)r^m} \times \cos m(\varphi - \varphi')\, d\varphi' \tag{6.71}$$

EXAMPLE 6.14. (Infinite Cylinder in a Uniform Electric Field).– A perfect, neutral conductor in the form of an infinitely long cylinder with axis Oz, and radius R, is placed in a uniform electric field directed along Ox, as seen in Figure 6.10. We wish to calculate the potential $V(r,\varphi)$ of the system outside the cylinder. It can be seen that the Ox axis is a symmetry axis for the system. That is, $V(r,\varphi) = V(r,-\varphi)$. A direct consequence of this restriction is that the B_m coefficients are all zero. Since the potential induced by the cylinder must not be infinite when $r \longrightarrow \infty$, the coefficients C_0 and C_m must also be eliminated. Consequently, the potential V at the point (r,φ) is:

$$V(r,\varphi) = -E_0 x + \frac{D_0}{2} + \sum_{m=1}^{\infty} \frac{D'_m}{r^m}\cos m\varphi = -E_0 r\cos\varphi + \frac{D_0}{2} + \sum_{m=1}^{\infty} \frac{D'_m}{r^m}\cos m\varphi \tag{6.72}$$

where $D'_m = A_m D_m$. $-E_0 x$ represents, here, the potential generating the external field E_0. Since the potential must be constant on the surface of the cylindrical conductor, that is, $V(R,\varphi) = V_0$, we obtain $V_0 = -E_0 R\cos\varphi + \frac{D_0}{2} + \sum_{m=1}^{\infty} \frac{D'_m}{R^m}\cos m\varphi$ or again $\left(\frac{D_0}{2} - V_0\right) + \left(\frac{D'_1}{R} - E_0 R\right)\cos\varphi + \sum_{m=2}^{\infty} \frac{D'_m}{R^m}\cos m\varphi = 0$; this results in the fact that $D_0 = 2V_0$, $D_1 = E_0 R^2$ and $D_m = 0$ $(m \geq 2)$, since the functions $\{\cos m\varphi\}_{m\geq 1}$ are independent. The final result is:

$$V(r > R,\varphi) = V_0 - E_0 r\cos\varphi + \frac{E_0 R^2}{r}\cos\varphi \tag{6.73}$$

and consequently:

$$\vec{\mathbf{E}} = -\vec{\nabla}V = E_0\,\vec{\mathbf{i}} + \frac{R^2 E_0}{r^2}\left(\cos\varphi\,\vec{\mathbf{e}}_r + \sin\varphi\,\vec{\mathbf{e}}_\varphi\right) \tag{6.74}$$

EXAMPLE 6.15. (Cylindrical Conductor with Conductivity γ_2 in a Uniform Field).– An infinitely long cylinder with radius R and conductivity γ_2, is placed in an infinite medium with conductivity γ_1. An electric field $E_0 \overrightarrow{\mathrm{i}}$ is suddenly activated at $t = 0$. This problem resembles the capacitor problem. We will therefore observe an accumulation of charges on the edges of the cylinder until they reach a stationary (equilibrium) state.

To solve the problem, we must find the potential inside and outside the cylinder and then relate these solutions (in each region) to each other through the boundary conditions at $r = R$. As previously, we will choose a solution of the type:

$$V(r, \varphi, t) = \begin{cases} \sum\limits_{m=1}^{\infty} A_m(t)\, r^m \cos m\varphi & r \leq R \\ -E_0 r \cos\varphi + \sum\limits_{m=1}^{\infty} \dfrac{C_m(t)}{r^m} \cos m\varphi & r > R \end{cases} \qquad [6.75]$$

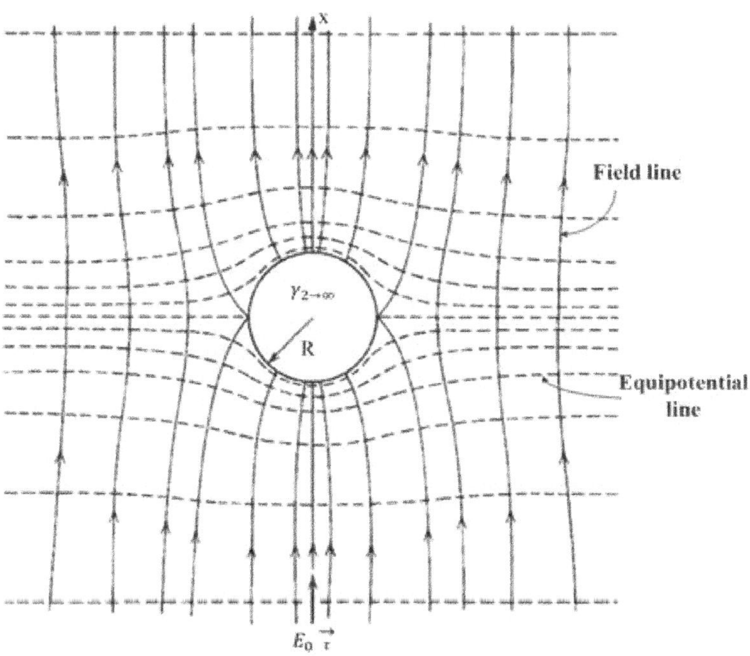

Figure 6.10. *Perfect, neutral, cylindrical conductor immersed into a uniform electric field $E_0 \overrightarrow{\mathrm{i}}$*

with the associated electric field:

$$\vec{\mathbf{E}}(t) = \begin{cases} -\sum_{m=1}^{\infty} mA_m(t)\, r^{m-1}\left[\cos m\varphi\, \vec{\mathbf{e}}_r - \sin m\varphi\, \vec{\mathbf{e}}_\varphi\right] & r \leq R \\ E_0\left(\cos\varphi\, \vec{\mathbf{e}}_r - \sin\varphi\, \vec{\mathbf{e}}_\varphi\right) + \sum_{m=1}^{\infty} \frac{mC_m(t)}{r^{m+1}} & \\ \left(\cos m\varphi\, \vec{\mathbf{e}}_r + \sin m\varphi\, \vec{\mathbf{e}}_\varphi\right) & r > R \end{cases} \quad [6.76]$$

Since the potential must be continuous at the boundary, we must have $\sum_{m=1}^{\infty} A_m(t)\, R^m \cos m\varphi = -E_0 r\cos\varphi + \sum_{m=1}^{\infty} \frac{C_m(t)}{R^m}\cos m\varphi$, which gives $A_1(t) = -E_0 + \frac{C_1(t)}{R^2}$ and $A_m(t) = \frac{C_m(t)}{R^{2m}}$ $(m \geq 2)$. These restrictions clearly show that the tangential component of the field $\vec{\mathbf{E}}$ is well conserved, that is, $E_\varphi(R^-, \varphi) = E_\varphi(R^+, \varphi)$. The charge is also conserved at the interface $J_r(R^+, \varphi) - J_r(R^-, \varphi) + \frac{\partial\sigma}{\partial t} = 0$, or again:

$$\gamma_1 E_r(R^+, \varphi) - \gamma_2 E_r(R^-, \varphi) + \frac{\partial}{\partial t}\left[\varepsilon_0 E_r(R^+, \varphi) - \varepsilon_0 E_r(R^-, \varphi)\right] = 0 \quad [6.77]$$

Using [6.76], we obtain $\left[\frac{R^2(\gamma_1-\gamma_2)E_0}{2\varepsilon_0} + \frac{C_1}{\tau} + \frac{dC_1}{dt}\right]\frac{\cos\varphi}{R^2} + \sum_{m=2}^{\infty}\left[\frac{C_m}{\tau} + \frac{dC_m}{dt}\right] \times \frac{m\cos m\varphi}{R^{m+1}} = 0$, which leads to:

$$\begin{aligned} &\frac{dC_1}{dt} + \frac{C_1}{\tau} + \frac{R^2(\gamma_1-\gamma_2)E_0}{2\varepsilon_0} = 0 \\ &\frac{dC_m}{dt} + \frac{C_m}{\tau} = 0 \qquad m \geq 2 \end{aligned} \quad [6.78]$$

where $\tau = \frac{2\varepsilon_0}{(\gamma_1+\gamma_2)}$ is the *relaxation time* of the system. The above equations admit the following solutions: $C_1(t) = -\frac{R^2(\gamma_1-\gamma_2)E_0}{\gamma_1+\gamma_2} + \mu_1 e^{-\frac{t}{\tau}}$ and $C_m(t) = \mu_m e^{-\frac{t}{\tau}}$ $(m \geq 2)$. The coefficients μ_1 and μ_m are determined by the initial conditions. We have: $C_m(0) = 0$ for $m \geq 1$. This implies that $\mu_1 = \frac{R^2(\gamma_1-\gamma_2)E_0}{\gamma_1+\gamma_2}$ and $\mu_m = 0$ for $m \geq 2$. It follows that:

$$\begin{aligned} &C_1(t) = -\frac{R^2(\gamma_1-\gamma_2)E_0}{\gamma_1+\gamma_2}\left(1 - e^{-\frac{t}{\tau}}\right) \\ &A_1(t) = -E_0 - \frac{(\gamma_1-\gamma_2)E_0}{\gamma_1+\gamma_2}\left(1 - e^{-\frac{t}{\tau}}\right) \\ &A_m(t) = C_m(t) = 0 \qquad m \geq 2 \end{aligned} \quad [6.79]$$

The surface charge density accumulated on the cylinder has the value:

$$\sigma(t) = \varepsilon_0 E_r(R^+, \varphi) - \varepsilon_0 E_r(R^-, \varphi) = \frac{2\varepsilon_0(\gamma_2-\gamma_1)E_0}{\gamma_1+\gamma_2}\left(1 - e^{-\frac{t}{\tau}}\right)\cos\varphi \quad [6.80]$$

The section of the cylinder defined with the coordinates $\left(-\frac{\pi}{2} \leq \varphi \leq \frac{\pi}{2}\right)$ is charged with one sign, while the other half $\left(\frac{\pi}{2} \leq \varphi \leq \frac{3\pi}{2}\right)$ is charged with the opposite sign. The net charge on the cylinder is, of course, zero. It can be seen that if $\gamma_1 = \gamma_2$, there is no charge accumulated on the surface of the cylinder. The electric field is therefore:

$$\vec{\mathbf{E}}(t) = \begin{cases} E_0 \vec{\mathbf{i}} + \dfrac{(\gamma_1 - \gamma_2) E_0}{\gamma_1 + \gamma_2} \left(1 - e^{-\frac{t}{\tau}}\right) \vec{\mathbf{i}} & r \leq R \\ E_0 \vec{\mathbf{i}} - \dfrac{R^2 (\gamma_1 - \gamma_2) E_0}{r^2 (\gamma_1 + \gamma_2)} \left(1 - e^{-\frac{t}{\tau}}\right) \left(\cos\varphi \vec{\mathbf{e}}_r + \sin\varphi \vec{\mathbf{e}}_\varphi\right) & r > R \end{cases} \quad [6.81]$$

When the stationary regime is established, the two formulas above are reduced to:

$$\sigma(t \longrightarrow \infty) = \frac{2\varepsilon_0 (\gamma_2 - \gamma_1) E_0}{\gamma_1 + \gamma_2} \cos\varphi \quad [6.82]$$

$$\vec{\mathbf{E}}(t \longrightarrow \infty) = \begin{cases} \dfrac{2\gamma_1 E_0}{\gamma_1 + \gamma_2} \vec{\mathbf{i}} & r \leq R \\ E_0 \vec{\mathbf{i}} - \dfrac{R^2 (\gamma_1 - \gamma_2) E_0}{r^2 (\gamma_1 + \gamma_2)} \left(\cos\varphi \vec{\mathbf{e}}_r + \sin\varphi \vec{\mathbf{e}}_\varphi\right) & r > R \end{cases}$$

In particular, when $\gamma_2 \longrightarrow \infty$ (the cylinder is a perfect conductor), we obtain:

$$\sigma(t \longrightarrow \infty) = 2\varepsilon_0 E_0 \cos\varphi$$

$$\vec{\mathbf{E}}(t \longrightarrow \infty) = \begin{cases} 0 & r \leq R \\ E_0 \vec{\mathbf{i}} + \dfrac{R^2 E_0}{r^2} \left(\cos\varphi \vec{\mathbf{e}}_r + \sin\varphi \vec{\mathbf{e}}_\varphi\right) & r > R \end{cases} \quad [6.83]$$

which is in agreement with [6.74]. Similarly, for $\gamma_2 \longrightarrow 0$ (i.e. the case of a perfect insulator), we find:

$$\sigma(t \longrightarrow \infty) = -2\varepsilon_0 E_0 \cos\varphi$$

$$\vec{\mathbf{E}}(t \longrightarrow \infty) = \begin{cases} 2E_0 \vec{\mathbf{i}} & r \leq R \\ E_0 \vec{\mathbf{i}} - \dfrac{R^2 E_0}{r^2} \left(\cos\varphi \vec{\mathbf{e}}_r + \sin\varphi \vec{\mathbf{e}}_\varphi\right) & r > R \end{cases} \quad [6.84]$$

If the cylinder is a perfect conductor, the internal electric field is zero, while the external electric field lines are radial in the neighborhood of the cylinder and extend to infinity, parallel to the Ox axis, as shown in Figure 6.10. On the other hand, when the cylinder is a perfect insulator, the external field lines must be tangential, because the incident normal current is zero and the internal field is twice as high as the applied field, as shown in Figure 6.11.

$$\text{perfectconductor:} \begin{cases} V(r,\varphi) = \text{const.} \Longrightarrow \left(\dfrac{r}{R} - \dfrac{R}{r}\right)\cos\varphi = \text{const.} \\ \dfrac{dr}{rd\varphi} = \dfrac{E_r}{E_\varphi} = \dfrac{1+\frac{R^2}{r^2}}{-1+\frac{R^2}{r^2}}\dfrac{\cos\varphi}{\sin\varphi} \\ \Longrightarrow \left(\frac{r}{R} + \frac{R}{r}\right)\sin\varphi = \text{const.} \end{cases} \quad [6.85]$$

$$\text{perfectinsulator:} \begin{cases} V(r,\varphi) = \text{const.} \Longrightarrow \left(\dfrac{r}{R} + \dfrac{R}{r}\right)\cos\varphi = \text{const.} \\ \dfrac{dr}{rd\varphi} = \dfrac{E_r}{E_\varphi} = \dfrac{1-\frac{R^2}{r^2}}{-1-\frac{R^2}{r^2}}\dfrac{\cos\varphi}{\sin\varphi} \\ \Longrightarrow \left(\frac{r}{R} - \frac{R}{r}\right)\sin\varphi = \text{const.} \end{cases} \quad [6.86]$$

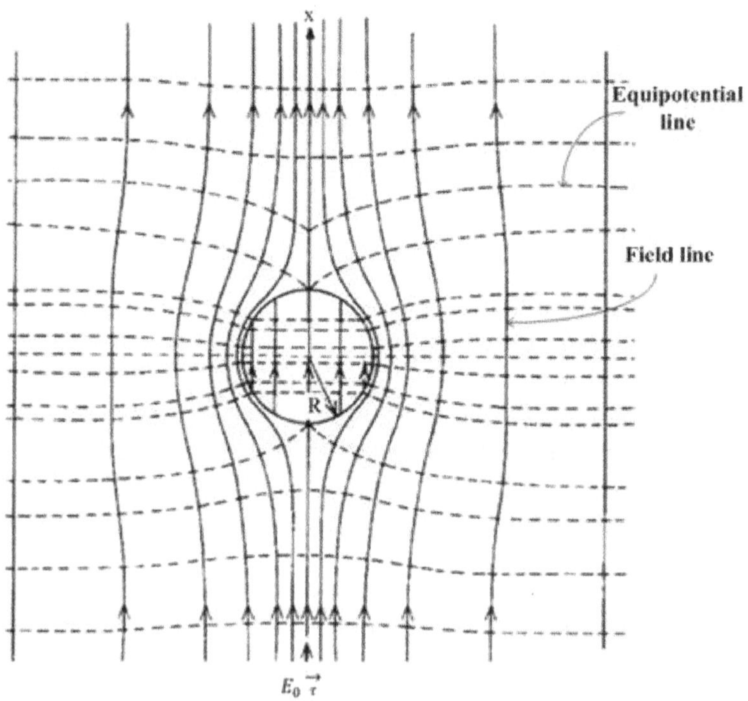

Figure 6.11. *Field and equipotential lines for a cylindrical insulator in a uniform field*

6.3.3. *Spherical coordinates with axial symmetry along* Oz

It is simpler to use spherical coordinates when the potential is specified on a spherical boundary, or when the system has spherical symmetry. Moreover, if the system is invariant by rotation around the Oz axis (axial symmetry), the Laplace equation is reduced to:

$$\frac{1}{r}\frac{\partial}{\partial r}\left(r^2\frac{\partial V}{\partial r}\right)+\frac{1}{r^2\sin\theta}\frac{\partial}{\partial\theta}\left(\sin\theta\frac{\partial V}{\partial\theta}\right)=0 \qquad [6.87]$$

Let us write a solution in the form $V(r,\theta) = R(r)\,\Theta(\theta)$. By substituting this solution into [6.87], we obtain $\frac{1}{R}\frac{d}{dr}\left(r^2\frac{dR}{dr}\right) = -\frac{1}{\Theta\sin\theta}\frac{d}{d\theta}\left(\sin\theta\frac{d\Theta}{d\theta}\right) \equiv l(l+1)$. The latter can be divided into two equations $\frac{d}{dr}\left(r^2\frac{dR}{dr}\right) = l(l+1)R$ and $\frac{1}{\sin\theta}\frac{d}{d\theta}\left(\sin\theta\frac{d\Theta}{d\theta}\right) = -l(l+1)\Theta$. The radial differential equation can also be written as $r^2\frac{d^2}{dr^2}(rR) = l(l+1)rR$ and admits the solution $R = Ar^l + \frac{B}{r^{l+1}}$, while the angular differential equation $\frac{1}{\sin\theta}\frac{d}{d\theta}\left(\sin\theta\frac{d\Theta}{d\theta}\right) + l(l+1)\Theta = 0$ can be transformed into a Legendre differential equation through the change of variable $x = \cos\theta$. This results in:

$$\frac{d}{dx}\left[\left(1-x^2\right)\frac{d\Theta}{dx}\right]+l(l+1)\Theta=0 \qquad [6.88]$$

Equation [6.88] admits the solutions $aP_l(x) + bQ_l(x)$, where $P_l(x)$ and $Q_l(x)$ are, respectively, the *Legendre functions of the first and second kind.* It is known that $|Q_l(\pm1)| = \infty$ (the Q_l part must therefore be eliminated from the solution) and that $|P_l(\pm1)| = \infty$ except when l is an integer. When l is an integer, the functions $P_l(x)$ are reduced to Legendre polynomials. Therefore, by virtue of the principle of superposition, the general solution is:

$$V(r,\theta)=\sum_{l=0}^{\infty}\left(A_l r^l+\frac{B_l}{r^{l+1}}\right)P_l(\cos\theta) \qquad [6.89]$$

EXAMPLE 6.16. (Perfect Spherical Conductor in a Uniform Electric Field).– Consider a perfect spherical conductor of radius R placed in a uniform electric field, $E_0\,\vec{k}$. Let us determine the potential and the associated electric field outside the conductor. This system is obviously invariant by rotation around Oz. The potential outside the sphere is given by the series [6.89] and satisfies the boundary condition $V(R^+,\theta) = V_0$. Additionally, we must have $V(r \longrightarrow \infty,\theta) = -E_0 z = -E_0 r\cos\theta$ (the electric field does not act on the sphere at very large distances). On the other hand, we write, for $r \longrightarrow \infty$, that $V(r,\theta) \simeq A_0 + A_1 r P_1(\cos\theta) + \sum_{l=2}^{\infty} A_l r^l P_l(\cos\theta) = A_0 +$

$A_1 r\cos\theta + \sum_{l=2}^{\infty} A_l r^l P_l(\cos\theta)$, which yields $A_0 = 0$, $A_1 = -E_0$ and $A_l = 0$ ($\forall l \geq 2$). In order to determine the coefficients B_l, let us use the boundary condition:

$$\begin{aligned} V_0 &= -E_0 R\cos\theta + \sum_{l=0}^{\infty} \frac{B_l}{R^{l+1}} P_l(\cos\theta) \\ &= \frac{B_0}{R} + \left(\frac{B_1}{R^2} - E_0 R\right)\cos\theta + \sum_{l=2}^{\infty} \frac{B_l}{R^{l+1}} P_l(\cos\theta) \end{aligned} \qquad [6.90]$$

Consequently, we obtain $B_0 = V_0 R$, $B_1 = E_0 R^3$ et $B_l = 0$ ($l \geq 2$). It finally comes that:

$$V(r,\theta) = V_0 + \left(\frac{R^3}{r^2} - r\right) E_0 \cos\theta \qquad [6.91]$$

The electric field induced by this potential outside the sphere is:

$$\begin{aligned} \vec{\mathbf{E}} &= -\vec{\nabla} V = \left(1 + \frac{2R^3}{r^3}\right) E_0 \cos\theta\, \vec{\mathbf{e}}_r + \left(\frac{R^3}{r^3} - 1\right) E_0 \sin\theta\, \vec{\mathbf{e}}_\theta \\ &= E_0\left(\cos\theta\, \vec{\mathbf{e}}_r - \sin\theta\, \vec{\mathbf{e}}_\theta\right) + E_0 \frac{R^3}{r^3}\left(2\cos\theta\, \vec{\mathbf{e}}_r + \sin\theta\, \vec{\mathbf{e}}_\theta\right) \\ &= E_0\left(\cos\theta\, \vec{\mathbf{e}}_r - \sin\theta\, \vec{\mathbf{e}}_\theta\right) + E_0 \frac{R^3}{r^3}\left[3\cos\theta\, \vec{\mathbf{e}}_r - \left(\cos\theta\, \vec{\mathbf{e}}_r - \sin\theta\, \vec{\mathbf{e}}_\theta\right)\right] \\ &= E_0 \vec{\mathbf{k}} + E_0 \frac{R^3}{r^3}\left[3\left(\vec{\mathbf{e}}_r \cdot \vec{\mathbf{k}}\right)\vec{\mathbf{e}}_r - \vec{\mathbf{k}}\right] \end{aligned} \qquad [6.92]$$

EXAMPLE 6.17. (Magnetic Scalar Potential of a Circular Loop).– Let us calculate the magnetic scalar potential, at any point in space (r,θ,φ), of a circular loop of center O, of radius R, located in the plane Oxy and traversed by a current I. As the system is invariant around Oz, the magnetic potential is governed by a series of type [6.89]. We need to consider two possibilities:

1) $r < R$: the term $\frac{B_l}{r^{l+1}}$ diverges when $r \longrightarrow 0$ and must therefore be eliminated from the expression of $V_m(r,\theta)$, that is, we must be limited to $V_m(r,\theta) = \sum_{l=0}^{\infty} A_l r^l P_l(\cos\theta)$. For $\theta = 0$, the magnetic potential is reduced to $V_m(z) = \sum_{l=0}^{\infty} A_l z^l$ (in this case, the vector $\vec{\mathbf{r}}$ lies along Oz, that is, $r = z$ and $P_l(1) = 1$). Further, it has already been seen that:

$$\begin{aligned} V_m(z) &= -\frac{I}{2}\frac{z}{\sqrt{z^2+R^2}} = -\frac{Iz}{2R}\left(1 + \frac{z^2}{R^2}\right)^{-1/2} \\ &= -\frac{I}{2}\sum_{n=0}^{\infty} \frac{(-1)^n (2n)!}{(n!)^2\, 2^{2n}} \left(\frac{z}{R}\right)^{2n+1} \end{aligned} \qquad [6.93]$$

which shows that $A_{2n+1} = -\frac{I}{2}\frac{(-1)^n(2n)!}{(n!)^2 2^{2n} R^{2n+1}}$ and $A_{2n} = 0$. By substituting these expressions in $V_m(r,\theta)$, the general solution will be:

$$V_m(r,\theta) = -\frac{I}{2}\sum_{n=0}^{\infty}\frac{(-1)^n(2n)!}{2^{2n}(n!)^2 R^{2n+1}} r^{2n+1} P_{2n+1}(\cos\theta) \quad [6.94]$$

and consequently:

$$\begin{aligned} B_r &= -\mu_0\frac{\partial V}{\partial r} = \frac{\mu_0 I}{2}\sum_{n=0}^{\infty}\frac{(-1)^n(2n+1)!}{2^{2n}(n!)^2 R^{2n+1}} r^{2n} P_{2n+1}(\cos\theta) \\ B_\theta &= -\mu_0\frac{1}{r}\frac{\partial V}{\partial\theta} = -\frac{\mu_0 I}{2}\sum_{n=0}^{\infty}\frac{(-1)^n(2n)!}{2^{2n}(n!)^2 R^{2n+1}} r^{2n}\sin\theta P'_{2n+1}(\cos\theta) \end{aligned} \quad [6.95]$$

The magnetic potential, V_m, can be expressed in terms of the variables r and z. For example, at expansion of the second order, we get:

$$V_m(r,z) = -\frac{I}{2}\left[\frac{z}{R} - \frac{z(5z^2-3r^2)}{4R^3} + \frac{3z(63z^4-70r^2z^2+15r^4)}{64R^5} + \cdots\right] \quad [6.96]$$

since $P_1(\cos\theta) = \cos\theta$, $P_3(\cos\theta) = \left(5\cos^3\theta - 3\cos\theta\right)/2$ and $P_5(\cos\theta) = \left(63\cos^5\theta - 70\cos^3\theta + 15\cos\theta\right)/8$.

2) $r > R$: the term $A_l r^l$ must be eliminated in this region, as it diverges for $r \longrightarrow \infty$. In other words, we must be limited to $V_m(r,\theta) = A_0 + \sum_{l=0}^{\infty}\frac{B_l}{r^{l+1}}P_l(\cos\theta)$. For $\theta = 0$, this reduces to $V_m(z) = A_0 + \sum_{l=0}^{\infty}\frac{B_l}{z^{l+1}}$. By comparing this series with:

$$V_m(z) = -\frac{I}{2}\frac{z}{\sqrt{z^2+R^2}} = -\frac{I}{2}\left[1+\left(\frac{R}{z}\right)^2\right]^{-1/2} = -\frac{I}{2}\sum_{n=0}^{\infty}\frac{(-1)^n(2n)!}{(n!)^2 2^{2n}}\left(\frac{R}{z}\right)^{2n} \quad [6.97]$$

it can be derived that $A_0 = -\frac{I}{2}$, $B_{2n} = 0$ and $B_{2n-1} = -\frac{I}{2}\frac{(-1)^n(2n)!R^{2n}}{(n!)^2 2^{2n}}$. It therefore follows that:

$$V_m(r,\theta) = \frac{I}{2}\left[-1 - \sum_{n=1}^{\infty}\frac{(-1)^n(2n)!R^{2n}}{(n!)^2 2^{2n}}\frac{1}{r^{2n}}P_{2n-1}(\theta)\right] \quad [6.98]$$

EXAMPLE 6.18. (Helmholtz Coil or Helmholtz Pair).– Let us calculate the magnetic potential and magnetic field generated by a pair of loops (of radius R), located at distance R from each other and parallel to the plane Oxy. The same current I flows through both loops in the same direction. For reasons of symmetry, it is assumed that the centers of the two turns are located at $z = -\frac{R}{2}$ and $z = \frac{R}{2}$, respectively, as indicated in Figure 6.12. On the one hand, the magnetic potential of the Helmholtz pair along the Oz axis is given by:

$$\begin{aligned} V_m(0,0,z) &= -\frac{I}{2}\left[\frac{z-\frac{R}{2}}{\sqrt{\left(z-\frac{R}{2}\right)^2+R^2}}+\frac{z+\frac{R}{2}}{\sqrt{\left(z+\frac{R}{2}\right)^2+R^2}}\right] \\ &= -\frac{8I}{5^{3/2}}\left[\frac{z}{R}-\frac{144}{5^4}\left(\frac{z}{R}\right)^5+\cdots\right] \end{aligned} \quad [6.99]$$

The expansion [6.99] is valid only between the two loops. Outside these two loops, we must use:

$$V_m(0,0,z) = -I\left[1-\frac{1}{2}\left(\frac{R}{z}\right)^2+\frac{15}{32}\left(\frac{R}{z}\right)^6+\cdots\right] \quad [6.100]$$

On the other hand, in the region between the two turns, the potential is represented by the series $V_m(r,\theta) = \sum_{l=1}^{\infty} A_l\left(\frac{r}{R}\right)^l P_l(\cos\theta)$, which reduces to $V_m(0,0,z) = \sum_l A_l\left(\frac{z}{R}\right)^l$ along the Oz axis. Comparing this result with [6.99], we obtain $A_0 = 0$, $A_1 = -\frac{8I}{5^{3/2}}$, $A_2 = A_3 = A_4 = 0$, $A_5 = -\frac{8I}{5^{3/2}} \cdot \frac{144}{5^4}$, etc. Consequently, it is found that:

$$V_m(r,\theta) = -\frac{8I}{5^{3/2}}\left[\frac{r}{R}P_1(\cos\theta)-\frac{144}{5^4}\left(\frac{z}{R}\right)^5 P_5(\cos\theta)+\cdots\right] \quad [6.101]$$

In cylindrical coordinates, this expression becomes:

$$V_m(r,z) = -\frac{8I}{5^{3/2}}\left[\frac{z}{R}-\frac{144}{625}\left(\frac{z^5}{R^5}+\frac{15}{8}\frac{zr^5}{R^5}-5\frac{z^3r^2}{R^5}\right)+\cdots\right] \quad [6.102]$$

and, consequently:

$$\begin{aligned} B_r(r,z) &= \frac{8\mu_0 I}{5^{3/2}} \cdot \frac{144}{125}\left(\frac{4rz^3-3zr^3}{2R^5}\right)+\cdots \\ B_z(r,z) &= \frac{8\mu_0 I}{5^{3/2}R}\left[1-\frac{144}{125}\left(\frac{z^4}{R^4}+\frac{3}{8}\frac{r^4}{R^4}-3\frac{z^2r^2}{R^4}\right)+\cdots\right] \end{aligned} \quad [6.103]$$

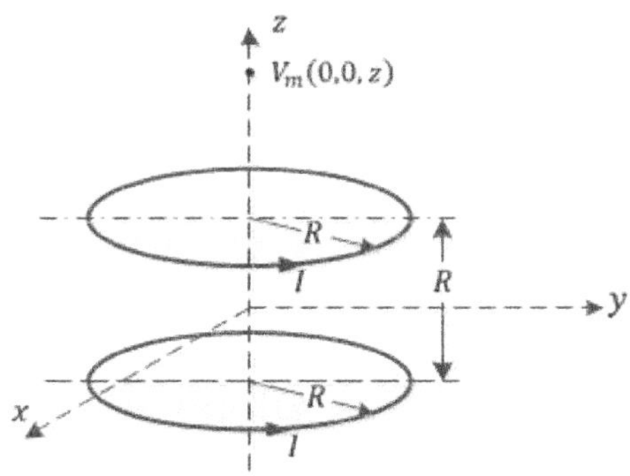

Figure 6.12. *Helmholtz coil*

EXAMPLE 6.19. (A Hollow Charged Sphere Rotating Around an Axis at a Uniform Angular Velocity).– Determine the magnetic field induced by a hollow uniformly charged sphere rotating around the axis Oz with angular velocity $\omega\overrightarrow{\mathbf{k}}$. The magnetic potential generated by the sphere is of the form $V_m^{\text{int}}(r,\theta)=\sum_{l=0}^{\infty}A_l r^l P_l(\cos\theta)$ for $r<R$ and $V_m^{\text{ext}}(r,\theta)=\sum_{l=0}^{\infty}\frac{B_l}{r^{l+1}}P_l(\cos\theta)$ for $r>R$. Consequently, the magnetic field produced by this potential is equal to $\overrightarrow{\mathbf{B}}^{\text{int}}=-\sum_{l=1}^{\infty}lA_l r^{l-1}P_l(\cos\theta)\times\overrightarrow{\mathbf{e}}_r+\sum_{l=0}^{\infty}A_l r^{l-1}\sin\theta P_l'(\cos\theta)\overrightarrow{\mathbf{e}}_\theta$ for $r<R$ and $\overrightarrow{\mathbf{B}}^{\text{ext}}=\sum_{l=1}^{\infty}(l+1)\frac{B_l}{r^{l+2}}\times P_l(\cos\theta)\overrightarrow{\mathbf{e}}_r+\sum_{l=0}^{\infty}\frac{B_l}{r^{l+2}}\sin\theta P_l'(\cos\theta)\overrightarrow{\mathbf{e}}_\theta$ for $r>R$. These fields $\overrightarrow{\mathbf{B}}^{\text{ext}}$ and $\overrightarrow{\mathbf{B}}^{\text{int}}$ obey the continuity equations:

$$\begin{aligned} &B_r^{\text{ext}}-B_r^{\text{int}}=0\\ &B_\theta^{\text{ext}}-B_\theta^{\text{int}}=\mu_0 J_\varphi \end{aligned} \qquad [6.104]$$

where $J_\varphi=\sigma\omega R\sin\theta$ (σ is the surface charge density). Equations [6.104] immediately give $\sum_{l=1}^{\infty}\left[\frac{(l+1)B_l}{R^{l+2}}+lA_l R^{l-1}\right]P_l(\cos\theta)=0$ and $\sum_{l=0}^{\infty}\left[\frac{B_l}{R^{l+2}}-A_l r^{l-1}\right]P_l'(\cos\theta)\sin\theta=\mu_0\rho\omega R\sin\theta$. It follows that:

$$\begin{aligned} &\frac{(l+1)B_l}{R^{l+2}}+lA_l R^{l-1}=0\ (l\geq 1)\\ &\frac{B_1}{R^3}-A_1=\mu_0\rho\omega R\\ &\frac{B_l}{R^{l+2}}-A_l r^{l-1}=0\ (l\geq 2) \end{aligned} \qquad [6.105]$$

which immediately leads to $A_1=-\frac{2R\sigma\omega}{3}$, $B_1=\frac{R^4\sigma\omega}{3}$ and $A_l=B_l=0$ for $l\geq 2$. The magnetic potential is therefore:

$$V_m(r,\theta) = \begin{cases} -\dfrac{2R\sigma\omega}{3} r\cos\theta = -\dfrac{2R\sigma\omega}{3} z & r < R \\ \dfrac{R^4\sigma\omega}{3r^2}\cos\theta & r > R \end{cases} \quad [6.106]$$

and consequently:

$$\vec{\mathbf{B}}(r > R) = \begin{cases} \dfrac{2\mu_0 R\sigma\omega}{3}\vec{\mathbf{k}} & r < R \\ \dfrac{2\mu_0 R^4\sigma\omega}{3r^3}\cos\theta\,\vec{\mathbf{e}}_r + \dfrac{\mu_0 R^4\sigma\omega}{3r^3}\sin\theta\,\vec{\mathbf{e}}_\theta & r > R \end{cases} \quad [6.107]$$

6.3.4. *Numerical 2D resolution*

In many cases, the geometry and boundary conditions are very complicated such that analytical solutions are not possible. In such cases, the Laplace equation must be solved numerically. In this section, we restrict ourselves to two-dimensional Cartesian solutions. It is obviously impossible to solve the Laplace equation numerically at any point in space. We restrict ourselves to a finite domain $\mathcal{D}$, which is regarded as a finite discrete 2D grid of size $M \times N$. It is already known that the Laplace equation admits a unique solution in a domain $\mathcal{D}$ if the value of V is fixed on the boundaries of the domain. In this case, the boundaries correspond to the contour of the grid, but can also include other points inside the grid on which the potential is fixed. By setting the boundary conditions, we can then use a finite difference method to solve the problem. The principle is exactly the same as for solving a differential equation using the Euler method: spatial derivatives of potential are approximated by the differences between the values of the potential between two neighboring points on the grid. The novelty compared to Euler's method is the iterative resolution: starting from an arbitrary initial value (value of V fixed arbitrarily and randomly at any point on the grid), the accuracy of the result can be improved by recurrence so that it conforms to Laplace's equation.

To illustrate this, let us take the two-dimensional grid $M \times N$, as depicted in Figure 6.13(a). The space between points grid is denoted by a: the coordinates of the point (I, J) of the grid are $(x_I, y_J) = (aI, aJ)$. The two-dimensional Laplace equation is written as: $\frac{\partial^2 V}{\partial x^2} + \frac{\partial^2 V}{\partial y^2} = 0$. Second derivatives can be written as finite differences using the finite expansion. Using Taylor's formula, it is possible to write:

$$V(x_I + a, y_J) = V(x_I, y_J) + a\frac{\partial V}{\partial x}(x_I, y_J) + \frac{1}{2}a^2\frac{\partial^2 V}{\partial x^2}(x_I, y_J) + O(a^3)$$

$$V(x_I - a, y_J) = V(x_I, y_J) - a\frac{\partial V}{\partial x}(x_I, y_J) + \frac{1}{2}a^2\frac{\partial^2 V}{\partial x^2}(x_I, y_J) + O(a^3)$$

[6.108]

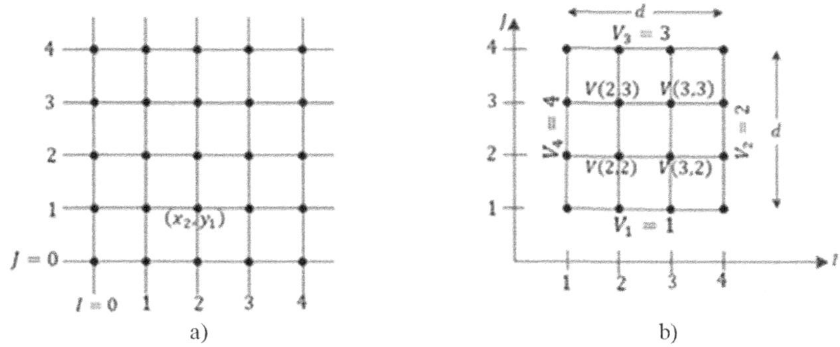

Figure 6.13. *(a) Grid in a two-dimensional domain $\mathcal{D}$. (b) The potential at the four internal points are found through successive iterations. The potential at any point is equal to the mean of the potentials of the four adjacent points*

For simplicity, we use the notation: $V[I,J] \equiv V(x_I, y_J)$. Equations [6.108] yield immediately:

$$\frac{\partial^2 V}{\partial x^2}(x_I, y_J) \simeq \frac{1}{a^2}\left(V[I+1,J] + V[I-1,J] - 2V[I,J]\right) \qquad [6.109]$$

Similarly, we also have:

$$\frac{\partial^2 V}{\partial y^2}(x_I, y_J) \simeq \frac{1}{a^2}\left(V[I,J+1] + V[I,J-1] - 2V[I,J]\right) \qquad [6.110]$$

Therefore, at any point (I,J) of the grid, the discrete Laplace equation is written as:

$$\begin{aligned}\Delta V[I,J] = \frac{1}{a^2}\big(&V[I+1,J] + V[I-1,J] + V[I,J+1] \\ &+V[I,J-1] - 4V[I,J]\big) = 0\end{aligned} \qquad [6.111]$$

This can also be expressed in the form:

$$V[I,J] = \frac{1}{4}\left(V[I+1,J] + V[I-1,J] + V[I,J+1] + V[I,J-1]\right) \qquad [6.112]$$

The potential at any point is therefore a mean of the potentials at neighboring points: in this way, it would be possible to construct the values of V at any point of the grid by recurrence, starting from the boundaries where the potential values are known.

EXAMPLE 6.20. (Grid with Four Internal Points).– Consider a grid with four internal points. Let us fix the potentials along the boundaries, as shown in Figure 6.13(b): $V_1 = V[I, J=1] = 1\ (\forall J)$, $V_2 = V[I=4, J] = 2\ (\forall J)$, $V_3 = V[I, J=4] = 3\ (\forall I)$ and $V_4 = V[I=1, J] = 4\ (\forall J)$. The initial values for the potential of all internal points are taken to be zero, that is, $V[2,2] = V[3,3] = V[2,3] = V[3,2] = 0$. The first iteration for $V[2,2]$ yields:

$$V[2,2] = \frac{1}{4}(V[2,1] + V[2,3] + V[1,2] + V[3,2]) = \frac{1}{4}(1+0+4+0) = 1.25 \quad [6.113]$$

By using this value of $V[2,2]$, for $V[3,2]$, we get:

$$\begin{aligned} V[3,2] &= \frac{1}{4}(V[2,2] + V[4,2] + V[3,1] + V[3,3]) \\ &= \frac{1}{4}(1.25+2+1+0) = 1.0625 \end{aligned} \quad [6.114]$$

Similarly, for $V[3,3]$:

$$\begin{aligned} V[3,3] &= \frac{1}{4}(V[3,2] + V[3,4] + V[2,3] + V[4,3]) \\ &= \frac{1}{4}(1.0625+3+0+2) = 1.5156 \end{aligned} \quad [6.115]$$

and $V[2,3]$:

$$\begin{aligned} V[2,3] &= \frac{1}{4}(V[2,2] + V[2,4] + V[1,3] + V[3,3]) \\ &= \frac{1}{4}(1.25+3+4+1.5156) = 2.4414 \end{aligned} \quad [6.116]$$

The procedure is then repeated for the four internal points, always using the most recent values of the potential. As the number of iterations increases, the potential values of the internal points approach the exact solutions.

Table 6.1 contains the first 10 iterations. It is clearly seen that the results obtained are very close to exact values.

	0	1	2	3	4	5
V [2, 2]	0	1.2500	2.1260	2.3777	2.4670	2.4911
V [3, 2]	0	1.0625	1.6604	1.9133	2.9770	1.9935
V [3, 3]	0	1.5156	2.2755	2.4409	2.4829	2.4952
V [2, 3]	0	2.4414	2.8504	2.9546	2.9875	2.9966
	6	**7**	**8**	**9**	**10**	**Exact**
V [2, 2]	2.4975	2.4993	2.4998	2.4999	2.5000	2.5000
V [3, 2]	1.9982	1.9995	1.9999	2.0000	2.0000	1.9771
V [3, 3]	2.4987	2.4996	2.4999	2.5000	2.5000	2.5000
V [2, 3]	2.9991	2.9997	2.9999	3.0000	3.0000	3.0299

Table 6.1. *The first ten iterations for the four internal points V[2,2], V[3,2], V[3,3] and V[2,3]*

According to [6.27], the value of the exact solution is:

$$V = \sum_{n=0}^{\infty} \frac{4}{(2n+1)\pi \sinh(2n+1)\pi} \left[\sin\frac{(2n+1)\pi y}{d} \left(V_3 \sinh\frac{(2n+1)\pi x}{d} + V_1 \sinh\frac{(2n+1)\pi(d-x)}{d} \right) + \sin\frac{(2n+1)\pi x}{d} \left(V_2 \sinh\frac{(2n+1)\pi y}{d} + V_4 \sinh\frac{(2n+1)\pi(d-y)}{d} \right) \right] \quad [6.117]$$

where $V_1 = 1$, $V_2 = 2$, $V_3 = 3$ and $V_4 = 4$. By restricting ourselves to the first harmonic, it comes that:

$$V[2,2] = V[3,3] = V\left(\frac{d}{3}, \frac{d}{3}\right) = V\left[\frac{2d}{3}, \frac{2d}{3}\right] \simeq \frac{10\sqrt{3}}{\pi \sinh \pi} \left(\sinh\frac{\pi}{3} + \sinh\frac{2\pi}{3} \right)$$

$$V[2,3] = V\left(\frac{d}{3}, \frac{2d}{3}\right) \simeq \frac{2\sqrt{3}}{\pi \sinh \pi} \left(7\sinh\frac{\pi}{3} + 3\sinh\frac{2\pi}{3} \right)$$

$$V[3,2] = V\left(\frac{2d}{3}, \frac{d}{3}\right) \simeq \frac{2\sqrt{3}}{\pi \sinh \pi} \left(7\sinh\frac{2\pi}{3} + 3\sinh\frac{\pi}{3} \right) \quad [6.118]$$

The above method can be applied to other boundaries forms and with a higher number of inner points.

6.3.5. *Solving in complex plane*

6.3.5.1. *Complex variable functions (reviews of results)*

The use of complex variable functions considerably simplifies the solution of the 2D Laplace equation. The introduction of a complex variable $z = x + iy$ in place of the real variable poses some problems, particularly with regard to the derivability and uniqueness of the definition.

EXAMPLE 6.21.– The function $f(z) = z^2$ poses no problem, since any point z in the complex plane is uniquely associated with another point z^2 (uniform function).

EXAMPLE 6.22.– The function $f(z) = \sqrt{z}$ is a multivalued function. For $z = re^{sfi\theta}$, we find $f(z) = \sqrt{r}e^{sfi\theta/2}$. This function is therefore said to have a *branching point* at $z = 0$, because if z completes a revolution around $z = 0$ to return to the same point (the same value of z), then the function changes sign (it no longer has the same value). For example, if we start at $z = z_0 = r_0 e^{sfi\theta_0}$, then $f(r_0, \theta_0) = \sqrt{r_0}e^{sfi\theta_0/2}$ is the initial value of the function. Rotating around the origin by 2π, the function takes the value $f(r_0, \theta_0 + 2\pi) = \sqrt{r_0}e^{i\theta_0/2}e^{i\pi} = -\sqrt{r_0}e^{i\theta_0/2} = -f(r_0, \theta_0)$. Similarly, a point located at infinity, $z \longrightarrow \infty$, is also a branching point. To see this, let us study the function $f\left(\frac{1}{z}\right) = r^{-1/2}e^{-i\theta/2}$. For this new function, $z = 0$ corresponds to $z \longrightarrow \infty$ in the original case. Let us make $f\left(\frac{1}{z}\right)$ rotate around this new origin. We therefore observe the same phenomenon, that is, the function has a double value. We then identify two branch points, one at $z = 0$ and the other at $z \longrightarrow \infty$. This type of singularity is not localized like a pole, but extends across the complex plane. Let us trace the half-line $[0, +\infty[$ along Ox, called the *branch line*, as shown in Figure 6.14(a). Imagine that we start at an initial point at $\theta_0 = 0^+$ (just above but infinitely close to the branch) and that there is a rotation (clockwise) around the origin, up to $\theta = 2\pi$ (just below but infinitely close to the branch line). We stop and go back to $\theta_0 = 0^+$. It is seen that the function regains its initial value as we have not crossed the branch line. If the branch line is crossed (independent of the distance from the origin), the value of the function changes. If we cross the branch line (independent of distance from coordinate origin), the value of the function will change. We say that by crossing the branch line, the function changes sheet (the collection of two sheets is called *Riemann surface*). Note that a function has as many Riemann sheets as there are possible values for a point.

EXAMPLE 6.23.– Let the function $f(z) = \sqrt{(z-a)(z-b)}$, where a and b are reals ($a < b$). Let us define the variables $z - a = r_a e^{i\theta_a}$ and $z - b = r_b e^{i\theta_b}$, which yields $f(z) = \sqrt{r_a r_b}e^{i(\theta_a + \theta_b)/2}$. Here are the different possibilities:

– $r_b < b - a$ (contour C_1 in Figure 6.14(b)): if we start from a point z_0 and make a complete turn around the point b, we see that $\theta_a \longrightarrow \theta_a$ and $\theta_b \longrightarrow \theta_b + 2\pi$; this shows that $f(z_0) \longrightarrow -f(z_0)$ (the function $f(z_0)$ changes sign). This confirms that b is a branch point. It would thus appear normal to trace the branch line between the two points a and b (the line segment $[a, b]$) or between b and a point located at ∞ (the half-line $[b, \infty[$).

$- r_b > b - a$ (contour C_2 in Figure 6.14(c)): in this case, if we do a complete turn around the point b, it is seen that $\theta_a \longrightarrow \theta_a + 2\pi$ and $\theta_b \longrightarrow \theta_b + 2\pi$. This implies that $f(z_0)$ does not change sign.

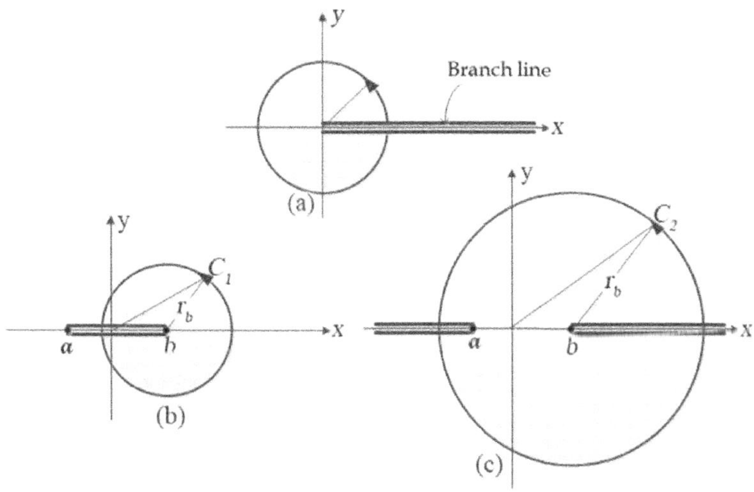

Figure 6.14. *(a) Branch line for the function $f(z) = \sqrt{z}$. (b and c) Different possibilities of branch lines for the function $f(z) = \sqrt{(z-a)(z-b)}$*

The same reasoning can be used for the point a. Generally, it is possible to define a branch line between a and b (the line segment $[a, b]$) or a branch line in two segments, the first between $]-\infty, a]$ and the second between $[b, +\infty[$. It can be seen that the contour C_1 crosses the branch line once, resulting in a change in the Riemann surface. However, the contour C_2 crosses the branch line twice, and the function returns to its initial value. ■

A complex function is written as: $f(z) = u(x, y) + iv(x, y)$, where $u(x, y)$ (respectively, $v(x, y)$) is the real (respectively, imaginary) part of f. This means that any point $z = x + iy$ of the complex plane is associated with a corresponding point $w = u + iv$ in the image plane, called the *plane of images* or *plane of f*, through f. In other words, the function f transforms $(x + iy)$ to $(u + iv)$. The following equations:

$$u = u(x, y) \qquad v = v(x, y) \qquad [6.119]$$

define a transformation between the z plane and the f plane. If to each point of the f plane corresponds one and only one point of the z plane, and conversely, we speak of a *one-to-one transformation*. In this case, a set of points in the z plane (such as a curve or region) is transformed into a set of points in the f plane (curve or region), and conversely. This is called a *mapping* (of the curve or region) in the z plane to the f plane. Here are a few examples.

EXAMPLE 6.24.– Let $\mathcal{D}$ be a rectangular region (in the z plane) bounded by $x = 0$, $y = 0$, $x = 2$ and $y = 1$. Let us determine the region $\mathcal{D}'$ obtained by mapping $\mathcal{D}$ through the transformation $w = \sqrt{2}\mathrm{e}^{\mathrm{i}\pi/4}z+1-2\mathrm{i} \Longrightarrow u+\mathrm{i}v = (1+\mathrm{i})(x+\mathrm{i}y)+1-2\mathrm{i}$ $\Longrightarrow u = x - y + 1$ and $v = x + y - 2$. The lines $x = 0$, $y = 0$, $x = 2$ and $y = 1$ are transformed into $u + v = 1$, $u - v = 3$, $u + v = 3$ and $u - v = 1$, respectively, as shown in Figure 6.15(a).

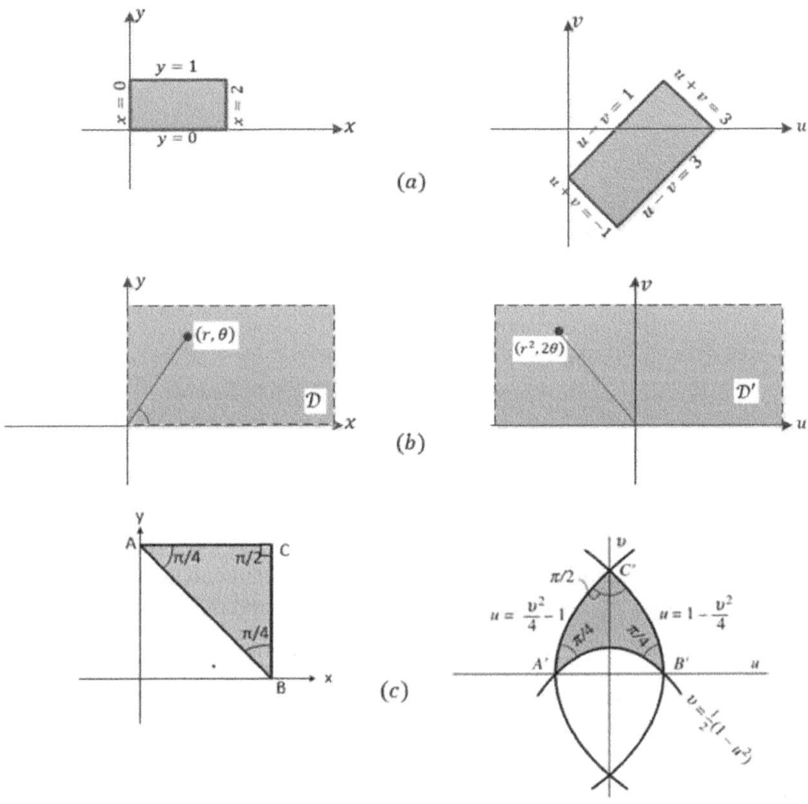

Figure 6.15. *(a) Mapping of the rectangle $\mathcal{D}$ through the transformation $w = \sqrt{2}\mathrm{e}^{\mathrm{i}\pi/4}z + 1 - 2i$. (b) Mapping of the first quadrant of the z plane using the transformation $w = z^2$. (c) Mapping of the region of the z plane that is bounded by the lines $x = 1$, $y = 1$ and $x + y = 1$ via the transformation $w = z^2$*

EXAMPLE 6.25.– Let us determine the region $\mathcal{D}'$ obtained by mapping through the transformation $w = z^2$: (a) $\mathcal{D}$ is the first quadrant of the z plane; (b) $\mathcal{D}$ is the region bounded by $x = 1, y = 1$ and $x + y = 1$. (a) Let $z = re^{i\theta} \Longrightarrow w = r^2 e^{i2\theta} \equiv \rho e^{i\phi} \Longrightarrow \rho = r^2$ and $\phi = 2\theta$. The z points whose coordinates are (r, θ) rotate from an angle θ. Consequently, all points in the first quadrant of the z plane (with $0 \leq \theta \leq \frac{\pi}{2}$) pass through the $0 \leq \phi \leq \pi$ region of the f plane (the upper half of the f plane), as shown in Figure 6.15(b). (b) $w = z^2 \Longrightarrow u + iv = x^2 - y^2 + 2ixy \Longrightarrow u = x^2 - y^2$ and $v = 2xy$. The lines $x = 1$, $y = 1$ and $x + y = 1$ are transformed, respectively, into $u = 1 - \frac{v^2}{4}$, $u = \frac{v^2}{4} - 1$ and $v = \frac{1}{2}\left(1 - u^2\right)$, as shown in Figure 6.15(c). ■

The function $f(z)$ is said to be *differentiable* at the point z if the ratio $\frac{f(z+\Delta z) - f(z)}{\Delta z}$ has a finite limit when $\Delta z \to 0$ (independent of the path taken). This limit is called *derivative* of $f(z)$ at the point z. It is denoted by $f'(z)$. If $f(z)$ is differentiable at the point $z = x + iy$, it must come that:

$$\frac{\partial u}{\partial x} = \frac{\partial v}{\partial y} \qquad \frac{\partial v}{\partial x} = -\frac{\partial u}{\partial y} \qquad [6.120]$$

These are the *Cauchy–Riemann conditions*. Inversely, if at the point (x, y), the functions $u(x, y)$ and $v(x, y)$ are differentiable and satisfy the conditions [6.120], then $f(z) = u(x, y) + iv(x, y)$ is differentiable at the point $z = x + iy$. The function $f(z)$ is said to be *analytic* at the point z, if it is differentiable at the point z and the derivative $f'(z)$ is continuous at the point z (differentiable in a certain neighborhood of the point z). The function $f(z)$ is said to be analytic in a domain $\mathcal{D}$, if it is analytic at all points in the domain. For any analytic function:

$$f'(z) = \frac{\partial u}{\partial x} + i\frac{\partial v}{\partial x} = \frac{\partial v}{\partial y} - i\frac{\partial u}{\partial y} = \frac{\partial u}{\partial x} - i\frac{\partial u}{\partial y} = \frac{\partial v}{\partial y} + i\frac{\partial v}{\partial x} \qquad [6.121]$$

Let $f(z)$ be an analytic function at the point z_0. Let us assume that $f'(z_0) \neq 0$. In the neighborhood of z_0, we write $w - w_0 \simeq f'(z_0)(z - z_0)$. $|f'(z_0)|$ appear here as the *dilatation coefficient* at z_0 through the mapping f: if $|f'(z_0)| > 1$ (respectively, $|f'(z_0)| < 1$), there is *dilatation* (respectively, *contraction*). The argument of $f'(z_0)$ is equal to the angle by which the tangent at z_0, of any smooth curve passing through z_0 (of the z plane), must be rotated to obtain the orientation of the tangent at the point $f(z_0)$ of the image of this curve in the f plane. Furthermore, if $\arg f'(z_0) > 0$ (respectively, $\arg f'(z_0) < 0$), the rotation is carried out clockwise (respectively, counterclockwise).

If the functions $u(x, y)$ and $v(x, y)$ have continuous second partial derivatives, it can be verified that:

$$\frac{\partial^2 u}{\partial x^2} + \frac{\partial^2 u}{\partial y^2} = \frac{\partial^2 v}{\partial x^2} + \frac{\partial^2 v}{\partial y^2} = 0 \qquad [6.122]$$

that is, they are *harmonic*. We can therefore conclude that both real and imaginary parts of an analytic function satisfy the two-dimensional Laplace equation. Any analytic function gives two solutions to the Laplace equation, suggesting that analytic functions can provide solutions to static potential problems. The functions u and v are also called *conjugate harmonic functions*. The essential points about these functions are outlined below.

– Let $f(z)$ be an analytic function in a region $\mathcal{D}$ and C a curve in $\mathcal{D}$. The integral of $f(z)$ along C can be reduced to the calculation of curvilinear integrals:

$$\int_C f(z)\,dz = \int_C (udx - vdy) + \mathrm{i}\int_C (vdx + udy) \tag{6.123}$$

In particular, for a closed curve C is closed, the integral of $f(z)$ along C is zero, since:

$$\begin{aligned}\oint_C f(z)\,dz &= \oint_C (udx - vdy) + \mathrm{i}\oint_C (vdx + udy) \\ &= -\int_{\mathcal{D}} \left(\frac{\partial v}{\partial x} + \frac{\partial u}{\partial y}\right) dxdy + \mathrm{i}\int_{\mathcal{D}} \left(\frac{\partial u}{\partial x} - \frac{\partial v}{\partial y}\right) dxdy = 0\end{aligned} \tag{6.124}$$

The integral $\int_{z_1}^{z_2} f(z)\,dz$ is therefore independent of the path connecting z_1 to z_2. This is *Cauchy's theorem*.

– If $f(z)$ is analytic inside C and on C, then:

$$f(z) = \frac{1}{2\pi\mathrm{i}} \oint_C \frac{f(z')}{z' - z} dz' \tag{6.125}$$

where z is a point inside C. This formula expresses the value of an analytic function at any interior point z of a domain $\mathcal{D}$ limited by a curve C in terms of its values over C. By deriving [6.125] n times, it comes that:

$$f^{(n)}(z) = \frac{n!}{2\pi\mathrm{i}} \oint_C \frac{f(z')\,dz'}{(z' - z)^{n+1}} \tag{6.126}$$

Equations [6.125] and [6.126] are called, respectively, the *first and second Cauchy formulas*.

– Let be a circle C with center O and radius R. Consider an analytic function $f(z)$ inside C and on C. If $z = re^{\mathrm{i}\theta}$ $(r < R)$ denotes a point inside C, then:

$$\begin{aligned}f\left(re^{\mathrm{i}\theta}\right) &= \frac{1}{2\pi\mathrm{i}} \oint_C \frac{f(z')}{z' - re^{\mathrm{i}\theta}} dz' \Longrightarrow f\left(re^{\mathrm{i}\theta}\right) \\ &= \frac{1}{2\pi} \int_0^{2\pi} \frac{f\left(Re^{\mathrm{i}\theta'}\right)\left(1 - \frac{r}{R} e^{-\mathrm{i}(\theta - \theta')}\right)}{1 + \frac{r^2}{R^2} - 2\frac{r}{R}\cos(\theta - \theta')} d\theta'\end{aligned} \tag{6.127}$$

In the same way, we also have:

$$0=\frac{1}{2\pi\mathrm{i}}\oint_C\frac{f(z')}{z'-\frac{R^2}{r}e^{\mathrm{i}\theta}}dz' \Longrightarrow 0=\frac{1}{2\pi}\int_0^{2\pi}\frac{f\left(Re^{\mathrm{i}\theta'}\right)\left(\frac{r^2}{R^2}-\frac{r}{R}e^{-\mathrm{i}(\theta-\theta')}\right)}{1+\frac{r^2}{R^2}-2\frac{r}{R}\cos(\theta-\theta')}d\theta' \qquad [6.128]$$

because $\frac{R^2}{r}e^{\mathrm{i}\theta}\notin C$. By subtracting [6.128] from [6.127], it finally comes that $f\left(re^{\mathrm{i}\theta}\right)=\frac{1}{2\pi}\int_0^{2\pi}\frac{(R^2-r^2)f\left(Re^{\mathrm{i}\theta'}\right)d\theta'}{R^2-2rR\cos(\theta-\theta')+r^2}$ and therefore:

$$\begin{aligned} u(r,\theta) &= \frac{1}{2\pi}\int_0^{2\pi}\frac{\left(R^2-r^2\right)u\left(R,\theta'\right)d\theta'}{R^2-2rR\cos(\theta-\theta')+r^2} \\ v(r,\theta) &= \frac{1}{2\pi}\int_0^{2\pi}\frac{\left(R^2-r^2\right)v\left(R,\theta'\right)d\theta'}{R^2-2rR\cos(\theta-\theta')+r^2} \end{aligned} \qquad [6.129]$$

These are *Poisson integral formulas for a circle.*

EXAMPLE 6.26.– Starting from [6.129], let us find the harmonic function, inside the circle of radius R, equal to u_0 on the boundary between θ_1 and θ_2 and 0 elsewhere:

$$\begin{aligned} u(r,\theta) &= \frac{1}{2\pi}\int_{\theta_1}^{\theta_2}\frac{\left(R^2-r^2\right)u_0 d\theta'}{R^2-2rR\cos(\theta'-\theta)+r^2} = \frac{u_0}{\pi}\tan^{-1}\left[\mu\tan\frac{\varphi'}{2}\right]\Bigg|_{\theta_1-\theta}^{\theta_2-\theta} \\ &= \frac{u_0}{\pi}\left\{\tan^{-1}\left[\mu\tan\frac{1}{2}(\theta_2-\theta)\right]-\tan^{-1}\left[\mu\tan\frac{1}{2}(\theta_1-\theta)\right]\right\} \end{aligned} \qquad [6.130]$$

where $\mu=\frac{R+r}{R-r}$. In particular, if the harmonic function takes, respectively, the constant values $u_0,u_1,\ldots,u_{n-1}$ over the intervals $]\theta_0,\theta_1]$, $]\theta_1,\theta_2],\ldots,]\theta_{n-1},\theta_n]$ with $\theta_0=0$ and $\theta_n=2\pi$, then:

$$\begin{aligned} u(r,\theta) &= \sum_{k=0}^{n-1}\frac{u_k}{\pi}\left\{\tan^{-1}\left[\mu\tan\frac{1}{2}(\theta_{k+1}-\theta)\right]-\tan^{-1}\left[\mu\tan\frac{1}{2}(\theta_k-\theta)\right]\right\} \\ &- \sum_{k=0}^{n}\frac{u_{k-1}-u_k}{\pi}\tan^{-1}\left[\mu\tan\frac{1}{2}(\theta_k-\theta)\right] \end{aligned} \qquad [6.131]$$

with the convention $u_{-1}=u_n=0$. ■

– Let $f(z)$ be an analytic function in the upper half-plane $\mathcal{P}(-\infty < x < \infty, y \geq 0)$. If $z = x + \mathrm{i}y$ is a point in this region, we write, according to Cauchy's formula, that:

$$f(z) = \lim_{R\to\infty}\left[\frac{1}{2\pi\mathrm{i}}\int_{-R}^{R}\frac{f(x')\,dx'}{x'-x-\mathrm{i}y}+\frac{1}{2\pi\mathrm{i}}\int_{C_R}\frac{f(z')}{z'-z}dz'\right] \quad [6.132]$$

where C_R is the upper boundary of the upper half-disk of radius R (the half-disk is large enough to contain z), as shown in Figure 6.16. Similarly, we also have:

$$0 = \lim_{R\to\infty}\left[\frac{1}{2\pi\mathrm{i}}\int_{-R}^{R}\frac{f(x')\,dx'}{x'-x+\mathrm{i}y}+\frac{1}{2\pi\mathrm{i}}\int_{C_R}\frac{f(z')}{z'-\overline{z}}dz'\right] \quad [6.133]$$

because $\overline{z} \notin \mathcal{P}(-\infty < x < \infty, y \geq 0)$. Subtracting [6.133] from [6.132], we find:

$$f(z) = \lim_{R\to\infty}\left[\frac{y}{\pi}\int_{-R}^{R}\frac{f(x')\,dx'}{(x'-x)^2+y^2}+\frac{\mathrm{i}Ry}{\pi}\int_{0}^{\pi}\frac{f\left(Re^{\mathrm{i}\theta'}\right)e^{\mathrm{i}\theta'}d\theta'}{\left(Re^{\mathrm{i}\theta'}-z\right)\left(Re^{\mathrm{i}\theta'}-\overline{z}\right)}\right] \quad [6.134]$$

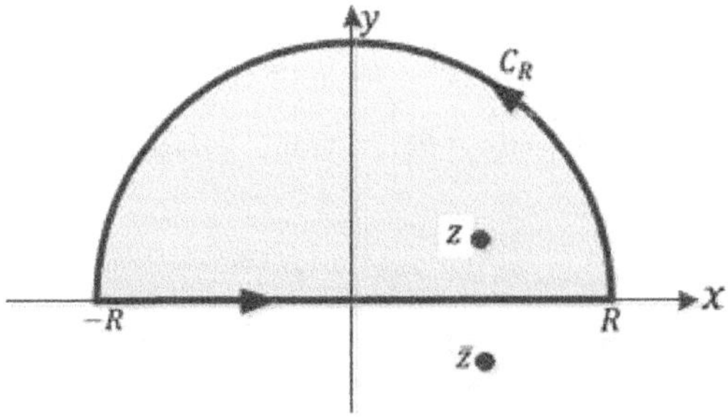

Figure 6.16. *Poisson integral formula for a half-plane*

The second term tends to zero when $R \to \infty$, because $\left|\frac{\mathrm{i}Ry}{\pi}\int_0^{\pi}\frac{f\left(Re^{\mathrm{i}\theta'}\right)e^{\mathrm{i}\theta'}d\theta'}{\left(Re^{\mathrm{i}\theta'}-z\right)\left(Re^{\mathrm{i}\theta'}-\overline{z}\right)}\right|$ $\leq \frac{RyM}{(R-|z|)^2} \longrightarrow 0$, where M is the maximum value of $|f(z')|$ on C_R.

Thus, we obtain $f(x+\mathrm{i}y) = \frac{y}{\pi}\int_{-\infty}^{\infty}\frac{f(x')dx'}{(x'-x)^2+y^2}$ and, therefore:

$$u(x,y) = \frac{y}{\pi}\int_{-\infty}^{\infty} \frac{u(x',0)\,dx'}{(x'-x)^2+y^2}$$
$$v(x,y) = \frac{y}{\pi}\int_{-\infty}^{\infty} \frac{v(x',0)\,dx'}{(x'-x)^2+y^2} \qquad [6.135]$$

This is the Poisson integral formula for a half-plane.

EXAMPLE 6.27.– From [6.135], let us determine the harmonic function in the upper half-plane equal to u_0 in the $]\alpha,\beta]$ interval of the Ox axis and 0 elsewhere, as illustrated in Figure 6.17:

$$u(x,y) = \frac{y}{\pi}\int_{\alpha}^{\beta} \frac{u_0 dx'}{(x'-x)^2+y^2} = \frac{u_0}{\pi}\left[\tan^{-1}\left(\frac{y}{x-\beta}\right) - \tan^{-1}\left(\frac{y}{x-\alpha}\right)\right] \qquad [6.136]$$

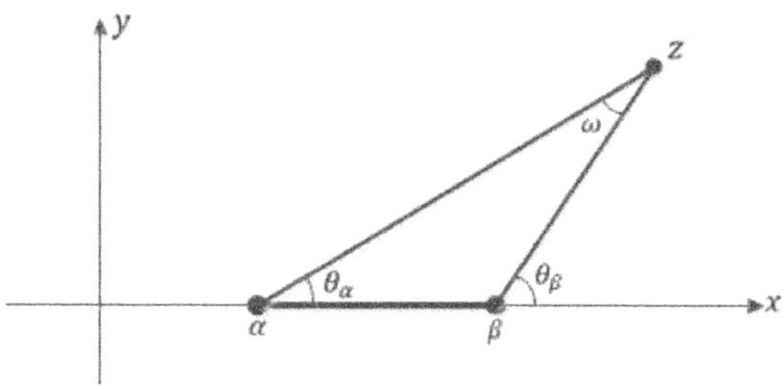

Figure 6.17. *Harmonic function in the upper half-plane using Poisson integral formula*

By introducing the angles θ_α and θ_β formed, respectively, by the vectors $z-\alpha$ and $z-\beta$ with the real axis Ox, we obtain:

$$u(x,y) = u_0\frac{\theta_\beta-\theta_\alpha}{\pi} \equiv u_0\frac{\omega}{\pi} \qquad [6.137]$$

Now let us find the harmonic function in the upper half-plane taking, respectively, on the segments $]-\infty,a_1]$, $]a_1,a_2]$, ..., $]a_n,+\infty[$ the constant values u_0, u_1, ..., u_n. The problem is solved by applying the formula [6.137] for each interval and then summing the results obtained (just note that $\theta_0 = 0$ and $\theta_{n+1} = \pi$):

$$u(x,y) = u_0\frac{\theta_1}{\pi} + u_1\frac{\theta_2-\theta_1}{\pi} + \cdots + u_n\frac{\pi-\theta_n}{\pi} = \frac{\theta_1}{\pi}(u_0-u_1)$$
$$+\frac{\theta_2}{\pi}(u_1-u_2) + \cdots + \frac{\theta_n}{\pi}(u_{n-1}-u_n) + u_n \qquad [6.138]$$

■

– Let $f(z)$ be an analytic function inside and on the circle of center a and radius R. For any point z inside C ($|z-a|<R$), we establish that:

$$\begin{aligned} f(z) &= \frac{1}{2\pi \mathrm{i}} \oint_C \frac{f(z')}{z'-z} dz' = \frac{1}{2\pi \mathrm{i}} \oint_C \frac{f(z')}{(z'-a)} \frac{1}{1-\left(\frac{z-a}{z'-a}\right)} dz' \\ &= \sum_{n=0}^{\infty} \frac{1}{2\pi \mathrm{i}} \oint_C \frac{f(z')(z-a)^n}{(z'-a)^{n+1}} dz' = \sum_{n=0}^{\infty} \frac{f^{(n)}(a)}{n!} (z-a)^n \end{aligned} \quad [6.139]$$

This is the *Taylor expansion*. If, in the neighborhood of a point z_0, an analytic function can be expanded into a Taylor series, we say that z_0 is an *ordinary point*. If this expansion is not possible, z_0 is said to be a *singular point*. A distinction is made between:

i) *Poles*: these are isolated singular points, around which $f(z)$ remains uniform and which are ordinary points for $\frac{1}{f(z)}$. The function $\frac{1}{z^2}$ has a double pole at its origin. The function $\frac{1}{z^2+1}$ has two simple poles: $z=-\mathrm{i}$ and $z=\mathrm{i}$.

ii) *Essential singular points*: these are isolated singular points around which $f(z)$ remains uniform, but which are still singular points for $\frac{1}{f(z)}$. The function $\exp\left(\frac{1}{z-a}\right)$ admits an essential singular point at the point $z=a$.

– Let $f(z)$ be an analytic function in a ring bounded from the outside by the circle C_R centered at the point a (a is a singular point for the function $f(z)$) and from the inside by the circle C_r, also centered at a. We also show that:

$$\begin{aligned} f(z) &= \frac{1}{2\pi \mathrm{i}} \oint_{C_R} \frac{f(z')}{z'-z} dz' - \frac{1}{2\pi \mathrm{i}} \oint_{C_r} \frac{f(z')}{z'-z} dz' \\ &= \frac{1}{2\pi \mathrm{i}} \oint_{C_R} \frac{f(z')}{(z'-a)} \frac{1}{1-\left(\frac{z-a}{z'-a}\right)} dz' + \frac{1}{2\pi \mathrm{i}(z-a)} \oint_{C_r} \frac{f(z')}{1-\left(\frac{z'-a}{z-a}\right)} dz' \\ &= \sum_{n=0}^{\infty} \frac{(z-a)^n}{2\pi \mathrm{i}} \oint_C \frac{f(z')}{(z'-a)^{n+1}} dz' + \sum_{n=1}^{\infty} \frac{(z-a)^{-n}}{2\pi \mathrm{i}} \oint_C \frac{f(z')}{(z'-a)^{-n+1}} dz' \\ &= \sum_{n=0}^{\infty} a_n (z-a)^n + \sum_{n=1}^{\infty} a_{-n} (z-a)^{-n} \end{aligned} \quad [6.140]$$

where:

$$\begin{aligned} a_n &= \frac{1}{2\pi i} \oint_C \frac{f(z')}{(z'-a)^{n+1}} dz' \quad \text{for } n \geq 0 \\ a_{-n} &= \frac{1}{2\pi i} \oint_C \frac{f(z')}{(z'-a)^{-n+1}} dz' \quad \text{for } n \geq 1 \end{aligned} \quad [6.141]$$

are *Laurent's coefficients*. The part $\sum_{n=0}^{\infty} a_n (z-a)^n$ (respectively, $\sum_{n=1}^{\infty} a_{-n} (z-a)^{-n}$) is called the *analytic part* (respectively, *main part*). If the main part has only a finite number of terms $\frac{a_{-1}}{z-a} + \cdots + \frac{a_{-n}}{(z-a)^n}$, the singular point $z=a$ is a pole of order n. On the other hand, if the main part has infinitely many terms, $z=a$ is an essential singularity.

– Let a be an isolated singular point of a function $f(z)$. From [6.141], we see that:

$$\oint_C f(z')\, dz' = 2\pi \mathrm{i} a_{-1} \qquad [6.142]$$

We therefore conclude that the integral $\oint_C f(z')\, dz'$ is expressed as a function of a single coefficient, namely, a_{-1}. a_{-1} is called the *residue* of $f(z)$ at a. We note that: $a_{-1} = \text{Res}[f(z), a]$. In particular, if a is a pole of order n, we verify that:

$$a_{-1} = \frac{1}{(n-1)!} \lim_{z \to a} \frac{d^{n-1}}{dz^{n-1}} \left[(z-a)^n f(z)\right] \qquad [6.143]$$

6.3.5.2. *Complex potential*

Consider an analytic function $f(z)$. We have already seen that the real part u and the imaginary part v of $f(z)$ satisfy the 2D Laplace equation. These functions can therefore play the role of static potentials. It is also easy to check that the curves $u(x,y) = c_1$ and $v(x,y) = c_2$, where c_1 and c_2 are constants, are perpendicular to each other, since $u\vec{\nabla} u \cdot \vec{\nabla} v = 0$. They always intersect at a point. These curves are represented by mutually orthogonal lines in the f plane, as shown in Figure 6.18. Vectors $\vec{\nabla} u$ and $\vec{\nabla} v$ can be used as a basis in the complex plane. If v represents the potential (or, more precisely, the potential lines), then u represents the field lines (the electric field). Consider a flat, infinite conductive plate held (in the Ouv plane) at potential V_0 along Ou. Above the Ou axis, the potential is of the form $V = V_0 - av$, where a is a constant. This potential can be thought of as the imaginary part of an analytical function $\Omega(u,v) = U(u,v) + \mathrm{i}V(u,v) = \Omega(f)$. Since $V = V_0 - av$, we deduce that:

$$\Omega(u,v) = \mathrm{i}V_0 - af = -au + \mathrm{i}(V_0 - av) \qquad [6.144]$$

The electric field induced by the potential V is $\vec{\mathbf{E}} = -\vec{\nabla} V = a\vec{\nabla} v$. We therefore conclude that the curvilinear components of the corresponding field are $E_u = 0$ and $E_v = a$.

EXAMPLE 6.28.– For $f(z) = z^2$, we get $\Omega(f) = \mathrm{i}V_0 - af = -a(x^2 - y^2) + \mathrm{i}(V_0 - 2axy)$ and, therefore:

$$u(x,y) = -a(x^2 - y^2) \quad v(x,y) = V_0 - 2axy \qquad [6.145]$$

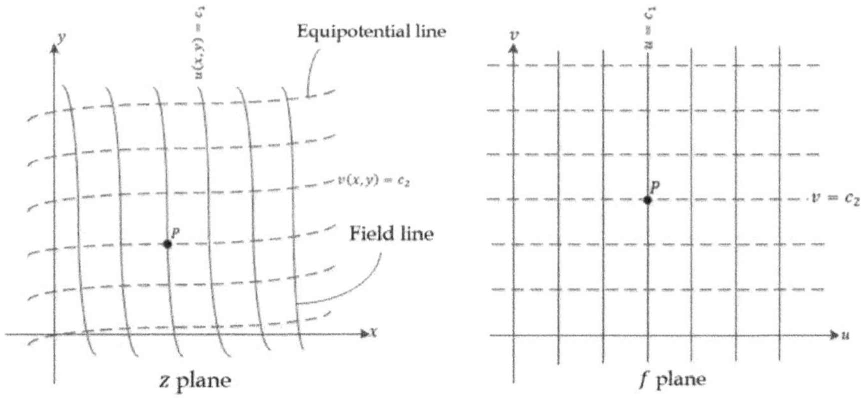

Figure 6.18. *Curve lines of an analytical function. For a color version of this figure, see www.iste.co.uk/abdesselam/electrodynamics1.zip*

The field and potential lines are produced, respectively, by $-a\left(x^2-y^2\right)=\text{const}$ and $V_0-2axy=\text{const}$, that is, by $u\left(x,y\right)=u$ and $v\left(x,y\right)=v$, as shown in Figure 6.19.

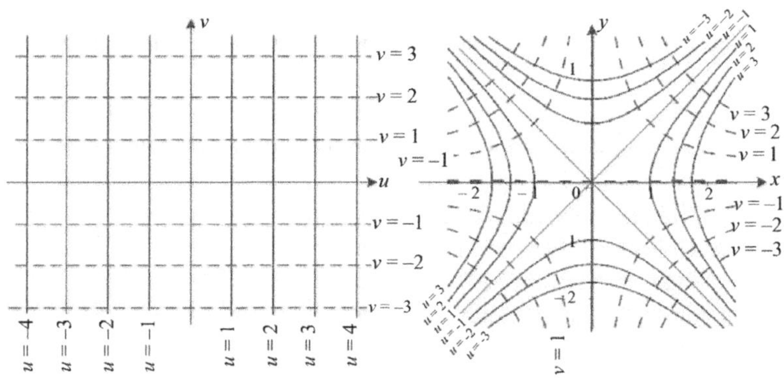

Figure 6.19. *Equipotentials and field lines produced by the analytical function $f\left(z\right)=z^2$. For a color version of this figure, see www.iste.co.uk/abdesselam/electrodynamics1.zip*

EXAMPLE 6.29.– Let us take $f = l\sin^{-1}\left(\frac{z}{l}\right)$ or, conversely, $z=l\sin f=l\left(\cosh v\times\ \sin u+\mathrm{i}\sinh v\cos u\right)$. This implies that $x=l\cosh v\sin u$ and $y=l\sinh v\cos u$. For this case, it is more difficult to express $\Omega\left(f\right)=\mathrm{i}V_0-af$ as a function of x and y by directly substituting the expression of f. However, this is not really necessary. For a point $\left(u,v\right)$ in the f plane, the corresponding point $\left(x,y\right)$

is easily identified. The equipotentials $v = $ const. are easy to obtain, simply by writing:

$$\frac{x^2}{l^2 \cosh^2 v} + \frac{y^2}{l^2 \sinh^2 v} = 1 \quad [6.146]$$

The equipotential is therefore an ellipse whose major and minor axes are $l \cosh v$ and $l \sinh v$, as shown in Figure 6.20. For high potentials, both $\cosh v$ and $\sinh v$ tend toward $\frac{e^v}{2}$, which means that the equipotentials tend to form circles. The field lines are produced by:

$$\frac{x^2}{l^2 \sin^2 u} - \frac{y^2}{l^2 \cos^2 u} = 1 \quad [6.147]$$

■

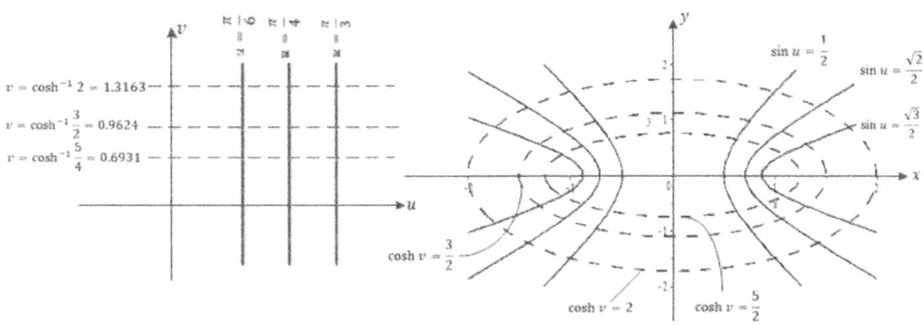

Figure 6.20. *Equipotentials (dotted blue) and field lines (red) generated by the analytical function $z = \sin^{-1}\left(\frac{z}{l}\right)$. For a color version of this figure, see www.iste.co.uk/abdesselam/electrodynamics1.zip*

From the above, it is obvious that there must be a harmonic function $U(x,y)$ conjugate to the electrostatic potential $V(x,y)$ measured in vacuum, so that the function:

$$\Omega(z) = U(x,y) + \mathrm{i}V(x,y) \quad [6.148]$$

is analytic in any region not occupied by charges (vacuum). We call $\Omega(z)$ the *complex electrostatic potential*, or simply the *complex potential*. The electrostatic field induced by the potential $V(x,y)$ is:

$$\mathbf{E}(z) = -\frac{\partial V}{\partial x} - \mathrm{i}\frac{\partial V}{\partial y} = -\frac{\partial V}{\partial x} - \mathrm{i}\frac{\partial U}{\partial x} = -\mathrm{i}\overline{\frac{d\Omega(z)}{dz}} = -\mathrm{i}\overline{\Omega'(z)} \quad [6.149]$$

The field magnitude is given by:

$$E = \left|\overrightarrow{\mathbf{E}}(z)\right| = \left|\overline{\Omega'(z)}\right| = |\Omega'(z)| \qquad [6.150]$$

The curves $U(x, y) = \text{const}$ and $V(x, y) = \text{const}$ are called *field lines* and *equipotential lines*, respectively.

EXAMPLE 6.30.– Let us calculate the potential at any point on a disk of radius R subjected on its boundary to the potential V_0 for $0 \leq \theta \leq \pi$ and $-V_0$ for $\pi < \theta < 2\pi$. The formula [6.131] immediately gives:

$$\begin{aligned} V(r, \theta) &= \frac{V_0}{\pi}\left(\tan^{-1}\left[\mu \tan\frac{1}{2}(\pi - \theta)\right] - \tan^{-1}\left[\mu \tan\frac{1}{2}(0 - \theta)\right]\right) \\ &- \frac{V_0}{\pi}\left(\tan^{-1}\left[\mu \tan\frac{1}{2}(2\pi - \theta)\right] - \tan^{-1}\left[\mu \tan\frac{1}{2}(\pi - \theta)\right]\right) \\ &= \frac{2V_0}{\pi}\left(\tan^{-1}\left[\mu \tan\frac{\theta}{2}\right] + \tan^{-1}\left[\frac{\mu}{\tan\frac{\theta}{2}}\right]\right) \\ &= \frac{2V_0}{\pi}\tan^{-1}\left[\frac{2\mu}{(1-\mu^2)\sin\theta}\right] = V_0\left(1 - \frac{2}{\pi}\tan^{-1}\left[\frac{(1-\mu^2)\sin\theta}{2\mu}\right]\right) \\ &= V_0\left(1 + \frac{2}{\pi}\tan^{-1}\left[\frac{2Rr\sin\theta}{R^2 - r^2}\right]\right) \end{aligned} \qquad [6.151]$$

where $\mu = \frac{R+r}{R-r}$. The identities $\tan^{-1}(a-b) + \tan^{-1}(a+b) = \tan^{-1}\frac{2a}{1-a^2+b^2}$ and $\tan^{-1}x + \tan^{-1}\frac{1}{x} = \frac{\pi}{2}$ are used here. This solution is the imaginary part of the complex potential:

$$\Omega(z) = V_0\left(\mathrm{i} + \frac{2}{\pi}\ln\left[\frac{R+z}{R-z}\right]\right) \qquad [6.152]$$

Indeed, we have:

$$\begin{aligned} \frac{R+z}{R-z} &= \sqrt{\frac{R^2 + r^2 + 2Rr\cos\theta}{R^2 + r^2 - 2Rr\cos\theta}}\exp\left\{\mathrm{i}\left(\tan^{-1}\left[\frac{r\sin\theta}{R + r\cos\theta}\right]\right.\right. \\ &\left.\left. + \tan^{-1}\left[\frac{r\sin\theta}{R - r\cos\theta}\right]\right)\right\} \\ &= \sqrt{\frac{R^2 + r^2 + 2Rr\cos\theta}{R^2 + r^2 - 2Rr\cos\theta}}\exp\left(\mathrm{i}\tan^{-1}\left[\frac{2Rr\sin\theta}{R^2 - r^2}\right]\right) \end{aligned} \qquad [6.153]$$

which gives $\ln\left[\frac{R+z}{R-z}\right] = \frac{1}{2}\ln\left(\frac{R^2+r^2+2Rr\cos\theta}{R^2+r^2-2Rr\cos\theta}\right) + \mathrm{i}\tan^{-1}\left[\frac{2Rr\sin\theta}{R^2-r^2}\right]$ and, consequently, $\Im\left(\ln\left[\frac{R+z}{R-z}\right]\right) = \tan^{-1}\left[\frac{2Rr\sin\theta}{R^2-r^2}\right]$. It follows that: $V(r,\theta) = \Im\left(V_0\left(\mathrm{i}+\ \frac{2}{\pi}\ln\left[\frac{R+z}{R-z}\right]\right)\right)$.

EXAMPLE 6.31.– Let us determine the potential at any point in the upper half-plane with the boundary condition $V(x,0) = \begin{cases} V_0 & x \geq 0 \\ -V_0 & x < 0 \end{cases}$. Formula [6.138] gives:

$$V(x, y \geq 0) = V_0\left[1 - \frac{2}{\pi}\tan^{-1}\left(\frac{y}{x}\right)\right] \quad [6.154]$$

This solution appears as the imaginary part of the complex potential:

$$\Omega(z) = V_0\left(\mathrm{i} - \frac{2}{\pi}\ln z\right) \quad [6.155]$$

The field and equipotential lines are produced by the curves $x^2 + y^2 = u^2$ (i.e. semicircles of radius u) and $\tan^{-1}\left(\frac{y}{x}\right) = v$ (i.e. straight lines $y = \tan vx$), as shown in Figure 6.21.

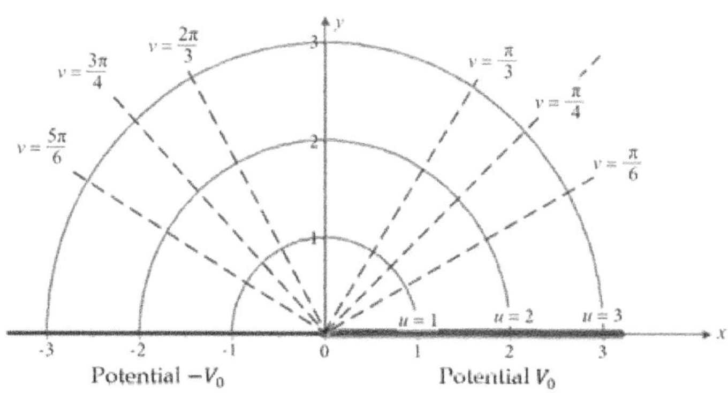

Figure 6.21. *Equipotential and field lines in a half-plane at potential V_0 for $x \geq 0$ and $-V_0$ for $x < 0$. For a color version of this figure, see www.iste.co.uk/abdesselam/electrodynamics1.zip*

6.3.5.3. *Complex form of Gauss' theorem*

Consider a closed curve C (oriented) in the z plane surrounding a line charge Q (per unit length perpendicular to the z plane). This curve can be seen as a closed ribbon of width one unit across the sheet (complex plane), as shown in Figure 6.22. In this case, Gauss' theorem allows us to write that:

$$\frac{Q}{\varepsilon_0} = \oint_C \left(\vec{\mathbf{E}} \cdot \vec{\mathbf{n}}\right) ds = \oint_C E_x dy - E_y dx = -\mathrm{i} \oint_C \left(E_x - \mathrm{i}E_y\right)\left(dx + \mathrm{i}dy\right) \quad [6.156]$$

because the circulation of $\vec{\mathbf{E}}$ around C is zero, that is, $\oint_C E_x dx + E_y dy = 0$. Consequently, we obtain:

$$\begin{aligned}\frac{Q}{\varepsilon_0} &= \mathrm{i}\oint_C \left(\frac{\partial V}{\partial x} - \mathrm{i}\frac{\partial V}{\partial y}\right)\left(dx + \mathrm{i}dy\right) = \oint_C \left(\frac{\partial U}{\partial x} + \mathrm{i}\frac{\partial V}{\partial x}\right)\left(dx + \mathrm{i}dy\right) \\ &= \oint_C \Omega'(z)\, dz\end{aligned} \quad [6.157]$$

which leads to:

$$Q = 2\pi \mathrm{i} \varepsilon_0 \sum \mathrm{Res}\left[\Omega'(z), \text{singularities}\right] \quad [6.158]$$

When the curve C contains no internal charge, the integral [6.157] is zero, that is, $\oint_C \Omega'(z)\, dz = 0$. In other words, the function $\Omega'(z)$ is analytic within C.

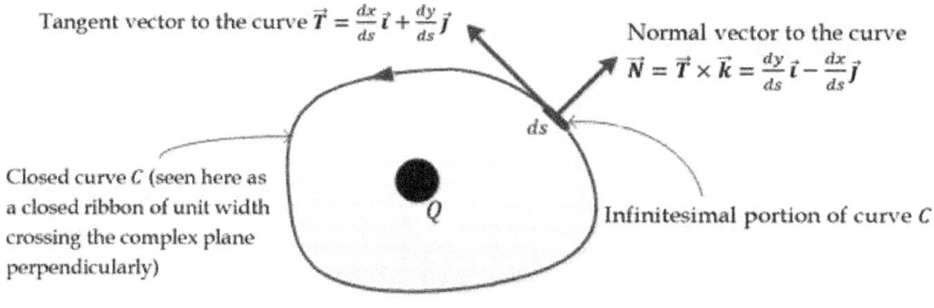

Figure 6.22. *Gauss' theorem in complex plane*

EXAMPLE 6.32.– (Example 6.32: Complex Potential of Line Charge).– Let us find the complex potential created by a line charge q (per unit length) perpendicular to z plane and located at the point $z = 0$. The induced electric field is radial. In other words, the normal component to any cylinder with radius r and axis through $z = 0$ is constant and equal to E_r, while the tangential component is zero. We can therefore write $\oint_C \left(\vec{\mathbf{E}} \cdot \vec{\mathbf{n}}\right) ds = E_r \oint_C ds = E_r 2\pi r = \frac{q}{\varepsilon_0}$, which gives $E_r = \frac{q}{2\pi\varepsilon_0 r}$. Since $E_r = -\frac{\partial V}{\partial r}$, we obtain:

$$V = -\frac{q}{2\pi\varepsilon_0} \ln r \quad [6.159]$$

(The additive integration constant is chosen to be zero.) This function appears as the imaginary part of the complex potential:

$$\Omega(z) = -\frac{q\mathrm{i}}{2\pi\varepsilon_0}\ln z \qquad [6.160]$$

Field and equipotential lines are described, respectively, by curves $\tan^{-1}\frac{y}{x} = u$ (straight line $y = \tan ux$) and $\ln r = \ln v$ (circle of radius v), as shown in Figure 6.23. If the line charge is located at $z = a$, formula [6.160] must be replaced by:

$$\Omega(z) = -\frac{q\mathrm{i}}{2\pi\varepsilon_0}\ln(z-a) \qquad [6.161]$$

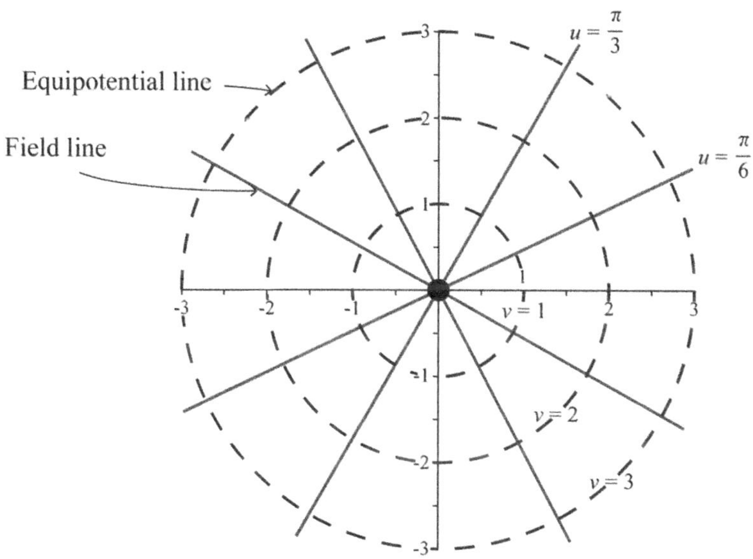

Figure 6.23. *Field and equipotential lines of a line charge placed at $z = 0$. For a color version of this figure, see www.iste.co.uk/abdesselam/electrodynamics1.zip*

EXAMPLE 6.33. (Complex Potential of a Set of Two Line Charges with Opposite Polarities).– Using the superposition principle, the complex potential of the two line charges of opposite polarities q and $-q$ located, respectively, at a and $-a$, is:

$$\Omega(z) = -\frac{q\mathrm{i}}{2\pi\varepsilon_0}\ln(z-a) + \frac{q\mathrm{i}}{2\pi\varepsilon_0}\ln(z+a) = \frac{q\mathrm{i}}{2\pi\varepsilon_0}\ln\left(\frac{z+a}{z-a}\right) \qquad [6.162]$$

The electrostatic potential corresponds to the imaginary part of the complex potential, that is, $V(x,y) = \Im\left[\frac{q\mathrm{i}}{2\pi\varepsilon_0}\ln\left(\frac{z+a}{z-a}\right)\right]$. We can clearly see that the potential is zero along Oy, that is, the ABC plane in Figure 6.24(a) is at the zero potential. This follows immediately from the fact that $z = \mathrm{i}y$ on the Oy axis and, consequently, $\Omega = -\frac{q\mathrm{i}}{2\pi\varepsilon_0}\ln\left(\frac{\mathrm{i}y+a}{\mathrm{i}y-a}\right) = -\frac{q\mathrm{i}}{2\pi\varepsilon_0}\ln\left(\frac{\mathrm{i}y-a}{\mathrm{i}y+a}\right) = -\frac{q\mathrm{i}}{2\pi\varepsilon_0}\ln\left(\frac{-\mathrm{i}y+a}{-\mathrm{i}y-a}\right) = \overline{\Omega}$. This implies that $\Im(\Omega) = 0$ on the Oy axis. We can therefore conclude that the potential due to an infinite plane ABC (located along Oy), held at zero potential, and a line charge q parallel to this plane (shown in Figure 6.24(b)) can be calculated from the formula [6.162]. By writing $z + a = r_1 e^{\mathrm{i}\theta_1}$ and $z - a = r_2 e^{\mathrm{i}\theta_2}$, we get $\Omega = U + \mathrm{i}V = \frac{q\mathrm{i}}{2\pi\varepsilon_0}\ln\left(\frac{r_1 e^{\mathrm{i}\theta_1}}{r_2 e^{\mathrm{i}\theta_2}}\right) = \frac{q}{2\pi\varepsilon_0}(\theta_2 - \theta_1) + \frac{\mathrm{i}q}{2\pi\varepsilon_0}\ln\left(\frac{r_1}{r_2}\right)$. This gives $U = \frac{q}{2\pi\varepsilon_0}(\theta_2 - \theta_1)$ and $V = \frac{q}{2\pi\varepsilon_0}\ln\left(\frac{r_1}{r_2}\right)$. The field and the equipotentials lines are produced, respectively, by $U = \frac{q}{2\pi\varepsilon_0}(\theta_2 - \theta_1) = \frac{q}{2\pi\varepsilon_0}\tan^{-1}u$ (u is a constant) and $V = \frac{q}{2\pi\varepsilon_0}\ln\left(\frac{r_1}{r_2}\right) = \frac{q}{2\pi\varepsilon_0}\tanh^{-1}v$ (v is a constant). Using the fact that $r_1 = \sqrt{(x+a)^2 + y^2}$, $r_2 = \sqrt{(x-a)^2 + y^2}$, $\theta_1 = \tan^{-1}\frac{y}{x+a}$ and $\theta_2 = \tan^{-1}\frac{y}{x-a}$, the equipotentials are written as follows:

$$\frac{\sqrt{(x+a)^2 + y^2}}{\sqrt{(x-a)^2 + y^2}} = e^{\tanh^{-1}v} \Longrightarrow \left[x - \frac{a}{v}\right]^2 + y^2 = \frac{a^2\left(1 - v^2\right)}{v^2} \qquad [6.163]$$

which, for a given value of v, is a circle centered on $\left(\frac{a}{v}, 0\right)$ and of radius $\frac{a\sqrt{1-v^2}}{|v|}$. These circles are drawn in blue in Figure 6.24(a). The field lines are given by $\tan^{-1}\frac{y}{x-a} - \tan^{-1}\frac{y}{x+a} = \tan^{-1}u$. Taking the tangent and simplifying, it comes that:

$$x^2 + \left[y + \frac{a}{u}\right]^2 = \frac{a^2\left(1 + u^2\right)}{u^2} \qquad [6.164]$$

which, for a given value of u, is a circle centered on $\left(0, -\frac{a}{u}\right)$ and of radius $\frac{a\sqrt{1+u^2}}{|u|}$. These circles, which pass through $(-a, 0)$ and $(a, 0)$, are represented in red in Figure 6.24(a).

6.3.5.4. *Capacitance*

If the curve C coincides with an equipotential $V(x,y) = V_1$, the total charge $Q_{\text{tot}}(V_1)$ contained within this curve is:

$$Q_{\text{tot}}(V_1) = \varepsilon_0 \oint_{V(x,y)=V_1} \left[\frac{\partial V}{\partial y}dx - \frac{\partial V}{\partial x}dy\right] \qquad [6.165]$$

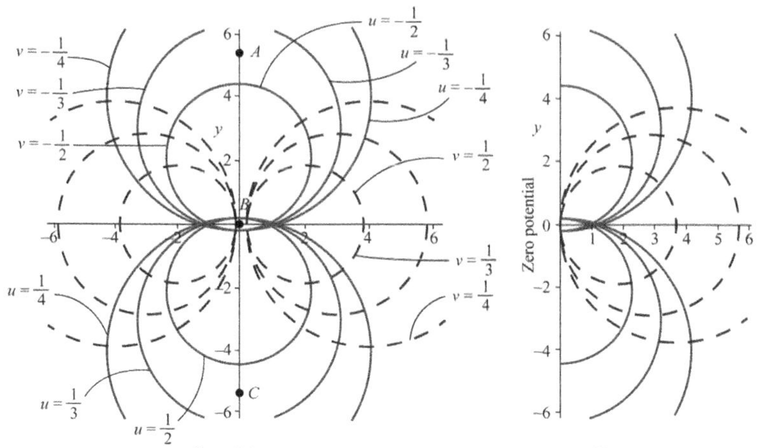

Case (a) Case (b)

Figure 6.24. *Complex potential created (a) by two line charges of opposite polarities, q and $-q$, located at a and $-a$, respectively, (b) by a line charge q placed perpendicular to a plane ABC held at zero potential. The equipotentials are represented in blue (dashed lines) and the field lines in red. For a color version of this figure, see www.iste.co.uk/abdesselam/electrodynamics1.zip*

In the portion between the points (x_a, y_a) and (x_b, y_b) of the equipotential, we can therefore define the charge:

$$Q = \varepsilon_0 \int_{(x_a,y_a)}^{(x_b,y_b)} \left[\frac{\partial V}{\partial y}dx - \frac{\partial V}{\partial x}dy\right]_{V(x,y)=V_1} \qquad [6.166]$$

Q can be interpreted here as the charge on the plate V_1 between the points (x_a, y_a) and (x_b, y_b). Using the Cauchy–Riemann conditions, it is also possible to write:

$$Q = \varepsilon_0 \int_{(x_a,y_a)}^{(x_b,y_b)} \left[\frac{\partial U}{\partial x}dx + \frac{\partial U}{\partial y}dy\right]_{V(x,y)=V_1} = \varepsilon_0 (U_2 - U_1) \qquad [6.167]$$

where U_1 and U_2 are, respectively, the field lines passing through (x_a, y_a) and (x_b, y_b), as illustrated in Figure 6.25. Clearly, the AB and CD portions of the V_1 and V_2 equipotentials carry the same charge. They can therefore be considered as the electrodes of a parallel-plate capacitor. Thus, the capacitance contained within the surface bounded by plates V_1, V_2, U_1 and U_2 is:

$$C_{12} = \frac{Q}{V_1 - V_2} = \varepsilon_0 \frac{U_2 - U_1}{V_1 - V_2} \qquad [6.168]$$

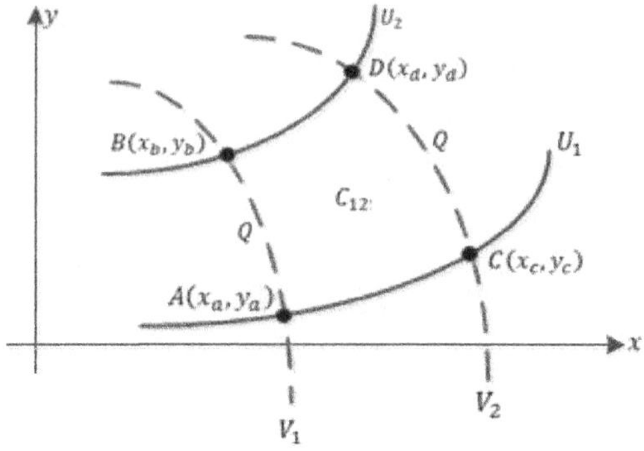

Figure 6.25. *(1) The charge measured on plates AB and CD is the same and equal to $\varepsilon_0 (U_2 - U_1)$. (2) The $ABCD$ element is regarded as a parallel-plate capacitor with capacitance $C_{12} = \varepsilon_0 \frac{U_2 - U_1}{V_1 - V_2}$. For a color version of this figure, see www.iste.co.uk/abdesselam/electrodynamics1.zip*

EXAMPLE 6.34.– A region is delimited by two infinitely long concentric cylindrical conductors of radii R_1 and R_2 ($R_2 > R_1$), carried at potentials V_1 and V_2 respectively, as shown in Figure 6.26. We wish to calculate the potential and capacitance of the capacitor formed by these two cylindrical conductors. We have already seen that the function V satisfying the Laplace equation in the region $R_1 \leq r \leq R_2$ is of the form $V(r) = A \ln r + B$ and must satisfy the boundary conditions $V(R_1) = V_1$ and $V(R_2) = V_2$, that is, $A \ln R_1 + B = V_1$ and $A \ln R_2 + B = V_2$. Therefore, it follows that $A = \frac{V_2 - V_1}{(R_2/R_1)}$ and $B = \frac{V_1 \ln R_2 - V_2 \ln R_1}{(R_2/R_1)}$. This gives $V(r) = \frac{V_2 \ln \frac{r}{R_1} - V_1 \ln \frac{r}{R_2}}{\ln(R_2/R_1)}$. This potential appears as the imaginary part of the complex potential:

$$\Omega(z) = \mathrm{i} \left(\frac{V_2 \ln \frac{z}{R_1} - V_1 \ln \frac{z}{R_2}}{\ln \frac{R_2}{R_1}} \right) \qquad [6.169]$$

which leads to $U(\theta) = -\frac{V_2 - V_1}{\ln(R_2/R_1)} \theta$. The total charge contained between the equipotentials V_1 and V_2 is:

$$Q = U_2 - U_1 = U(2\pi) - U(0) = -2\pi \frac{V_2 - V_1}{\ln \frac{R_2}{R_1}} \qquad [6.170]$$

and consequently, the capacitance of the region bounded by the two cylinders is equal to:

$$C_{12} = \varepsilon_0 \frac{U_2 - U_1}{V_1 - V_2} = \frac{2\pi\varepsilon_0}{\ln \frac{R_2}{R_1}} \qquad [6.171]$$

which agrees with the results we have already obtained.

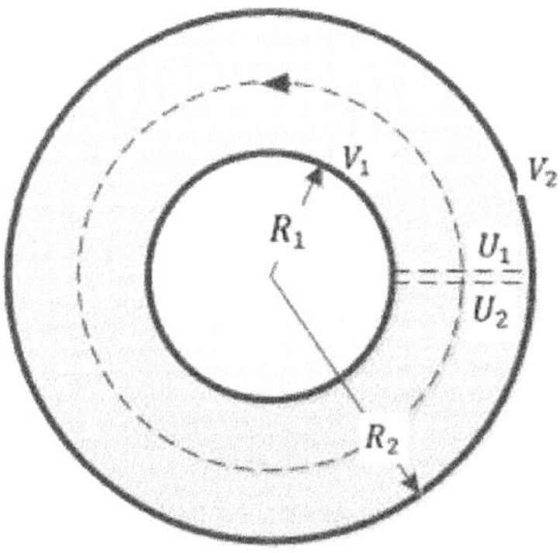

Figure 6.26. *Capacitance of a region bounded by two concentric cylinders of radii R_1 and R_2 taken to the potentials V_1 and V_2*

EXAMPLE 6.35.– Two cylindrical conductors of the same radius R, centers A and B, and separated by a distance D ($D > R_1 + R_2$), are taken to potentials $-V_0$ and V_0, respectively, as shown in Figure 6.27. This system is equivalent to two line charges of opposite polarities, $-q$ and q, placed at points A' and B', respectively. These lines produce circular equipotentials whose centers are $\left(\frac{a}{\tanh\left(\frac{2\pi\varepsilon_0 V}{q}\right)}, 0\right)$ and radii $\frac{a}{\left|\sinh\left(\frac{2\pi\varepsilon_0 V}{q}\right)\right|}$ (Example 6.33). By choosing $V = V_0$ (i.e. placing ourselves on V_0), it can be derived that $\frac{D}{2} = \frac{a}{\tanh\left(\frac{2\pi\varepsilon_0 V_0}{q}\right)}$ and $R = \frac{a}{\sinh\left(\frac{2\pi\varepsilon_0 V_0}{q}\right)}$. As a result:

$$q = \frac{2\pi\varepsilon_0 V_0}{\cosh^{-1}\frac{D}{2R}} \qquad [6.172]$$

Therefore, the capacitance of the system is:

$$C = \frac{\text{charge}}{\text{Potential difference}} = \frac{q}{2V_0} = \frac{\pi\varepsilon_0}{\cosh^{-1}\frac{D}{2R}} \qquad [6.173]$$

6.3.5.5. *Conformal transformation*

Through transformation [6.119], an infinitesimal region $\mathcal{D}$ in the z plane is transformed into a region $\mathcal{D}'$ in the f plane. If $\Delta x\Delta y$ and $\Delta u\Delta v$ denote the areas of these regions, then:

$$\begin{aligned} \lim_{\Delta x,\Delta y\longrightarrow 0}\frac{\Delta u\Delta v}{\Delta x\Delta y} &= \left|\frac{\partial(u,v)}{\partial(x,y)}\right| = \left|\frac{\partial u}{\partial x}\frac{\partial v}{\partial y} - \frac{\partial u}{\partial y}\frac{\partial v}{\partial x}\right| \\ &= \left(\frac{\partial u}{\partial x}\right)^2 + \left(\frac{\partial v}{\partial x}\right)^2 = |f'(z)|^2 \end{aligned} \qquad [6.174]$$

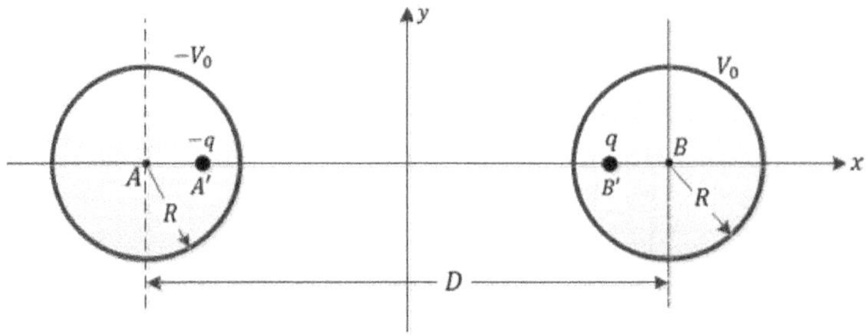

Figure 6.27. *Capacitance of a system of two identical, parallel cylinders, separated by a distance D and taken to the potentials V_0 and $-V_0$*

This is the *Jacobi transformation.* It therefore follows that the transformation is one-to-one in the regions where $f'(z) \neq 0$. The solutions of equation $f'(z) = 0$ are called *critical points*. Suppose that the point (x_0, y_0) is transformed into a point (u_0, v_0) by the transformation [6.119], while the curves C_1 and C_2 (intersecting at (x_0, y_0)) are transformed, respectively, into C_1' and C_2' (intersecting at (u_0, v_0)). When the transformation preserves the angle measured at the point (x_0, y_0) between C_1 and C_2, in amplitude and direction (i.e. equal to the angle at the point (u_0, v_0) between C_1' and C_2'), we will say that the transformation is *conformal* in (x_0, y_0), as shown in Figure 6.28. A transformation that preserves the amplitudes of the angles but not necessarily the direction is said to be *isogonal*. In general, if $f(z)$ is analytic and $f'(z) \neq 0$ in a region $\mathcal{D}$, then the transformation is conformal at all points in $\mathcal{D}$. To prove this, consider an analytic function $f(z)$ in z_0 with $f'(z_0) \neq 0$. By this transformation, the tangent at z_0 to any curve C (in the z plane) passing through z_0 turns through the angle $\alpha = \arg f'(z_0)$. When we move from z_0 to $z_0 + \Delta z_0$ along C, the image moves along C' from w_0 to $w_0 + \Delta w_0$. If C is described by the parametric equation $z(t)$ (here t is the time parameter), then the derivatives $\frac{dz}{dt}$ and $\frac{dw}{dt}$ represent the tangent vectors at the corresponding points in C and C'. Since $\frac{dw}{dt} = \frac{dw}{dz}\frac{dz}{dt} = f'(z)\frac{dz}{dt}$, we obtain, at the points z_0 and w_0, $\left.\frac{dw}{dt}\right|_{w_0} = f'(z_0)\left.\frac{dz}{dt}\right|_{z_0}$. By writing that $\left.\frac{dw}{dt}\right|_{w_0} = \rho_0 e^{i\phi_0}$, $f'(z_0) = Re^{i\alpha}$ and $\left.\frac{dz}{dt}\right|_{z_0} = r_0 e^{i\theta_0}$, we found $\rho_0 e^{i\phi_0} = Rr_0 e^{i(\alpha+\theta_0)}$ and, consequently:

$$\phi_0 = \theta_0 + \alpha = \theta_0 + \arg f'(z_0) \qquad [6.175]$$

So, we deduce $\phi_0^1 = \theta_0^1 + \arg f'(z_0)$ for C_1 and $\phi_0^2 = \theta_0^2 + \arg f'(z_0)$ for C_2. It follows that:

$$\Delta\phi_0 = \phi_0^2 - \phi_0^1 = \theta_0^2 - \theta_0^1 = \Delta\theta_0 \qquad [6.176]$$

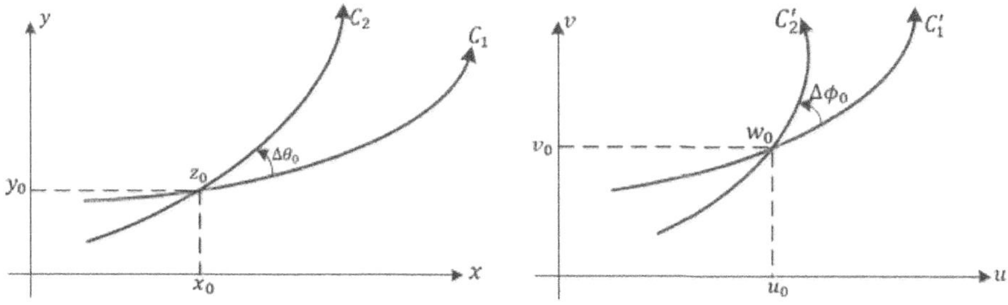

Figure 6.28. *Conservation of the amplitude and direction of the angle by a conformal transformation*

The angle between C_1 and C_2 is therefore preserved. Note that if $f'(z_0) = 0$, then α is not determined.

6.3.5.6. *Riemann's mapping theorem*

Let C be a curve delimiting a region $\mathcal{D}$ in the z plane and C' a circle of radius unit radius and center $(0,0)$ enclosing a region $\mathcal{D}'$ in the f plane. Riemann showed that there exists an analytic function $f(z)$ in $\mathcal{D}$ that uniquely maps to each point of $\mathcal{D}$ a point of $\mathcal{D}'$ and to each point of C a point of C'. This is the *Riemann mapping theorem*, as shown in Figure 6.29. This function $f(z)$ contains arbitrary constants that can be determined by mapping the center of C' to a given point on $\mathcal{D}$ and a given point on C' to a given point on C. The Riemann theorem can be extended to a region contained between two curves C_1 and C_2, one inside the other. This region is transformed into a region bounded by two concentric circles. In addition, any related region (which is not the entire complex plane), for example, the upper half-plane, can be transformed in a conformal manner into a unit disk. The invariants of the transformation, that is, the solutions of equation $z = f(z)$, are called *fixed points*. For example, the fixed points of the transformation $w = z^2$ are the solutions of $z^2 = z$, that is, $z = 0$ and $z = 1$. Here are the simplest transformations:

1) *Translation*: $w = z + \beta$, where $\beta \in \mathbb{C}$. By this transformation, the figures are translated in the direction of the β vector.

2) *Rotation*: $w = e^{i\theta_0} z$, where $\theta_0 \in \mathbb{R}$. In this transformation, the figures rotate about an angle θ_0. If $\theta_0 > 0$ (respectively, $\theta_0 < 0$), the rotation is clockwise (respectively, counterclockwise).

3) *Stretching*: $w = az$, where $a \in \mathbb{R}$. In this transformation, the figures are stretched (respectively, contracted) if $a > 1$ (respectively, $0 < a < 1$).

4) *Inversion*: $w = 1/z$.

More complex transformations can be defined from these simple transformations.

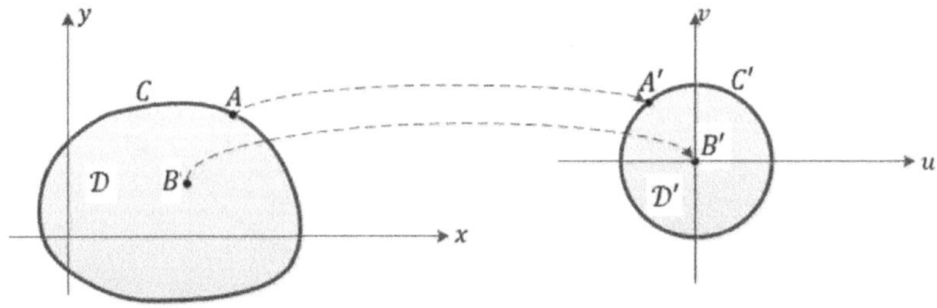

Figure 6.29. *Riemann mapping theorem: if $\mathcal{D}$ is a simply connected region of a complex plane, distinct from the plane, then there exists an analytic, one-to-one mapping between $\mathcal{D}$ and the unit disk $\mathcal{D}'$*

– *Linear transformation*: $w = \alpha z + \beta$, where $\alpha, \beta \in \mathbb{C}$. By writing $\alpha = ae^{i\theta_0}$, we can see that a linear transformation is a combination of translation, rotation and stretching.

– *Bilinear or homographic transformation*:

$$w = \frac{\alpha z + \beta}{\gamma z + \delta} \qquad [6.177]$$

$\alpha\delta - \beta\gamma \neq 0$ is called a *bilinear or homographic transformation.* This is a combination of several simple transformations: translation, rotation, stretching and inversion. Indeed, we write:

$$w = \frac{\alpha}{\gamma} + \frac{\beta\gamma - \alpha\delta}{\gamma(\gamma z + \delta)} = \lambda + \frac{\mu}{z + \nu} \qquad [6.178]$$

where $\lambda = \frac{\alpha}{\gamma}$, $\mu = \frac{\beta\gamma - \alpha\delta}{\gamma^2}$ and $\nu = \frac{\delta}{\gamma}$. Transformation [6.177] has the property of transforming circles in the z plane into circles in the f plane (circles of infinite radius, i.e. straight lines, are included here). Let us give the formulas (without proving them) that allow us to calculate the images of lines and circles by the transformation [6.177]:

i) at the lines $\mathrm{Re}\,(az) = b$, which do not pass through the point $z = -\frac{\delta}{\gamma}$ $\left(b \neq -\mathrm{Re}\left(a\frac{\delta}{\gamma}\right)\right)$, correspond to the circles $|w - w_0| = \rho_0$, where $w_0 = \frac{2\alpha b\overline{\gamma} + \alpha\overline{\delta}\overline{a} + \beta\overline{\gamma}a}{2b|\gamma|^2 + 2\mathrm{Re}(\gamma\overline{\delta}\overline{a})}$ and $\rho_0 = \left|\frac{\alpha}{\gamma} - w_0\right| = \left|\frac{(\alpha\delta - \beta\gamma)a}{2b|\gamma|^2 + 2\mathrm{Re}(\gamma\overline{\delta}\overline{a})}\right|$;

ii) at the lines $\mathrm{Re}\,(az) = -\mathrm{Re}\left(a\frac{\delta}{\gamma}\right)$, which pass through the point $z = -\frac{\delta}{\gamma}$, correspond to the lines $\mathrm{Re}\left(\frac{\alpha\delta - \beta\gamma}{\gamma^2}a\overline{w}\right) = \mathrm{Re}\left(\frac{\alpha\delta - \beta\gamma}{\gamma^2}a\frac{\overline{\alpha}}{\overline{\gamma}}\right)$;

iii) at the circles $|z - z_0| = r$, which do not pass through the point $z = -\frac{\delta}{\gamma}$ $\left(r \neq \left|z_0 + \frac{\delta}{\gamma}\right|\right)$, correspond to the circles $|w - w_0| = \rho_0$, where $w_0 = \frac{(\alpha z_0+\beta)(\overline{\gamma z_0}+\overline{\delta})-\alpha\overline{\gamma}r^2}{|\gamma z_0+\delta|^2-|\gamma|^2 r^2}$ and $\rho_0 = \frac{r|\alpha\delta-\beta\gamma|}{\left||\gamma z_0+\delta|^2-|\gamma|^2 r^2\right|}$;

iv) at the circles $|z - z_0| = \left|z_0 + \frac{\delta}{\gamma}\right|$, which pass through the point $z = -\frac{\delta}{\gamma}$, correspond to the lines $\mathrm{Re}\left(\frac{\alpha\delta-\beta\gamma}{\gamma(\gamma z_0+\delta)}\overline{w}\right) = \frac{|\alpha\delta-\beta\gamma|^2+2\mathrm{Re}\left(\gamma(\alpha z_0+\beta)\overline{(\alpha\delta-\beta\gamma)}\right)}{2|\gamma(\gamma z_0+\delta)|^2}$.

EXAMPLE 6.36.– Let us find the image of the line $y = x + 2$ (also written as $\mathrm{Re}\left(\left(-1 - \mathrm{i}\right)\left(x + \mathrm{i}y\right)\right) = 2$, i.e. $a = -\mathrm{i} - 1$ and $b = 2$) by the transformation $w = \frac{z+1}{z-1}$ ($\alpha = \beta = \gamma = 1$ and $\delta = -1$). As the straight line does not pass through the point $z = 1$, it is transformed into a circle with center $w_0 = \frac{2-\mathrm{i}}{3}$ and radius $\rho_0 = \frac{|\mathrm{i}+1|}{3} = \frac{\sqrt{2}}{3}$. ■

Note that [6.177] transforms three distinct points of the z plane (one of which may be at infinity) into three distinct points of the f plane. If z_1, z_2, z_3 and z_4 are distinct points, then the quantity $\frac{(z_4-z_1)(z_2-z_3)}{(z_2-z_1)(z_4-z_3)}$ is called the *cross ratio* of z_1, z_2, z_3 and z_4. This ratio is invariant under the bilinear transformation and can be used to obtain specific bilinear transformations that connect three points to three others. Let us try to find a bilinear transformation that connects points z_1, z_2 and z_3 to points w_1, w_2 and w_3. If w_k corresponds to z_k, for $k = 1, 2, 3$, we can write: $w - w_k = \frac{\alpha z+\beta}{\gamma z+\delta} - \frac{\alpha z_k+\beta}{\gamma z_k+\delta} = \frac{(\alpha\delta-\beta\gamma)(z-z_k)}{(\gamma z+\delta)(\gamma z_k+\delta)}$, that is, $w - w_1 = \frac{(\alpha\delta-\beta\gamma)(z-z_1)}{(\gamma z+\delta)(\gamma z_1+\delta)}$ and $w - w_3 = \frac{(\alpha\delta-\beta\gamma)(z-z_3)}{(\gamma z+\delta)(\gamma z_3+\delta)}$. By replacing w by w_2 and z by z_2, we obtain $w_2 - w_1 = \frac{(\alpha\delta-\beta\gamma)(z_2-z_1)}{(\gamma z_2+\delta)(\gamma z_1+\delta)}$ and $w_2 - w_3 = \frac{(\alpha\delta-\beta\gamma)(z_2-z_3)}{(\gamma z_2+\delta)(\gamma z_3+\delta)}$. Hence:

$$\frac{(w - w_1)(w_2 - w_3)}{(w - w_3)(w_2 - w_1)} = \frac{(z - z_1)(z_2 - z_3)}{(z - z_3)(z_2 - z_1)} \qquad [6.179]$$

EXAMPLE 6.37.– By choosing $z_1 = 0$, $z_2 = -\mathrm{i}$, $z_3 = -1$, $w_1 = \mathrm{i}$, $w_2 = 1$, $w_3 = 0$, we arrive at the transform $w = -\mathrm{i}\left(\frac{z+1}{z-1}\right)$. ■

Let z_0 be any point in the upper half-plane (denoted here by $\mathcal{D}$). The function:

$$w = e^{\mathrm{i}\theta_0}\left(\frac{z - z_0}{z - \overline{z_0}}\right) \qquad [6.180]$$

transforms this half-plane biunivocally into a circle $\mathcal{D}'$ of unit radius $|w| = 1$. Each point on the Ox axis is transformed into a point on the periphery of the circle. The constant θ_0 is determined by mapping a particular point on Ox to a given point on the circle, as shown in Figure 6.30. In fact, we can clearly see that $|w| = \left|e^{i\theta_0}\left(\frac{z-z_0}{z-\overline{z_0}}\right)\right| = \left|\frac{z-z_0}{z-\overline{z_0}}\right|$. Since $|z - z_0| \leq |z - \overline{z_0}|$ (because z and z_0 are located in the upper half-plane), then $|w| \leq 1$.

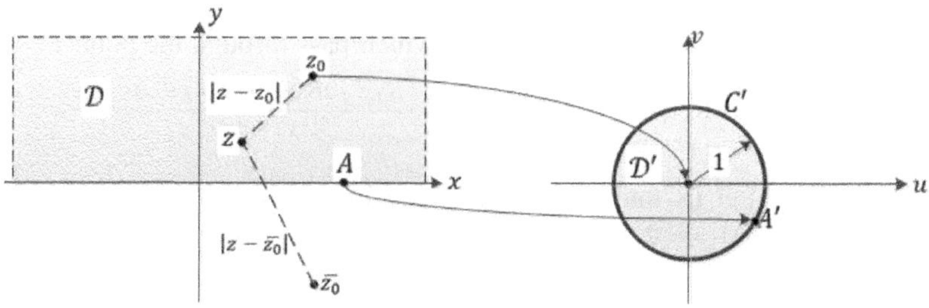

Figure 6.30. *Mapping the upper half-plane into a circle of unit radius*

EXAMPLE 6.38.– Let us find the transformation [6.180] that maps the upper half of the z plane to a circle of unit radius in the f plane, so that $z = i$ and $z \longrightarrow \infty$ are represented by $w = 0$ and $w = -1$, respectively. Starting from $w = e^{i\theta_0}\left(\frac{z-z_0}{z-\overline{z_0}}\right)$, we obtain $0 = e^{i\theta_0}\left(\frac{i-z_0}{i-\overline{z_0}}\right)$ and $-1 = \lim_{z\longrightarrow\infty} e^{i\theta_0}\left(\frac{z-z_0}{z-\overline{z_0}}\right) = e^{i\theta_0}$. This gives $z_0 = i$ and $e^{i\theta_0} = -1$. Thus, the transformation will be: $w = -1\left(\frac{z-i}{z+i}\right) = \left(\frac{i-z}{z+i}\right)$.

– *Joukowski transformation*: the mapping of the upper half-plane onto the outside of the semicircle of radius a, shown in Figure 6.31, is achieved by the *Joukowski transformation*:

$$f(z) = \frac{z}{a} + \sqrt{\frac{z^2}{a^2} - 1} \quad \Longleftrightarrow \quad z = \frac{a}{2}\left(w + \frac{1}{w}\right) \qquad [6.181]$$

Indeed, the points of the segment $[N, Q]$ are transformed into $\left(u = \frac{x}{a}, v = \sqrt{1 - \frac{x^2}{a^2}}\right)$, that is, into points located on the upper semicircle of equation $u^2 + v^2 = 1$. The points of the segments $]-\infty, N[\cup]Q, +\infty[$ are transformed into $\left(u = \frac{x}{a} + \sqrt{\frac{x^2}{a^2} - 1}, v = 0\right)$, that is, to points on $]-\infty, N'[\cup]Q', +\infty[$. Furthermore, any point $w = \rho e^{i\theta}$ ($\rho > 1$ and $0 < \theta < \pi$), located outside the circle, is the image of the point $\left(x = \frac{a}{2}\left[\rho + \frac{1}{\rho}\right]\cos\theta, y = \frac{a}{2}\left[\rho - \frac{1}{\rho}\right]\sin\theta\right)$ of the upper half-plane, because $\frac{a}{2}\left[\rho - \frac{1}{\rho}\right]\sin\theta > 0$.

– *Schwarz–Christoffel transformation*: let us choose $n - 1$ points $x_1, \ldots, x_{n-1}$ on the Ox axis and take $x_n = +\infty$. We wish to construct a function $f(z)$ which transforms the real axis into a polygon $calD'$ with vertices w_i, as shown in

Figure 6.32. We will write $w_i = f(x_i)$ for $i = 1, 2, \ldots, n-1$ and $w_n = f(+\infty)$. It is thus evident that when z varies on Ox, $\arg f'(z)$ must jump each time that we pass through one of the x_i values. To establish the function f that satisfies the above statement, it is sufficient to take:

$$f'(z) = A(z - x_1)^{-\beta_1}(z - x_2)^{-\beta_2} \cdots (z - x_{n-1})^{-\beta_{n-1}} \quad [6.182]$$

where $A \in \mathbb{C}$, and, consequently:

$$\arg f'(z) = \arg A - \sum_{i=1}^{n-1} \beta_i \arg(z - x_i) \quad [6.183]$$

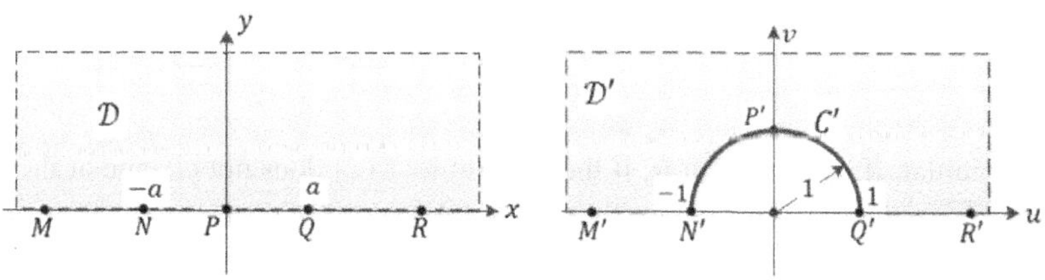

Figure 6.31. *Mapping the upper half-plane onto the outside of a circle of radius a using the Joukowski transformation*

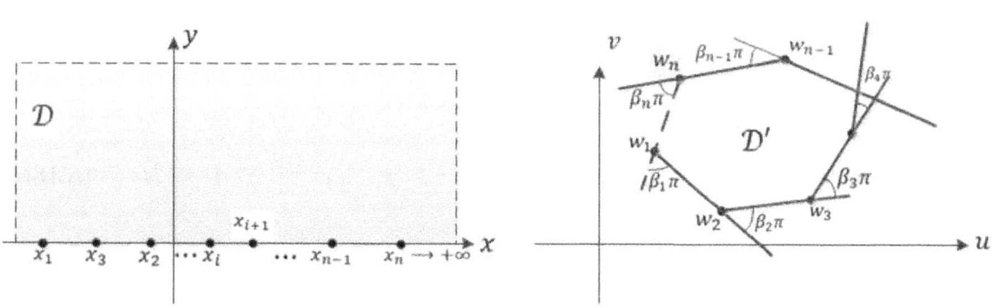

Figure 6.32. *Mapping the upper half-plane into a polygon of angle $\alpha_k\pi = \pi - \beta_k\pi$ by the Schwarz–Christoffel transformation*

We can see that when z is real, with $z < x_1$, $\arg(z - x_i) = \pi$ for $i = 1, \ldots, n-1$. If z now passes through the interval $]x_1, x_2[$, then $\arg(z - x_1) = 0$ and $\arg(z - x_i) = \pi$ for $i = 2, \ldots, n-1$. It follows that the argument of $f'(z)$ increases by $-\beta_1\pi$ as z passes through the point x_1. Since $\theta = \arg f'(z)$ represents the angle between the tangent to $f'(z)$ and Ox, we can conclude that the tangent to $f'(z)$ changes as we pass through x_1. This reasoning is valid for all of the x_i points, that is, every time that we pass through an x_i point, the argument of $f'(z)$ changes by $-\beta_i\pi$. The angles $\beta_i\pi$ are the external angles of the polygon described by $w = f(z)$ when z varies over the real axis and lie in $]-\pi,\pi]$. It must just be noted that we had to write $f'(z) = K(z-x_1)^{-\beta_1}\cdots(z-x_{n-1})^{-\beta_{n-1}}(z-x_n)^{-\beta_n}$. Putting $K = \frac{A}{(-x_n)^{-\beta_n}}$, the right-hand term becomes $f'(z) = A(z-x_1)^{-\beta_1}\cdots(z-x_{n-1})^{-\beta_{n-1}}\left(\frac{x_n-z}{x_n}\right)^{-\beta_n}$. It can clearly be seen that the term $\left(\frac{x_n-z}{x_n}\right)^{-\beta_n}$ tends to 1 when $x_n \longrightarrow \infty$. Assuming that the sides of the polygon do not intersect each other, that the orientation is in the positive direction, and that w_n is the image of $z=\infty$, we find:

$$\sum_{i=1}^{n} \pi - \beta_i\pi = (n-2) \Longrightarrow \beta_n\pi = 2\pi - \pi\sum_{i=1}^{n-1}\beta_i \Longrightarrow \sum_{i=1}^{n}\beta_i = 2 \quad [6.184]$$

In particular, if $\beta_n = 0$, that is, if the tangent to $f(z)$ does not change at the point w_n, then $\sum_{i=1}^{n-1}\beta_i = 2$. This means that w_n is not a vertex. Consequently, the polygon has only $n-1$ vertices $w_1,\ldots,w_{n-1}$. Similarly, by choosing $w_n = w_1$, the polygon is closed. Integrating [6.182], we obtain:

$$f(z) = A\int(z-x_1)^{-\beta_1}(z-x_2)^{-\beta_2}\cdots(z-x_{n-1})^{-\beta_{n-1}}\,dz + B \quad [6.185]$$

where $B \in \mathbb{C}$. This is the *Schwarz–Christoffel transformation.* This function is analytic in the upper half-plane, except for the x_i points. Depending on the interior angles of the polygon, it is also possible to write that:

$$f(z) = A\int(z-x_1)^{\frac{\alpha_1}{\pi}-1}(z-x_2)^{\frac{\alpha_2}{\pi}-1}\cdots(z-x_{n-1})^{\frac{\alpha_{n-1}}{\pi}-1}\,dz + B \quad [6.186]$$

Here are two important observations:

i) The Schwarz–Christoffel function transforms each point of the upper half-plane z into a point inside the polygon. Since we already know that the upper half-plane can be transformed into a disk of unit radius, we need only prove that this function transforms this unit disk into a polygon. Let us check that each point a of the polygon corresponds to one, and only one, point z_0 of the disk of unit radius, so that $f(z_0) = a$. Since a is inside the polygon, Cauchy's first formula means that $\frac{1}{2\pi i}\oint_{\text{polygon}}\frac{dw}{w-a} = 1$. Now, $w - a = f(z) - a$, which gives $\frac{1}{2\pi i}\oint_{\text{circle}}\frac{f'(z)}{f(z)-a}dz = 1$. Since $f(z)-a$ is analytic inside the circle, we can conclude that $f(z)-a$ admits a single zero inside the circle, that is, $f(z_0) = a$.

ii) Let us find the transformation that maps a disk of unit radius to a polygon. On the one hand, the transformation that maps the upper half-plane to a polygon is given by [6.185]. On the other hand, the transformation that maps the upper half-plane to the disk of unit radius is $\xi = e^{i\theta}\frac{z-z_0}{z-\overline{z_0}}$. It follows that $z = \frac{\overline{z_0}\xi - z_0 e^{i\theta}}{\xi - e^{i\theta}}$. If the points $x_1, \ldots, x_{n-1}, x_n = +\infty$ become $\xi_1, \ldots, \xi_{n-1}, \xi_n$ on the circle of radius unity, we get $z - x_k = \frac{(z_0-\overline{z_0})(\xi-\xi_k)e^{i\theta}}{(\xi-e^{i\theta})(\xi_k-e^{i\theta})}$, for $k = 1, \ldots, n$. We also find that $dz = \frac{(z_0-\overline{z_0})e^{i\theta}}{(\xi-e^{i\theta})^2}d\xi$. Substituting these results in [6.185], we get:

$$f(\xi) = A'\int (\xi-\xi_1)^{-\beta_1}(\xi-\xi_2)^{-\beta_2}\cdots(\xi-\xi_n)^{-\beta_n}\,d\xi + B \qquad [6.187]$$

where $A' = \frac{A(\xi_1-e^{i\theta})^{\beta_1}(\xi_2-e^{i\theta})^{\beta_2}\cdots(\xi_n-e^{i\theta})^{\beta_n}}{e^{i\theta}(z_0-\overline{z_0})}$.

EXAMPLE 6.39.– Let us determine the transformation that maps the upper half-plane to the infinite sector of angle $\frac{\pi}{m}$ shown in Figure 6.33. Using the Schwarz–Christoffel formula [6.186], we write $f(z) = A\int z^{\frac{\pi/m}{\pi}-1}dz + B = A'z^{\frac{1}{m}} + B$. The boundary conditions $f(0) = 0$ and $f(1) = 1$ give $B = 0$ and $A' = 1$. Consequently, it follows that:

$$f(z) = z^{\frac{1}{m}} \qquad [6.188]$$

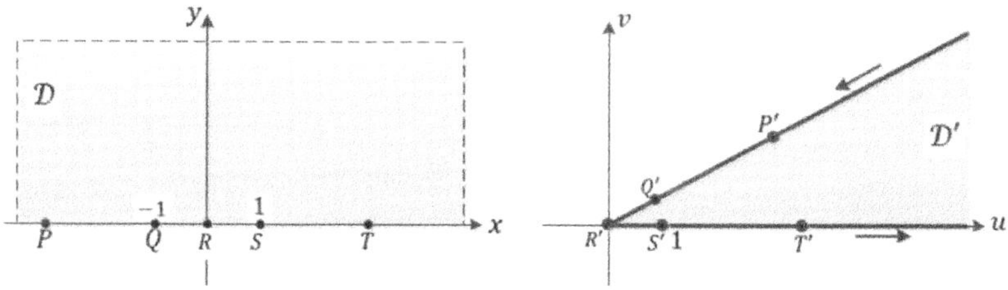

Figure 6.33. *Mapping the upper half-plane into an infinite sector of angle π/m*

EXAMPLE 6.40.– Let us find the transformation that maps the upper half-plane to the semi-infinite band of width $2b$, as shown in Figure 6.34(a). The points P, Q, S and T are transformed into P, Q', S' and T', respectively. $P'Q'S'T'$ is a limit case of a triangle with two vertices in Q' and S'. The third vertex P' or T' is at infinity. Since the interior angles at points Q' and S' are equal to $\frac{\pi}{2}$, we obtain $f'(z) = A(z+1)^{\frac{\pi/2}{\pi}-1}(z-1)^{\frac{\pi/2}{\pi}-1} = \frac{A}{\sqrt{z^2-1}}$. It follows that $f(z) = A\int\frac{dz}{\sqrt{z^2-1}} + B = A\cosh^{-1}z + B$. The boundary conditions $f\cosh^{-1}(1) = b$ and $f\cosh^{-1}(-1) = -b$ impose the restrictions $b = A\cosh^{-1}(1) + B = B$ and $-b = A\cosh^{-1}(-1) +$

$B = \mathrm{i}\pi A + B$. This gives $B = b$ and $A = \frac{2\mathrm{i}b}{\pi}$. Therefore, we get:

$$f(z) = b + \frac{2\mathrm{i}b}{\pi} \cosh^{-1} z \qquad [6.189]$$

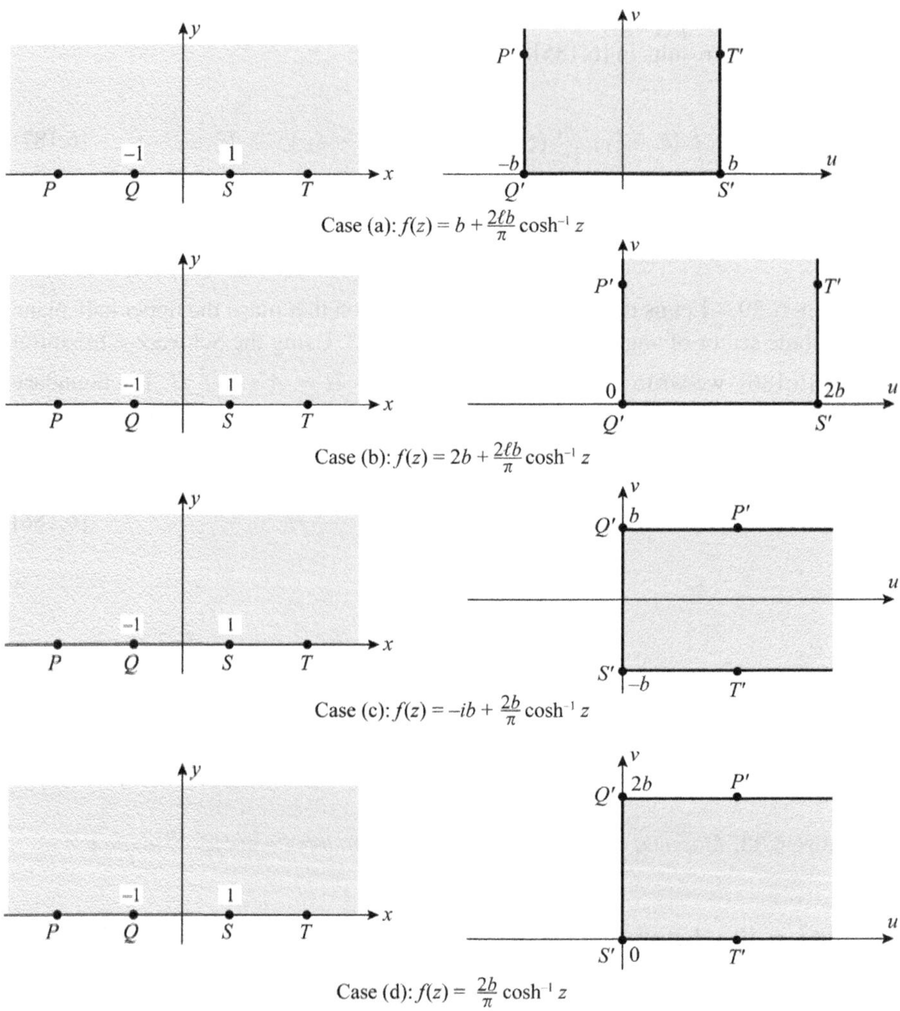

Figure 6.34. *Mapping the upper half-plane into a semi-infinite $2b$-wide band*

If the band extends between 0 and $2b$ instead of $-b$ and b, as shown in Figure 6.34(b), the previous formula must be replaced by $f(z) = 2b + \frac{2ib}{\pi}\cosh^{-1} z$. Similarly, if the semi-infinite band extends along the Ou axis between 0 and $+\infty$ with a finite width between $-b$ and b (respectively, 0 and $2b$) along the $0v$ axis, as shown in Figure 6.34(c) (respectively, Figure 6.34(d)), we deduce that $f(z) = \frac{2b}{\pi}\cosh^{-1} z - ib$ (respectively, $f(z) = \frac{2b}{\pi}\cosh^{-1} z$).

EXAMPLE 6.41.– Let us determine the transformation that maps the two regions in Figure 6.35. In terms of interior angles, the Schwarz–Christoffel transformation is written as: $f(z) = A\int (z-0)^{\frac{3\pi/2}{\pi}-1}(z-1)^{\frac{3\pi/2}{\pi}-1} dz + B = A\int\sqrt{\frac{z-1}{z}} + B$. Using the variable change $z = \cosh^2\theta$, we get $f(z) = 2A\int \sinh^2\theta d\theta + B = A(\sinh\theta\cosh\theta - \theta) + B$. The boundary conditions $f(0) = 0$ and $f(1) = bi$ give $B = bi$ and $A = \frac{2b}{\pi}$. We therefore obtain:

$$f(z) = \frac{2b}{\pi}\left(\sqrt{z(z-1)} - \cosh^{-1}\sqrt{z}\right) + bi \qquad [6.190]$$

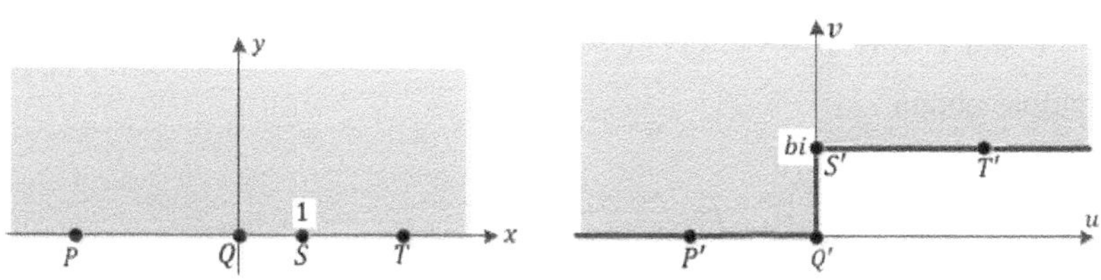

Figure 6.35. *Cartography produced by the transformation* $f(z) = \frac{2b}{\pi}\left(\sqrt{z(z-1)} - \cosh^{-1}\sqrt{z}\right) + bi$

EXAMPLE 6.42.– The mapping of the upper half-plane into an infinite strip (along Ou) of width a, as shown in Figure 6.36, is performed by the function:

$$f(z) = \frac{a}{\pi}\ln z \qquad [6.191]$$

In fact, if $z = \rho e^{i\theta}$, then $w = \frac{a}{\pi}\ln r + i\frac{a\theta}{\pi}$. This gives $u = \frac{a\ln r}{\pi}$ and $v = \frac{a\theta}{\pi}$. The points on the half-line $[0, +\infty[$ on the Ox axis are transformed into $\left(u = \frac{a\ln x}{\pi}, v = 0\right)$, that is, into points on the line $v = 0$, while the points on the half-line $[-\infty, 0[$ of Ox are transformed into $\left(u = \frac{a\ln|x|}{\pi}, v = a\right)$, that is, into points on the $v = a$ line. This transformation can also be obtained using the Schwarz–Christoffel method. The width band a is regarded as a polygon with one of its vertices at $+\infty$. As the sides of the polygon are parallel, we can see that $\alpha = 0$.

It follows that $\frac{df}{dz} = Az^{0-1} = \frac{A}{z}$ and, consequently, $f(z) = A\ln z + B$. Fixing $f(1) = 0$ and $f(-1) = \mathrm{i}a$, we find $A = \frac{a}{\pi}$ and $B = 0$. Thus: $f(z) = \frac{a}{\pi}\ln z$.

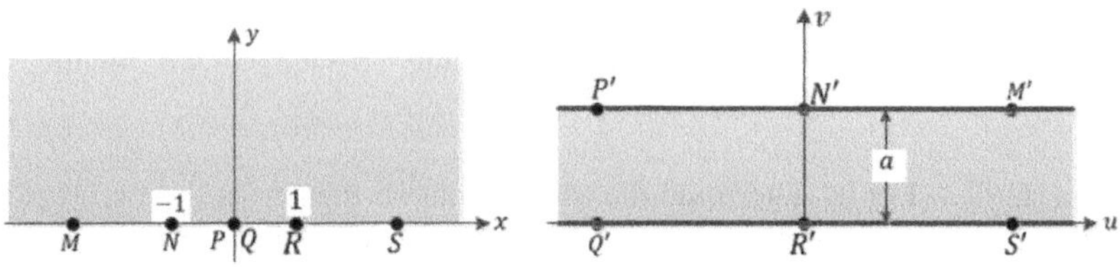

Figure 6.36. *Mapping the upper half-plane into an infinite band of width a*

EXAMPLE 6.43.– Let us calculate the function that maps the upper half-plane into a triangle, as shown in Figure 6.37. According to the Schwarz–Christoffel transformation, we have: $f'(z) = Az^{\frac{\alpha}{\pi}-1}(z-1)^{\frac{\beta}{\pi}-1} = Kz^{\frac{\alpha}{\pi}-1}(1-z)^{\frac{\beta}{\pi}-1}$. This yields $f(z) = K\int_0^z \xi^{\frac{\alpha}{\pi}-1}(1-\xi)^{\frac{\beta}{\pi}-1}d\xi + B$. Since $f(0) = 0$ and $f(1) = 1$, we get $B = 0$ and $1 = K\int_0^1 \xi^{\frac{\alpha}{\pi}-1}(1-\xi)^{\frac{\beta}{\pi}-1}d\xi = K\frac{\Gamma(\frac{\alpha}{\pi})\Gamma(\frac{\beta}{\pi})}{\Gamma(\frac{\alpha+\beta}{\pi})}$, i.e. $K = \frac{\Gamma(\frac{\alpha+\beta}{\pi})}{\Gamma(\frac{\alpha}{\pi})\Gamma(\frac{\beta}{\pi})}$. We therefore obtain:

$$f(z) = \frac{\Gamma\left(\frac{\alpha+\beta}{\pi}\right)}{\Gamma\left(\frac{\alpha}{\pi}\right)\Gamma\left(\frac{\beta}{\pi}\right)}\int_0^z \xi^{\frac{\alpha}{\pi}-1}(1-\xi)^{\frac{\beta}{\pi}-1}d\xi \qquad [6.192]$$

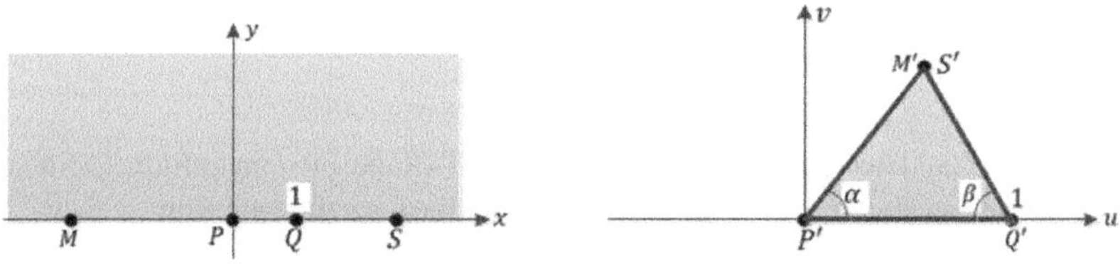

Figure 6.37. *Mapping the upper half-plane into a triangle*

EXAMPLE 6.44.– Let us determine the transformation that maps the upper half-plane to the rectangle $M'N'Q'S'$, as shown in Figure 6.38. Let $\frac{1}{k}$ be the coordinate of the point S, where $0 < k < 1$. According to the Schwarz–Christoffel transformation, the transformation we are looking for is $f(z) = A\int_0^z \frac{dt}{\sqrt{(t^2-1)(t^2-\frac{1}{k^2})}} + B = A'\int_0^z \frac{dt}{\sqrt{(1-t^2)(1-k^2t^2)}} + B$. Since $f(0) = 0$, we get $B = 0$. To calculate A', let us use the fact that $f(1) = K$ and $f\left(\frac{1}{k}\right) = K + \mathrm{i}K'$. The result is $K = A'\int_0^1 \frac{dt}{\sqrt{(1-t^2)(1-k^2t^2)}}$ and

$K + \mathrm{i}K' = A' \int_0^{\frac{1}{k}} \frac{dt}{\sqrt{(1-t^2)(1-k^2t^2)}}$. The second equation gives $K' = A' \int_1^{\frac{1}{k}} \frac{dt}{\sqrt{(t^2-1)(1-k^2t^2)}}$. For simplicity, the dimensions of the rectangle, K and K', will be chosen so that $A' = 1$, that is, $K = \int_0^1 \frac{dt}{\sqrt{(1-t^2)(1-k^2t^2)}}$ and $K' = \int_1^{\frac{1}{k}} \frac{dz}{\sqrt{(t^2-1)(1-k^2t^2)}}$. The mapping of the half-plane into a rectangle is thus performed by the function:

$$f(z) = \int_0^z \frac{dt}{\sqrt{(1-t^2)(1-k^2t^2)}} = \mathrm{sn}^{-1}(z, k) \qquad [6.193]$$

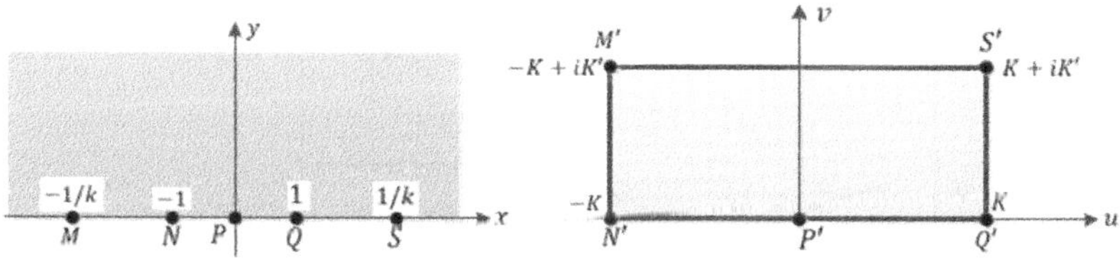

Figure 6.38. *Mapping of the upper half-plane into a rectangle*

This is the *inverse Jacobi function.*

EXAMPLE 6.45.– Let us write the transformation that maps the upper half-plane (Figure 6.39(a)) into the polygon of Figure 6.39(c). The Schwarz–Christoffel transformation gives $\frac{df}{dZ} = A(Z+1)^{\frac{\pi+\alpha\pi}{\pi}-1} Z^{0-1} (Z-1)^{\frac{\pi+\alpha\pi}{\pi}-1} = A\frac{(Z^2-1)^\alpha}{Z}$. The result is: $f(Z) = A\int_{-1}^{Z} \frac{(t^2-1)^\alpha}{t} dt + \frac{\mathrm{i}a}{2}$, because $f(-1) = \frac{\mathrm{i}a}{2}$. The second condition $f(1) = -\frac{\mathrm{i}a}{2}$ yields $A = \frac{-\mathrm{i}a}{\int_{-1}^{1} \frac{(t^2-1)^\alpha}{t} dt} = \frac{-\mathrm{i}a}{\int_{\mathrm{i}\pi}^{0} (e^{2t}-1)^\alpha dt}$. Therefore, we get:

$$f(Z) = -\mathrm{i}a \frac{\int_{\mathrm{i}\pi}^{\ln Z} (e^{2t}-1)^\alpha dt}{\int_{\mathrm{i}\pi}^{0} (e^{2t} \quad 1)^\alpha dt} \qquad [6.194]$$

For $\alpha = 1$ (see Figure 6.39(d)), we obtain the well-known mapping:

$$f(Z) = \frac{a}{\pi}\left(\ln Z \quad \frac{Z^2}{2} + \frac{1}{2} - \frac{\mathrm{i}\pi}{2}\right) \qquad [6.195]$$

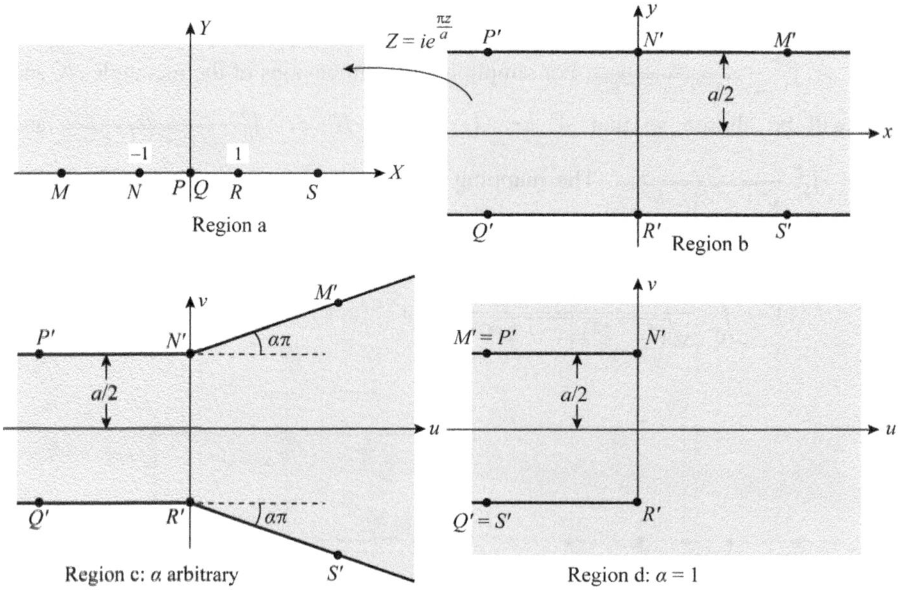

Figure 6.39. *Mapping the semi-plane into a Rogowski capacitor*

If we wish to transform the region shown in Figure 6.39(b) directly into the region shown in Figure 6.39(d), Z is simply replaced by $\mathrm{i}e^{\frac{\pi z}{a}}$:

$$f(z) = \frac{a}{2\pi}\left(e^{\frac{2\pi z}{a}} + \frac{2\pi z}{a} + 1\right) \qquad [6.196]$$

6.3.5.7. *Dirichlet and Neumann problems using conformal transformations*

Dirichlet and Neumann problems can be solved in any connected region $\mathcal{D}$ that can be conformally mapped by an analytic function onto a disk of unit radius or an upper half-plane. This can be done by following the below steps:

1) using the mapping function to transform the boundary problem for the region $calD$ into a corresponding problem for the disk of unit radius or the upper half-plane;

2) solving the problem for the disk of unit radius or the upper half-plane;

3) using the inverse mapping function to find the solution in the region $\mathcal{D}$.

The important theorems used to carry out this process are given below:

i) Let $w = f(z)$ be a one-to-one analytical function defined between the region $\mathcal{D}$ and the unit disk or the upper half-plane. If $f'(z) \neq 0$ at any point in $\mathcal{D}$, that is, the transformation is conformal in $\mathcal{D}$, then the function $f(z)$ admits a unique inverse function $z = f^{-1}(w)$.

ii) Let $V(x, y)$ be a harmonic function in $\mathcal{D}$. If the region $\mathcal{D}$ is mapped on $\mathcal{D}'$ by the conformal transformation $w = f(z)$, then the transformed function $V(u, v) = V(x(u, v), y(u, v))$ is also harmonic in $\mathcal{D}'$. Indeed, we verify that:

$$\begin{aligned}
\frac{\partial V}{\partial x} &= \frac{\partial V}{\partial u}\frac{\partial u}{\partial x} + \frac{\partial V}{\partial v}\frac{\partial v}{\partial x} \\
\frac{\partial V}{\partial y} &= \frac{\partial V}{\partial u}\frac{\partial u}{\partial y} + \frac{\partial V}{\partial v}\frac{\partial v}{\partial y} \\
\frac{\partial^2 V}{\partial x^2} &= \frac{\partial V}{\partial u}\frac{\partial^2 u}{\partial x^2} + \frac{\partial u}{\partial x}\left[\frac{\partial u}{\partial x}\frac{\partial^2 V}{\partial u^2} + \frac{\partial v}{\partial x}\frac{\partial^2 V}{\partial u \partial v}\right] + \frac{\partial V}{\partial v}\frac{\partial^2 v}{\partial x^2} \\
&\quad + \frac{\partial v}{\partial x}\left[\frac{\partial u}{\partial x}\frac{\partial^2 V}{\partial u \partial v} + \frac{\partial v}{\partial x}\frac{\partial^2 V}{\partial v^2}\right] \\
\frac{\partial^2 V}{\partial y^2} &= \frac{\partial V}{\partial u}\frac{\partial^2 u}{\partial y^2} + \frac{\partial u}{\partial y}\left[\frac{\partial u}{\partial y}\frac{\partial^2 V}{\partial u^2} + \frac{\partial v}{\partial y}\frac{\partial^2 V}{\partial u \partial v}\right] + \frac{\partial V}{\partial v}\frac{\partial^2 v}{\partial y^2} \\
&\quad + \frac{\partial v}{\partial y}\left[\frac{\partial u}{\partial y}\frac{\partial^2 V}{\partial u \partial v} + \frac{\partial v}{\partial y}\frac{\partial^2 V}{\partial v^2}\right]
\end{aligned} \qquad [6.197]$$

It follows from these results that:

$$\begin{aligned}
&\frac{\partial^2 V}{\partial x^2} + \frac{\partial^2 V}{\partial y^2} = \frac{\partial V}{\partial u}\left(\frac{\partial^2 u}{\partial x^2} + \frac{\partial^2 u}{\partial y^2}\right) + \frac{\partial V}{\partial v}\left(\frac{\partial^2 v}{\partial x^2} + \frac{\partial^2 v}{\partial y^2}\right) \\
&+2\frac{\partial^2 V}{\partial u \partial v}\left[\frac{\partial u}{\partial x}\frac{\partial v}{\partial x} + \frac{\partial u}{\partial y}\frac{\partial v}{\partial y}\right] + \frac{\partial^2 V}{\partial u^2}\left[\left(\frac{\partial u}{\partial x}\right)^2 + \left(\frac{\partial u}{\partial y}\right)^2\right] \\
&+\frac{\partial^2 V}{\partial v^2}\left[\left(\frac{\partial v}{\partial x}\right)^2 + \left(\frac{\partial v}{\partial y}\right)^2\right] = |f'(z)|^2\left[\frac{\partial^2 V}{\partial u^2} + \frac{\partial^2 V}{\partial v^2}\right]
\end{aligned} \qquad [6.198]$$

because $\frac{\partial^2 u}{\partial x^2} + \frac{\partial^2 u}{\partial y^2} = \frac{\partial^2 v}{\partial x^2} + \frac{\partial^2 v}{\partial y^2} = \frac{\partial u}{\partial x}\frac{\partial v}{\partial x} + \frac{\partial u}{\partial y}\frac{\partial v}{\partial y} = 0$ and $\left(\frac{\partial u}{\partial x}\right)^2 + \left(\frac{\partial u}{\partial y}\right)^2 = \left(\frac{\partial v}{\partial x}\right)^2 + \left(\frac{\partial v}{\partial y}\right)^2 = \left(\frac{\partial u}{\partial x}\right)^2 + \left(\frac{\partial v}{\partial x}\right)^2 = \left|\frac{\partial u}{\partial x} + \mathrm{i}\frac{\partial v}{\partial x}\right|^2 = |f'(z)|^2$. Since $\frac{\partial^2 V}{\partial x^2} + \frac{\partial^2 V}{\partial y^2} = 0$ and $f'(z) \neq 0$, it can therefore be derived that $\frac{\partial^2 V}{\partial u^2} + \frac{\partial^2 V}{\partial v^2} = 0$.

iii) If $V(x, y) = a$ (a is constant) on a part of the $\mathcal{D}$ boundary, then $V(u, v) = V(x(u, v), y(u, v)) = a$ on its image on the $\mathcal{D}'$ boundary. Similarly, if the normal derivative of $V(x, y)$ is zero, that is, $\frac{\partial V}{\partial n} = 0$ on part of the boundary of $\mathcal{D}$, then the normal derivative of $V(u, v)$ is zero on its image on the boundary of $\mathcal{D}'$.

Let us look at some examples to illustrate the above statements.

EXAMPLE 6.46. (Harmonic Function Inside a Disk).– Let us find the harmonic function inside a disk of radius R taking on its boundary the values V_0 for $0 \leq \theta \leq \pi$

and 0 for $pi < \theta < 2\pi$. To do this, the disk can be mapped into the upper half-plane by the conformal transformation $w = \mathrm{i}\left(\frac{R-z}{R+z}\right)$ (or, conversely, $z = R\left(\frac{\mathrm{i}-w}{\mathrm{i}+w}\right)$), as illustrated in Figure 6.40. Under this transformation, the arcs MNP and PQS are transformed, respectively, into $M'N'P'$ and $P'Q'S'$ of axis Ou. The boundary values $V = 0$ on arc MNP and $V = V_0$ on arc PQS become $V = 0$ on $M'N'P'$ and $V = V_0$ on $P'Q'S'$, respectively. The problem therefore reduces to finding a harmonic function in the upper half of the f plane, with boundary values V_0 for $u \geq 0$ and 0 for $u < 0$. According to the formula [6.138], this problem admits the solution:

$$V(u, v) = V_0 - \frac{V_0}{\pi} \tan^{-1}\left(\frac{v}{u}\right) \qquad [6.199]$$

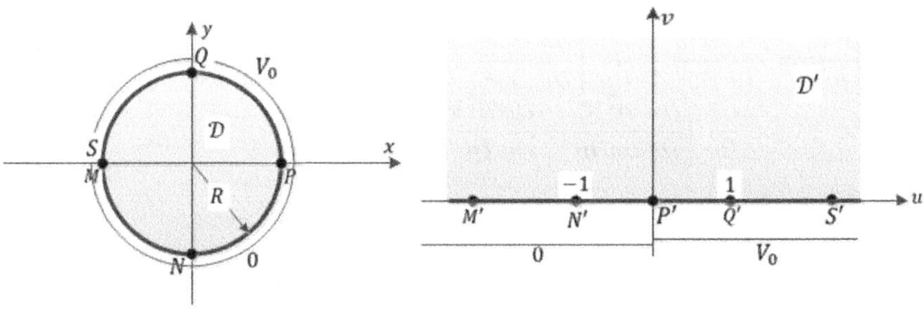

Figure 6.40. *Harmonic function in a disk with radius R with the boundary conditions $V(R, \theta) = \begin{cases} V_0 & \text{for } 0 \leq \theta \leq \pi \\ 0 & \text{for } \pi < \theta < 2\pi \end{cases}$. For a color version of this figure, see www.iste.co.uk/abdesselam/electrodynamics1.zip*

From the conformal transformation $w = \mathrm{i}\frac{R-z}{R+z}$, we obtain $u = \frac{2Ry}{(R+x)^2+y^2}$ and $v = \frac{R^2-x^2-y^2}{(R+x)^2+y^2}$. Substituting these results into the expression for V, we find:

$$V(x, y) = V_0 - \frac{V_0}{\pi} \tan^{-1}\left(\frac{2Ry}{R^2 - x^2 - y^2}\right) \qquad [6.200]$$

which can also be written in polar coordinates as follows:

$$V = V_0 - \frac{V_0}{\pi} \tan^{-1}\left(\frac{2rR\sin\theta}{R^2 - r^2}\right) \qquad [6.201]$$

This result can be obtained directly from formula [6.131].

EXAMPLE 6.47. (Harmonic Function within a Semi-infinite Band of Width a).– Let us find the harmonic function in the region $\mathcal{D}$, shown in Figure 6.41, which takes the boundary values indicated along its borders. This region is mapped into the upper half-plane by the conformal transformation $w = \sin\left(\frac{\pi z}{a}\right)$. This gives $u = \sin\left(\frac{\pi x}{a}\right)\cosh\left(\frac{\pi y}{a}\right)$ and $v = \cos\left(\frac{\pi x}{a}\right)\sinh\left(\frac{\pi y}{a}\right)$. The boundary conditions V_0 for $\left(x = -\frac{a}{2}, y > 0\right)$, 0 for $\left(-\frac{a}{2} \leq x \leq \frac{a}{2}, y = 0\right)$ and $2V_0$ for $\left(x = \frac{a}{2}, y > 0\right)$ are transformed, respectively, into V_0 for $-\infty < u < -1$, 0 for $-1 \leq u \leq 1$ and $2V_0$ for $1 < u < \infty$. The problem can now be easily solved in the f plane. Using [6.138], we obtain:

$$V(u,v) = \frac{2V_0}{\pi}\tan^{-1}\left(\frac{v}{u-1}\right) - \frac{V_0}{\pi}\tan^{-1}\left(\frac{v}{u+1}\right) - 2V_0 \qquad [6.202]$$

and, consequently:

$$\begin{aligned} V(x,y) = \frac{2V_0}{\pi}\tan^{-1}&\left(\frac{\cos\left(\frac{\pi x}{a}\right)\sinh\left(\frac{\pi y}{a}\right)}{\sin\left(\frac{\pi x}{a}\right)\cosh\left(\frac{\pi y}{a}\right)-1}\right) \\ - \frac{V_0}{\pi}\tan^{-1}&\left(\frac{\cos\left(\frac{\pi x}{a}\right)\sinh\left(\frac{\pi y}{a}\right)}{\sin\left(\frac{\pi x}{a}\right)\cosh\left(\frac{\pi y}{a}\right)+1}\right) - 2V_0 \end{aligned} \qquad [6.203]$$

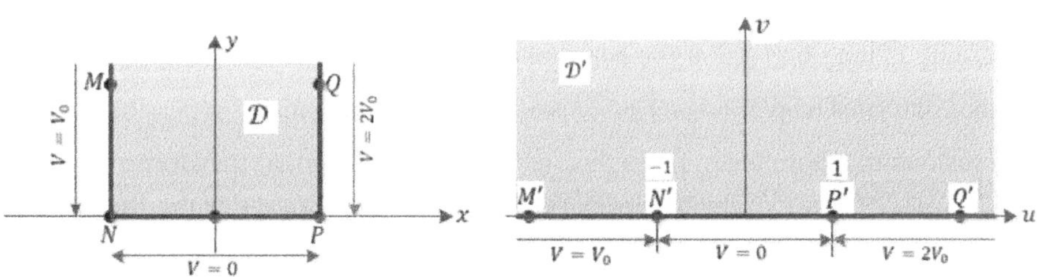

Figure 6.41. *Harmonic function in a semi-infinite region of width a directed in the positive direction of the Oy axis, with boundary conditions V_0 for $\left(x = -\frac{a}{2}, y > 0\right)$, 0 for $\left(-\frac{a}{2} \leq x \leq \frac{a}{2}, y = 0\right)$ and $2V_0$ for $\left(x = \frac{a}{2}, y > 0\right)$. For a color version of this figure, see www.iste.co.uk/abdesselam/electrodynamics1.zip*

This is the solution requested in the semi-infinite band.

EXAMPLE 6.48. (Electric Field Inside a Plane Capacitor). Let us calculate the electrostatic potential inside a plane capacitor formed by two parallel infinite plates separated by a distance a and held at potentials $\pm V_0$, as shown in Figure 6.42. We have already seen that the transformation used to map the region $\mathcal{D}$ into $\mathcal{D}'$ is $w = \mathrm{i}e^{\frac{\pi z}{a}}$. Thus, we obtain $u = -e^{\frac{\pi x}{a}}\sin\left(\frac{\pi y}{a}\right)$ and $v = e^{\frac{\pi x}{a}}\cos\left(\frac{\pi y}{a}\right)$. The boundary conditions are transformed, as shown in Figure 6.42. In the f plane, the solution is

$V(u,v) = \frac{V_0}{\pi}\left[\tan^{-1}\left(\frac{v}{u}\right) - 0\right] - \frac{V_0}{\pi}\left[\pi - \tan^{-1}\left(\frac{v}{u}\right)\right] = \frac{2V_0}{\pi}\tan^{-1}\left(\frac{v}{u}\right) - V_0$ and, consequently, $V(x,y) = \frac{2V_0}{\pi}\tan^{-1}\left(\frac{\cos\left(\frac{\pi y}{a}\right)}{-\sin\left(\frac{\pi y}{a}\right)}\right) - V_0 = \frac{2V_0}{\pi}\tan^{-1}\left(\frac{\sin\left(\frac{\pi}{2}+\frac{\pi y}{a}\right)}{\cos\left(\frac{\pi}{2}+\frac{\pi y}{a}\right)}\right) - V_0 = \frac{2V_0}{a}y$. The electric field induced in the capacitor is $\vec{\mathbf{E}} = -\vec{\nabla}V = -\frac{2V_0}{a}\vec{\mathbf{j}}$. The field lines and equipotential lines are shown in Figure 6.42.

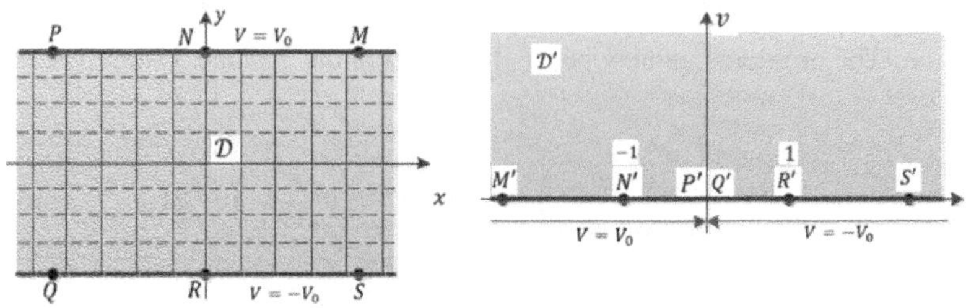

Figure 6.42. *Harmonic function inside a planar capacitor with boundary conditions V_0 for $\left(-\infty < x < \infty, y = -\frac{a}{2}\right)$ and $-V_0$ for $\left(-\infty < x < \infty, y = -\frac{a}{2}\right)$: field lines and equipotential lines are represented by solid red lines and dotted blue lines, respectively. For a color version of this figure, see www.iste.co.uk/abdesselam/electrodynamics1.zip*

EXAMPLE 6.49.– (Electric Field Near the Edges of a Plane Capacitor).– The electric field inside a planar capacitor is assumed to be practically uniform. However, this field uniformity is not observed at the edges and a special calculation is required. Let us use the conformal map $z = \frac{a}{2\pi}\left(e^{\frac{2\pi w}{a}} + \frac{2\pi w}{a} + 1\right)$, which makes it possible to map both regions in Figure 6.43. To do this, let us use the conformal transformation $z = \frac{a}{2\pi}\left(e^{\frac{2\pi w}{a}} + \frac{2\pi w}{a} + 1\right)$, which maps the two regions of Figure 6.43. In the f plane, the solution (harmonic function) of the problem is $V(u,v) = \frac{2V_0}{a}v$. This function appears as the imaginary part of the complex potential $\Omega = \frac{2V_0}{a}w$ or, conversely, $w = \frac{a}{2V_0}\Omega$. It therefore follows that:

$$z = \frac{a}{2\pi}\left[e^{\frac{\pi\Omega}{V_0}} + \frac{\pi\Omega}{V_0} + 1\right] \qquad [6.204]$$

This is the solution to the problem in the z plane. To obtain the field lines and equipotential lines, it is sufficient to separate the real and imaginary parts in [6.204]:

$$\begin{aligned} x &= \frac{a}{2\pi}\left[e^{\frac{\pi U}{V_0}}\cos\left(\frac{\pi V}{V_0}\right) + \frac{\pi U}{V_0} + 1\right] \\ y &= \frac{a}{2\pi}\left[e^{\frac{\pi U}{V_0}}\sin\left(\frac{\pi V}{V_0}\right) + \frac{\pi V}{V_0}\right] \end{aligned} \qquad [6.205]$$

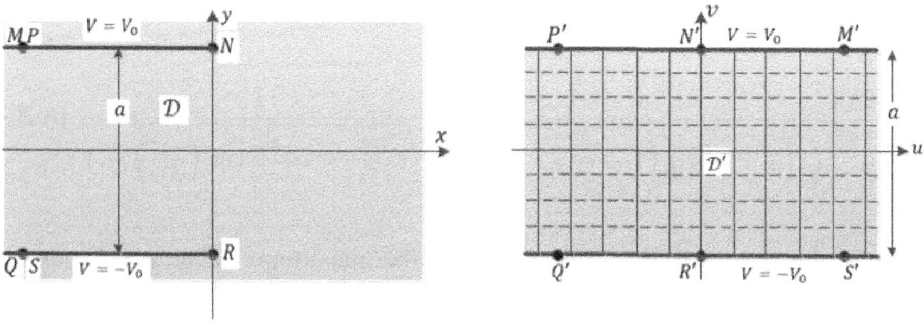

Figure 6.43. *Harmonic function at the edges of a plane capacitor. For a color version of this figure, see www.iste.co.uk/abdesselam/electrodynamics1.zip*

The field lines and equipotential lines are shown in Figure 6.44. According to [6.149], the electric field intensity has the value:

$$\mathbf{E} = -\mathrm{i}\frac{\overline{d\Omega}}{dz} = -\frac{\mathrm{i}}{\frac{dz}{d\Omega}} = -\mathrm{i}\frac{2V_0}{a}\frac{1}{1+e^{\frac{\pi\Omega}{V_0}}} \qquad [6.206]$$

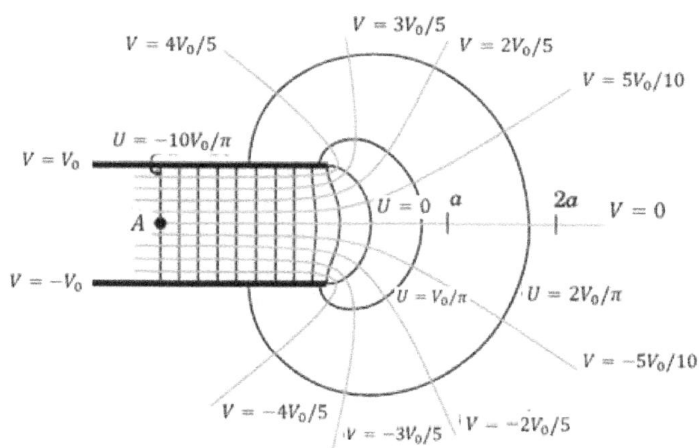

Figure 6.44. *Field lines (bold) and equipotential lines (thin) at the edges of a planar capacitor*

Inside the capacitor, that is, for points z in proximity to A, $\overline{\Omega}$ tends toward $-\infty$; this results in $\mathbf{E} \simeq -\mathrm{i}\frac{2V_0}{a}$ (the field is practically uniform in this region). As we approach the edges of the capacitor (points N and R), we see that $\overline{\Omega} \longrightarrow \pm \mathrm{i}V_0$ and, as a consequence, the field intensity increases indefinitely. Let us follow the variation in the value of the field intensity:

$$|\mathbf{E}| = \frac{2V_0}{a}\frac{1}{\sqrt{\left(1+e^{\frac{\pi\Omega}{V_0}}\right)\left(1+e^{\frac{\pi\bar{\Omega}}{V_0}}\right)}} = \frac{2V_0}{a}\frac{1}{\sqrt{e^{\frac{2\pi U}{V_0}}+2e^{\frac{\pi U}{V_0}}\cos\left(\frac{\pi V}{V_0}\right)+1}} \quad [6.207]$$

along the equipotential lines. To calculate the maximum value of $|\mathbf{E}|$ on the equipotential line, we simply determine the minimum value with respect to U for a fixed V of the expression under the square root. Taking the derivative with respect to U equal to zero, we obtain $e^{\frac{\pi U}{V_0}} + \cos\left(\frac{\pi V}{V_0}\right) = 0$. This condition is not satisfied when the cosine is positive, that is, in the region where $\frac{\pi|V|}{V_0} < \frac{\pi}{2}$ or $|V| < \frac{V_0}{2}$. Along the equipotential lines located in this region, the intensity of the field $|\mathbf{E}|$ varies monotonically with no maximum or minimum. For $V = \pm\frac{V_0}{2}$, the maximum of $|\mathbf{E}| = \frac{2V_0}{a}\frac{1}{\sqrt{e^{\frac{2\pi U}{V_0}}+1}}$ is reached for $U \longrightarrow -\infty$, that is, at the left edge of the capacitor. The equipotentials $\pm\frac{V_0}{2}$ are described by the equations $x = \frac{a}{2\pi}\left[\frac{\pi U}{V_0}+1\right]$ and $y = \pm\frac{a}{2\pi}\left[e^{\frac{\pi U}{V_0}} + \frac{\pi}{2}\right]$. As a result, $y = \pm\left(\frac{a}{4} + \frac{ae^{-\pi}}{2\pi}e^{\frac{2\pi x}{a}}\right)$. The zone where $|V| \le \frac{V_0}{2}$ is called the *Rogowski zone*. For $\frac{V_0}{2} < |V| \le V_0$, the equation $|\mathbf{E}| = \frac{2V_0}{a}\frac{1}{\sqrt{1-e^{\frac{2\pi U}{V_0}}}}$ reaches its maximum for $U = 0$, that is, at the right edge of the capacitor.

EXAMPLE 6.50.– (Line Charge between Two Earthed Planes).– Conformal transformations can also be applied in other situations. Consider two parallel, infinite planes separated by a distance a and maintained at zero potential (earthed). A linear charge q is located between the planes at a distance b from the bottom plane, as shown in Figure 6.45. Let us determine the potential at any point between the two planes. In Figure 6.45, ABC and DEF represent the two planes (in the z plane) connected to ground. The line charge q is located at point bi. We already know that the transformation $w = e^{\frac{\pi z}{a}}$ maps the region between the two planes ABC and DEF into the upper half-plane bounded by the plane $A'B'C'D'E'F'$ held at zero potential. The charge q now moves to position $e^{\frac{i\pi b}{a}}$. To reproduce the same boundary conditions, simply place another charge line $-q$ at the point $\overline{e^{\frac{i\pi b}{a}}} = e^{-\frac{i\pi b}{a}}$ and disconnect the plane $A'B'C'D'E'F'$ from the ground. This new system will produce the complex potential $\Omega = -\frac{iq}{2\pi\varepsilon_0}\ln\left(w - e^{\frac{i\pi b}{a}}\right) + \frac{iq}{2\pi\varepsilon_0}\ln\left(w - e^{-\frac{i\pi b}{a}}\right) = \frac{iq}{2\pi\varepsilon_0}\ln\left(\frac{w-e^{\frac{-i\pi b}{a}}}{w-e^{\frac{i\pi b}{a}}}\right)$ in the f plane. Therefore:

$$\Omega = \frac{iq}{2\pi\varepsilon_0}\ln\left(\frac{e^{\frac{\pi z}{a}} - e^{\frac{-i\pi b}{a}}}{e^{\frac{\pi z}{a}} - e^{\frac{i\pi b}{a}}}\right) \quad [6.208]$$

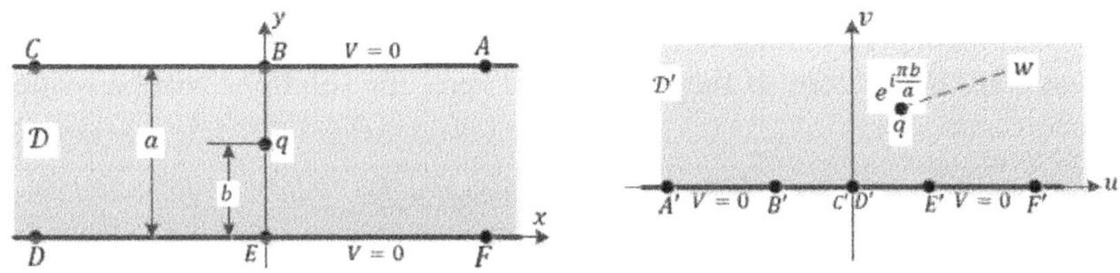

Figure 6.45. *Potential created by a charge line located between two infinite planes connected to earth*

This is the solution in the z plane. The electrostatic potential corresponds to the imaginary part of Ω:

$$\begin{aligned} V(x,y) &= \frac{q}{2\pi\varepsilon_0}\Re\left[\ln\left(\frac{e^{\frac{\pi z}{a}} - e^{\frac{-i\pi b}{a}}}{e^{\frac{\pi z}{a}} - e^{\frac{i\pi b}{a}}}\right)\right] \\ &= \frac{q}{4\pi\varepsilon_0}\ln\left[\left(\frac{e^{\frac{\pi z}{a}} - e^{\frac{-i\pi b}{a}}}{e^{\frac{\pi z}{a}} - e^{\frac{i\pi b}{a}}}\right)\left(\frac{e^{\frac{\pi \bar{z}}{a}} - e^{\frac{i\pi b}{a}}}{e^{\frac{\pi \bar{z}}{a}} - e^{\frac{-i\pi b}{a}}}\right)\right] \\ &= \frac{q}{4\pi\varepsilon_0}\ln\left[\frac{\cosh\left(\frac{\pi x}{a}\right) - \cos\left(\frac{\pi(b+y)}{a}\right)}{\cosh\left(\frac{\pi x}{a}\right) - \cos\left(\frac{\pi(b-y)}{a}\right)}\right] \end{aligned} \qquad [6.209]$$

This solution is obviously zero when $y = 0$ and $y = a$.

6.4. 3D Laplace equation

6.4.1. *3D Cartesian coordinates*

When the potential depends on three coordinates (x, y, z), we must find a solution in the form of a product $V(x,y,z) = X(x)Y(y)Z(z)$. Substituting this solution into the Laplace equation and dividing by XYZ gives $\frac{1}{X}\frac{d^2X}{dx^2} + \frac{1}{Y}\frac{d^2Y}{dy^2} + \frac{1}{Z}\frac{d^2Z}{dz^2} = 0$: three terms, each entirely a function of a single coordinate. This equality is verified unless each term is separately equal to a constant, that is:

$$\begin{aligned} \frac{1}{X}\frac{d^2X}{dx^2} &= -\mathsf{k}_x^2 \\ \frac{1}{Y}\frac{d^2Y}{dy^2} &= -\mathsf{k}_y^2 \\ \frac{1}{Z}\frac{d^2Z}{dz^2} &= \mathsf{k}_z^2 = \mathsf{k}_x^2 + \mathsf{k}_y^2 \end{aligned} \qquad [6.210]$$

In the last equation, the sign of k_z^2 is changed, as the sum of the squares of the separation constants is zero. If they are all non-zero, the solutions will be written as follows:

$$\begin{aligned} X(x) &= A_1 \sin \mathsf{k}_x x + A_2 \cos \mathsf{k}_x x \\ Y(y) &= B_1 \sin \mathsf{k}_y y + B_2 \cos \mathsf{k}_y y \\ Z(z) &= C_1 \sinh \mathsf{k}_z z + C_2 \cosh \mathsf{k}_z z = D_1 e^{\mathsf{k}_z z} + D_2 e^{-\mathsf{k}_z z} \end{aligned} \quad [6.211]$$

The separation constants k_x, k_y and k_z are assumed to be real here. Consequently, the dependencies in x and y are trigonometric, while the dependence in z is hyperbolic (or exponential). However, if one of them is imaginary, we must convert the corresponding solution from trigonometric to hyperbolic, and vice versa. Since the sum of the squares of the separation constants must be zero, at least one of the solutions must be trigonometric and at least one must be hyperbolic. The remaining solution is then trigonometric or hyperbolic, depending on the boundary conditions. If the separation constants are all equal to zero, the solutions reduce to:

$$\begin{aligned} X(x) &= A_1 x + A_2 \\ Y(y) &= B_1 y + B_2 \\ Z(z) &= C_1 z + C_2 \end{aligned} \quad [6.212]$$

When a single constant is zero, for example, $\mathsf{k}_z = 0$, we must take $Z(z) = C_1 z + C_2$. For the other two solutions, one must be a trigonometric and the other hyperbolic.

6.4.2. *Cylindrical coordinates*

If we wish to highlight the properties of the harmonic function on a cylinder of revolution with axis Oz, it is best to switch to cylindrical coordinates. In this case, the Laplace equation is written as:

$$\frac{1}{r}\frac{\partial}{\partial r}\left(r\frac{\partial V}{\partial r}\right) + \frac{1}{r^2}\frac{\partial^2 V}{\partial \varphi^2} + \frac{\partial^2 V}{\partial z^2} = 0 \quad [6.213]$$

Let us find a solution in the form of a product, $V(r, \varphi, z) = R(r)\Phi(\varphi)Z(z)$. Therefore, it follows that $\frac{1}{rR}\frac{d}{dr}\left(r\frac{dR}{dr}\right) + \frac{1}{r^2\Phi}\frac{d^2\Phi}{d\varphi^2} = -\frac{1}{Z}\frac{d^2 Z}{dz^2} \equiv -\mathsf{k}^2$. This gives:

$$\begin{aligned} &\frac{d^2 Z}{dz^2} - \mathsf{k}^2 Z = 0 \\ &\frac{1}{rR}\frac{d}{dr}\left(r\frac{dR}{dr}\right) + \frac{1}{r^2\Phi}\frac{d^2\Phi}{d\varphi^2} = -\mathsf{k}^2 \end{aligned} \quad [6.214]$$

The first equation of the system [6.214] admits the solution:

$$Z(z) = \begin{cases} A\sinh \mathsf{k}z + B\cosh \mathsf{k}z \ \left(\text{or } Ae^{\mathsf{k}z} + Be^{-\mathsf{k}z}\right) & \mathsf{k} \neq 0 \\ A'z + B' & \mathsf{k} = 0 \end{cases} \qquad [6.215]$$

The second differential equation can also be rewritten as $\frac{r}{R}\frac{d}{dr}\left(r\frac{dR}{dr}\right) + \mathsf{k}^2 r^2 = -\frac{1}{\Phi}\frac{d^2\Phi}{d\varphi^2} \equiv n^2$. The latter equation is also separated as follows:

$$\begin{aligned} &\frac{d^2\Phi}{d\varphi^2} + n^2\Phi = 0 \\ &r\frac{d}{dr}\left(r\frac{dR}{dr}\right) + \mathsf{k}^2 r^2 R = n^2 R \end{aligned} \qquad [6.216]$$

The first differential equation admits the solution:

$$\Phi(\varphi) = \begin{cases} C\sin n\varphi + D\cos n\varphi & n \neq 0 \\ C'\varphi + D' & n = 0 \end{cases} \qquad [6.217]$$

Since $\Phi(\varphi + 2\pi) = \Phi(\varphi)$ ($(r, \varphi + 2\pi, z)$ and (r, φ, z) represent the same point on the cylinder), n must be an integer and $C' = 0$. The resulting differential equation for the radial part:

$$r\frac{d}{dr}\left(r\frac{dR}{dr}\right) + \left(\mathsf{k}^2 r^2 - n^2\right) R = 0 \qquad [6.218]$$

is the Bessel equation and has the solutions:

$$R(r) = \begin{cases} EJ_n(\mathsf{k}r) + FN_n(\mathsf{k}r) & \text{for } \mathsf{k} \neq 0 \\ E'r^n + \dfrac{F'}{r^n} & \text{for } \mathsf{k} = 0 \text{ et } n \neq 0 \\ E''\ln r + F'' & \text{for } \mathsf{k} = 0 \text{ et } n = 0 \end{cases} \qquad [6.219]$$

where $J_n(\mathsf{k}r)$ and $N_n(\mathsf{k}r)$ are, respectively, the *Bessel functions of the first kind* and *Neumann functions* (or *Bessel functions of the second kind*) of order n. Note that if k is purely imaginary, R satisfies the modified Bessel equation, $r\frac{d}{dr}\left(r\frac{dR}{dr}\right) - \left(n^2 + \mathsf{k}^2 r^2\right) R = 0$, which admits as solutions the modified Bessel functions $I_n(\mathsf{k}r)$ and $K_n(\mathsf{k}r)$. The functions $K_n(\mathsf{k}r)$ and $N_n(\mathsf{k}r)$ diverge for $r \longrightarrow 0$ and are therefore excluded from problems where the region includes $r = 0$, while $I_n(\mathsf{k}r)$ diverges for $r \longrightarrow \infty$ and are therefore excluded from all outside solutions. The complete solution therefore takes the following form:

$$
\begin{aligned}
&V(r,\varphi,z) = (E\ln r + F)(Az+B) + \sum_{n=1}^{\infty}\left(E_n r^n + \frac{F_n}{r^n}\right)\times \\
&(C_n \sin n\varphi + D_n \cos n\varphi)(A_n z + B_n) + \\
&\sum_{n=1}^{\infty}\sum_{\mathsf{k}}\left[E_{\mathsf{k},n} J_n(\mathsf{k}r) + F_{\mathsf{k},n} N_n(\mathsf{k}r)\right]\left[C_{\mathsf{k},n}\sin n\varphi + D_{\mathsf{k},n}\cos n\varphi\right]\times \\
&\left[A_{\mathsf{k},n}\sinh \mathsf{k}z + B_{\mathsf{k},n}\cosh \mathsf{k}z\right] + \sum_{n=1}^{\infty}\sum_{\mathsf{k}}\left[E'_{\mathsf{k},n} I_n(\mathsf{k}r) + F'_{\mathsf{k},n} K_n(\mathsf{k}r)\right]\times \\
&\left[C'_{\mathsf{k},n}\sin n\varphi + D'_{\mathsf{k},n}\cos n\varphi\right]\left[A'_{\mathsf{k},n}\sin \mathsf{k}z + B'_{\mathsf{k},n}\cos \mathsf{k}z\right]
\end{aligned}
\quad [6.220]
$$

EXAMPLE 6.51. (General Solution Method).– Let us find the potential inside a cylinder of length L and radius a. We will assume that the lower base of the cylinder is grounded (i.e. at zero potential), while the upper base and lateral surface of the cylinder are held at potentials $V_L(r,\varphi)$ and $V_a(\varphi,z)$, respectively, as shown in Figure 6.46. As in the case of Cartesian coordinates, we divide the problem into two distinct steps (or three, when the potential of the lower base is not zero). We first solve the Laplace equation, assuming in the first stage that the potential of the upper base is zero and $V_a(r,\varphi)$ on the lateral surface, then assuming that the potential of the lateral surface is zero and that of the upper base is $V_L(r,\varphi)$ in the second stage. The general solution that simultaneously satisfies both boundary conditions is the sum of the two solutions:

i) *Solution with zero potential on the lower and upper bases of the cylinder* ($V(r,\varphi,0) = V(r,\varphi,L) = 0$): to avoid an infinite solution at $r = 0$, the terms $\ln r$, $\frac{1}{r^n}$, $K_n(\mathsf{k}r)$ and $N_n(\mathsf{k}r)$ must be eliminated from the solution. The fact that the solution is zero at $z = 0$ and $z = L$ clearly shows that $A = B = A_n = B_n = A_{\mathsf{k},n} = B_{\mathsf{k},n} = 0$. These same conditions also eliminate the term $\cos \mathsf{k}z$, that is, $B'_{\mathsf{k},n} = 0$, and require that $\mathsf{k} = \frac{m\pi}{L}$ to obtain a non-trivial solution, that is, $A'_{\mathsf{k},n} \neq 0$. The potential is then:

$$
V_1(r,\varphi,z) = \sum_{m,n=1}^{\infty}\left[A''_{m,n}\sin n\varphi + B''_{m,n}\cos n\varphi\right] I_n\left(\frac{m\pi r}{L}\right)\sin\frac{m\pi z}{L} \quad [6.221]
$$

and reduces, for $r = a$, to $V_a(\varphi,z) = \sum_{m,n}^{\infty}\left[A''_{m,n}\sin n\varphi + B''_{m,n}\cos n\varphi\right]\times I_n\left(\frac{m\pi a}{L}\right)\sin\frac{m\pi z}{L}$. We can recognize here the Fourier double series of the function $V_a(\varphi,z)$. The coefficients $A''_{m,n}$ and $B''_{m,n}$ are obtained from the integral equations:

$$
\begin{Bmatrix} A''_{m,n} \\ B''_{m,n} \end{Bmatrix} = \frac{2}{\pi L I_n\left(\frac{m\pi a}{L}\right)}\int_0^{2\pi}\int_0^{L} V_a(\varphi,z)\begin{Bmatrix}\sin n\varphi \\ \cos n\varphi\end{Bmatrix}\sin\frac{m\pi z}{L}\,d\varphi dz \quad [6.222]
$$

and consequently:

$$V_1(r,\varphi,z) = \sum_{m,n=1}^{\infty} \frac{2I_n\left(\frac{m\pi r}{L}\right)\sin\frac{m\pi z}{L}}{\pi L I_n\left(\frac{m\pi a}{L}\right)} \times \int_0^{2\pi}\int_0^L V_a(\varphi',z')\cos n(\varphi-\varphi')\sin\frac{m\pi z'}{L}d\varphi' dz' \qquad [6.223]$$

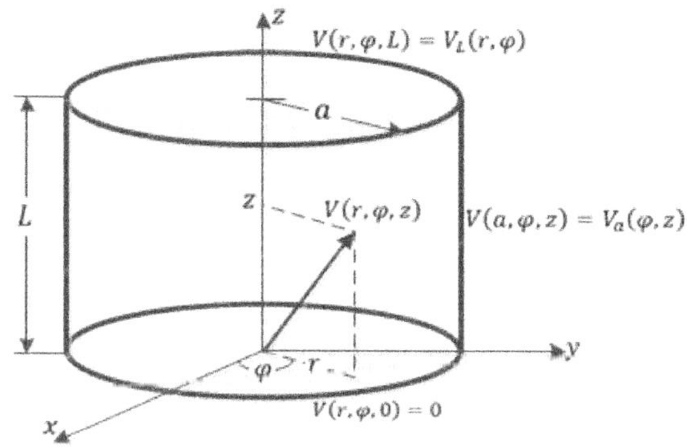

Figure 6.46. *Solution to the Laplace equation in cylindrical coordinates with the boundary conditions: zero potential on the lower base, a potential $V_L(r,\varphi)$ on the upper base, and potential $V_a(\varphi,z)$ on the lateral surface*

ii) *Solution with zero potential on the lateral surface* ($V(a,\varphi,z) = 0$): we must also eliminate the terms $\ln r$, $\frac{1}{r^n}$, $K_n(\mathrm{k}r)$ and $N_n(\mathrm{k}r)$ here. Similarly, $F(Az+B)$, $E_n r^n$ and I_n admit no zeros and must therefore be deleted, as their inclusion would prevent V from canceling at $r = a$ for all φ and z. The solution therefore reduces to $V_2(r,\varphi,z) = \sum_{\mathrm{k},n} J_n(\mathrm{k}r)\left[C''_{\mathrm{k},n}\sin n\varphi + D''_{\mathrm{k},n}\cos n\varphi\right]$ $[A_{\mathrm{k},n}\sinh \mathrm{k}z + B_{\mathrm{k},n}\cosh \mathrm{k}z]$.

Since $V_2(r,\varphi,z)$ is canceled for $z = 0$, it can be derived that $B_{\mathrm{k},n} = 0$. Similarly, the condition that $V_2(r,\varphi,z)$ is zero for $r = a$ results in the fact that $\mathrm{k}a$ must be a root of J_n. Let $\rho_{n,m}$ denote the mth root of J_n. This results in the fact that $\mathrm{k}_m = \frac{\rho_{n,m}}{a}$. Therefore, the required solution is:

$$V_2(r,\varphi,z) = \sum_{n,m} J_n\left(\frac{\rho_{n,m}r}{a}\right)[C''_{m,n}\sin n\varphi + D''_{m,n}\cos n\varphi]\sinh\left(\frac{\rho_{n,m}z}{a}\right) \qquad [6.224]$$

and for $z = L$, it is reduced to:

$$V_L(r,\varphi) = \sum_{n,m} J_n\left(\frac{\rho_{n,m} r}{a}\right)\left[C''_{m,n}\sin n\varphi + D''_{m,n}\cos n\varphi\right]\sinh\left(\frac{\rho_{n,m} L}{a}\right) \quad [6.225]$$

Let us multiply [6.225] by $J_l\left(\frac{\rho_{l,p} r}{a}\right)\left\{\begin{array}{c}\sin l\varphi\\ \cos l\varphi\end{array}\right\} r dr d\varphi$ and then integrate over φ and r, respectively, between $[0.2\pi[$ and $[0.a]$:

$$\int_0^a\int_0^{2\pi} V_L(r,\varphi,z) J_l\left(\frac{\rho_{l,p} r}{a}\right)\left\{\begin{array}{c}\sin l\varphi\\ \cos l\varphi\end{array}\right\} r dr d\varphi = \frac{\pi a^2}{2}\left\{\begin{array}{c}C''_{p,l}\\ D''_{p,l}\end{array}\right\}\times \left[J_{l+1}(\rho_{l,p})\right]^2 \sinh\left(\frac{\rho_{l,p} L}{a}\right) \quad [6.226]$$

where we have used the orthogonality relations:

$$\int_0^a J_l\left(\frac{\rho_{l,m} r}{a}\right) J_l\left(\frac{\rho_{l,p} r}{a}\right) r dr = \frac{1}{2}a^2\left[J_{l+1}(\rho_{l,p})\right]^2 \delta_{m,p} \quad [6.227]$$

Therefore, it comes that:

$$\left\{\begin{array}{c}C''_{p,l}\\ D''_{p,l}\end{array}\right\} = \frac{2\int_0^a\int_0^{2\pi} V_L(r,\varphi) J_l\left(\frac{\rho_{l,p} r}{a}\right)\left\{\begin{array}{c}\sin l\varphi\\ \cos l\varphi\end{array}\right\} r dr d\varphi}{\pi a^2\left[J_{l+1}(\rho_{l,p})\right]^2\sinh\left(\frac{\rho_{l,p} L}{a}\right)} \quad [6.228]$$

$$V_2(r,\varphi,z) = \sum_{n,m}\frac{2J_n\left(\frac{\rho_{n,m} r}{a}\right)\sinh\left(\frac{\rho_{n,m} z}{a}\right)}{\pi a^2\left[J_{n+1}(\rho_{n,m})\right]^2\sinh\left(\frac{\rho_{n,m} L}{a}\right)}\times \int_0^a\int_0^{2\pi} V_L(r',\varphi') J_n\left(\frac{\rho_{n,m} r'}{a}\right)\cos n(\varphi-\varphi') r' dr' d\varphi' \quad [6.229]$$

EXAMPLE 6.52. (Coaxial Cylinders).– The system shown in Figure 6.47 consists of two coaxial cylinders, one above the other, with radii a and b ($a < b$). The cylinder with radius a (the bottom one) is considered a perfect conductor. The entire system is placed above a plane connected to the earth. The plane at $z = 0$ and the outer ring of the cylinder with radius a are therefore at zero potential, while a constant potential V_0 is imposed along the ring of radius b. No electric current can flow through the surfaces at $z = L$ and $z = L - d$. As this problem does not depend on the angle φ, we will try solutions with a zero separation constant:

$$V(r,z) = \left\{\begin{array}{ll} A_1 z + B_1 z\ln r + C_1\ln r + D_1 & \text{for } L-d < z \leq L\\ A_2 z + B_2 z\ln r + C_2\ln r + D_2 & \text{for } 0 \leq z \leq L-d\end{array}\right. \quad [6.230]$$

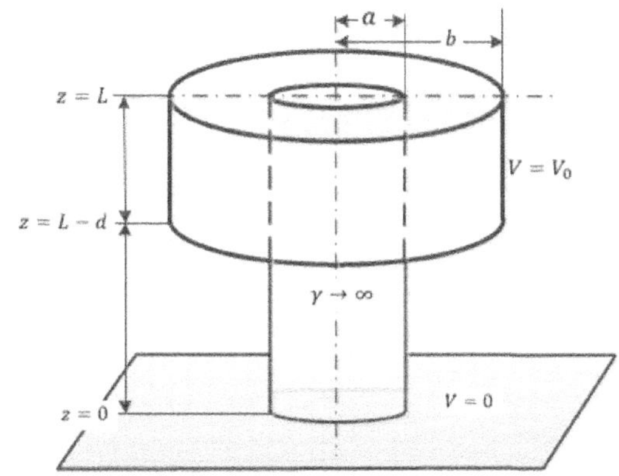

Figure 6.47. *A cylindrical conductor with finite conductivity is mounted on a perfect cylindrical conductor post placed on a perfectly conductive plane connected to the ground*

Applying the boundary conditions of the problem, we obtain:

$$
\begin{aligned}
&V(r,0) = 0 \Longrightarrow C_2 \ln r + D_2 = 0 \Longrightarrow C_2 = D_2 = 0 \\
&V(a,z) = 0 \Longrightarrow \begin{cases} A_2 + B_2 \ln a = 0 \\ A_1 + B_1 \ln a = 0 \\ C_1 \ln a + D_1 = 0 \end{cases} \\
&V(b, z > L-d) = V_0 \Longrightarrow \begin{cases} A_1 + B_1 \ln b = 0 \\ C_1 \ln b + D_1 = V_0 \end{cases} \\
&V\left(r,(L-d)^-\right) = V\left(r,(L-d)^+\right) \Longrightarrow \begin{cases} (L-d)\,A_2 = (L-d)\,A_1 + D_1 \\ (L-d)\,B_2 = (L-d)\,B_1 + C_1 \end{cases}
\end{aligned}
\quad [6.231]
$$

Solving the system [6.231] gives $A_1 = B_1 = 0$, $C_1 = \frac{V_0}{\ln(b/a)}$, $D_1 = -\frac{V_0 \ln a}{\ln(b/a)}$, $A_2 = -\frac{V_0 \ln a}{(L-d)\ln(b/a)}$ and $B_2 = \frac{V_0}{(L-d)\ln(b/a)}$. Finally, we find:

$$
V(r,z) = \begin{cases} \dfrac{V_0 \ln(r/a)}{\ln(b/a)} & \text{for } L-d < z \leq L \\ \dfrac{V_0 z \ln(r/a)}{(L-d)\ln(b/a)} & \text{for } 0 \leq z \leq L-d \end{cases} \quad [6.232]
$$

$$
\vec{\mathbf{E}} = \begin{cases} -\dfrac{V_0}{r\ln(b/a)}\vec{\mathbf{e}}_r & \text{for } L-d < z \leq L \\ -\dfrac{V_0}{(L-d)\ln(b/a)}\left(\ln\dfrac{r}{a}\vec{\mathbf{k}} + \dfrac{z}{r}\vec{\mathbf{e}}_r\right) & \text{for } 0 \leq z \leq L-d \end{cases} \quad [6.233]
$$

In the region $0 \leq z \leq L - d$, the field lines are described by $\frac{dr}{dz} = \frac{E_r}{E_z} = \frac{z}{r \ln \frac{r}{a}}$. This gives:

$$z^2 = r^2 \left[\ln \frac{r}{a} - \frac{1}{2} \right] + \text{const.} \qquad [6.234]$$

which are shown in Figure 6.48. In the region $L - d < z \leq L$, the field lines and equipotential lines are linear.

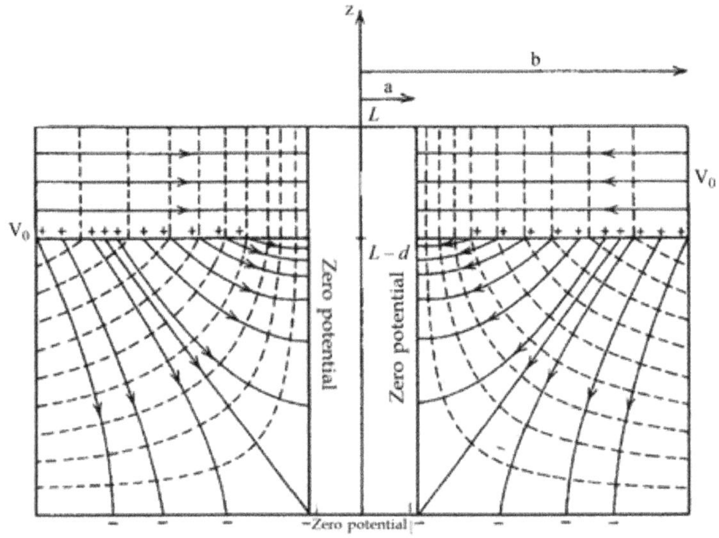

Figure 6.48. *Field and equipotential lines inside cylinders*

6.4.3. *Spherical coordinates*

If we want to reflect the properties of the harmonic function on a sphere with center O or on a cone of revolution with vertex O, it is convenient to switch to spherical coordinates. Let us find the bounded solutions (continuous, one-to-one and finite throughout the domain of variation of the variables $0 \leq \theta \leq \pi$ and $0 \leq \varphi < 2\pi$) of the Laplace equation in spherical coordinates. In this coordinate system, the Laplace equation is written as:

$$\Delta V = \frac{1}{r^2} \frac{\partial}{\partial r} \left(r^2 \frac{\partial V}{\partial r} \right) + \frac{1}{r^2 \sin \theta} \frac{\partial}{\partial \theta} \left(\sin \theta \frac{\partial V}{\partial \theta} \right) + \frac{1}{r^2 \sin^2 \theta} \frac{\partial^2 V}{\partial \varphi^2} = 0 \qquad [6.235]$$

When the solution depends only a single spatial coordinate, the determinant equations and solutions for each of the three coordinates are:

$$\frac{d}{dr}\left(r^2\frac{dV(r)}{dr}\right)=0 \Longrightarrow V(r)=\frac{A_1}{r}+A_2 \qquad [6.236]$$

$$\frac{d}{d\theta}\left(\sin\theta\frac{dV(\theta)}{d\theta}\right)=0 \Longrightarrow V(\theta)=B_1\ln\left(\tan\frac{\theta}{2}\right)+B_2 \qquad [6.237]$$

$$\frac{d^2V}{d\varphi^2}=0 \Longrightarrow V(\varphi)=C_1\varphi+C_2 \qquad [6.238]$$

The first solution represents the potential of a point-like charge. The new solutions are those that depend only on θ and φ.

EXAMPLE 6.53. (Cones of Revolution).– Two identical cones of revolution with the same vertex O and the same generating line have a potential difference V_0, as shown in Figure 6.49. Let us find the electric potential and the electric field. Since this problem does not depend on the variables r and φ, let us try a solution of the type $V(\theta)=B_1\ln\left(\tan\frac{\theta}{2}\right)+B_2$. According to the boundary conditions imposed on the system, $V(\alpha)=-\frac{V_0}{2}$ and $V(\pi-\alpha)=-\frac{V_0}{2}$, we deduce that $\frac{V_0}{2}=B_1\ln\tan\frac{\alpha}{2}+B_2$ and $-\frac{V_0}{2}=-B_1\ln\tan\frac{\alpha}{2}+B_2$. This gives $B_1=\frac{V_0}{2\ln\left(\tan\frac{\alpha}{2}\right)}$ and $B_2=0$. Thus, the electric potential and electric field are written as:

$$V(\theta)=\frac{V_0}{2}\frac{\ln\left(\tan\frac{\theta}{2}\right)}{\ln\left(\tan\frac{\alpha}{2}\right)} \qquad [6.239]$$

$$\vec{\mathbf{E}}=-\vec{\nabla}V=-\frac{V_0}{2r\sin\theta\ln\left(\tan\frac{\alpha}{2}\right)}\vec{\mathbf{e}}_\theta \qquad [6.240]$$

■

Let us try to find the solutions of [6.235] in the general case using the separation-of-variables method, that is, taking $V(r,\theta,\varphi)=R(r)Y(\theta,\varphi)$. By substituting this in [6.235], we get $\frac{1}{R}\frac{d}{dr}\left(r^2\frac{dR}{dr}\right)=-\frac{1}{\sin\theta Y}\frac{\partial}{\partial\theta}\left(\sin\theta\frac{\partial Y}{\partial\theta}\right)-\frac{1}{\sin^2\theta Y}\frac{\partial^2 Y}{\partial\varphi^2}\equiv\mu$, where μ is a constant. This last equation can be separated into two parts:

$$\begin{aligned}&\frac{d}{dr}\left(r^2\frac{dR}{dr}\right)=\mu R\\&\frac{1}{\sin\theta}\frac{\partial}{\partial\theta}\left(\sin\theta\frac{\partial Y}{\partial\theta}\right)+\frac{1}{\sin^2\theta}\frac{\partial^2 Y}{\partial\varphi^2}+\mu Y=0\end{aligned} \qquad [6.241]$$

The second equation of the system [6.241] is also solved by the separation-of-variables method. To do this, let us write $Y(\theta,\varphi)=\Theta(\theta)\Omega(\varphi)$.

Consequently, we get $\frac{\sin\theta}{\Theta(\theta)}\frac{d}{d\theta}\left(\sin\theta\frac{d\Theta(\theta)}{d\theta}\right)+\mu\sin^2\theta=-\frac{\Omega''(\varphi)}{\Omega(\varphi)}\equiv\nu$, where ν is a constant. This gives rise to the following differential equations:

$$\sin\theta\frac{d}{d\theta}\left(\sin\theta\frac{d\Theta\left(\theta\right)}{d\theta}\right)+\mu\sin^2\theta\Theta\left(\theta\right)=\nu\Theta\left(\theta\right) \quad [6.242]$$

$$\Omega''\left(\varphi\right)+\nu\Omega\left(\varphi\right)=0 \quad [6.243]$$

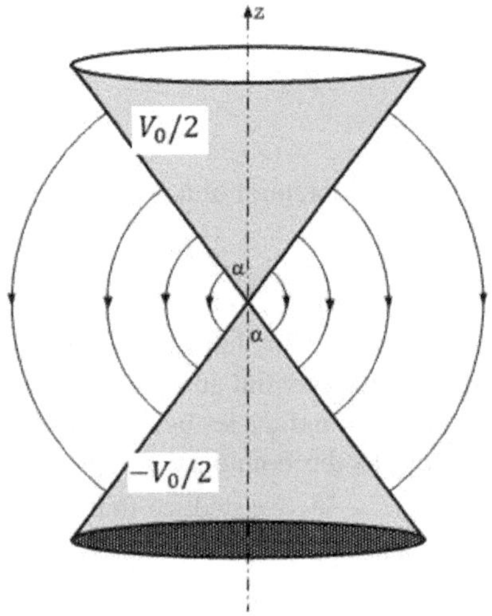

Figure 6.49. *Two cones of revolution with the same vertex and same generating line, maintained at the potentials* $V_0/2$ *and* $-V_0/2$*, respectively*

The last equation admits the solutions:

$$\Omega\left(\varphi\right)=\begin{cases}Ae^{\mathrm{i}\sqrt{\nu}\varphi} & \text{pour } \nu\neq 0\\ A'\varphi+B & \text{pour } \nu=0\end{cases} \quad [6.244]$$

For $V\left(r,\theta,\varphi+2\pi\right)$ and $V\left(r,\theta,\varphi\right)$ to coincide, $\sqrt{\nu}$ must be a relative integer, that is, $\nu=m^2$, and $A'=0$. This gives two linearly independent solutions (m is taken as positive), $\Omega_m\left(\varphi\right)=\frac{1}{\sqrt{2\pi}}e^{im\varphi}$ and $\Omega_{-m}\left(\varphi\right)=\frac{1}{\sqrt{2\pi}}e^{-im\varphi}$. Functions:

$$\Omega_m\left(\varphi\right)=\frac{1}{\sqrt{2\pi}}e^{im\varphi}\quad\left(m=0,\pm1,\pm2,\ldots\right) \quad [6.245]$$

satisfy the orthogonality relations:

$$\int_0^{2\pi}\overline{\Omega_m\left(\varphi\right)}\Omega_{m'}\left(\varphi\right)d\varphi=\delta_{mm'} \quad [6.246]$$

Let us turn to the differential equation [6.242]. Posing $\cos\theta \equiv x$, it reduces to the *associated Legendre differential equation*:

$$\frac{d}{dx}\left[\left(1-x^2\right)\frac{d\Theta}{dx}\right]+\left(\mu-\frac{m^2}{1-x^2}\right)\Theta=0 \qquad [6.247]$$

which admits as solutions the *associated Legendre functions of the first and second kind*, $P_l^m(x)$ and $Q_l^m(x)$, respectively, where μ is related to l by the relation $\mu = l(l+1)$. $Q_l^m(\pm 1) = \pm\infty$. $P_l^m(x)$ also diverges for $x \longrightarrow \pm 1$, unless l is an integer. $P_l^0(x) \equiv P_l(x)$ are *Legendre polynomials*. Consequently, we take (one-to-one and bounded solution):

$$\Theta_{lm}(x) = C_{lm}P_l^m(x) = C_{lm}(-1)^l\frac{\left(1-x^2\right)^{\frac{m}{2}}}{l!2^l}\frac{d^{l+m}}{dx^{l+m}}\left(1-x^2\right)^l \qquad [6.248]$$

where $l \geq 0$ and $-l < m \leq l$. C_{lm} are the normalization constants. Clearly, the functions $\Theta_{lm}(x)$ satisfy the orthogonality relations that follow from the properties of the associated Legendre polynomials, $\int_{-1}^{1}\Theta_{lm}(x)\,\Theta_{l'm}(x)\,dx = C_{lm}C_{l'm}\int_{-1}^{1}P_l^m(x)\,P_{l'}^m(x)\,dx = \frac{2C_{lm}^2}{2l+1}\frac{(l+m)!}{(l-m)!}\delta_{ll'}$, which gives $C_{lm} = \sqrt{\frac{2l+1}{2}\frac{(l-m)!}{(l+m)!}}$. We therefore conclude that the second equation of system [6.241] admits, for $\mu = l(l+1)$ and $\nu = m^2$, bounded solutions of the form:

$$\begin{aligned} Y_m^l(\theta,\varphi) &= \sqrt{\frac{2l+1}{4\pi}\frac{(l-m)!}{(l+m)!}}e^{im\varphi}P_l^m(\cos\theta) \\ \overline{Y_m^l(\theta,\varphi)} &= \sqrt{\frac{2l+1}{4\pi}\frac{(l-m)!}{(l+m)!}}e^{-im\varphi}P_l^m(\cos\theta) = (-1)^m\,Y_{-m}^l(\theta,\varphi) \end{aligned} \qquad [6.249]$$

where $-l \leq m \leq l$. The functions $Y_m^l(\theta,\varphi)$ are called *spherical harmonics of degree l and order m*. They satisfy the orthogonality relations:

$$\int_{\Omega} Y_m^l(\theta,\varphi)\,\overline{Y_{m'}^{l'}(\theta,\varphi)}\sin\theta d\theta d\varphi = \delta_{ll'}\delta_{mm'} \qquad [6.250]$$

Spherical harmonic functions verify the relation known as the *addition formula for spherical harmonics*:

$$\sum_{m=-l}^{l} Y_m^l(\theta_1,\varphi_1)\,\overline{Y_m^l(\theta_2,\varphi_2)} = \frac{2l+1}{4\pi}P_l(\cos\theta) \qquad [6.251]$$

where θ represents the angle measured between the directions (θ_1, φ_1) and (θ_2, φ_2), $\cos\theta = \cos\theta_1 \cos\theta_2 + \sin\theta_1 \sin\theta_2 \cos(\varphi_1 - \varphi_2)$. Let us return to the differential equation $\frac{d}{dr}\left(r^2 \frac{dR}{dr}\right) - l(l+1)R = 0$ (μ here has the value $l(l+1)$). This equation has the solution $R(r) = Ar^l + \frac{B}{r^{l+1}}$, where A and B are constants. By virtue of additivity, the general solution of [6.235] is:

$$V(r,\theta,\varphi) = \sum_{l=0}^{\infty} \sum_{m=-l}^{l} \left[A_{lm} r^l + \frac{B_{lm}}{r^{l+1}}\right] Y_m^l(\theta,\varphi) \quad [6.252]$$

The functions $r^l Y_m^l(\theta,\varphi)$ are used to solve boundary problems within a spherical domain, while the functions $r^{-l-1} Y_m^l(\theta,\varphi)$ are used to solve boundary problems outside the domain.

– *Inside the sphere*: let us write the solution of [6.235] inside a sphere of radius R with the boundary condition $V(R,\theta,\varphi) = V_R(\theta,\varphi)$ (potential on the surface of the sphere). The desired solution will be:

$$\begin{aligned} V(r,\theta,\varphi) &= \sum_{l=0}^{\infty} \sum_{m=-l}^{l} A_{lm} \left(\frac{r}{R}\right)^l Y_m^l(\theta,\varphi) \\ A_{lm} &= \int_{\theta'=0}^{\pi} \int_{\varphi'=0}^{2\pi} V_R(\theta',\varphi') \overline{Y_m^l(\theta',\varphi')} \sin\theta' d\theta' d\varphi' \end{aligned} \quad [6.253]$$

where we made use of the orthogonality relations [6.250]. This can also be written in the form of an integral:

$$\begin{aligned} V(r,\theta,\varphi) &= \int_0^{\pi} \int_0^{2\pi} \sin\theta' d\theta' d\varphi' V_R(\theta',\varphi') \\ &\quad \times \left[\sum_{l=0}^{\infty} \sum_{m=-l}^{l} \left(\frac{r}{R}\right)^l \overline{Y_m^l(\theta',\varphi')} Y_m^l(\theta,\varphi)\right] \\ &= \int_0^{\pi} \int_0^{2\pi} \sin\theta' d\theta' d\varphi' V_R(\theta',\varphi') \\ &\quad \times \left[\sum_{l=0}^{\infty} \frac{2l+1}{4\pi} \left(\frac{r}{R}\right)^l P_l(\cos\phi)\right] \end{aligned} \quad [6.254]$$

with $\cos\phi = \cos\theta\cos\theta' + \sin\theta\sin\theta'\cos(\varphi - \varphi')$. To calculate the sum in square brackets, we will use the Legendre polynomial generating function $\sum_{l=0}^{\infty}(2l+1)t^l \times P_l(\zeta) = 2\sqrt{t}\frac{d}{dt}\left[\sqrt{t}\sum_{l=0}^{\infty} t^l P_l(\zeta)\right] = 2\sqrt{t}\frac{d}{dt}\left[\frac{\sqrt{t}}{\sqrt{1-2t\zeta+t^2}}\right] = \frac{1-t^2}{(1-2t\zeta+t^2)^{3/2}}$.

We therefore find:

$$V(r,\theta,\varphi)=\frac{R\left(R^2-r^2\right)}{4\pi}\int_0^{\pi}\int_0^{2\pi}\frac{V_R(\theta',\varphi')\sin\theta' d\theta' d\varphi'}{\left(R^2-2rR\cos\phi+r^2\right)^{3/2}} \qquad [6.255]$$

– *Outside the sphere*: solving equation [6.235] outside the sphere (with the same boundary conditions) leads to the solution:

$$\begin{aligned} V(r,\theta,\varphi) &= \sum_{l=0}^{\infty}\sum_{m=-l}^{l} B_{lm}\left(\frac{R}{r}\right)^{l+1} Y_m^l(\theta,\varphi) \\ B_{lm} &= \int_0^{\pi}\int_0^{2\pi} V_R(\theta',\varphi')\,\overline{Y_m^l(\theta',\varphi')}\sin\theta' d\theta' d\varphi' \end{aligned} \qquad [6.256]$$

It therefore follows that:

$$\begin{aligned} V(r,\theta,\varphi) &= \int_0^{\pi}\int_0^{2\pi} V_R(\theta',\varphi') \\ &\times\left[\sum_{l=0}^{\infty}\sum_{m=-l}^{l}\left(\frac{R}{r}\right)^{l+1}\overline{Y_m^l(\theta',\varphi')}Y_m^l(\theta,\varphi)\right]\sin\theta' d\theta' d\varphi' \\ &= \frac{1}{4\pi}\int_0^{\pi}\int_0^{2\pi} V_R(\theta',\varphi') \\ &\times\left[\sum_{l=0}^{\infty}(2l+1)\left(\frac{R}{r}\right)^{l+1} P_l(\cos\phi)\right]\sin\theta' d\theta' d\varphi' \\ &= \frac{R\left(r^2-R^2\right)}{4\pi}\int_0^{\pi}\int_0^{2\pi}\frac{V_R(\theta',\varphi')}{\left(r^2-2rR\cos\phi+R^2\right)^{3/2}}\sin\theta' d\theta' d\varphi' \end{aligned} \qquad [6.257]$$

where $\cos\phi=\cos\theta\cos\theta'+\sin\theta\sin\theta'\cos(\varphi-\varphi')$.

EXAMPLE 6.54.– Let us determine the potential inside and outside the sphere for $V_R(\theta,\varphi)=V_0\sin^2\theta\cos 2\varphi$. Let us first note that:

$$V_R(\theta,\varphi)=\frac{1}{2}V_0\sin^2\theta\left(e^{i2\varphi}+e^{-i2\varphi}\right)=V_0\sqrt{\frac{8\pi}{15}}\left(Y_2^2(\theta,\varphi)+Y_2^{-2}(\theta,\varphi)\right) \quad [6.258]$$

– *Inside the sphere*: let $r = R$ in [6.253]:

$$V_R(\theta,\varphi) = \sum_{l=0}^{\infty}\sum_{m=-l}^{l} A_{lm} Y_m^l(\theta,\varphi) \equiv V_0\sqrt{\frac{8\pi}{15}}\left(Y_2^2(\theta,\varphi) + Y_2^{-2}(\theta,\varphi)\right) \quad [6.259]$$

It comes immediately that $A_{2,2} = A_{2,-2} = V_0\sqrt{\frac{8\pi}{15}}$. It therefore follows that the solution inside the sphere will be:

$$V(r \le R,\theta,\varphi) = V_0\sqrt{\frac{8\pi}{15}}\frac{r^2}{R^2}\left(Y_2^2(\theta,\varphi) + Y_2^{-2}(\theta,\varphi)\right) = V_0\frac{r^2}{R^2}\sin^2\theta\cos 2\varphi \quad [6.260]$$

– *Outside the sphere*: at $r = R$, the external solution [6.256] reduces to $\sum_{l=0}^{\infty}\sum_{m=-l}^{l} B_{lm} Y_m^l(\theta,\varphi) \equiv V_0\sqrt{\frac{8\pi}{15}}\left(Y_2^2(\theta,\varphi) + Y_2^{-2}(\theta,\varphi)\right)$. This means that: $B_{2,2} = B_{2,-2} = V_0\sqrt{\frac{8\pi}{15}}$. Outside, the potential is:

$$V(r > R,\theta,\varphi) = V_0\sqrt{\frac{8\pi}{15}}\frac{R^3}{r^3}\left(Y_2^2(\theta,\varphi) + Y_2^{-2}(\theta,\varphi)\right) = V_0\frac{R^3}{r^3}\sin^2\theta\cos 2\varphi \quad [6.261]$$

EXAMPLE 6.55. (Charged Sphere on Its Surface).– Let us find the potential inside and outside a sphere of radius R charged on its surface. We will take $\sigma(\theta,\varphi) = \sigma_0\sin^2\theta\cos 2\varphi$. The presence of $\sigma(\theta,\varphi)$ means that the normal component of the electric field is discontinuous by $\frac{\sigma}{\varepsilon_0}$ on the surface of the sphere, while the tangential components are continuous, that is:

$$\begin{aligned} \left.\frac{\partial V}{\partial r}\right|_{r=R^-} - \left.\frac{\partial V}{\partial r}\right|_{r=R^+} &= \frac{\sigma}{\varepsilon_0} \\ \left.\frac{\partial V}{\partial \theta}\right|_{r=R^-} - \left.\frac{\partial V}{\partial \theta}\right|_{r=R^+} &= 0 \\ \left.\frac{\partial V}{\partial \varphi}\right|_{r=R^-} - \left.\frac{\partial V}{\partial \varphi}\right|_{r=R^+} &= 0 \end{aligned} \quad [6.262]$$

where V is $\sum_{l=0}^{\infty}\sum_{m=-l}^{l} A_{lm} r^l Y_m^l(\theta,\varphi)$ for $r \le R$ and $\sum_{l=0}^{\infty}\sum_{m=-l}^{l} B_{lm} r^{-l-1} \times Y_m^l(\theta,\varphi)$ for $r > R$. Substituting these expressions into the first equation of [6.265], we obtain:

$$\sum_{l=0}^{\infty}\sum_{m=-l}^{l}\left(lC_{lm}R^{l-1}+\frac{l+1}{R^{l+2}}D_{l,m}\right)Y_m^l(\theta,\varphi)=\frac{\sigma_0}{\varepsilon_0}\sqrt{\frac{8\pi}{15}}\left(Y_2^2(\theta,\varphi)+Y_2^{-2}(\theta,\varphi)\right)$$

[6.263]

Therefore, it follows that: $A_{lm}=-\frac{(l+1)B_{l,m}}{lR^{2l+1}}$ for $(l,m)\neq(2,\pm2)$ and $2A_{2,\pm2}+\frac{3B_{2,\pm2}}{R^4}=\frac{\sigma_0}{\varepsilon_0}\sqrt{\frac{8\pi}{15}}$. The other two continuity equations of the system [6.262] require that $A_{lm}=\frac{B_{l,m}}{R^{2l+1}}$, $\forall(l,m)$. The simultaneous resolution of these restrictions gives $A_{2,\pm2}=\frac{\sigma_0}{5\varepsilon_0 R}\sqrt{\frac{8\pi}{15}}$ and $B_{2,\pm2}=\frac{\sigma_0 R^4}{5\varepsilon_0}\sqrt{\frac{8\pi}{15}}$. The other constants are zero. Therefore, the potential is:

$$V(r,\theta,\varphi)=\begin{cases}\frac{\sigma_0 r^2}{5\varepsilon_0 R}\sqrt{\frac{8\pi}{15}}\left(Y_2^2(\theta,\varphi)+Y_2^{-2}(\theta,\varphi)\right)=\frac{\sigma_0 r^2}{5\varepsilon_0 R}\sin^2\theta\cos2\varphi & r\leq R\\ \frac{\sigma_0 R^4}{5\varepsilon_0 r^3}\sqrt{\frac{8\pi}{15}}\left(Y_2^2(\theta,\varphi)+Y_2^{-2}(\theta,\varphi)\right)=\frac{\sigma_0 R^4}{5\varepsilon_0 r^3}\sin^2\theta\cos2\varphi & r>R\end{cases}$$

[6.264]

6.4.4. *Parabolic cylindrical coordinates*

In parabolic cylindrical coordinates, the Laplace equation is written as follows:

$$\frac{1}{u^2+v^2}\left(\frac{\partial^2V}{\partial u^2}+\frac{\partial^2V}{\partial v^2}\right)+\frac{\partial^2V}{\partial z^2}=0 \qquad [6.265]$$

Separating the variables u, v and z by taking $V(u,v,z)=F(u)G(v)Z(z)$, substituting in [6.265] and dividing by $F(u)G(v)Z(z)$, we find $\frac{1}{u^2+v^2}\left(\frac{F''}{F}+\frac{G''}{G}\right)=-\frac{Z''}{Z}\equiv-\mathrm{k}^2$, where k is a constant. We can also write $\frac{1}{u^2+v^2}\left(\frac{F''}{F}+\frac{G''}{G}\right)=-\frac{Z''}{Z}\equiv\mathrm{k}^2$, that is, replace k by $-i$k. We will restrict ourselves to the first convention. This last equation splits into two equations:

$$\begin{aligned}&Z''-\mathrm{k}^2Z=0\longrightarrow\qquad Z(z)=\begin{cases}A\sinh\mathrm{k}z+B\cosh\mathrm{k}z & \text{for } \mathrm{k}\neq0\\ A'z+B' & \text{for } \mathrm{k}=0\end{cases}\\&\frac{F''}{F}+\mathrm{k}^2u^2=-\frac{G''}{G}-\mathrm{k}^2v^2\equiv\mu\end{aligned} \qquad [6.266]$$

where μ is another constant. The second equation in turn yields:

$$\begin{aligned} &F'' + \left(-\mu + \mathsf{k}^2 u^2\right) F = 0 \\ &G'' + \left(\mu + \mathsf{k}^2 v^2\right) G = 0 \end{aligned} \qquad [6.267]$$

These are *Hermite differential equations*. To see this, let us make a scale change $u_1 = \alpha u$ and $v_1 = \beta v$: $\frac{d^2F}{du_1^2} + \left(-\frac{\mu}{\alpha^2} + \frac{\mathsf{k}^2}{\alpha^4} u_1^2\right) F = 0$ and $\frac{d^2G}{dv_1^2} + \left(\frac{\mu}{\beta^2} + \frac{\mathsf{k}^2}{\beta^4} v_1^2\right) G = 0$. Choosing $\alpha = \mathsf{i}\sqrt{2\mathsf{ik}}$ and $\beta = \sqrt{2\mathsf{ik}}$, we deduce that $\frac{d^2F}{du_1^2} + \left(\frac{\mu}{2\mathsf{ik}} - \frac{1}{4}u_1^2\right) F = 0$ and $\frac{d^2G}{dv_1^2} + \left(\frac{\mu}{2\mathsf{ik}} - \frac{1}{4}v_1^2\right) G = 0$. These equations admit bounded solutions, unless $\frac{\mu}{2\mathsf{ik}} = n + \frac{1}{2}$, that is, $\mu = 2\mathsf{ik}n + \mathsf{ik}$. The solutions are, respectively, $e^{-u_1^2/4} H_n(u_1)$ and $e^{-v_1^2/4} H_n(v_1)$, that is, $e^{\mathsf{ik}u^2/2} H_n\left(\mathsf{i}\sqrt{2\mathsf{ik}}u\right)$ and $e^{-\mathsf{ik}v^2/2} H_n\left(\sqrt{2\mathsf{ik}}v\right)$, where $H_n(z)$ are *Hermite polynomials*. Thus, the general solution will be:

$$\begin{aligned} V(u, v, z) = &\sum_{n=0}^{\infty} \int \left(A_{\mathsf{k},n} \sinh \mathsf{k}z + B_{\mathsf{k},n} \cosh \mathsf{k}z\right) e^{\frac{\mathsf{ik}\left(u^2 - v^2\right)}{2}} \\ &H_n\left(\mathsf{i}\sqrt{2\mathsf{ik}}u\right) H_n\left(\sqrt{2\mathsf{ik}}v\right) d\mathsf{k} \end{aligned} \qquad [6.268]$$

By adopting the second convention ($\mathsf{k} \longrightarrow -\mathsf{ik}$), we obtain:

$$\begin{aligned} V(u, v, z) = &\sum_{n=0}^{\infty} \int \left(A'_{\mathsf{k},n} \sin \mathsf{k}z + B'_{\mathsf{k},n} \cos \mathsf{k}z\right) e^{\frac{\mathsf{k}\left(u^2 - v^2\right)}{2}} \\ &H_n\left(\mathsf{i}\sqrt{2\mathsf{k}}u\right) H_n\left(\sqrt{2\mathsf{k}}v\right) d\mathsf{k} \end{aligned} \qquad [6.269]$$

6.4.5. *Paraboloidal coordinates*

In parabolic coordinates, the Laplace equation is written as:

$$\frac{1}{u^2 + v^2}\left[\frac{1}{u}\frac{\partial}{\partial u}\left(u\frac{\partial V}{\partial u}\right) + \frac{1}{v}\frac{\partial}{\partial v}\left(v\frac{\partial V}{\partial v}\right)\right] + \frac{1}{u^2 v^2}\frac{\partial^2 V}{\partial \phi^2} = 0 \qquad [6.270]$$

Let us write $V(u, v, \phi) = F(u) G(v) \Phi(\phi)$. Inserting this expression in [6.270], we deduce $\frac{u^2 v^2}{u^2+v^2}\left[\frac{1}{uF}\frac{d(uF')}{du} + \frac{1}{vG}\frac{d(vG')}{dv}\right] = -\frac{\Phi''}{\Phi} \equiv \mathsf{k}^2$, where k is a constant.

This last equation yields two equations:

$$\begin{aligned}
&\Phi'' + \mathrm{k}^2\Phi = 0 \Longrightarrow \Phi(\phi) = \begin{cases} A\sin \mathrm{k}\phi + B\cos \mathrm{k}\phi & \text{for } \mathrm{k} \neq 0 \\ A'\phi + B' & \text{for } \mathrm{k} = 0 \end{cases} \\
&\frac{1}{uF}\frac{d(uF')}{du} - \frac{\mathrm{k}^2}{u^2} = -\frac{1}{vG}\frac{d(vG')}{dv} + \frac{\mathrm{k}^2}{v^2} \equiv \mu \\
&\Longrightarrow \begin{cases} F'' + \dfrac{1}{u}F' - \left(\mu + \dfrac{\mathrm{k}^2}{u^2}\right)F = 0 \\ G'' + \dfrac{1}{v}G' + \left(\mu - \dfrac{\mathrm{k}^2}{v^2}\right)G = 0 \end{cases}
\end{aligned} \qquad [6.271]$$

where μ is another constant. In order for $V(u, v, \phi + 2\pi)$ and $V(u, v, \phi)$ to coincide ($\Phi(\phi)$ is periodic), k must be an integer, that is, $\mathrm{k} = n$, and $A' = 0$. By making the scale change $u_1 = \sqrt{\mu}u$ and $v_1 = \sqrt{\mu}v$, we get:

$$\begin{aligned}
&\frac{d^2F}{du_1^2} + \frac{1}{u_1}\frac{dF}{du_1} - \left(1 + \frac{n^2}{u_1^2}\right)F = 0 \\
&\frac{d^2G}{dv_1^2} + \frac{1}{v_1}\frac{dG}{dv_1} + \left(1 - \frac{n^2}{v_1^2}\right)G = 0
\end{aligned} \qquad [6.272]$$

Here, we recognize the *Bessel equations*. The solutions are $F(u) = C_nI_n(\sqrt{\mu}u) + D_nK_n(\sqrt{\mu}u)$ and $G(v) = E_nJ_n(\sqrt{\mu}v) + F_nN_n(\sqrt{\mu}v)$. Thus, the global solution of the problem is:

$$\begin{aligned}
V(u, v, \phi) = &\sum_{n=0}^{\infty}\int d\mu\left[C_{n,\mu}I_n(\sqrt{\mu}u) + D_{n,\mu}K_n(\sqrt{\mu}u)\right] \times \\
&\left[E_{n,\mu}J_n(\sqrt{\mu}v) + F_{n,\mu}N_n(\sqrt{\mu}v)\right]\left[A_{n,\mu}\sin n\phi + B_{n,\mu}\cos n\phi\right]
\end{aligned} \qquad [6.273]$$

EXAMPLE 6.56. (Elliptical Coordinates).– In elliptical coordinates, the Laplace equation is written as follows:

$$\begin{aligned}
&\frac{1}{a^2(\sinh^2\xi + \sin^2\eta)}\left[\frac{1}{\sinh\xi}\frac{\partial}{\partial\xi}\left(\sinh\xi\frac{\partial V}{\partial\xi}\right) + \frac{1}{\sin\eta}\frac{\partial}{\partial\eta}\left(\sin\eta\frac{\partial V}{\partial\eta}\right)\right] \\
&+\frac{1}{a^2\sinh^2\xi\sin^2\eta}\frac{\partial^2V}{\partial\phi^2} = 0
\end{aligned} \qquad [6.274]$$

For simplicity's sake, let us assume that the potential V is independent of the coordinates η and ϕ. This leads us to a solution that depends only on ξ. The potential $V(\xi)$ must satisfy the reduced equation $\frac{d}{d\xi}\left(\sinh\xi\frac{dV}{d\xi}\right) = 0$. This leads to the fact that $\frac{dV}{d\xi} = \frac{\mathsf{k}}{\sinh \xi}$ and, consequently, $V(\xi) = \mathsf{k} \ln \tanh \frac{\xi}{2} + C$, where k and C are constants. Since the potential is zero for very high values of ξ, the constant C is zero because $V(\infty) = 0 = \lim_{\xi \longrightarrow \infty} \mathsf{k} \ln \tanh \frac{\xi}{2} + C = C$. Moreover, if the potential is V_0 on the surface of the ellipsoid defined by $\xi = \xi_0$, we deduce that $V_0 = \mathsf{k} \ln \tanh \frac{\xi_0}{2}$, that is, $\mathsf{k} = \frac{V_0}{\ln \tanh \frac{\xi_0}{2}}$. It follows that:

$$V(\xi) = V_0 \frac{\ln \tanh \frac{\xi}{2}}{\ln \tanh \frac{\xi_0}{2}} \quad [6.275]$$

The electric field induced by the potential $V(\xi)$ is therefore:

$$\begin{aligned} \vec{\mathbf{E}} &= -\frac{1}{a\sqrt{\sinh^2 \xi + \sin^2 \eta}} \frac{\partial V}{d\xi} \vec{\mathbf{e}}_\xi \\ &= \frac{V_0}{2a \ln \tanh \frac{\xi_0}{2} \sqrt{\sinh^2 \xi + \sin^2 \eta}} \frac{\tanh^2 \frac{\xi}{2} - 1}{\tanh \frac{\xi}{2}} \vec{\mathbf{e}}_\xi \end{aligned} \quad [6.276]$$

To obtain the charge contained inside the ellipsoid, we need to integrate over the surface $\xi = \xi_0$:

$$\begin{aligned} \frac{Q}{\varepsilon_0} &= \oint_{\text{Surf.}\xi=\xi_0} \vec{\mathbf{E}} \cdot d\vec{\mathbf{S}} = \int_{\eta=0}^{\pi} \int_{\phi=0}^{2\pi} E_{\xi_0} h_\eta h_\phi d\eta d\phi \\ &= \int_{\eta=0}^{\pi} \int_{\phi=0}^{2\pi} E_{\xi_0} a^2 \sqrt{\sinh^2 \xi_0 + \sin^2 \eta} \sinh \xi_0 \sin \eta d\eta d\phi \\ &= -\frac{V_0 a}{\ln \tanh \frac{\xi_0}{2}} \int_{\eta=0}^{\pi} \int_{\phi=0}^{2\pi} \sin \eta d\eta d\phi = -\frac{4\pi a V_0}{\ln \tanh \frac{\xi_0}{2}} \end{aligned} \quad [6.277]$$

The potential V_0 is therefore related to the charge Q contained inside the ellipsoid as follows: $V_0 = -\frac{Q}{4\pi\varepsilon_0 a} \ln \tanh \frac{\xi_0}{2}$. We get:

$$\begin{aligned} V(\xi) &= -\frac{Q}{4\pi\varepsilon_0 a} \ln \tanh \frac{\xi}{2} \\ \vec{\mathbf{E}} &= \frac{Q}{8\pi\varepsilon_0 a^2} \frac{1}{\sqrt{\sinh^2 \xi + \sin^2 \eta}} \frac{1 - \tanh^2 \frac{\xi}{2}}{\tanh \frac{\xi}{2}} \vec{\mathbf{e}}_\xi \end{aligned} \quad [6.278]$$

EXAMPLE 6.57.– In elliptic cylindrical coordinates, the Laplace equation can be written as follows:

$$\frac{1}{a^2 \left(\cosh^2 \xi - \cos^2 \varphi\right)} \left[\frac{\partial^2 V}{\partial \xi^2} + \frac{\partial^2 V}{\partial \varphi^2}\right] + \frac{\partial^2 V}{\partial z^2} = 0 \quad [6.279]$$

Let $V(\xi,\varphi,z) = \Xi(\xi)\,\Phi(\varphi)\,Z(z)$: $\frac{1}{a^2(\cosh^2\xi-\cos^2\varphi)}\left[\frac{\Xi''}{\Xi}+\frac{\Phi''}{\Phi}\right] = -\frac{Z''}{Z} \equiv -\lambda^2$, where λ is a constant. On the one hand, this gives $Z''-\lambda^2 Z=0 \Longrightarrow Z(z) = A\sinh\lambda z + B\cosh\lambda z$ for $\lambda\neq 0$ and $Z(z)=A'z+B'$ for $\lambda=0$. On the other hand, we get $\frac{\Xi''}{\Xi}+\lambda^2a^2\cosh^2\xi = -\frac{\Phi''}{\Phi}+\lambda^2a^2\cos^2\varphi = \mu$, where μ is another constant. This leads to $\Xi''+\left[\frac{1}{2}\lambda^2a^2\cosh(2\xi)-\mu+\frac{1}{2}\lambda^2a^2\right]\Xi = 0$ and $\Phi''+\left[\mu+\frac{1}{2}\lambda^2a^2-a^2\lambda^2\cos(2\varphi)\right]\Phi = 0$. These equations admit Mathieu's sine and cosine functions as solutions. The parameters must be restricted to obtain bounded and periodic solutions in ϕ.

References

Abbel, M.L. and Braselton, J.P. (2005). *Maple by Example*. Elsevier Academic Press, Cambridge.

Abdesselam, B. (2007). *Introduction à la mécanique analytique*. Ellipses, Paris.

Ait-Ameur, K. (2016). *Physique, les rayonnements électromagnétiques : résumé de cours et problèmes corrigés*. Ellipses, Paris.

Alastuey, A., Margo, M., Pujol, P. (2008). *Physique et outils mathématiques : méthodes et exemples*. EDP Sciences/CNRS Editions, Paris.

Alfvèn, H. (1950). *Cosmical Electrodynamics*. Oxford University Press, Oxford.

Arfken, G. (1985). *Mathematical Methods for Physicists*. Academic Press, Cambridge.

Arnold, V.I. (1980). *Ordinary Differential Equations*. MIT Press, Cambridge.

Baker, B.B. and Copson, E.E. (2003). *Mathematical Theory of Huygens's Principale*. American Mathematical Society, Providence.

Bérest, P. (1997). *Calcul des variations : application à la mécanique et à la physique*. Ellipses, Paris.

Berthon, A. (2011). *Electromagnétisme-équations intégrales de l'électromagnétisme, applications aux problèmes axisymétriques*. Ellipses, Paris.

Bitsadze, A.V. (1980). *Equations of Mathematical Physics*. Editions Mir, Moscow.

Bitsadze, A.V. and Kalinichenko, D.F. (1980). *A Collection of Problems on the Equations of Mathematical Physics*. Edition Mir, Moscow.

Bok, J. and Hulin-Jung, N. (1975). *Ondes électromagnétiques relativité*. Herman, Paris.

Boyce, W.E. and Diprima, R.C. (2009). *Elementary Equations and Boundary Value Problems*. Wiley, New York.

Brewster, H.D. (2010). *Electromagnetism*. Oxford Book Company, Oxford.

Brillouin, L. (1960). *Wave Propagation and Group Velocity*. Academic Press, Cambridge.

Butkov, E. (1968). *Mathematical Physics*. Addison-Wesley, Boston.

Cahn, S.B. and Nadgorny, B.E. (1994). *A Guide to Physics Problems. Part 1: Mechanic, Relativity and Electrodynamics*. Kluwer Academic Publishers, Alphen aan den Rijn.

Cardone, F. and Mignani, R. (2004). *Energy and Geometry: An Introduction to Deformed Special Relativity*. World Scientific, Singapore.

Cohen-Tannoudji, C. (2004). *Atoms in Electromagnetic Fields*. World Scientific, Singapore.

Cordier, D. (2007). *Cours de physique – électromagnétisme – volume 2 : phénomènes d'induction et ondes électromagnétiques – cours et exercices corrigés – licence*. Dunod, Paris.

Cornille, P. (2003). *Advanced Electromagnetism and Vacuum Physics*. World Scientific, Singapore.

Coulomb, C.A. (1884). *Collection de mémoires relatifs à la physique 3*. Gauthier-Villars, Paris.

Courant, R. and Hilbert, D. (1989). *Methods of Mathematical Physics*. Wiley, New York.

D'Auria, R. and Trigiante, M. (2016). *From Special Relativity to Feynman Diagrams: A Course of Theoretical Particle Physics for Beginners*. Springer, Berlin.

Debye, P. (1945). *Polar Molecules*. Dover Publications, New York.

Dirac, P.A.M. (1958). *The Principles of Quantum Mechanics*. Oxford University Press, Oxford.

Durand, E. (1953). *Electrostatique et magnétostatique*. Masson et Cie, Paris.

Edminister, J.A. (1979). *Schaum's Outline of Theory and Problems of Electromagnetics*. McGraw-Hill, New York.

Einstein, A., Lorentz, H.A., Minkowski, H., Weyl, H. (1952). *The Principale of Relativity*. Dover Publications, New York.

Eom, H.J. (2004). *Electromagnetic Wave Theory of Boundary-Value Problems: An Advanced Course on Analytical Methods*. Springer, Berlin.

Faraday, M. (1965). *Experimental Researches in Electricity*. Dover Publications, New York.

Feynman, R.P., Leighton, R.B., Sands, M. (1979). *Le cours de physique de Feynman — électromagnétisme 1 et 2*. Inter-Editions, Paris.

Fitzpatrick, R. (2008). *Maxwell's Equations and the Principles Electromagnetism*. Infinity Science Press LLC, Sydney.

Foot, C.J. (2005). *Atomic Physics*. Oxford University Press, Oxford.

Friedman, B. (2011). *Principales and Techniques of Applied Mathematics*. Dover Publications, New York.

Friedrich, H. (2004). *Theoretical Atomic Physics*. Springer, Berlin.

Fröhlich, H. (1949). *Theory of Dielectrics*. Oxford University Press, Oxford.

Fujimoto, M. (2007). *Physics of Classical Electromagnetism*. Springer, Berlin.

Funaro, D. (2008). *Electromagnetism and the Structure of Matter*. World Scientific, Singapore.

Goldstein, H. (1950). *Classical Mechanis*. Addison Wesley, Boston.

Grimes, D.M. and Grimes, G.A. (2004). *The Electromagnetic Origin of Quantum Theory of Light*. World Scientific, Singapore.

Guru, B. and Hiziroglu, H.R. (2009). *Electromagnetic Fields Theory Fundamentals*. Cambridge University Press, Cambridge.

Hairer, E. and Wanner, G. (2006). *Analyse complexe et séries de Fourier*. Université de Genève, Geneva.

Halpern, A. and Erlbach, E. (1998). *Theory and Problems of Beginning Physics II: Waves, Electromagnetism, Optics and Modern Physics*. McGraw-Hill, New York.

Heitler, W. (2010). *Quantum Theory of Radiation*. Dover Publications, New York.

Henley, E.M. and Garcia, A. (2007). *Subatomic Physics*. World Scientific, Singapore.

Hertz, H. (1962). *Electric Waves Being Researches on the Propagation of Electric Action With Finite Velocity Through Space*. Dover Publications, New York.

Hoffmann, J.D. (2001). *Numerical Methods for Engineers and Scientists*. Marcel Dekker, New York.

Iwanenko, D. and Sokolow, A. (1953). *Klassische Feltheorie*. Akademic-Verlag, Berlin.

Jackson, J.D. (1962). *Classical Electrodynamics*. Wiley, New York.

Jeans, J.H. (2009). *Mathematical Theory of Electricity and Magnetism*. Cambridge University Press, Cambridge.

Jeffereys, H. and Jeffreys, B.S. (2000). *Methods of Mathematical Physics*. Cambridge University Press, Cambridge.

Johnson, W.R. (2007). *Atomic Structure Theory: Lectures on Atomic Physics*. Springer, Berlin.

Jordan, E.C. and Balman K.G. (1968). *Electromagnetic Waves and Radiating Systems*. Prentice Hall, Englewood Cliffs.

Kamal, A.A. (2010). *1000 Solved Problems in Modern Physics*. Springer, Berlin.

Kay, D.C. (2011). *Theory and Problems of Tensor Calculus*. McGraw-Hill, New York.

Kibble, T.W.B. and Berkshire, F.H. (2004). *Classical Mechanics*. Imperial College Press, London.

King, R.W.P. and Wu, T.T. (1959). *Scattering and Diffracton of Wave*. Harvard University Press, Cambridge.

Landau, L.D. and Lifchitz, E.M. (1962). *Classical Theory of Fields*. Addison-Wesley, Boston.

Landau, L.D., Lifshitz, E.M., Pitaevskii, L.P. (1984). *Electrodynamics of Continuous Media*. Addison-Wesley, Boston.

Le Bella, C. (1998). *Des phénomènes critiques aux champs de jauge : une introduction aux méthodes et aux applications de la théorie quantique des champs*. EDP Sciences, Les Ulis.

Lichnerowicz, A. (1955). *Théories relativistes de la gravitation et de l'électromagnétisme, relativité générale et théories unitaires*. Masson et Cie, Paris.

Lorentz, H.A. (1952). *Theory of Electrons*. Dover Publications, New York.

Mathey, P. (2017). *Electromagnétisme : ondes et propagation guidée, cours, 130 QCM et exercices corrigés, exemples d'applications*. Dunod, Paris.

Maxwell, J.C. (1954). *Treatise on Electricity and Magnetism*. Dover Publications, New York.

Moliton, A. (2004). *Applications de l'électromagnétisme dans les milieux matériels*. Hermès-Lavoisier, Paris.

Morin, D. (2008). *Introduction to Classical Mechanics with Problems and Solutions*. Cambridge University Press, Cambridge.

Neff, H.P. (1991). *Introductory Electromagnetics*. Wiley, New York.

Nikiforov, A. and Ouvarov, V. (1983). *Fonctions spéciales de la physique mathématique*. Editions Mir, Moscow.

Nogarède, B. (2005). *Electrodynamique appliquée : bases et principes physiques de l'électrotechnique – cours et exercices corrigés licence 3ème année, master, écoles d'ingénieurs*. Dunod, Paris.

Okun, L.B. (2008). *Energy and Mass in Relativity Theory*. World Scientific, Singapore.

Panofsky, W. and Philips, M. (1955). *Classical Electricity and Magnetism*. Addison-Wesley, Boston.

Pauli, W. (1981). *Theory of Relativity*. Dover Publications, New York.

Pérez, J.P.H., Carles, R., Fleckinger, R. (2020). *Electromagnétisme : fondements et applications, avec 300 exercices et problèmes résolus*, 4th edition. Dunod, Paris.

Provost, J.P. and Vallée, J.P. (2011). *Les maths en physique: la physique à travers le filtre des mathématiques avec éléments d'analyse numérique*. Dunod, Paris.

Raab, R.E. and De Lange, O.L. (2004). *Multipole Theory in Electromagnetism: Classical, Quantum, and Symmetry Aspects, with Applications*. Oxford Science Publications, Oxford.

Rax, J.M. (2017). *Electromagnétisme : milieux, structures et énergie*. De Boeck, Brussels.

Reed, M. and Barry, S. (1980). *Methods of Modern Mathematical Physics*. Academic Press, Cambridge.

Reitz, J.R., Milford, F.J., Christy, R.W. (2008). *Foundations of Electromagnetic Theory*. Addison-Wesley, Boston.

Reuse, F.A. (2012). *Electrodynamique : 125 exercices avec solutions — bachelor, licence, classes prépas*. PPUR, Lausanne.

Riley, K.F., Hobson, M.P., Bence, S.J. (2012). *Mathematical Methods for Physics and Engineering*. Cambridge University Press, Cambridge.

Rudnicki, K. (2000). *Gravitation, Electromagnetism and Cosmology: Toward a New Synthesis*. Apeiron Books, Montreal.

Samarskii, A.A. and Vabishchevich, P.N. (2007). *Numerical Methods for Solving Inverse Problems of Mathematical Physics*. De Gruyter, Berlin.

Schultz, B. (1980). *Geometrical Methods of Mathematical Physics*. Cambridge University Press, Cambridge.

Sénéchal, D. (2007). *Electromagnétisme avancé*. Université de Sherbrooke, Sherbrooke.

Sommerfeld, A. (1952). *Electrodynamics*. Academic Press, Cambridge.

Spiegel, M.R. (1968). *Mathematical Handbook of Formulas and Tables*. McGraw-Hill, New York.

Spiegel, M.R. (1974). *Fourier Analysis with Applications to Boundary Value Problems*. McGraw-Hill, New York.

Thirring, W. (1997). *Classical Mathematical Physics, Dynamical Systems and Field Theories*, 3rd edition. Springer, Berlin.

Van Bladel, J. (2007). *Electromagnetic Fields*, 2nd edition. Wiley and Sons, New York.

Vanderlinde, J. (2005). *Classical Electromagnetic Theory*, 2nd edition. Kluwer Academic Publishers, Dordrecht.

Verweij, M.D., Van Den Berg, P.M., Blok H. (2006). *Electromagnetic Waves: An Introductory Course*, 2nd edition. VSSD, Delft.

Wrede, R.C. and Spiegel, M.R. (2010). *Advanced Calculus*. McGraw-Hill, New York.

Zhang, M. (1979). *Electromagnetic Field Theory: A Problem Solving Approach*. Wiley, New York.

Index

A, B

C

D

E

Other titles from

iSTE

in

Waves

2024

BERNARD Jean-Michel L.
Advanced Studies in the Mathematical Theory of Scattering (Waves and Scattering Set – Volume 3)

JUVÉ Daniel, GALLAND Marie-Annick, CLAIR Vincent
Acoustics of Fluid Media 1: Principles and Applications

RÉVEILLAC Jean-Michel
Synthesizers and Subtractive Synthesis 1: Theory and Overview
Synthesizers and Subtractive Synthesis 2: Application and Practice

2022

GONTRAND Christian
Electromagnetism: Links to Special Relativity

LACAZE Pierre Camille, LACROIX Jean-Christophe
Nanotechnology and Nanomaterials for Energy

RÉVEILLAC Jean-Michel
Recording and Voice Processing 1: History and Generalities
Recording and Voice Processing 2: Working in the Studio

ROYER Daniel, VALIER-BRASIER Tony
Elastic Waves in Solids 1: Propagation

SAKHO Ibrahima
Nuclear Physics 1: Nuclear Deexcitations, Spontaneous Nuclear Reactions

VALIER-BRASIER Tony, ROYER Daniel
Elastic Waves in Solids 2: Radiation, Scattering, Generation

2021

DAHOO Pierre-Richard, LAKHLIFI Azzedine
Infrared Spectroscopy of Symmetric and Spherical Top Molecules for Space Observation 1
(Infrared Spectroscopy Set – Volume 3)
Infrared Spectroscopy of Symmetric and Spherical Top Molecules for Space Observation 2
(Infrared Spectroscopy Set – Volume 4)

2020

DANIELE Vito G., LOMBARDI Guido
Scattering and Diffraction by Wedges 1: The Wiener-Hopf Solution - Advanced Applications
(Waves and Scattering Set – Volume 1)
Scattering and Diffraction by Wedges 2: The Wiener-Hopf Solution - Advanced Applications
(Waves and Scattering Set – Volume 2)

SAKHO Ibrahima
Introduction to Quantum Mechanics 2: Wave-Corpuscle, Quantization & Schrödinger's Equation

2019

BERTRAND Pierre, DEL SARTO Daniele, GHIZZO Alain
The Vlasov Equation 1: History and General Properties

DAHOO Pierre-Richard, LAKHLIFI Azzedine
Infrared Spectroscopy of Triatomics for Space Observation
(Infrared Spectroscopy Set – Volume 2)

RÉVEILLAC Jean-Michel
Electronic Music Machines: The New Musical Instruments

ROMERO-GARCÍA Vicente, HLADKY-HENNION Anne-Christine
Fundamentals and Applications of Acoustic Metamaterials: From Seismic to Radio Frequency
(Metamaterials Applied to Waves Set – Volume 1)

SAKHO Ibrahima
Introduction to Quantum Mechanics 1: Thermal Radiation and Experimental Facts Regarding the Quantization of Matter

2018

SAKHO Ibrahima
Screening Constant by Unit Nuclear Charge Method: Description and Application to the Photoionization of Atomic Systems

2017

DAHOO Pierre-Richard, LAKHLIFI Azzedine
Infrared Spectroscopy of Diatomics for Space Observation
(Infrared Spectroscopy Set – Volume 1)

PARET Dominique, HUON Jean-Paul
Secure Connected Objects

PARET Dominque, SIBONY Serge
Musical Techniques: Frequencies and Harmony

RÉVEILLAC Jean-Michel
Analog and Digital Sound Processing

STAEBLER Patrick
Human Exposure to Electromagnetic Fields

2016

ANSELMET Fabien, MATTEI Pierre-Olivier
Acoustics, Aeroacoustics and Vibrations

BAUDRAND Henri, TITAOUINE Mohammed, RAVEU Nathalie
The Wave Concept in Electromagnetism and Circuits: Theory and Applications

PARET Dominique
Antennas Designs for NFC Devices

PARET Dominique
Design Constraints for NFC Devices

WIART Joe
Radio-Frequency Human Exposure Assessment

2015

PICART Pascal
New Techniques in Digital Holography

2014

APPRIOU Alain
Uncertainty Theories and Multisensor Data Fusion

JARRY Pierre, BENEAT Jacques N.
RF and Microwave Electromagnetism

LAHEURTE Jean-Marc
UHF RFID Technologies for Identification and Traceability

REMY Jean-Gabriel, LETAMENDIA Charlotte
LTE Standards

REMY Jean-Gabriel, LETAMENDIA Charlotte
LTE Services

SAVAUX Vincent, LOUËT Yves
MMSE-based Algorithm for Joint Signal Detection, Channel and Noise Variance Estimation for OFDM Systems

THOMAS Jean-Hugh, YAAKOUBI Nourdin
New Sensors and Processing Chain

TING Michael
Molecular Imaging in Nano MRI

VALIÈRE Jean-Christophe
Acoustic Particle Velocity Measurements using Laser: Principles, Signal Processing and Applications

VANBÉSIEN Olivier, CENTENO Emmanuel
Dispersion Engineering for Integrated Nanophotonics

2013

BENMAMMAR Badr, AMRAOUI Asma
Radio Resource Allocation and Dynamic Spectrum Access

BOURLIER Christophe, PINEL Nicolas, KUBICKÉ Gildas
Method of Moments for 2D Scattering Problems: Basic Concepts and Applications

GOURE Jean-Pierre
Optics in Instruments: Applications in Biology and Medicine

LAZAROV Andon, KOSTADINOV Todor Pavlov
Bistatic SAR/GISAR/FISAR Theory Algorithms and Program Implementation

LHEURETTE Eric
Metamaterials and Wave Control

PINEL Nicolas, BOURLIER Christophe
Electromagnetic Wave Scattering from Random Rough Surfaces: Asymptotic Models

SHINOHARA Naoki
Wireless Power Transfer via Radiowaves

TERRE Michel, PISCHELLA Mylène, VIVIER Emmanuelle
Wireless Telecommunication Systems

2012

BÉCHERRAWY Tamer
Electromagnetism: Maxwell Equations Wave Propagation and Emission

BOUCHET Olivier
Wireless Optical Communications

BOUTIER Alain
Laser Metrology in Fluid Mechanics: Granulometry, Temperature and Concentration Measurements

LALAUZE René
Chemical Sensors and Biosensors

LE MENN Marc
Instrumentation and Metrology in Oceanography

PICART Pascal, LI Jun-chang
Digital Holography

SAGUET Pierre
Numerical Analysis in Electromagnetics: The TLM Method

2011

BÉCHERRAWY Tamer
Mechanical and Electromagnetic Vibrations and Waves

BESNIER Philippe, DÉMOULIN Bernard
Electromagnetic Reverberation Chambers

GOURE Jean-Pierre
Optics in Instruments

GROUS Ammar
Applied Metrology for Manufacturing Engineering

LE CHEVALIER François, LESSELIER Dominique, STARAJ Robert
Non-standard Antennas

2010

BEGAUD Xavier
Ultra Wide Band Antennas

GUNIOT Vincent
Wave Propagation in Fluids: Second Edition – Models and Numerical Techniques

MARAGE Jean-Paul, MORI Yvon
Sonar and Underwater Acoustics

2009

BOUDRIOUA Azzedine
Photonic Waveguides

BRUNEAU Michel, POTEL Catherine
Materials and Acoustics Handbook

DE FORNEL Frédérique, FAVENNEC Pierre-Noël
Measurements using Optic and RF Waves

FRENCH COLLEGE OF METROLOGY
Transverse Disciplines in Metrology

2008

CHARRUAU Stéphane
Electromagnetism and Interconnections: Advanced Mathematical Tools for Computer-aided Simulation

FILIPPI Paul J.T.
Vibrations and Acoustic Radiation of Thin Structures

LALAUZE René
Physical Chemistry of Solid-Gas Interfaces

2007

KUNDU Tribikram
Advanced Ultrasonic Methods for Material and Structure Inspection

PLACKO Dominique
Fundamentals of Instrumentation and Measurement

RIPKA Pavel, TIPEK Alois
Modern Sensors Handbook

2006

BALAGEAS Daniel, FRITZEN Claus-Peter, GÜEMES Alfredo
Structural Health Monitoring

BOUCHET Olivier, SIZUN Hervé, BOISROBERT Christian, DE FORNEL Frédérique, FAVENNEC Pierre-Noël
Free-Space Optics

BRUNEAU Michel, SCELO Thomas
Fundamentals of Acoustics

FRENCH COLLEGE OF METROLOGY
Metrology in Industry

GUILLAUME Philippe
Music and Acoustics

GUYADER Jean-Louis
Vibration in Continuous Media

RIGNEAULT Hervé, LOURTIOZ Jean-Michel, DELALANDE Claude, LEVENSON Juan Ariel
Nanophotonics

Printed and bound by CPI Group (UK) Ltd, Croydon, CR0 4YY

02/02/2026

14817344-0001